Advanced Vehicles and Infrastructure Systems

Advanced Vehicles and Infrastructure Systems

Computer Applications, Control and Automation

Christopher O. Nwagboso
Vehicle Systems Research Centre
Faculty of Technology
Bolton Institute, Bolton, UK

JOHN WILEY & SONS
Chichester • New York • Weinheim • Brisbane • Singapore • Toronto

National (01243) 779777
International (+44) 1243 779777

e-mail (for orders and customer service enquiries): cs-books@wiley.co.uk

Visit our Home Page on http://www.wiley.co.uk
or
http://www.wiley.com

Other Wiley Editorial Offices

John Wiley & Sons, Inc., 605 Third Avenue,
New York, NY 10158-0012, USA

VCH Verlagsgesellschaft mbH, Pappelallee 3,
D-69469 Weinheim, Germany

Jacaranda Wiley Ltd, 33 Park Road, Milton,
Queensland 4064, Australia

John Wiley & Sons (Asia) Pte Ltd, 2 Clementi Loop #02-01,
Jin Xing Distripark, Singapore 0512

John Wiley & Sons (Canada) Ltd, 22 Worcester Road,
Rexdale, Ontario M9W 1L1, Canada

Library of Congress Cataloging-in-Publication Data

Nwagboso, Christopher O.
Advanced vehicles and infrastructure systems : computer applications, control, and automation / Christopher O. Nwagboso.
p. cm.
Includes bibliographical references and index.
ISBN 0 471 5645 7
1. Intelligent Vehicle Highway Systems. 2. Motor vehicles — Automatic control. 3. Motor vehicles — Automatic location systems. 4. Automobile driving — Automation. 5. Transport Telematics I. Title.
TE228.3.N85 1997 96-27886
388.3'12 — DC20 CIP

British Library Cataloguing in Publication Data

A catalogue record for this book is available from the British Library

ISBN 0 471 95645 7

Typeset in 10.5/12.5pt Sabon by Laser Words, Madras, India
Printed and bound in Great Britain by Bookcraft (Bath) Ltd
This book is printed on acid-free paper responsibly manufactured from sustainable forestation, for which at least two trees are planted for each one used for paper production.

Contents

Preface

There can only be few people in our society who are not aware of the problems caused by road vehicles and the general transportation system. Therefore, it is appropriate to ask oneself if we are now in the position to own an advanced or intelligent vehicle capable of, e.g. detecting if the driver has illegally a high alcohol content in the blood, and therefore is able to inform the driver not to drive the vehicle. If the driver persists to attempt to drive, further warning can be given to the driver, and after an elapsed time, the vehicle can demobilize itself. On the other hand, if the driver is technical enough to reactivate the vehicle, the vehicle will be left with no alternative but to inform the police through its radio or digital communication system, so that the driver will be gently persuaded not to drive the vehicle. Under extreme conditions, the vehicle utilizes its intelligence to vertically or horizontally eject the driver from the vehicle and invoke its autonomous algorithm before he/she causes any accident. My view is that the fundamental technology required to make road vehicles more intelligent exists, but the reliability of these technologies must be proven.

Coming back to the historical perspective, the world's first automobile vehicle was built by Nicholas Joseph Cugnot of Lorraine in 1769. This was a large, heavy, steam-powered tricycle that ran for 20 minutes at about 3.2 km/h, carrying four passengers. This vehicle ran without any notable road infrastructure. Since this early innovation, there have been scores of notable inventions (see Chapter 4) which have contributed to advances in road vehicles. This has resulted in an increasing demand for vehicles, but as this demand increases, so does the volume of the vehicles produced.

The increasing volume of road traffic due to expanding world economy and personal wealth has meant that road infrastructure expansion is pursued, but there is limit to how far and by how much the expansion can be pursued. This has meant that in the recent times, the road infrastructure has seen overcrowding, resulting in inefficient, unsafe and environmentally damaging transport systems. An efficient transportation system is regarded by the user as a necessity; also the user expects a transportation system that is safe, secure and with minimal environmental impact. Innovative solutions to the problems caused by the transport system are needed.

The recent technological advancement to solving these problems provide for the continuing development of advanced vehicle and infrastructure systems.

Within the transport industry, it is now known that the technological revolution is progressively taking place in the area of vehicle systems and infrastructure technology. This revolution includes the development in information technology, telecommunications and electronics in road vehicles and traffic systems. This provided the necessary automation required to reduce congestion, maximize performance, increase road safety, reduce environmental pollution and provide an extension of the office and home activities based on an efficient communication system. The technological development has been expedited by various major research and development in Europe, the USA and Japan. For example, in Japan, the faster rate of development in these technologies has resulted in the development of a third generation of navigation and route guidance systems, while some other countries are still in the first and second generation.

As the technological development continues, there is an urgent need for present and future engineers to be aware of these developments. This motivated me to develop this book to include the information and experience of my international colleagues who are experts in their field. This book is therefore written with the aim that engineers and managers in the industry, academics and students in both senior undergraduate and postgraduate in automotive, electronics, transportation, computing, communication and systems engineering will benefit from this book.

This book is divided into six parts, with each part containing a number of chapters grouped into common theme. Part 1 deals with the development in intelligent transport systems. Part 2 contains three chapters that discuss the dynamics of intelligent vehicles. Part 3 presents four chapters dealing with the technology and development of autonomous vehicle systems. In Part 4 the technique and development of vehicle navigation and guidance system is discussed. Part 5 contains three chapters that discuss aspects of transport informatics systems. Finally, Part 6 contains five chapters that discusses the development of infrastructure system consisting of systems for automated traffic control and management.

C.O. Nwagboso

List of Contributors

Y. Akuta,
Honda Research and Development Co. Ltd, Chassis Research Division, Haga-gun, Tochigi-Ken, Japan

J. Alizon,
S.M.T.I Laboratoire d'eletronique, Universite Blaise Pascal, Claemont 2 France

G. Ambrosino,
AUTOMA, Sistemi di Automazione Industriale Scri, Genova, Italy

M.G.H. Bell,
Department of Civil Engineering, The University of Newcastle Upon Tyne, UK

M. Bielli,
AUTOMA, Sistemi di Automazione Industriale Scri, Genova, Italy

M. Boero,
AUTOMA, Sistemi di Automazione Industriale Scri, Genova, Italy

F. Collange,
S.M.T.I Laboratoire d'eletronique, Universite Blaise Pascal, Claemont 2 France

B. De St-Laurent,
CETE Medierranee, Europroject-Transport Avances, France

B.T. Fijalkowski,
Institute of Automotive Vehicle and Combustion Engines, Politechnika Kraowaska IM Tradeus, Krakow, Poland

J. Gallice,
S.M.T.I Laboratoire d'eletronique, Universite Blaise Pascal, Claemont 2 France

B.F. Heinrich,
Vehicle Engineering, Chrysler Corporation, Michigan, USA

G. Heuser,
Department of Safety Research and Development, TUV Rheinland eV, Koln, Germany

N.B. Hounsell,
Department of Civil Engineering, University of Southampton, UK

J.C. Hsu,
Department of Power Mechanical Engineering, National Tsing Hua University, Taiwan

R. Hurteau,
Department of Electrical and Computer Engineering, Ecole Polytechnique de Montreal, Montreal, Canada

M. Irgens,
Department of Knowledge Based Systems, University of Oslo, Norway

K. Kanayama,
Research and Development Centre, OMRON Corporation, Kyoto, Japan

K. Kin,
Honda Research and Development Co. Ltd,

Chassis Research Division, Haga-gun, Tochigi-Ken, Japan

C. Krogh,
Department of Philosspphy, University of Oslo, Norway

Y. Laperriere,
Noranda Technology Centre, Mining Automation Group, Canada

Y. Loyaerts,
Inspecteur General Des Ponts Et Chausse'es, Ministere Wallon De L'Equipment Et Des Transports, Bruxells, Belgium

P.G. Luebbers,
Department of Computer Science and Engineering, Florida Atlantic University, Florida, USA

T. Naito,
Research and Development Centre, OMRON Corporation, Kyoto, Japan

C.O. Nwagboso,
Vehicle Systems Research Centre(VSRC), School of Engineering, Bolton Institute, Bolton, UK

A.S. Pandya,
Department of Computer Science and Engineering, Florida Atlantic University, Florida, USA

A.M. Parkes,
Institute for Transport Studies, University of Leeds

J. Radford,
Defence Research Agency, Ministry of Defence, Malvern, UK

A. Rourke,
Department of Civil Engineering, The University of Newcastle Upon Tyne, UK

S. Shiraishi,
Honda Research and Development Co. Ltd, Chassis Research Division, Haga-gun, Tochigi-Ken, Japan

M. St-Amant,
C. C. A. R. M., Ecole Polytechnique de Montreal, Montreal, Canadan

A. Stephen,
Transport Research Laboratory, Crowthorne, UK

H. Traettebers,
Department of Knowledge Based Systems, University of Oslo, Norway

L. Trassoudaine,
S.M.T.I Laboratoire d'eletronique, Universite Blaise Pascal, Claemont 2 France

H. Trier,
Department of Safety Research and Development, TUV Rheinland eV, Koln, Germany

K.I. Trovato,
Philips Laboratories, North America Philips Corporation, New York, USA

M. Varga,
Defence Research Agency, Ministry of Defence, Malvern, UK

S.R. Vishnubhotla,
Department of Computer Science and Engineering, Oakland University, Michigan, USA

R.H. Wu,
Department of Power Mechanical Engineering, National Tsing Hua University, Taiwan

O. Yamanoto,
Honda Research and Development Co. Ltd, Chassis Research Division, Haga-gun, Tochigi-Ken, Japan

E.C. Yeh,
Department of Power Mechanical Engineering, National Tsing Hua University, Taiwan

Acknowledgments

I am grateful to my national and international colleagues who have directly or indirectly contributed to this book. My deepest gratitude is due to Ms Ann-Marie Halligan and Mr Andrew McLeod of John Wiley & Sons Ltd for their persistence in making this book a reality.

PART 1

INTELLIGENT TRANSPORTATION SYSTEMS

1

Introduction to Intelligent Transport Systems

1.1 INTRODUCTION

A collection and combination of new and advanced technologies are being prototyped or pilot tested in various form throughout the world to realize the most advanced vehicles and infrastructure systems. For the sake of commonality in definition, advanced vehicle and infrastructure systems are considered in this chapter as the central part of intelligent transport systems (ITS). The trend is towards the development of ITS. By 1998, it is estimated that over US$1.7 billion will have been spent on transportation research and projects in the European Union in order to approach the vision of ITS. In the USA, expenditure on advanced vehicle and infrastructure systems will have reached US$10 billion by the year 2000. With motorways stretched to full capacity, traffic congestion at an all time high, safety a prerequisite and the need to communicate becoming ever important for the advanced vehicle user, it is therefore important that continuous development is made in the field of ITS.

It is expected that for the first decade of the twenty-first century, advanced vehicles which form part of the ITS will have advanced traffic management systems, car platooning, electronic toll collection systems, vehicle communication, navigation and location-based technology with global positioning systems and a cellular link of some type (see Figure 1.1). It is also expected that such a vehicle will have as part of its integrated system, an intelligent cruise control system, advanced collision avoidance, interactive driving, enhanced vision, advanced driver information system and a personalized security system. As most of these systems are discussed in some detail

Advanced Vehicles and Infrastructure Systems, Edited by C.O. Nwagboso

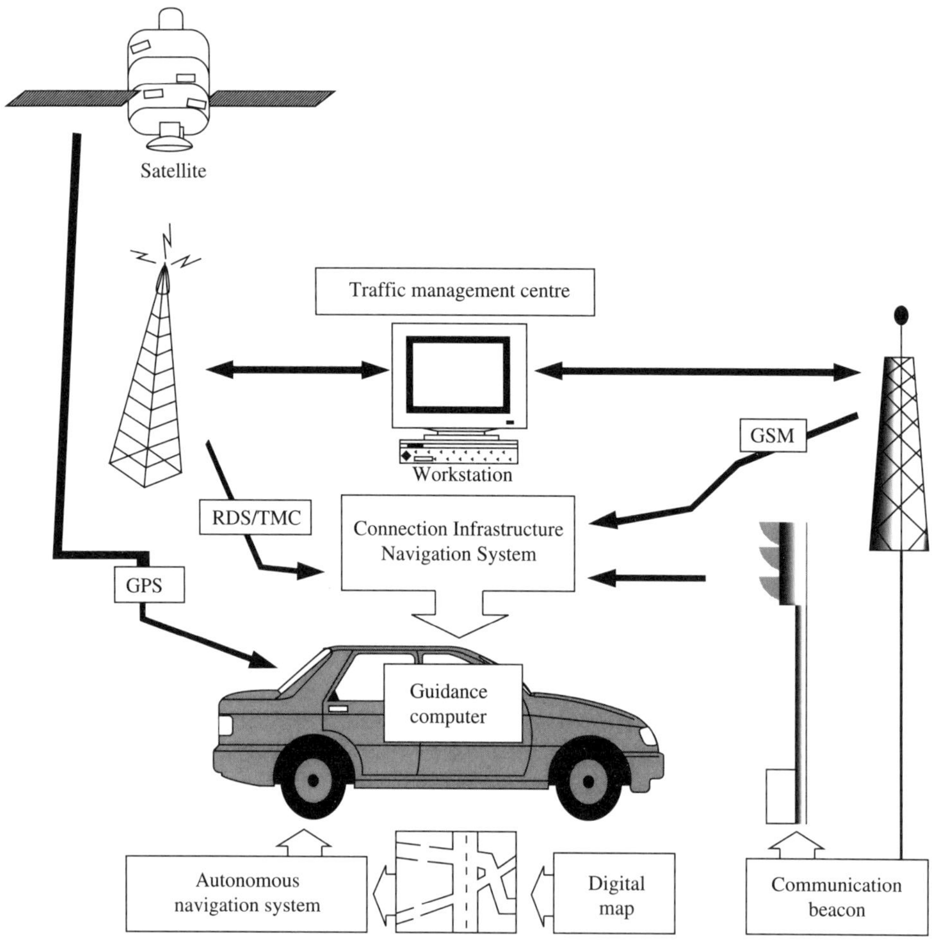

Figure 1.1. Vehicle-Infrastructure communication

in the following chapters, this introductory chapter will concentrate on providing a basic understanding of ITS.

1.2 INTELLIGENT TRANSPORT SYSTEMS

ITS is that field of transportation concerned with the intelligent connection of perception to action integrated with existing transportation systems. The ITS is a mechanism which intelligently serves the need to travel to effect access. The user is concerned with the overall efficiency of the end-to-end journey rather than just the stages within it. Within an ITS, the user of the system becomes an "informed traveller", the transport network (infrastructure) in which the ITS operates becomes environmentally "clear zone" and the vehicle becomes an advanced vehicle. There

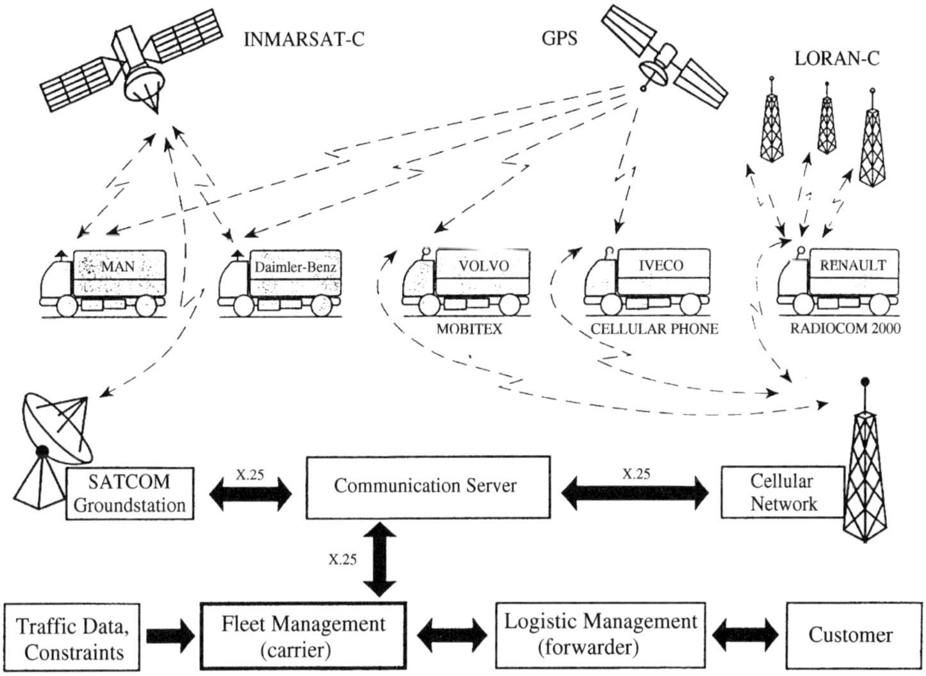

Figure 1.2. Fleet Vehicle–Vehicle–infrastructure communication (PROMETHEUS)

are three essential developments which form the foundation of ITS, these are the development in microelectronics, computers and telecommunication. Computers involving normal computing algorithms to fuzzy logic and neural networks will automate and be used to control some of the functions of the micro-electronic devices onboard a vehicle, while the telecommunications will enable the advanced vehicle to become an integrated part of the advances in the information superhighway. This will enable communication between vehicle and vehicle, vehicle and office, office and vehicle, vehicle and home, home and vehicle, vehicle and traffic centres throughout. Some of these communication systems have already been implemented in road transport trial projects as shown in Figure 1.2.

1.2.1 European Union Initiatives

The first concept of ITS developed in Europe was known as advanced transport telematics (ATT) and relied on a decentralized or autonomous transportation approach. In the 1980s vehicle manufacturers began to realize the limitations imposed by the increasing volume of vehicles in the European transport network. These limitations were due to urban congestion, increased cost from time devoted to travelling, accidents, safety, security, environmental pollution and the need to utilize the time spent on driving more efficiently. Subsequently, manufacturers envisioned the possibility of a new transport system based on the application and integration of electronics, computers and telecommunications to reduce these limitations. This gave birth to a research programme entitled "*PRO*gra*M*me for a *E*uropean *T*raffic with a *H*ighest Efficiency and *U*nprecedented *S*afety", *PROMETHEUS*. This research programme began in 1986, with an 8-year life span. In 1987, it was established as an European Union backed EUREKA project (EU-45) investigating improvements in safety and efficiency on the roads. This research programme being industrially led was designed to increase the competitiveness of the European automotive industry in the world market of ITS.

The initial stage of the programme (1986–87) was a definition phase, which provided the topics of research and the basis for partnership and consortium. There were initially 15 European vehicle manufacturers, but at the later stages, included other research projects sponsored by various other research institutions. In 1988, the launch phase of the programme was implemented, with the first project demonstrations at the programme board member meeting in early 1989, and in January 1989 the research phase of the programme was implemented. At that stage, the annual funding was about US$113 million with two-thirds provided by industry, and one-third provided by national ministries of technology, research and industry.

In order to provide better organization of the programme, PROMETHEUS was supported by three subprogrammes, namely:

1. *PRO-CAR*. In-vehicle systems, such as, vehicle monitoring and intelligent driving aids (e.g., radar collision avoidance, in-vehicle navigation systems and tyre monitoring, etc.).

2. *PRO-ROAD*. Communication systems for vehicle to roadside infrastructure systems. Examples of projects in this subprogramme included the development of route guidance and information systems integrated to traffic conditions and other road infrastructural development.

3. *PRO-NET*. Vehicle to vehicle communication systems for accident warning to authorities.

The three subprogrammes were further subdivided into four working groups:

1. *Pro-Art*: identifying and substantiating the requirements for artificial intelligence, expert systems and software.

2. *Pro-Chip*: micro-electronics systems research and development for information systems (e.g., information for forces, acceleration and pressures, etc.).

3. *Pro-Com*: definition of standards, structures and architectures of communication systems.

4. *Pro-Gen*: developing and analysing how systems will be integrated with traffic management systems.

The groups included academics and research institutions with the responsibility for the above thematic research into various supporting areas, and assist the relevant industries to develop hardware and software, and help to co-ordinate and meet the requirements of the Common European Demonstrators (CEDs).

PROMETHEUS led to the development of several demonstrator programmes called CEDs; each of demonstrator formed to represent a near prototype stage of research containing a particular application area focused to meet customers requirements. Systems such as, collision avoidance, vision enhancement, short range communication, route guidance, formed part of these near prototypes. These resulted in the demonstration by manufacturers of prototype systems in road-worthy advanced vehicles. In most of the development, the prototype served as technological demonstrators, showing the practicability of taking systems originally developed in other domains such as military and aircraft/aerospace technology, and then integrate them into land transport. Members of the consortium within these CEDs programmes are shown in Table 1.1.

In June 1988, a parallel programme entitled DRIVE (Dedicated Road Infrastructure for Vehicle Safety in Europe) was formally adopted by the European Union as a research and development programme in transport telematic and informatics. The objectives of the programme were defined as the improvement of road safety, maximization of transport efficiency and contribution to environmental quality. DRIVE envisaged a common European road transport environment in which drivers are better informed and intelligent vehicles communicate and cooperate with road infrastructure and transport network itself. This entails the following:

(1) the assessment, research and development of a range of road transport informatics technologies;

(2) the evaluation of strategic choices of candidate systems; and

(3) a significant amount of European Union standardization and harmonization

The outcome of the programme was aimed at achieving the following results:

(1) the identification of the best choice of systems on the basis of economic and technical criteria;

(2) the best strategy for their implementation;

(3) the specification of performance and compatibility standards which will enable industry to develop the necessary equipment, components and systems;

(4) the provision of directives and guidelines to which industrial products and intelligent European road transport infrastructures should conform; and

(5) the design and implementation of pilot schemes to assess the performance of equipment and systems.

Table 1.1. Industrial/research institute in common european demonstrators

CEDs	Vehicle manufacturers	Industrial and research establishments	Project areas
1	Jaguar, Fiat, Renault, Matra, PSA, Saab, Volvo	GEC-Marconi, Gilardini, Husat, Hella, Magneti, Marelli, Pilkington, Ultralux, Valeo	Vision Enhancement
2.1	Fiat, MAN, Porsche, Volvo, VW		Friction monitoring and vehicle dynamics
2.2	BMW, MAN, Porsche, Volvo, VW		Lane keeping support
2.3	BMW, Opel, VW		Visibility range monitoring Driver status monitoring
2.4	Ford	Husat	
3	BMW, Daimler-Benz, Fiat, Jaguar, PSA, Renault	GEC-Marconi, Lucas, Thomson-CFS	Collision avoidance
4	BMW, Daimler-Benze, Fiat, Jaguar, Matra, Opel, PSA, Renault, Saab, Volvo	Bosch, Matra CAP Systems, Thomson-CFS	Cooperative driving
5	BMW, Daimler-Benze, Fiat, Jaguar, MAN, Opel, Porsche, PSA, Renault, Saab, Volvo, VW	Bosch, Celsius Tech, DASA, GEC-Marconi, Husat, Leica, Lucas, Pilkington	Autonomous intelligent cruise control
6	BMW	IBR, Draxlmaier	Automatic emergency call
7	Daimler-Benze, Ford, MAN, Volvo	AA, Debis, GEC-Marconi, IBM, MAN Technologies, Nokia, Spectronic	Fleet management
9	BMW, Daimler-Benze, Fiat, Opel, PSA, Renault, Saab, Volvo, VW	EGT, Husat, Motorola. Philips, Sepa	Dual mode route guidance
10	BMW, Daimler-Benze, Fiat, Ford, MAN, Opel, PSA, Renault, Saab, Volvo, VW	IBM, Nokia	Traffic and travel information

Between 1989 and 1992 the DRIVE I programme produced 71 projects from 224 telematic-related proposals, involving 67 manufacturers. The DRIVE programme brought together road users, research institutions, providers of broadcasting and telecommunication services, industry and road transportation authorities.

Phase 1 of the programme was initially limited to 3 years (1989–92), specifically for exploring options in transport telematics and informatics. As most of the vision of the programme was realized, what started as a vision became a reality and subsequently the second phase of the programme was initiated, and this was entitled

DRIVE II in ATT. This subsequent programme was reduced to 57 projects and the programme lasted from 1991 to 1994.

There are seven areas of research and particular attention was given to validation of the R&D results achieved so far in other European projects such as PROMETHEUS, EUROPOLIS,* CARMINAT† and COST.‡ The seven areas of interest are as follows:

(1) demand management;
(2) traffic and travel information;
(3) integrated urban traffic management;
(4) integrated inter-urban management;
(5) driver assistance and cooperative driving;
(6) freight and fleet management; and
(7) public transport management.

The projects developed within these areas were focused on preparing for implementing and exploiting the ATT system leading to the integrated road transport environment.

The technologies required to implement ITS in Europe are to a large extent available, but these technologies are varied and have been developed by different organizations within Europe. To make the development of ITS a reality, the European Union Commission appealed to the major players in the ITS to create a public–private organization in which participants can work together to make ITS a reality. As a direct response to this appeal, on 13 November 1991, the European Road Transport Telematic Implementation Coordination Organization (ERTICO) was formed. Its purpose was to contribute to the co-ordination of the ongoing ITS activities in Europe and to provide support and guidelines for their transfer from research and experimentation to practical application

ERTICO brings together industry, public and private infrastructure operators, public authorities and users. It has over 26 members, including: ACEA, AIT, Eurotunnel, Fiat, Ford of Europe, France-Telecom, PSA, Polis, Renault, Rijkswaterstaat, Rover, Swedish National Road Administration, UTIP, SECAP, IRITECHNA, GEC Traffic, Volvo, Philips, Daimler-Benz, IBM, VDA, RWS, Melta, Department of Transport (UK) and the European Conference of Ministers of Transport (EMTC).

In order to achieve its objectives, ERTICO has the following defined tasks:

1. To conduct strategic planning for ITS implementation.
2. To analyse needs and market of ITS.

* EUROPOLIS = an European telematic project

† CARMINAT = EUREKA project providing comprehensive in-vehicle information system

‡ COST = European Cooperation in the Field of Scientific and Technical Research

3. To conduct ITS engineering resulting in common functional specifications.
4. To follow and assess continued research results.
5. To follow and assess field trials.
6. To promote common specifications, laws and regulations.
7. To initiate and participate in implementation projects.
8. To assist in arranging consortia and financing

Although the PROMETHEUS and DRIVE I and II programmes were finalized in 1994, further programmes such as the 4th FRAMEWORK are presently in progress to expedite systems integration, development and exploitation based on user requirements and cost-effectiveness. Individual state governments have also initiated their national programmes which will implement the technology developed in these programmes. In Germany the initiative entitled MOTIVE has started, while in the UK associations such as PROMOTE and ITS Focus were formed to expedite exploitation and implementation of the systems, and initiatives such as the UK EPSRC Land Transport programme, the UK Office of Science and Technology Foresight Challenge initiative 1996 and the UK DTI/DoT Surface Transport LINK programme announced in February of 1996 will generate research programmes that will contribute to the development of ITS.

1.2.2 US Initiatives

A 1986 traffic study of 29 major US cities, estimated that US$24 billion a year was lost in those cities because of traffic congestion. In the USA, some 41 000 people are killed in road accidents every year and further 5 million people are injured at a total cost of US$70 billion. The total cost of road network congestion is estimated at US$100 billion with the inefficient movement of vehicles resulting in depressed productivity, waste energy and increased vehicle exhaust emission. Therefore, in 1991, with an initial budget of US$660 million, the USA launched a programme initiative which was initially entitled *I*ntelligent *V*ehicle–*H*ighway *S*ystems (IVHS). This led to the establishment of IVHS America.

IVHS America is a utilized and chartered Federal Advisory Committee, with agencies and organizations consisting of various levels of government, the private sector, academic institutions and transport organizations. The main role of IVHS America is to identify the most appropriate technologies and strategies for the future development of IVHS. This includes:

(1) programme planning;
(2) systems architecture;
(3) establishment of standards and protocol;
(4) institutional and legal issues;

(5) national IVHS information clearing house;

(6) international liaison; and

(7) public awareness and education.

During late 1994, the IVHS board approved a name change and therefore, the programme name was changed to Intelligent Transport Systems of America (ITS America). The pressure to adopt the term ITS started from a growing feeling within the USA transportation community that IVHS did not encompass all the transportation modes that need to be addressed. In particular, the programme is under pressure to spend more resources on advanced technology for public transit systems. Others feel that the programme should include technology to assist passengers and freight to make transfers among different transport modes, forming a seamless transportation system.

The ITS programme is a 6-year federally funded R&D programme with a greater proportion of the funding coming from the private sector. It was seen as one of the outcomes of the US Congressional Act passed in December 1991 which is the Intermodal Surface Transportation Efficiency Act (ISTEA) of 1991. This act allowed for the allocation of US$660 million, from 1992 to 1997, in order to relieve congestion and improve safety on the road. This act, which includes the Intelligent Highway Systems Act, stressed the importance of energy efficiency, movement of people and goods, and implementation of new systems. The ITS R&D programme has objectives which are broadly in line with that of ISTEA'91. However, the act allocated significant funds specially towards the development of ITS projects.

The objectives of the ITS programme is to use innovation and technology to fulfil national transportation goals of safety and efficiency, and to meet the transportation needs of the future. The long-term strategic plan covers a 20-year span with two important goals: improvement of transport safety and a reduction in network congestion. The programme is administered through the federal Department of Transport, but much of the programme definition, direction and review are provided by the Intelligent Transport Systems Society of America.[1]

There are six areas of interest identified in the strategic plan, these are:

(1) advanced travel and traffic management systems (ATTMS);

(2) electronics payment systems (EPS);

(3) emergency management systems (EMS);

(4) advanced vehicle safety systems (AVSS);

(5) commercial vehicle operations (CVO);

(6) advanced public transportation systems (APTS).

In 1994, the first detailed programme definition was published and this included projects already under way and project outlines awaiting suitable parties and proposals. In it approximately 300 activities were organized under the following headings:

1. ITS user services (research projects and operational tests).
2. National compatibility planning.
3. Deployment studies.
4. Programme assessments and other related projects.

The six areas of interest previously discussed and the sub-areas consisting of 28 user services form the ITS structure shown in Figure 1.3. In addition to the six areas,

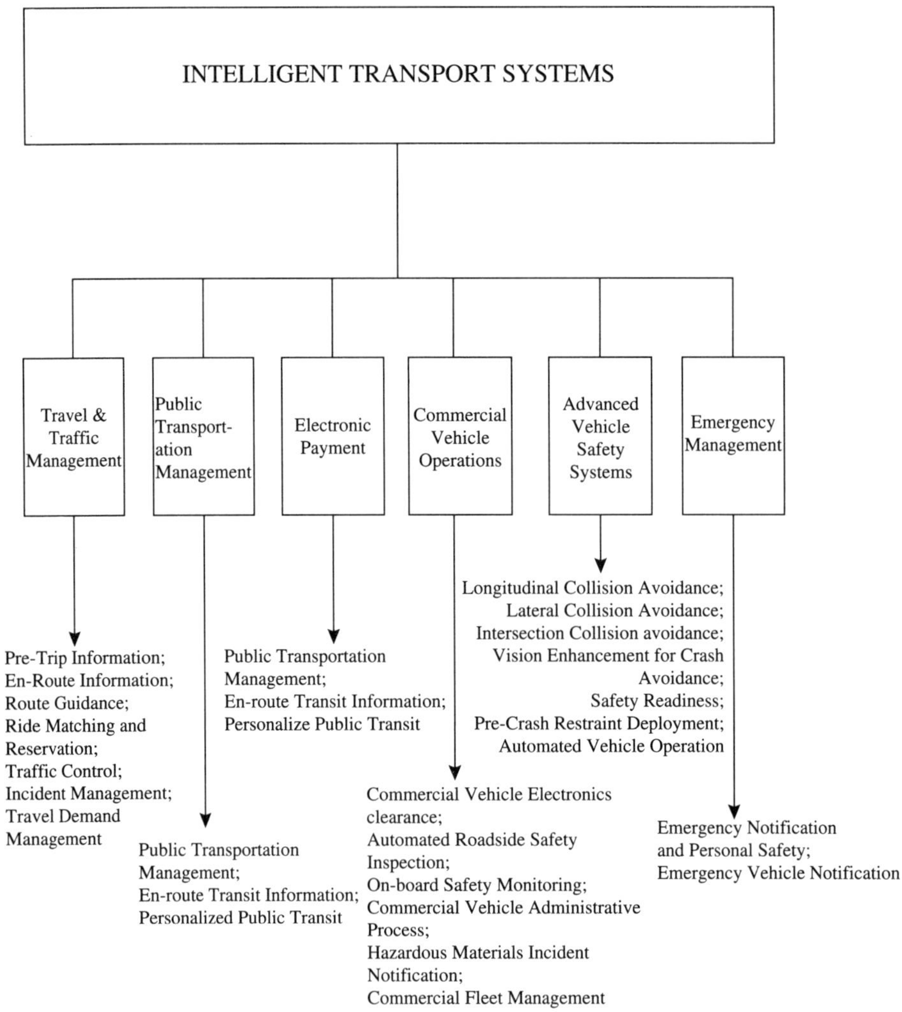

Figure 1.3. ITS Programme area structure

there are a number of projects that cut across these areas and are grouped into other categories.

It is expected that the ITS programme will evolve towards more development of applicable technologies, user requirement understanding, national and international standardization and cost-effectiveness issues (ITS will be paid for principally by the users of its products and services). It is expected that what happens to ITS development, practices and legislation, may have important implications for the rest of the world.

1.2.3 Japanese initiatives

In the 1970s Japan identified that traffic problems such as congestion and increased road death (16 500 in 1970) were occurring due to the fast industrial growth and rapid growth in cities. During the middle 1970s considerable effort was expended to improve the situation by developing new multi-lane highways and traffic control systems. To develop systems further which are capable of solving the Japanese traffic problem and the same developing intelligent systems for vehicles, a Japanese ITS initiative was started in the 1980s, although Japanese researchers have been investigating a range of advanced transportation technologies since the 1960s. The recent initiatives have resulted in several projects being set-up by national agencies, universities, research institutes and private companies. Some of the major projects include:

(1) advanced mobile traffic information and communication systems (AMTICS);

(2) road–automobile communication systems (RACS);

(3) vehicle information and communication system (VICS); and

(4) super smart vehicle system (SSVS).

Advanced mobile traffic information and communication systems

AMTICS is sponsored by the National Police Agency and supported by the Ministry of Post and Telecommunications. There are approximately 60 private companies participating in the project through the Japan Traffic Management Technology Association. AMTICS is designed for route guidance and information in urban areas and it employs in-vehicle navigation system that uses dead reckoning with map matching techniques. Instead of using roadside beacons, AMTICS uses teleterminals, which is similar in operation to cellular radios, for communication between the navigation systems and the traffic control centre. The basic concept of AMTICS is shown in Figure 1.4.

The project was launched in 1987, and pilot tests of this system were first carried out in Tokyo in 1988 and later in 1990. The system is capable of displaying on screen all traffic information collected at traffic control and surveillance centres managed by the police in 74 cities. AMTICS provides the benefits of being able to display in real time, the vehicle current location, route and also information on traffic congestion,

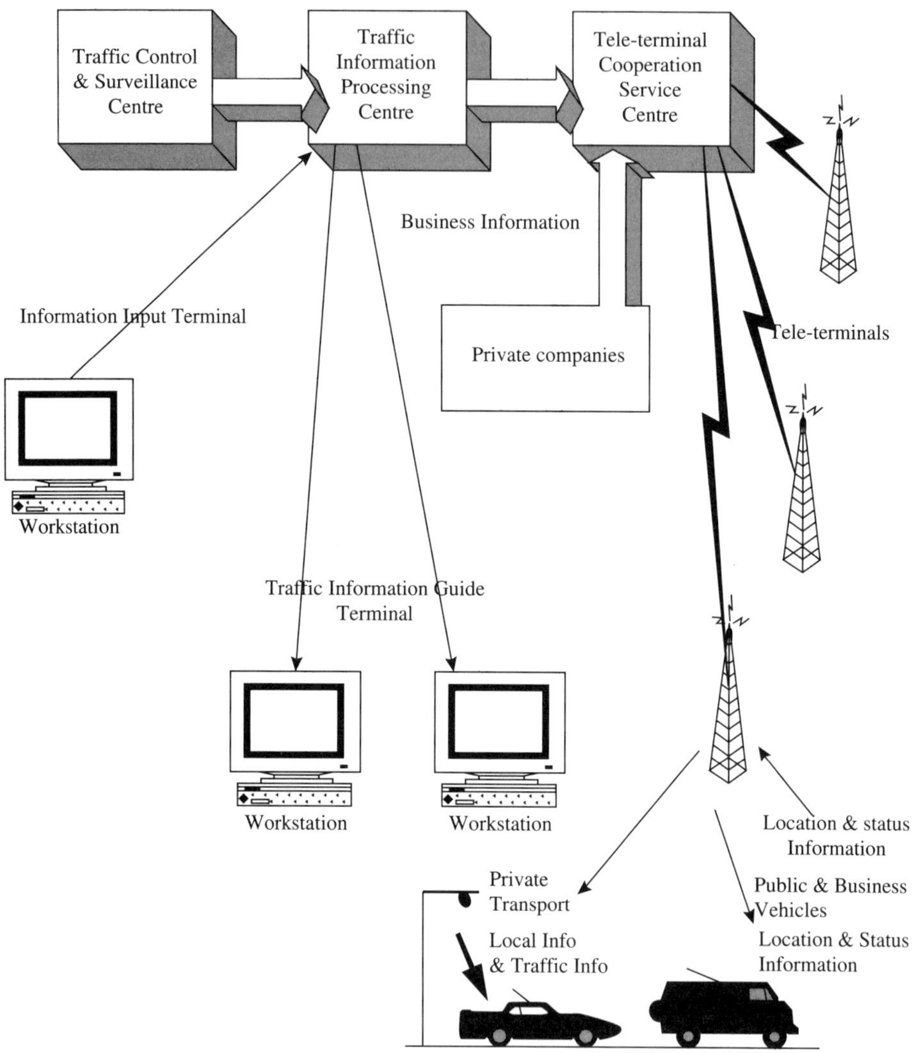

Figure 1.4. Basic Systems concept of advanced mobile traffic information and communication systems (AMTICS)

regulations, road works and parking. The AMTICS system is capable of offering static and dynamic information, and the typical information that can be issued by the system is listed in Table 1.2.

Road–automobile communication systems

RACS is a joint programme involving the Public Works Research Institute of the Ministry of Construction and 25 private companies. It is co-ordinated by the

Table 1.2. Typical information through AMTICS

Type of information	Description of information	Form	Sources
Dynamic	Congestion, road works accidents, events emergency information weather, parking others (e.g. route guidance)	Basic	Supplied through teleterminal
Static	Road networks, traffic regulations, parking areas, general information (e.g. city boundaries), Location of major facilities, service and repair stations, sight-seeing, residential maps, etc.	Service	Supplied from on-board storage medium, e.g. CD-ROM

National Police Agency and the Ministry of Construction. The programme which is funded by 2 billion Yen has the objective to establish a roadside beacon-based driver navigation and communication systems. Tests of the system started in March 1987 in areas covering parts of Tokyo and Yokohama City, and ended in 1989. The system developed from this programme has three major subsystems:

(1) navigation/route guidance, it has dead reckoning and CD-ROM, with roadside beacons locating errors;

(2) traffic information capable of providing congestion, accident and parking spaces information based on real-time roadside beacon signals;

(3) individual communication system including vehicle identification, monitoring of vehicle location and communication data such as telephones and faxes.

The roadside beacon used for the system has a typical specification shown in Table 1.3, and it operates on the microwave band with a communication zone of about 70 m long by 15 m wide. The system which operates with a number of different beacons has its first beacon, called the location beacon, transmit its location, configuration of the next intersection and the destination of the roads branching from that point of the road. The second beacon called the information beacon,

Table 1.3. Beacon system specification of the RACS

Antenna	Microstrip antenna
Transmission frequency	2.5378 GHz
Reception frequency	2.59 GHz
Transmission power	19 mW
Modulation method	BPSK
Transmission speed	512 kbps

transmit road and traffic information as well as location data, and it is connected to the control centre of the RACS. An individual communication beacon can be used, this functions as the location beacon and the information beacon, and also allows for personal two-way communication.

Vehicle information and communication system

VICS started in Japan in 1991 with more than 200 private sector companies involved. It is aimed at better integration of the functions of AMTICS and RACS project to form this new VICS project. This was made possible because both AMTICS and RACS were aiming to provide traffic information to the driver, the main difference being in the technical approach used in providing the telecommunication link. The strategic plan of the VICS programme requires the National Police Agency, Ministry

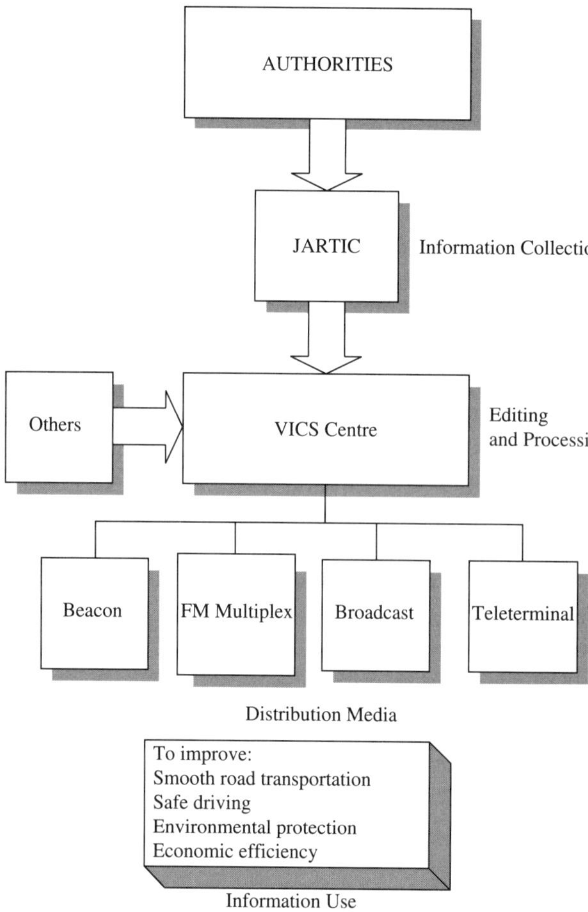

Figure 1.5. Basic systems concept of vehicle information and communication system (VICS)

Table 1.4. Issued information from the VICS beacon system

Category	Content
Present position information	Standardized co-ordination of the location where a location is installed Direction of road Name of place where beacon is installed Class and the route number of a road Name of road Names of the start/end of road Mileage to the beacon
Intersection information	Name of intersection Distance between a beacon and an intersection Intersection type Traffic regulations at intersection Composition of cross-section at an intersection
Guide sign information	Name of destination of crossing roads Distance between destinations and an intersection Type of road and road name Name of crossing roads Class of roads which cross the crossing roads and their route number Major facilities along crossing roads Location of major facilities Distance of major facilities from intersection Services at facilities
Warning information	Type Supplement to a warning Distance to a warning point from a beacon

of Construction and Ministry of Post and Telecommunications to work together in order to arrive at the objective of the programme. The programme which aims to inform the driver about road traffic conditions through the use of the radio communication medium, requires that the National Police Agency expand their traffic control centres and analyse the traffic within the network, and the Ministry of Construction operates a system of roadside communication beacons. The two government agencies are allowed to operate parallel systems for supplying data to vehicles, the Ministry of Construction will achieve this through its roadside beacons and the National Police Agency by some wide area broadcast system, such as the FM sideband. The Ministry of Post and Telecommunications also participates in these activities.

As shown in Figure 1.5 a central information collection centre is provided. This receives the traffic information from existing sources, such as the VICS optical and radio beacon, and interprets and displays information at specific locations on the roads, teleterminals and FM radio. Typical information provided by VICS is listed in Table 1.4.

In June 1993, a demonstration of the system began in Tokyo, and in November 1993, 500 people had taken part in the demonstration, which used about 45 vehicles. Generally, VICS has set standards on the future use of mobile communications in Japan covering digital cellular radio, FM subcarrier broadcasting, beacons and package media.

Super smart vehicle system

In August 1990, Japan started a new initiative entitled Super Smart Vehicle System (SSVS). This programme aims to promote the research and development of ITS technologies for the next 20–30 years, particularly those concerned with highway safety. The following issues are addressed in the SSVS projects:

(1) analysis of the fundamental characteristics of in-vehicle systems based on technological capabilities;

(2) investigation of human–machine interfaces and human factor issues;

(3) evaluation of user requirement in terms of safety, efficiency and convenience;

(4) technical, social and economic issues concerned with the development and implementation of ITS technologies;

(5) policies concerning technology research, development and evaluation.

Six areas of system's technology considered under this programme are:

(1) intelligent intersection systems;
(2) intelligent physical transportation information systems;
(3) intelligent environment recognition systems;
(4) active driver assistance systems;
(5) ultraminicar systems;
(6) cooperative driving systems.

Intelligent intersection systems On road junctions, roundabout junctions and other hazardous road links, it will be possible to use an intelligent in-vehicle system to allow vehicles to communicate with monitoring devices mounted around the road intersection to inform the driver about potential hazards and the traffic conditions on the intersection. This system would be very useful on hidden junctions and at poor visibility.

Intelligent physical transportation information systems The system proposed in this project will be responsible for managing the physical transportation using the information from all the stake holders within the transportation provision establishment. Information which can be processed will include public service vehicle booking and active control of the transportation system. It is specified within the project that intercity transportation would be performed by autonomous vehicles in a platoon/convoy and linked to a leading vehicle.

Intelligent environment recognition systems This is an adaptive control system which is capable of recognizing environmental conditions and will be able to take driving task decisions. The system will effect automatic driving and guidance strategy based on the perceived environmental conditions.

Active driver support systems The research is focused towards the provision of in-vehicle sensors/detectors and receivers that will allow the driver to monitor the driving conditions and receive broadcast data. In an emergency condition, the system will be capable of advising the driver of the hazardous conditions and accidents.

Ultraminicar systems This is aimed at developing vehicle of ultralight weight and reduced size with high strength designed to be driven under an automatic guidance system. This can be a test-bed for developing novel composite materials capable of a very high-force/weight ratio.

Cooperative driving systems The vehicle with the cooperative driving system is capable of recognizing its location and the driving conditions. It will have the ability to communicate with other vehicles (vehicle–vehicle communication) and infrastructure systems (vehicle–infrastructure–vehicle communication).

Universal traffic management system (UTMS)

In 1971, the Japanese central government (National Police Agency) initiated the UTMS programme aimed at developing new traffic control systems with driving assistance systems, such as information systems, route guidance, public transport systems and pollution monitoring and advisory systems. Research projects have been developed in the programme to provide vehicle to vehicle communication and vehicle to infrastructure communication using optical beacons. The optical beacons use near-infra-red light projection on to the road to use reflection from the vehicle as the basis of monitoring the traffic flow and providing communication to the vehicle.

As shown in Table 1.5 the programme development is divided into five distinct areas with the integrated traffic control system providing the central management and control of the functions of the five areas. Project development in this programme is staged in a 5-year phase with funding provided by the central and local government. Estimated funding levels are in the range of US$1 billion per year. Optical beacons are being installed in 500 vehicles in Yokohama and Kawasaki on an experimental basis with the objective of having a fully operational system in late 1996 and early 1997.

In Europe, USA and Japan, commercialization and exploitation of the subsystems developed in these programmes are in progress. This is more the case in Japan, where for example, autonomous navigation systems are now offered as part of new in-car options, and this has proven to be popular with users. It is estimated that approximately 12 000 vehicles per month are being sold in Japan equipped with navigation systems at approximately US$3000 per unit. On the whole, the current market for the navigation system is one-half million units in Japan plus 10 000 split between the USA and Europe. In the USA, the majority of the market is currently: commercial vehicles, rental cars, and other selected cars. The predicted annual market in Japan is 3 million units with a market value of US$490 million. Enablers of the market in Japan are the lack of street names, and the numbers and landmarks being historical rather than sequential. The USA and European market enablers will be the real-time traffic information.

Table 1.5. Functional features of UTMS

Areas of development	Purpose of systems	Function of systems
Integrated traffic control system (ITCS)	The aim of the system is to develop intelligent systems which achieve the automatic generation of traffic signal parameters and advanced information processing. Infra-red vehicle detectors and other advanced sensors are incorporated in the system	Installation of infra-red vehicle detector Installation of U-type traffic signal controllers for centralized control
Advanced mobile information systems (AMIS)	To provide drivers with sophisticated traffic information	Provides road and traffic information on congestion, accidents, road works, etc. Provides travel time to destination Provides flexible traffic information
Mobile operation control systems (MOCS)	To give support for efficient transport management of trucks and other transport industry vehicles	Provides information on travelling location Provides travelling time Provides information on the state of the vehicle Provides traffic and regional information Provides security information
Dynamic route guidance systems (DRGS)	To advise the driver on the optimum route to get to a given destination	Provides the route selection information Provides travel times to destination Provide route guidance information Provide security information
Public transport priority systems (PTPS)	To improve the efficiency and conveniences of public transport, as well as ensuring transit priority of buses and other high occupancy vehicles	Gives transit priority to public transport vehicles Provides travel times Provides travelling information Sets up and clears bus lane, etc
Environmental protection management systems	To reduce traffic pollution for environmental preservation	Advanced signal control methods Provides and displays the recommended speed Provides traffic pollution information Provides alternative routes to avoid entering heavily polluted areas

1.3 ELEMENTS OF INTELLIGENT TRANSPORT SYSTEMS

As new and different private enterprise enters into the design and development of the elements and devices that make up ITS, it becomes difficult for one book to cover all the elements, especially where the development is still dynamic and fast progressing. However, it is essential to note that ITS can be divided into several component parts. These will either belong to in-vehicle systems, or part of an infrastructure of the transport system. The infrastructure is used to define any part of the ITS not resident in the vehicle, but which may or may not interact with the vehicle. For in-vehicle and infrastructure systems to operate, it is likely that some form of sensor will be required.[2] We will now discuss some of the elements of ITS which are parts of the programme areas discussed earlier in this chapter. Other elements not presented in this chapter, such as those involved in vehicle control tasks, are discussed in subsequent chapters.

1.3.1 Traffic Management and Control

This can be categorized into three areas, namely, urban traffic management, inter-urban traffic management and demand management.

Urban traffic management has several sub-functions, part of which includes, traffic control, ramp metering, variable message sign (VMS), traffic light management weather monitoring, pollution monitoring, long distance route guidance, limited access regulation, public transport network management, service and emergency vehicle management, etc. There are several of these systems in operation or being developed in different parts of the world, examples of which include:

(1) Sydney co-ordinated adaptive traffic systems (SCATS), Australia;
(2) split cycle offset optimization technique (SCOOT), UK;
(3) optimization policies for adaptive control (OPAC), USA;
(4) universal traffic management system (UTMS), Japan;
(5) advanced traffic control systems (ATC), Japan.

Demand management seeks to influence traffic demands during a given time so that such a demand does not exceed the effective ability of the transport/traffic network to accommodate the requirements in terms of flow, delays, energy consumption, exhaust pollution, safety and intrusion. This can be achieved through policies which seek to alter transport systems user behaviour in terms of time, space or mode in order not to exceed network capacity. Various demand management techniques are being explored and tested in different part of the European Union. Techniques such as constraints on private cars traffic (e.g. car-free inner cities, speed limitations, zone access control systems, etc.) are normally supported by incentives to use public transport or multimodal systems, such as park and ride.

1.3.2 Public Transport and Freight Operation Systems

There are a number of developments which form the growing subsystems in public transport and freight operation systems. The trends of development required to support and progressively realize better public transport and freight operation systems over time are shown in Figure 1.6.

Some systems such as automatic vehicle location or tracking and bus detection and priority at traffic signals are some of these developments.

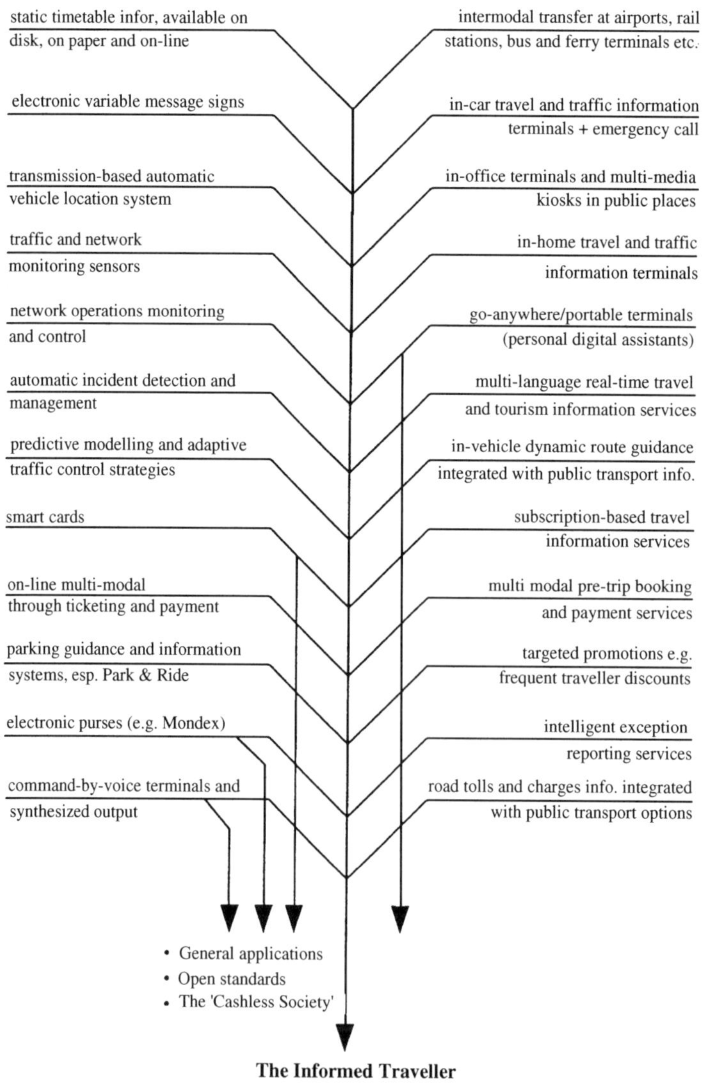

Figure 1.6. Development track of public transport system for the informed traveller[6]

Automatic vehicle location (AVL) are vehicle tracking systems which employ hardware equipment such as odometers, roadside beacons or GPS receivers; voice and data communication such as radio data systems (RDS); and on-board and ground equipment. The technology of AVL has been in existence since the 1980s, but in recent times, there has been a transition from odometers and beacons towards definite continuous systems such as GPS for vehicle tracking.[3] AVL systems are designed to provide vehicle location information to a central control, rather than to the vehicle driver. This system can be used for freight monitoring, especially when integrated with automatic identification and electronic data interchange systems.

Bus detection and priority at traffic signals are systems which are capable of detecting passenger service vehicles in order to prioritize them in traffic network. The systems are well established and they can be integrated into other functional systems of the vehicles. The detection technologies used include inductive loops, microwave and infra-red transponders and radar. This can be linked with AVL and a systems management strategy can be used to determine which class of passenger service vehicle gets a specified level of priority within an integrated urban traffic control system.

1.3.3 Mobile Telecommunications Systems

Mobile communication is an essential part of ITS which provides the link between the driver/vehicle to the external systems. Essential information on the existing mobile communication is provided in further chapters of this book, but three methods of communication are introduced as follows:

1. *Cellular radio system* is a communication system which is capable of supplying medium range two way communication link needed for traffic and transport data exchange. For detailed discussions of cellular radio system readers may want to refer to Ott.[4] There are ITS trials that have been tested in Europe based on this system. In Germany a project entitled RHAPIT was tested in the Rhein/Main area, while in Paris and Gothenburg ELITE and TANGO projects respectively were tested under the SOCRATES (System Of Cellular RAdio for Traffic Efficiency and Safety) project.[5] Further development of the SOCRATES Project to extend it within the European GSM standard for digital mobile telephone communication is in its final stages.

2. *Radio data system–traffic message channel (RDS-TMC)* is a subcarrier system which enables information to be conveyed to the driver via the normal VHF-FM vehicle radio broadcasting systems. This can be used to provide the motorist with the information on road conditions, traffic and weather conditions. RDS-TMC is an extension of the normal radio broadcasting RDS system which will enable traffic messages to be carried digitally and silently as a one-way transmission without interfering with the normal vehicle radio programme.

3. *Roadside beacon systems* is part of the infrastructure within the transportation network. It is a roadside electronic device for a two-way communication of data to and from vehicles within the network. It has transmitter and receiver capability which may employ infra-red, inductive loop or microwave signal. For communications between vehicle and roadside beacon a 5.8 GHz two-way microwave link is strongly favoured. Within the PROMETHEUS programme, GEC Marconi and ANT-Bosch have developed this type of system in conjunction with Jaguar Cars. The car antenna is in this case mounted in the top centre of the windscreen behind the rear view mirror and the roadside antennae mounted on traffic-light poles or overhead gantries. The system is known as TRICS (traffic and road information communication system).

TRICS sends and receives code numbers rather than text characters. This takes less air time and is necessary for pan-European operation. For example, the message "FOG" could be represented by digital code number 10010 so whatever the country, a fog warning would always be conveyed over the microwave link by the code 10010. So whenever such code is received in the car, the receiving unit will translate the code into the driver's own language.

The communication message from the beacon to the car fits into a time slot of 28.8 ms. During this time a vehicle travelling at 50 km/h moves less than half a metre so the message will be received several times, enabling it to be checked for accuracy. This helps to overcome the problems of the signal being reflected from nearby buildings and passing vehicles.

Several other field trials have been carried out in other countries, for traffic, traveller, automatic debiting and route guidance information. An example is the Siemens Euro-Scout system, which uses an infra-red communication link.

1.3.4 Debiting and Smart Card Systems

As we move towards transport network decongestion, optimizing and managing the demands made to the network, it will become essential to integrate network payment systems (electronic toll collection system) with ITS. Automatic debiting systems or electronic toll collection are systems that enable a vehicle to communicate with a roadside infrastructural device and register the presence of that vehicle as it passes through the designated area. Automatic debiting systems for highways are now being applied in several cities, such as urban cordon tolling experimented in Norway and interurban debiting used in Italy or in some French motorway network. About US$23 billion of tolls are currently being collected worldwide, with the USA accounting for 16%, Japan for 64% and Europe for 20%. Some of the systems use a short-range communication frequency allocated to ITS functions. Parking control and zone access control can also be implemented using these technologies. Enforcement of the tolling system can be effected by automatic vehicle classification which can use tags or image processing techniques. The basic concept of automatic debiting systems which permits non-stop toll collection is shown in Figure 1.7. Further extensions to standard debiting systems include two-way transactions, which also can allow transaction to be progressed at high vehicle speed of more than 150 km/h. Having a smart card in the vehicle would allow a vehicle driver to drive past the toll collection point at high speed without stopping to pay the toll manually.

Smart cards are being considered as a single universal card which can be used to pay for tolls, parking, mass transit and other host of services. The smart card can have all

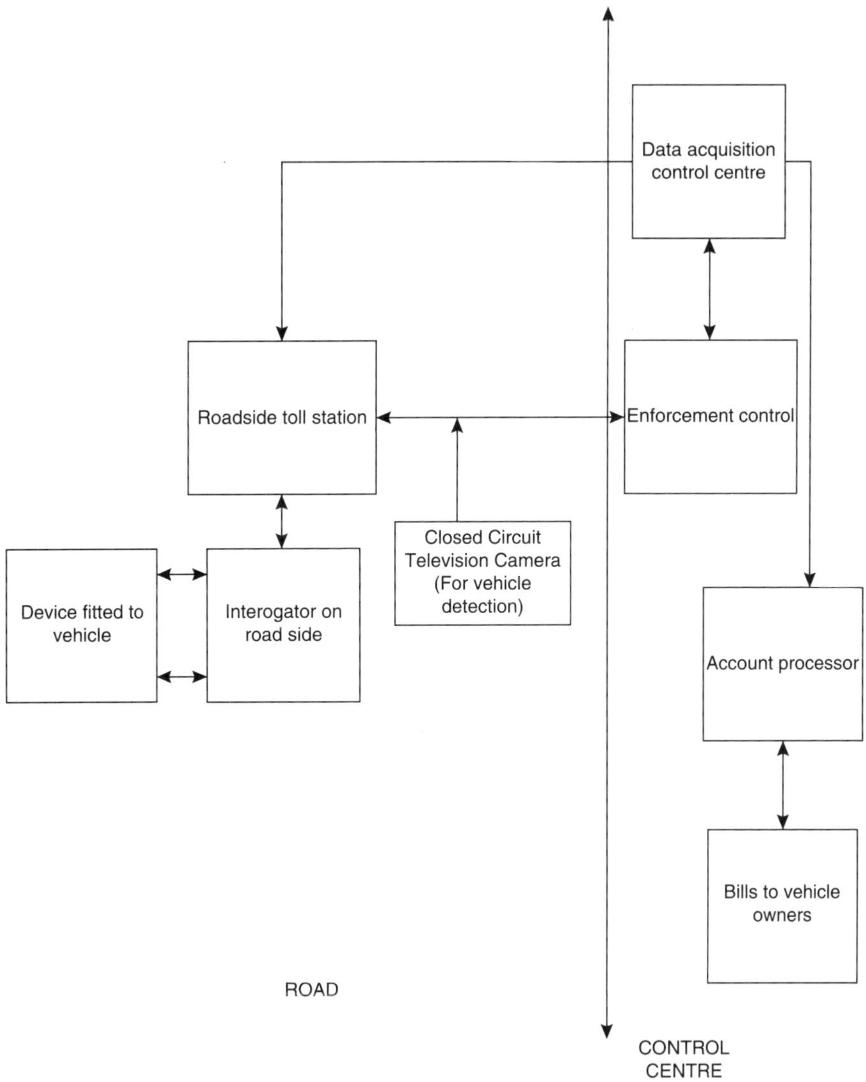

Figure 1.7. Basic concept of an automatic debiting system

the essential personal information of the car, the driver and the network transactions. Specific transport masks for smart cards and magnetic cards are being standardized within Europe for toll collection. Multilane systems of toll collection and two-way automatic debiting involving possible writing on the driver's card were successfully tested in Europe. Dynamic ticketing using contactless cards is also possible for fare collection in public transport, although there are other fare collection systems that can be used (e.g. electronic ticket machines).

1.3.5 User Information Systems

The user of the ITS network needs to be better informed, in the description contained in the UK Technology Foresight,[6] he or she should be an "informed traveller". The user information system is a medium through which the transport user is presented with information about the vehicle and its environment within the transport network. The information provided to the user allows the user to make judgements about the prevailing network conditions. There are various systems currently being developed to enable the user of ITS to be the informed traveller, two examples of these are.

Passenger information systems

These provide the necessary information about the transport to the user. There are several types of the systems that are being developed, these include; onboard information displays, information displays at the vehicle stops and videotext services, etc. Some of these systems provide information concerning the current location of the vehicle or the waiting time to the next vehicle. Some of the European and North American systems can deliver precalculated routes and may be linked with a geographical information system.

Future interactive information terminals

These are futuristic systems which may be in-vehicle or at the vehicle stops, such as, bus stops. They are proposed to extend the capability of the current user information system to include the use of natural language, multimedia, use of personal digital assistant, multimodal information for route guidance and subsequently linked to AVL systems.

1.3.6 Advanced Vehicle Safety Systems

As shown in Figure 1.8, these are systems of hardware and software dedicated to increasing the safety and efficiency of vehicle operation.

The in-vehicle systems that are required to assist the driver to avoid accident under traffic conditions, and are classified as advanced vehicle safety systems user services are as follows.

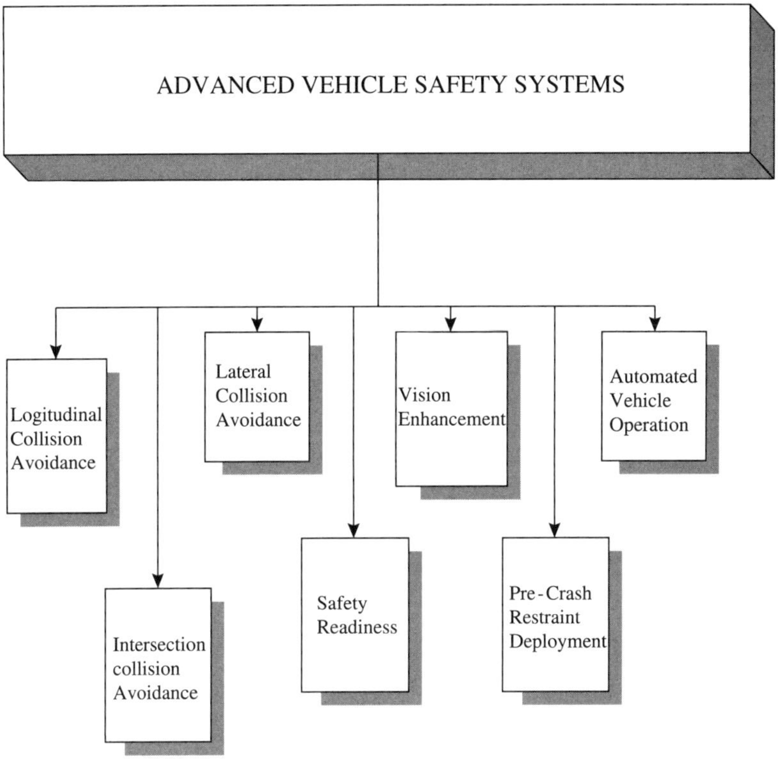

Figure 1.8. Components of Advanced Vehicle Safety Systems

Longitudinal collision avoidance

Vehicle crash when the vehicles are moving in a parallel path of travel (i.e. X-axis of motion) prior to collision, or those which involve a stationary vehicle can be reduced by integrating sensory systems on the vehicle. This type of sensory system can dictate such crashes under a safe time limit, and subsequently inform the driver or automatically initiates an avoidance action.

Lateral collision avoidance

Side impact on vehicles, along its Y-axis of motion, is a common occurrence in traffic accidents. This can be vehicle to infrastructure or vehicle to vehicle collision. Sensory systems can be used to augment the driver's ability to avoid this type of side impact or lateral collision.

Intersection collision avoidance

This is the type of sensory-assisted system which will provide a warning of possible collision when a vehicle violates the road priority of another vehicle to the point of causing collision or when the right of way at a road intersection is not clear as to cause collision.

Vision enhancement

Driver visibility can be reduced at night, under foggy conditions and during poor weather conditions. Poor driving visibility conditions have the risk of causing traffic accidents. To help in the reduction of the number of collisions caused by poor visibility, sensory systems which enhance driver vision are being developed for collision avoidance.

Safety readiness

These are in-vehicle sensory systems that will help to give advanced warning about conditions such as vehicle component failures, driver impairment and poor infrastructure conditions. These systems are being researched for inclusion into vehicles.

Pre-crash restraint deployment

Systems linked to collision avoidance systems can be used for detecting an imminent collision and subsequently deploying other safety devices, such as air bags or active/intelligent side-impact bars.

Automated vehicle operation

These are systems that are capable of providing a fully automated driving control. Autonomous systems that will provide hands and/or feet free driving operation with the aim of improving safety and efficiency are being developed.

1.4 INTELLIGENT TRANSPORT SYSTEMS USER REQUIREMENTS

The overall approach to the development on ITS must be application driven, but above all, it must be firmly grounded on the needs of the users. The users of transportation systems have been instrumental to the various development in ITS. A private individual requires a transportation system to satisfy basic needs such as; safety, comfort, socially acceptability, convenience, security and privacy. At the same time, the commercial user requires a transport system to be reliable, cost-effective, flexible, secure, safe and adaptable. Society as whole needs to travel without delay due to traffic congestion, but expects to use a transport system with a lower environmental impact.

In all the areas that are being researched and developed for ITS, it is not only the technology that will determine the exploitation and implementation of the systems, but rather what the user requires and how much they are willing to pay. There are a range of users of ITS products which include the driver, fleets operators, individuals, public transport passengers, local authorities, network operators, etc. As these groups will inevitably be influencing the "take-up" factor of technology, it is important to understand their requirement profiles.

1.4.1 Public Transport

Passenger transport systems which are used for public mobility will be an important area of transportation that will receive the benefits of the recent development in ITS. The users of this class of transportation should be encouraged to use it, but also more and new users should be motivated to use public transport. To make this possible it must be an essential part of any ITS research and development strategy to understand that public transport users have unique requirements. Some of these requirements include:

(1) increased certainty and improved control over their journey;
(2) high quality of service and flexibility that can compete with private cars;
(3) a transferable and integrated fare system;
(4) a comprehensive passenger information system;
(5) timetable reliability and regularity within the network;
(6) increased personal security (e.g. in suburban buses, subway/undergrounds, etc.);
(7) easy access to multimodal services (e.g. transferability of tickets, park and ride, etc.).

1.4.2 Freight Operators

The requirement of the freight operators is that which enables the operation of the vehicle freight to meet customers and organizational needs at the right time and low cost. Some of the freight operators', requirements include:

(1) ITS tools for better freight operation;
(2) systems for driver and organizational information (e.g. communication systems, transit facilitation, route guidance, etc.);
(3) systems for monitoring of freight within the network (e.g. tracking systems, etc.);
(4) standardization of data interchange protocol for exchange of freight management information (e.g. trans-boarder crossing facility, common invoicing system, etc.);

(5) traffic information and better weather forecasting systems;

(6) safety and security systems.

1.4.3 Fleet/Freight Drivers

In working to make the road transport systems more intelligent, it is important to understand the needs of the fleet driver, whose main occupation is driving to move passengers or goods from end-to-end of a journey. For fleet/freight driver, driving is a continuing recognition–judgement–action. For the driver to carry out this human–machine interface task and at the same time deliver the contents of the vehicle at the given time and destination, driver requirement analysis would include:

(1) in-vehicle ITS devices (e.g. driver information system, communication, safety, etc.);

(2) navigation and route guidance;

(3) better intermodal facilities;

(4) traffic and weather information.

1.4.4 Private/Individual Users

In a democratic world the needs of individuals are paramount to the development and exploitation of ITS products and subsystems. Whether the individual is in his/her private vehicle or public transport, there are basic user requirements which should be considered, some of which are:

(1) current and forecasts of the traffic and weather conditions;

(2) route information (e.g. hazards, emergencies maps/navigation, etc.);

(3) safety and security systems;

(4) toll payment information within the network;

(5) information on other transport modes, timetables, park and ride, fares, etc.;

(6) personal communication system;

(7) manoeuvring aids (e.g. for parking);

(8) freedom from breakdowns and accidents.

1.4.5 Network Operators

The cities, local authorities and toll companies have certain requirements which are based on what ITS product, services and technologies should provide. Examples of these requirements are:

(1) security and safety systems (e.g. vehicle identification and tracking, etc.);
(2) network integrity and reliability;
(3) automatic traffic and transport data acquisition;
(4) integrated transport and traffic information system;
(5) models and systems for transport emission monitoring and forecasting within the network;
(6) systems and tools for better enforcement of transport policies;
(7) communication between traffic control centre;
(8) tools to improve network use efficiency.

In this section examples of ITS user requirements have been listed, but readers must be aware that the lists are only used as examples. There are in fact several factors which have not been listed, especially with the issues of systems reliability and risks as the user sees it. The user's requirements in terms of cost–benefit and risk–benefit modelling is an important area of user requirement analysis which is currently receiving research interest.[7]

1.5 CONCLUSIONS

The aim of this chapter was to introduce to the reader aspects of the historical issues that contribute to ITS development. It was also intended that the reader who may be new to the technology involved in ITS will utilize this chapter for the necessary background information on ITS. Some of the technologies that have been covered are still going through development. It is evident from the systems presented that ITS technologies are developing at great pace. If the majority of the systems discussed were fully implemented, the transportation of the future will be very different from what we have at the moment.

Most of the basic technologies for ITS have been tested. There should now be a combined effort to prove their reliability, cost-effectiveness, low-risk factors and harmonize standards throughout the world transport communities.

REFERENCES

1 Sheild TR. Programme and technology of IVHS-America. *Communication Techniques in Road Traffic Informatics, Proceedings of the DACAR Conference, Rome and Brussels*. Amsterdam: Bakkenist Management Consultants, 1991.
2 Nwagboso CO. *Automotive Sensory Systems*. London Chapman & Hall, 1993.
3 Scorer T. The current status of the securicor datatrak system. *2nd International Conference on Road Vehicle Automation*, 11–13 September 1995.

4 Ott GD. Vehicle location in cellular radio systems. *IEE Transactions on Vehicular Technology* 1977; **26**, (1) 43–60.
5 Catling I, Op de Beek F, Casimir C, Mannings R, Zijderland F, Zechnall W, Hellaker J. SOCRATES: system of cellular radio for traffic efficiency and safety. *Advanced Telematics in Road Transport: Proceedings of the DRIVE Conference, Brussels.* Amsterdam: Elsevier, 1991.
6 Office of Science and Technology. *Technology Foresight (Transport): Progress Through Partnership*. London, 1995.
7 Godfrey D, Nwagboso C, Ball R, Davies D. *IMI Land Transport Telematic Programme Recommendation for Focused Call.* Engineering, Physical and Science Research Council Report, UK, 1996.

2

Microprocessor-based Control Technique for Vehicle Performance Optimization

2.1 INTRODUCTION

This chapter introduces some important recent developments in auto electronics that will enhance the performance and the intelligence of the modern vehicle. Performance of a vehicle is judged by parameters such as economy, safety, reliability, comfort and ease of use. Intelligence of the vehicle is judged by how well the environment in and around the vehicle is being monitored dynamically by the sensory system, whether prudent decisions are being taken by the computer system based on the measurements of the sensory system, and finally how well the actuators are responding to the commands being issued by the computer system. The electronic system broadly consists of the sensory system within the vehicle, the analogue-to-digital (A/D) converters that convert the measured values into the digital form, the multiplexers and the data paths that will bring these data to the computer system, the computer systems itself, the demultiplexers that will take the digital values from the computer system to the actuators through the digital-to-analogue (D/A) converters and finally the actuators that will enforce the decisions of the computer system. The above parameters are entirely technical in nature. There are other parameters such as the economic and reliability parameters. Ultimately, the vehicle must be cost-effective to attract the customer, and the design should be reliable so that the vehicle maintenance and repair will not place an undue burden on the customer. Most of the research and development work in the automobile

Advanced Vehicles and Infrastructure Systems, Edited by C.O. Nwagboso

technology is constrained by the economic, safety, reliability and regulatory parameters. Manufacturers are very hesitant to increase the number of electronic gadgets within the vehicle, unless it is proven that such an increase will result in increasing the system intelligence proportionately, and the electronic system makes optimum use of the new devices. Manufacturers have their bitter experiences when some designs which are technically sound, have failed in the market-place, because the product is not cost-effective.

Reliability is another important area that eventually decides the reputation of a vehicle. Most vehicles will operate reliably in the first 2 or 3 years. Problems start from then onwards. Major repairs such as transmission failures or motor breakdown are not very common problems. Most repairs are simpler in nature, and in many cases the very sensors intended to indicate certain failures themselves will develop errors, causing a great deal of confusion. Vehicles must have reliable components and there must be a watchdog diagnostic network that can give technicians information about the probable causes of the failure. It is very likely that future vehicles will have some of these features. In recent years, automobile manufacturers have improved in the area of reliability substantially, mainly due to the international competition. Also, multiprocessor-based systems are being increasingly used to monitor various sensor outputs and control the vehicle performance dynamically.[1–8]

The topics in this chapter are organized as follows. In the next section, all preliminaries are discussed. They include all the topics that are relevant to the auto electronics such as the sensors, intelligence in a sensor, sensory network, the intelligence of the network, concepts of multiplexers and demultiplexers in the context of automobiles, A/D and D/A converters and the resolution of the data, and single bus multiprocessor architectures. In Section 2.3, parameters of vehicle control and diagnostics are discussed. In Section 2.4, simple vehicles are considered and a single bus architecture for the engine control and diagnostics is presented. In Section 2.5, a single bus architecture with some degree of fault-tolerance for the real-time control is presented for the complex vehicles. In both the cases multiprocessor architectures are designed for the dynamic control of the various engine, body, ride, and safety parameters and diagnostics. Dynamic parameters such as fuel/air mixture, spark timing, etc., are controlled by executing the control routines. Control of the static body parameters and the dynamic parameters are implemented by using a single bit serial bus. The concept of diagnostics is emphasized both for the static and dynamic parameters. Finally, in Section 2.6, some design ideas of the future vehicles are expressed.

2.2 PRELIMINARIES

All components of a typical automobile electronic system are explained in this section. The importance of each component is discussed in relation to the overall system performance. The topics include the sensors and actuators[9,10] and their intelligence, A/D and D/A converters, multiplexing and demultiplexing[11,12] and finally the multiprocessor architectures.[13–15]

2.2.1 Sensors and Actuators

Sensors are instruments that will measure various parameters such as temperature, pressure, fluid levels and fuel/air mixers. Positional sensors will measure linear and rotary positions, speed sensors will measure speed, acceleration sensors measure acceleration, oxygen sensors will measure the oxygen content and are used in the engine and in the exhaust system, etc. In the simplest form, a sensor is like a binary switch, i.e. it will indicate yes or no. The oil and fuel lights we see on the dash boards are examples of such binary sensors. These sensors represent the smallest amount of intelligence. There are more complicated sensors that actually measure the quantity of a parameter within the tolerance limits and will convert the measured value into an analogue electrical signal. These latter type sensors are costlier than the former type, and they carry more intelligence in their measurements. Both of these sensor types can be broadly classified as contacting and non-contacting types. Some sensors such as the temperature and oil/fuel pressure gauges are exclusively of contacting types. Contacting sensors suffer from wear and tear over a period of time. In the case of automobiles, where most cars do not stay on the roads beyond 15 years, contacting sensors are reasonably reliable. In recent years, there is a substantial improvement in the design of these sensors. Interested readers may refer to volume I of this series,[10] which is an excellent reference for the various types of sensors in the automobiles. Most non-contacting sensors are of visual type and have some kind of camera. An image of the object under measurement is taken, and is converted into electrical signals. These sensors have better longevity as they do not make contact with the object, but the built-in camera facility makes them costly. Most of the present-day automobiles do not use these non-contacting sensors. These sensors may have larger applications in the future vehicles. A sensory system is a network of the sensor units. The data measured by the sensors are communicated to a computer system which processes the data, understands the situation, and sends commands to the actuators to take appropriate action. The intelligence of a sensory system therefore has two components, the intelligence of the individual sensors and the intelligence embedded in the topology of the sensory network.

Sensors are used at many places in the automobiles. For example, engines have a set of sensors to measure various engine parameters; there are sensors in the air-conditioning unit, anti-lock brakes, power steering, exhaust units, transmission unit, suspension system, anti-skid mechanisms, etc. Most of them are temperature or pressure transducers. Sensors that carry larger amounts of intelligence are called smart sensors. These smart sensors will be used increasingly in the future designs. An actuator on the other hand is the mechanism that takes the action. In most cases, the signal sent by the computer will magnetize an electromagnet, which in turn will pull or push mechanical parts such as leavers to implement the action.

2.2.2 A/D and D/A Converters

The electrical signals measured by the sensors being analogue (continuous) signals, it is necessary to convert them into the digital form before sending them to the computer system for processing. An A/D converter simply accepts an analogue signal as the input and will produce the equivalent digital signal as the output. There is an important concept know as the resolution in the A/D conversion. For example, assume that all the signals are between 0 and 5 V. If we use only 1 byte (i.e. 8 bits) to store the equivalent digital value, then each digital count differs from the previous count by an amount equal to 256th of 5 V. This is the resolution for this unit. Suppose we wish to measure smaller quantities than this, then we should go for 2 bytes to store the digital value. This will give better resolution by a factor of two, but will double the size of all data paths, multiplexers, etc. For an application, the designer will calculate during the design phase, whether the single byte resolution is adequate or not. For automobiles, single byte resolution is considered adequate.

The D/A conversion is the reverse processes. When the computer system monitors the sensory values, it will use these values to compute some built-in control routines and generates commands in the form of digital values. These values are converted into continuous signals and are given to the actuators. Actuators will enforce the action determined by the computer system. There are many processors and even some complicated sensors and actuators, that have built-in A/D and D/A converters.

2.2.3 Multiplexing, Demultiplexing and Buses

In this section, we will discuss the concepts of multiplexing, demultiplexing and the concepts of serial and parallel buses. When data are coming from various sources such as the sensors, and all of them should reach the same destination such as the computer system, a combinational unit is used to route the data from each source to the destination, one at a time. This unit is called a multiplexer and the unit that performs the reverse process, a demultiplexer. The bus is a collection of wires along which binary signals can flow. Each bus will have a register to serve as the source of the data. Buses are classified as parallel and serial. Suppose, the 8 bits of data should reach the destination from the source, all the eight values can reach the destination at the same time in one clock pulse along eight wires. This makes the bus a parallel bus. On the other hand, if there is only one wire in the bus, it takes eight clock pulses for the 8 bit data to reach the destination. This is called the serial bus. Serial transmissions also include some other bits such as the start and stop bits, parity bit, etc. While the serial transmission is slow, it takes only one wire, for the data. In automobiles, serial buses are used most frequently (called the twisted pair system) due to the low cost and simplicity of wiring. The sacrifice in speed is affordable in the case of automobiles as the required response time is not so acute. We will first illustrate multiplexers and demultiplexers[12] with examples and will then illustrate a typical serial transmission.

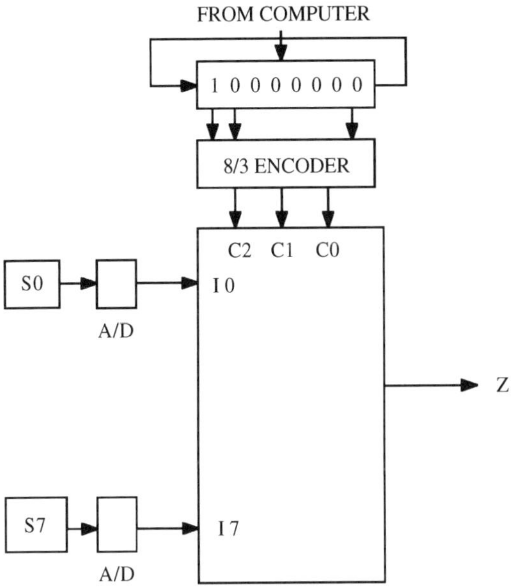

Figure 2.1. Multiplexer unit with polling/selecting

Consider the multiplexer unit with polling/selecting scheme given in Figure 2.1. The multiplexer has eight input data lines $I0, \ldots, I7$, each data line Ij coming from a sensor Sj through an A/D converter. $C2$, $C1$ and $C0$ are the three control lines and can have a value between 0 and 7. If the value on these three control lines is j, then Ij will be connected to the output Z. The polling register has a 1 in only one of the eight places and this position (0–7) is converted into a binary value by the 8/3 encoder. This encoded value becomes the value on $C2$, $C1$ and $C0$. The one bit in the polling register is rotated, so that each sensor value is connected to the output Z, which in turn is connected to the computer system. This way the computer can receive the measured values from all the sensors.

The operation of a demultiplexer is the exact opposite and is shown in Figure 2.2. The input I comes in one line from the computer system. The demultiplexer has eight outputs $L0, \ldots, L7$. The control lines $C2$, $C1$ and $C0$ in Figure 2.2, will decide which output is connected to the input. The eight outputs $L0, \ldots, L7$ go to the actuators $A0, \ldots, A7$ through a D/A converter as shown in Figure 2.2. In each case, the computer selects one actuator using similar polling mechanism. It is important to realize that the computer system may not decide to go through the sequence of actuators in any regular order. At any time, the computer may load a new value into the polling register to activate a specific actuator. This may happen in the case of sensors also. But it is more common for the computer system to maintain a regular order while monitoring the sensor values than while effecting the actuators.

Figure 2.3(a) describes a typical serial transmission. All the local modules are connected to the system, and each module has an address. The transmission starts

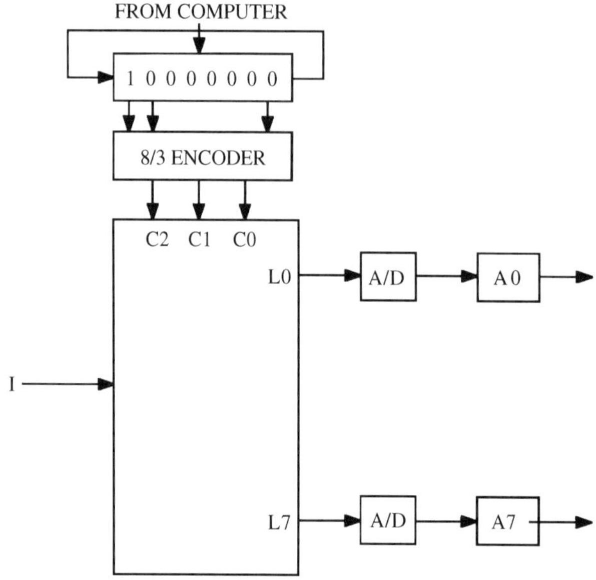

Figure 2.2. A demultiplexer unit with polling/selecting

(b)

Mode	Operation
0 0	Test
0 1	Activate operation
1 0	Deactivate Operation
1 1	Unused

Figure 2.3. (a) Serial data format. (b) Mode definition

with a start bit, followed by the address of the unit. The two-bit mode will dictate whether the transmission is for testing, activation of an operation or deactivation of an operation. The last mode configuration in Figure 2.3(b) is unused and may be used for some specific transmission. All the data are transmitted serially this way from the local modules to the computer and vice versa. As mentioned earlier, the degradation in the speed is acceptable for most automobile applications.

2.2.4 Multiprocessor Architecture

Multiprocessor architectures[13–15] has been an important research area in computer architecture for the last 15 years. The objective is to build parallel architectures using many smaller processors and distributed memory modules so that all or most of the processors can simultaneously execute codes, increasing the system throughput. Single bus architectures are some of the earliest and simplest objects of study.

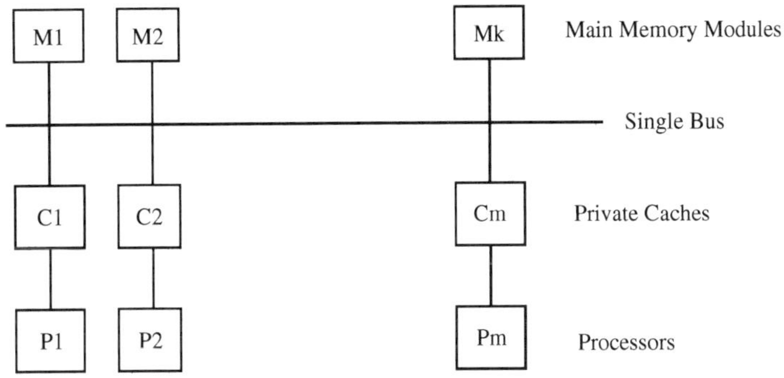

Figure 2.4. A single bus multiprocessor system with private caches

Figure 2.4 illustrates such a single bus architecture. The main memory is subdivided into many modules. There is one bus, through which all the processors and memory modules communicate with each other one pair at a time. The organization of Figure 2.4 uses private caches associated with each processor. There are other architectures such as the multiple bus architectures, which are more powerful than the single bus architectures, but these architectures are too costly for the automobile applications. Also, the single bus architectures are adequate for the computational needs of most automobiles, including the newer designs in the foreseeable future. In this section, we introduce some terminology and will introduce the typical function of a single bus architecture. The architectures for simple and complex vehicles are presented in Sections 2.4 and 2.5, respectively.

In Figure 2.4, $M1, M2, \ldots, Mk$ are k memory modules, $P1, P2, \ldots, Pm$ are m processors and $C1, C2, \ldots, Cm$ are the m cache memory units, each associated with a processor. Blocks from any memory module may be loaded into any cache at a given time and we shall assume the set-associative block placement policy with the least recently used (LRU) block replacement policy. Each processor may also have its own private memory and is not shown in Figure 2.4.

When a processor wishes to read a data item in a block, and if the most recent copy of the data is found in the cache, then it is called a read-hit. It is a read-miss, either if the block itself is not in the cache or if the data in the block are not the most recent. Similarly, it is a write-miss if either a block is not found in the cache when a processor wants to write into the block, or if there is no write access for the block. If the block is present in the cache with a write access, then it is called a write-hit. In case of a write-miss because the block is not in the cache, the block must be brought into the cache from a memory module and then updated. At the same time, if the same block is in other caches, then the data in those blocks will not be current after the present write operation. Also, the data in the original memory module will not be current. Caches having old data copies is called the cache coherency problem and the system must somehow invalidate the old copies to avoid reading erroneous data. There are two policies to handle the cache coherency problem. The first one is called the write-through policy. In this policy, whenever a block is updated in a cache, the new data are also written immediately into the main memory and the copies of this block in the other caches are invalidated. This policy is not very efficient as the computer must immediately update the block in the main memory irrespective of whether this block is needed by another processor or not. There is another policy called the write-back policy. In this, only the block in the cache is updated. The copies in the main memory and other caches are simply invalidated. If the same block is being updated in the same cache repeatedly (which is usually the case), then invalidation is necessary only for the first time. This block is updated in the main memory, only when another processor encounters a read-miss for the same block. We will use this policy for our architecture in Sections 2.4 and 2.5.

Single bus architectures reach saturation by the time there are about 15 processors in the system.[13] Beyond this point, the traffic along the single bus becomes so crowded, any addition of new processors will not enhance the performance. Fortunately, for the automobile applications, the computation power with 15 processors is very adequate even under the presence of the most complex sensory network.

2.3 VEHICLE CONTROL AND DIAGNOSTICS

Vehicles can be broadly classified into three categories: (i) small vehicles (all automobiles, vans and pick-up trucks, etc.); (ii) medium-sized vehicles (large trucks and small aircraft); and (iii) large vehicles (large aircraft, spacecraft, etc.). The concept of employing microprocessor-based systems to monitor sensory data dynamically to optimize performance, is common to all the vehicle categories. However, the precise processor architecture to be used is a function of many vehicle parameters both from the economy and speed points of view. For example, in smaller automobiles, the number of sensors effecting the performance decisions are relatively fewer. Consequently, the processor controlling engine operations can be interrupted in real-time by the sensors, without adversely effecting the control. So, for smaller vehicles, simple architectures using a few microprocessor would be adequate. A larger number of microprocessors are required in complex vehicles. Presently, most of these microprocessors control various operations in isolation. Microprocessor architectures using two or four microprocessors are being used only for engine control in some vehicles. This section discusses both dynamic performance control and diagnostic parameters to be used in deciding the appropriate microprocessor architectures for different types of vehicles.

2.3.1 Control Parameters

Some of the most important parameters to be considered while deciding upon the processor architecture are listed below.

1. *Vehicle speed.* The real-life operating speed range of the vehicle is a very important parameter. This in turn decides how fast the processor system should respond to the changes in the sensor measurements. For automobiles, trucks, etc., this is not a very serious problem. For aircraft and spacecraft, this issue is very critical. Controlling an operation within a time frame poses the most serious challenge for both hardware and software real-time computations in the spacecraft.

2. *Vehicle size.* The physical size of the vehicle is another parameter to be considered. The level of sophistication in a vehicle remaining constant, the complexity of sensory network and the required processing power of the computer system depends on the physical size of the vehicle. For example, we need a more complicated network for a truck than for a car, to implement the same parameters of ride and safety.

3. *Intelligence in the sensory system being planted.* The level of intelligence supported by the sensory system being planted in the vehicle is another important parameter. Ten years ago, sensors were used only for the most important functions and were very simple in design. The usage of sensors has increased tremendously during the last 10 years.[16] Nowadays, many drivers are not even aware of the list of sensors they have in their automobiles. There is a marked improvement in the areas of traction control,[17] anti-lock brakes[18] etc. More intelligent sensory systems invariably involve large numbers of and more sophisticated sensors. This will have a direct bearing on the computer system and it's architecture.

2.3.2 Controls

Controls are of two kinds: (i) driver initiated, and (ii) sensory system initiated. Driver-initiated controls are very simple controls such as head lights on/off, wind shield wipers, etc. It is also necessary to incorporate diagnostics with them also. For example, if one or both lights got burned out, then the driver should get a message. These diagnostics are already available on some automobiles. Sensory-system-initiated controls are more complicated. For example, a sensor unit reads data and sends it to the computer in digital form. The reading indicates either an abnormality (as in the case of skidding) or a changed dynamic situation (as in the case when brakes are applied).

No matter what the reason for the change might be, the process will read the data, proccss it and initiate a corrective operation (as in the case of anti-skid control) or simply enforce the required control on the engine (such as reducing fuel to engine in the case of braking). The time of response depends on the nature of control. Most controls of abnormal situations must be immediate. These controls are given the highest priority by the computer system.

2.3.3 Diagnostics

Every automobile owner realizes the hardships caused by failures. It is more frustrating when the failure persists after the repair. Diagnostics is a very important area from the time, money and safety viewpoint.[19] Proper utilization of the intelligence in the sensory system can be used to store more precise diagnostic information, in the memory. This information is down-loaded into the diagnostic unit of the mechanic, to give precise diagnostic information during the repair. Theoretically, it is a good idea to increase the complexity of the sensory network inside the automobile for the sake of the diagnostics. Currently, the cost factors do not permit such an increase in the complexity of the sensory system. By proper planning in the design phase, some of the redundancies in the sensors also may be used to improve the diagnostics.

Body control diagnostics

They deal with displaying diagnostic messages regarding various body controls, such as front/rear lights, oil level, fuel level, AC unit, heater, etc. Diagnostics are not very important for certain body controls such as windshield wiper, etc.

Engine control diagnostics

Diagnostics in engine control are more important than the body control diagnostics. The computer executes control routines in real-time using the sensory data and controls the engine functions. These routines are stored in ROM and correspond to the fault-free assumptions. One or more sensors may fail to send the data to the engine computer. This may be due to a failure in the sensor itself or in the A/D converter, loose connections in the wires that bring these values to the computer, or in the input/output (I/O) ports of the computer. In the electronic industry, I/O ports are blamed for the loose connections more than any other unit. In the presence of faults on input lines, the computer should continue execution of the control routines with default values. This will result in degraded performance in ride comfort and fuel economy. However, if the error is in the safety parameters such as the anti-lock brakes, anti-skid system, etc., the system should sound an alarm to the driver and display the list of the malfunctioning units.

2.4 SMALL VEHICLES

In this section all parameters relevant to small vehicles are analysed. A single bit serial bus-based architecture is given to control body and other static parameters. This is further extended into a multiprocessor architecture with shared memory modules for the control of dynamic parameters and diagnostics.

2.4.1 Analysis of the Parameters

The parameters of a modern automobile can be broadly classified into the following four categories: (i) ride parameters; (ii) engine parameters; (iii) safety parameters; and (iv) body and miscellaneous parameters. From the design point of view, it is also convenient to classify them as the static and dynamic parameters. Static parameters are those, whose changes will not effect the engine performance immediately. Most of the body parameters except the brakes and traction parameters are considered static. Also oil/fuel warning lights can be considered static, because there is nothing the computer system can do to remedy these situations.

1. *Ride parameters.* Cruise control, acceleration and deceleration, anti-lock brakes, transmission, steering and traction control fall into this category. Some of them are also safety parameters. The main objective of studying these parameters is to improve the comfort in ride.

2. *Engine parameters.* All the parameters measured within the engine fall into this class. All the temperature and pressure measurements, fuel–air mixture measurements, measurements of the oxygen content in the exhaust, throttle valve switch, spark time, Environmental Protection Agency measurements, etc., are the engine parameters. Controlling these engine parameters will result in better fuel economy, safer and more reliable engine performance, and a reduction of pollutants in the emissions.

3. *Safety parameters.* Measurements of proximity to the vehicle in the front, anti-lock brakes, proper operation of the power door and power window systems, locking the vehicle in the parking position when the vehicle is parked and locking the gear system in the drive while driving so that the gear will not slip into the reverse position, and traction control, are the main safety parameters.

4. *Body and miscellaneous parameters.* All the internal warning lights, head lights, brake lights, burglar alarm, speedometer, internal diagnostic lights, seat belt light/alarm, etc., fall into these categories.

The control routines that are executed in real-time are designed taking the interrelationships between these parameters into account. The control of any action is a function of some of these parameters. The overall control of the automobile is a set of such controls with interrelationships. All these interrelationships should be represented mathematically and then transformed into the control routines. The quantitative relationships depend on the type of the vehicle. Some such relationships are listed below:

1. When the proximity sensors measure the presence of another vehicle in the front being fast approached by this vehicle, brakes are applied, fuel to the engine is cut off and the brake lights are turned on.

2. Any time the brakes are applied, the brake lights should be on, and fuel to the engine is reduced automatically.
3. When the cruise control is engaged, constant speed within tolerance is achieved by controlling the fuel supply.
4. While taking a sharp turn, if the traction control realizes that the vehicle will be dangerously unstable, then the engine power is automatically reduced.
5. When the cruise control is on and if the coast is engaged then fuel to engine is reduced, and the brake lights are turned on.
6. While driving on the slippery roads, the computer system realizes that the roads are hazardous when some of the sensor measurements do not confirm with the expected profile of the values. The computer system goes into an emergency handling mode, and enforces the anti-skidding operations giving them the highest priority.

Let $f_1, f_2, \ldots, f_k$ be the control functions to be computed in the real-time for the dynamic control where

$f_1 = f_1(p_{11}, p_{12}, \ldots)$

$f_2 = f_2(p_{21}, p_{22}, \ldots)$

$f_k = f_k(p_{k1}, p_{k2}, \ldots)$, where $p_{ij}$is a reading of sensor j effecting function f_i.

When the computer fails to receive the reading from a sensor due to a fault, the computer upon recognizing the missing parameter, issues a fake default value for the parameter and continues with the computation.

2.4.2 Control of Body Parameters

Most of the body parameters (with the exception of anti-lock brakes and traction control) are static and there is no real-time data processing involved to operate on them. For example, when the driver pushes a button (such as to turn on the head lights), the body control processor (BCP) will simply have to send the signal to the lights. As wiring should always be minimized, a single bit serial bus as shown in Figure 2.5 is used, for control and diagnostics. $U1, \ldots, Uk$ are the static units being controlled.

While communicating with any unit, the BCP will first send the unit identification number, and only one unit will succeed in decoding it and will get connected to the serial bus (see Figure 2.3). The next two bits known as the mode bits, will decide whether the present operation is a test, normal activation of an operation, or a normal deactivation of a previously activated operation. A value is then sent to the unit. This value may correspond to the various actions or to a diagnostic test pattern. The BCP will monitor the returned value. If an error is indicated, it will store that

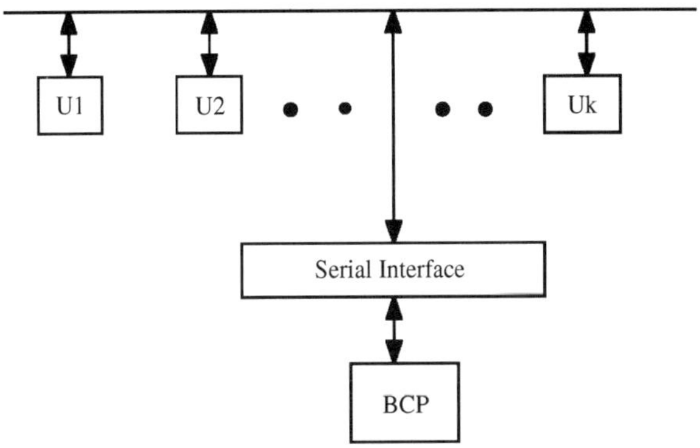

Figure 2.5. Serial communication between body control processor (BCP) and body control units

information in its memory and may also display it to the driver. More mode bits may be used if there are more than four modes, using the $\log_2 n$ formula, where n is the total number of modes.

2.4.3 Control of Engine/Dynamic Parameters

This is the more important part of the control. The computer monitors the sensor readings and computes dynamically in real-time, and decides the optimal control strategy at that point in time. It is valid to think of these strategies as functions of the various sensor readings. So, in the absence of sensor failures, the same routine will be executed for the optimal control strategy. In the presence of a sensor failure, the computer will execute the control routine with default values. Timing is a very important factor for the dynamic engine control. In the anti-lock brakes and traction control, they are very critical. When a processor is selected as the engine control processor (ECP), it must be fast enough to compute and respond in the worst circumstances. In most of the modern automobiles, at least two processor-based systems are being used for these very critical controls.

Consider the architecture given in Figure 2.5. One processor is devoted for the body control operations and for the central scheduling. Processors $p1$, $p2, \ldots, pm$ are the processors that will execute the control routines for the real-time dynamic control. The central scheduler monitors all the sensor readings and sends the control signals to the various actuators after the computation. $M1, \ldots, Mk$ are the shared memory modules. The shared memory modules will have both ROM and RAM memories. In any control routine, all the code that can be executed in parallel are stored in different memory modules.

As parallel software should be developed simultaneously with the multiprocessor architecture, we assume that the control routines are written maximizing parallelism. When the central scheduler decides to execute a control routine, it will allocate different idle processors, one each to execute a parallel section of the code. If all the processors are busy, then the request for a processor will interrupt a processor currently executing a lower priority job. When a processor should wait for the results obtained by executing a different section of the code by another processor, the present processor will wait. The scheduler has the knowledge of where (in memory) results are being stored. Final results are computed taking these intermediate values from the memory.

The central scheduler plays the most important part in this architecture and it is advisable to make this unit fault tolerant. If one of the processors *p*1, ..., *pm*, fail, the system can still operate at a reduced efficiency. But if the central unit fails, then the whole system would fail.

2.5 COMPLEX VEHICLES

In this section, the control of complex vehicles is presented. These vehicles (such as trucks) are very huge and involve many controls for loading and unloading, manipulation of the wheel parts, etc., besides all the other controls discussed in Sections 2.3 and 2.4. Consequently, there is a more severe demand for the real-time commuting power. The architecture of Figure 2.6, is modified for the complex vehicles as given in Figure 2.7. In the previous section, it is observed that a fault in the central scheduler will make the whole system inoperative. The modified architecture shown in the Figure 2.7, will not have this disadvantage, as different processors can take the place of the scheduler. However, this requires a more complicated built-in protocol. As a result, faults in individual processors will not make the system totally inoperative. The system operates with a corresponding degradation in performance.

The architecture of Figure 2.7 is a single shared bus architecture. This can support up to 15 processors if enough instruction cache space is available inside the processors, and the degree of data sharing is not too severe,[13] a condition easily satisfied in the automobile applications.

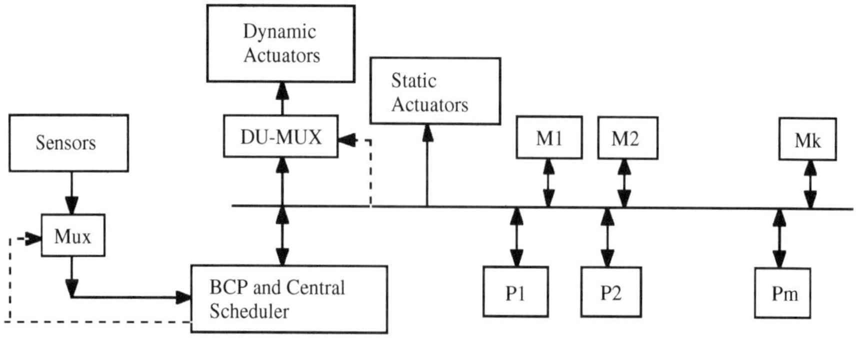

Figure 2.6. Central scheduler-based architecture for simple vehicles. BCP, body control processor

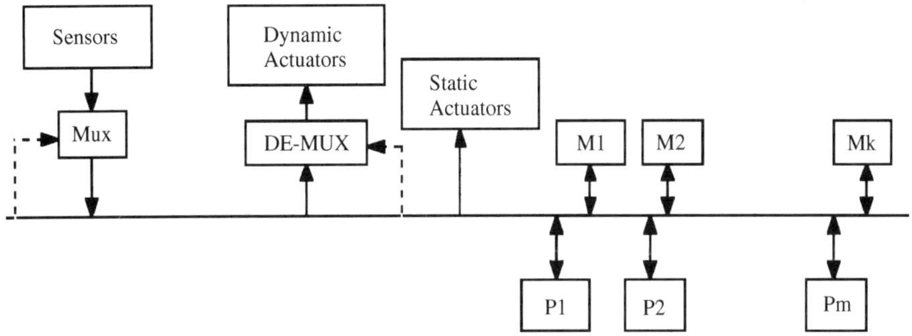

Figure 2.7. Microprocessor architecture for complex vehicles

2.5.1 Description of the System

The novel idea of this architecture is that one processor is used as a central processor (CP). The CP is responsible for monitoring the sensors and then scheduling the appropriate job to an available processor. When the CP becomes inactive, another processor becomes the CP. This new CP executes a default program to keep the vehicle running, and a diagnostic program to test all the sensor units. If everything in the system is fault-free, except the old CP, then the new CP executes the normal program to keep the vehicle running in the normal mode. As one of "m" processors will not be available for operation, the performance of the system will be degraded. The amount of degradation depends on the value of "m". Larger the value of "m", smaller will be the amount of degradation. After the old CP becomes faulty or inactive, a message will be displayed on the dash board to let the driver know that a processor is faulty. The ID number of the faulty processor will also be stored with a diagnostic message in a fixed area in the memory. This will help the mechanic to identify the faulty processor.

2.5.2 Operation of the System

The sensors are connected to the shared bus through a multiplexer. A few critical sensors may be permanently connected to some dedicated lines of the shared bus. Thus, these signals are always available to all the processors. The actuators are connected to the shared bus by a demultiplexer and the communication between the central module and the local modules is always through a serial line.

A processor status register (PSR) is maintained in the memory to show the status of the processors. This register is shared by all the processors. The number of bits in the PSR is equal to the number of processors in the system. If bit i of the register is one then processor i is busy, otherwise, processor i is free.

The CP monitors all the sensors. Whenever a program segment has to be executed (depending on the value of the sensor signals) the CP looks at the PSR to find a free processor.

1. Let processor j be free at that time. The CP interrupts processor j and sends the starting address of the program segment to be executed through the shared bus. The CP also sets bit j of the PSR. Processor j reads the address of the program segment from the bus and then starts executing the program. The last instruction of any program segment will be a Return From Interrupt instruction. Thus, any program segment is considered as an interrupt service routine. Processor j resets bit j of the PSR after completing execution of the program segment.

2. Let no processor be free. In this case the CP interrupts a processor which is executing a program segment of the lowest priority. Let this interrupted processor be processor n. Processor n reads the address from the shared bus and stores the address in its service queue. The queue is maintained in the memory. When processor n is done with the previous program segment, it reads the address of the next program segment from the service queue, and then executes that program segment. At the end of execution of this program segment, processor n resets bit n of the PSR if its service queue is empty.

The CP also checks (periodically) all the other processors to see if they are operational or not. A line of the bus called the CP active line (CPA) is used exclusively for the purpose of indicating whether the CP is active or not. An active signal is available on this line as long as the CP is active. Whenever no active signal is available on this line, a watchdog timer becomes active. If the active signal does not come back within some predefined time interval then the watchdog timer interrupts the next processor in the hierarchy to be the new CP. This hierarchy was predefined during initialization. If the interrupted processor does not send the active signal through the CPA within some predefined time, then the watchdog timer interrupts the next processor in the hierarchy to become the CP. This type of interrupt is the highest priority interrupt. After a processor becomes the new CP, it executes a default program to keep the vehicle running. This new CP then goes through some diagnostic procedure or it may schedule another processor to go through some diagnostic procedure. After executing the diagnostic procedure, if everything is found to be operational except the old CP, then the new CP runs the normal operation.

Different program segments being executed concurrently by different processors might require mutual communication, such as data sharing. This is implemented using the shared memory concept. Fixed areas of memory are reserved for all such data shared between program segments. For example, consider two program segments A and B, being processed concurrently by two processors. Suppose B needs to use a data set produced by A. The CP when scheduling A and B will first invalidate the data in the memory area reserved for the data sharing between A and B. This is done by using a single bit flag. Whenever A completes the computation, it will place computed data items in these locations and will validate the data. If B requires the data after data are stored by A, then there is no problem. If B needs the data while

A is still computing, then B sees that the data in these locations is invalid and waits until A completes its computation, stores the data in these locations, and turns on the validation flag.

2.6 FUTURE AUTOMOBILES

Currently, automobile manufacturers are unable to implement some of the technological developments already available, due to the prohibitive cost factors. But the future automobiles invariably will have to implement them in their designs and hopefully the cost factors will come down. Some of the features of the future vehicles are listed below. Non-touching sensors such as cameras are not being widely used presently. These sensors can visualize the changing environment around the vehicle, and can take corrective actions in situations such as a collision. Also, in the event of skidding, the slipping of the vehicle from the appropriate future course in the next few instances of time can be detected. Once the computer system recognizes this, a corrective action can be initiated by the computer to minimize the severity of skidding. Making this control a closed-loop control is costly, but is justifiable in view of the severity of the skidding phenomenon.

In the present-day technology, sensors are not being used widely for diagnosis. Sensors are being appended at different positions by the electronic engineers after the engine design phase. They have not become lower-level ingredients of the design, except possibly within some parts of the engine. Currently, sensors are being viewed exclusively as the means of feedback for the important vehicle control operations. In future, sensors will be integrated with the design at a much lower level. Sensors will have dual functions: control operations and different forms of the diagnostic functions, such as diagnosing the engine operations, diagnosing other sensors, etc. Engine designs become more modular and there will be diagnostic structures within the modules implemented with the help of the built-in sensory networks. Diagnostic strategies will be embedded in the system taking advantage of the existing redundancies in the intelligence of the sensors and the sensory system. Automobiles will be designed with the capabilities to take prudent decisions by monitoring the traffic signals, the yellow and white lines on the roads, etc. There will be a substantial increase in the voice controls, and interaction between the driver and the vehicle will become lot more friendlier. Auto-diagnostics is an area that will see a great improvement and technicians will be provided precise repair information by plug-in-type auto-diagnostic units. In spite of all these advances, the computation needs of the vehicle operation will not increase beyond a single bus architecture.

2.7 CONCLUSIONS

The control of static and dynamic parameters and diagnostics are presented in this chapter for modern automobiles. Two types of architecture are presented, one for smaller vehicles and the other one for complex vehicles such as trucks. The problem of dynamic real-time control in larger vehicles is more challenging both in view of the speed of response required and the complexity of the sensory system, whose readings are the inputs to the control routines. This problem becomes more complex for aircraft and spacecraft and requires a special study. The present multiprocessor technology is very adequate for all increased computational demands caused by utilization of more extensive sensory systems in future automobiles. In space vehicles, we may have to go into multibus architectures as the real-time computational needs are very critical in such vehicles.

REFERENCES

1 Bishel RA. *Integration of Truck Electronics: A Look at the 90's*, Automotive Engineering, February 1988, **96**, Warrendale, Pennsylvania, 115–119.

2 *Automotive Electronics II*, Society of Automotive Engineers, 1975, SP-393, Warrendale, Pennsylvania.

3 Sood AK, Fahs A, Henein NA. *A Real-Time Microprocessor Based System for Engine Deficiency Analysis*. IEEE Workshop on Automotive Applications of Microprocessors, 1982, Washinton, D.C., 111–118.

4 Flis TJ. *The Use of Microprocessors for Electronic Engine Control*, IEEE Workshop on Automotive Applications of Microprocessors, 1982, Washinton, D.C., 3–19.

5 Wist AO, Meiksin ZH. *Electronic Design of Microprocessor-Based Instruments and Control Systems*, Prentice-Hall, 1986, Englewood Cliffs, New Jersey.

6 Gant GC, Alves GJ. *Progress in Electronic Control of Large Diesel Engines*, The 11th Annual Fall Technical Conf. of the American Society of Mechanical Engineers, Internal Combustion Engine Division, Dearborn, Michigan, October 15–18, 1989, **9**, 67–77.

7 Bazari Z. *The Transient Performance Analysis of a Turbocharged Vehicle Diesel Engine with Electronic Fueling Control*, Electronic Diesel Engine Controls: New Developments, SP-819, Society of Automotive Engineers, Warrendale, Pennsylvania, 1990, 51–62.

8 Majeed KN. *Dual Processor Automotive Controller*, IEEE Workshop on Automotive Applications of Electronics, October 19, 1988, Dearborn, Michigan, 39–44.

9 *Sensors and Actuators*, Society of Automotive Engineers, Warrendale, Pennsylvania, SP-737, 1988.

10 Nwagboso C. *Automotive Sensory Systems*, **1**, Chapman & Hall, Bolton, England, 1993.

11 Uramoto T, Inomata T. *Development of Multiplex Wiring Systems in Automobiles, Multiplex Technology*, Society of Automotive Engineers, Warrendale, Pennsylvania, SP-742, 1988, 49–54.

12 Mano MM. *Computer Engineering: Hardware Design*, Prentice-Hall, Englewood Cliffs, New Jersey, 1988.

13 Archibald J, Baer J-L. *Cache Coherency Protocols: Evaluation Using a Multiprocessor Simulation Model*, ACM Transactions on Computer Systems, **4**(4), November 1986, 273–298.

14 Vishnubhotla SR, Mahmud SM. *A Centralized Multiprocessor-Based Control to Optimize Performance in Complex Vehicles*, IEEE Workshop on Automotive Applications of Electronics, October 19, 1988, Dearborn, Michigan, 52–56.

15 DeCegama AL. *Parallel Processing Architectures and VLSI Hardware*, **1**, Published by Prentice-Hall, Englewood Cliffs, New Jersey, 1989.
16 *Engine and Driveline Control Systems*, Society of Automotive Engineers, SP-739, Warrendale, Pennsylvania, March 1988.
17 *ABS Traction Control*, Society of Automotive Engineers, SP-744, Warrendale, Pennsylvania, March 1988.
18 Sachs K, Fridman A. *In-plant Testing of Anti-lock Brake Systems*, Society of Automotive Engineers Tech. Paper Series, No. 880257, International Congress and Exposition, February 29-March 3, 1988, Detroit, Michigan.
19 Kiencke U, Dais S, Litschel M, Unruh J. *Error Handling Strategies for Automobile Networks*, Multiplex Technology, Published by Society of Automotive Engineers, SP-742, Warrendale, Pennsylvania, 1988, 19–28.

3

Automotive Electronic Trends and their Influence on Intelligent Transport Systems*

3.1 INTRODUCTION

Electronics has been effectively used in today's vehicles to enhance their performance, reduce emissions, improve economy, and to increase driver and occupant safety, comfort and convenience. Its use has been primarily market driven, its implementation has been evolutionary rather than revolutionary and in some cases its success has been very sensitive to driver acceptance. Intelligent vehicle–highway systems (IVHS) in vehicle equipment is expected to follow the same trends.

IVHS is seen as a means to improve highway system efficiency. Expectations are that it will lead to less highway construction, less congestion, lower auto emissions, less fuel consumption and improved safety. How IVHS will accomplish these goals, and to what extent societal and individual benefits will be realized is yet to be determined.

In its ultimate form IVHS will be a marriage of the "smart highway" and "smart vehicle", along with its driver, into an integrated vehicle/highway system. The "smart highway" system will be needed before IVHS can become a reality. "Smart vehicles" exist today; however, the additional electronic equipment required in the vehicle to work effectively with the "smart highway" needs to be defined. Whether or not the

* The name "intelligent vehicle highway systems" (IVHS) has now been changed to "intelligent transport systems (ITS)"

Advanced Vehicles and Infrastructure Systems, Edited by C.O. Nwagboso

car buyer will perceive sufficient value in IVHS to buy and use the related in-vehicle electronic equipment is a key issue addressed in this chapter.

A variety of in-vehicle IVHS offerings will be required to satisfy the different driver needs and preferences for traffic assistance. These offerings may span from simple highway advisory radios (HAR) to very sophisticated route guidance systems. To justify the billions of dollars required for the "smart highway", affordable in-vehicle systems must be available to allow most, if not all, drivers to access the traffic advisory system.

One objective of this chapter is to review the automotive electronic trends for the past 30 years and to determine the forces behind the rapid expansion of electronics in our modern day vehicles. Some of the lessons learned during this period which could benefit the implementation of IVHS technologies are highlighted. Numerous examples of Chrysler Corporation automotive electronic systems are cited to illustrate past experiences and industry trends which may influence IVHS development.

The chapter will discuss the current trends in automotive electronics (as viewed by the author) that may influence the future implementation of related in-vehicle IVHS equipment. It will conclude with a discussion of the issues that need to be addressed to ensure driver acceptance, thought to be necessary for the success of IVHS.

3.2 AUTOMOTIVE ELECTRONICS TRENDS

Figure 3.1 provides an overview of the past and anticipated industry trends in automotive electronics over a 40-year period. It subdivides the automotive

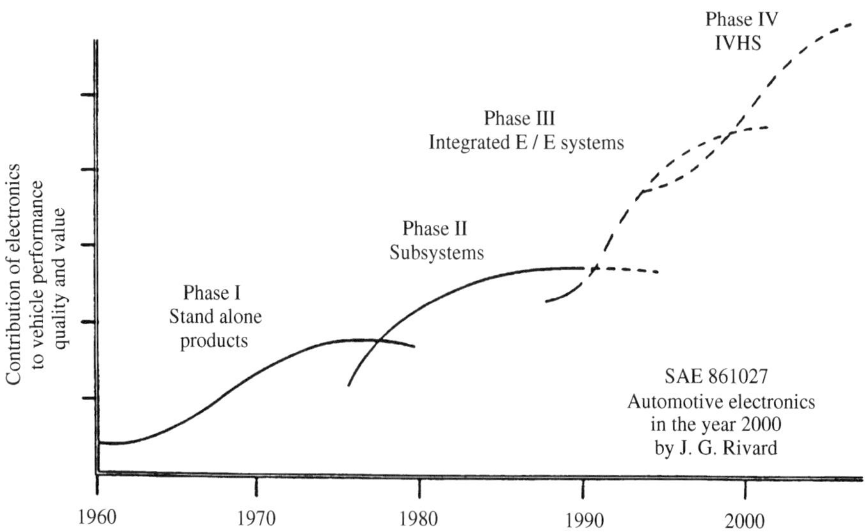

Figure 3.1. Automotive electronics evolution

electronics evolution into four phases.[1] The first two phases, occurring over the past 30 years, have been essentially completed.

During this period, the auto industry experienced a tremendous growth in automotive electronics to the extent of creating a whole new industry. A continuation of its growth will occur where car buyers perceive the value of automotive electronic systems and are willing to pay the additional cost.

3.2.1 Historical Trends

The automotive electronics growth has been primarily market driven, its implementation has been evolutionary rather than revolutionary; and in some cases, its success has been very sensitive to driver acceptance. It has been heavily influenced by the need to improve vehicle reliability and performance, meet legislative regulations, and reduce vehicle cost while improving occupant safety, convenience and comfort. These market forces have precipitated the following major changes in the auto industry over the past 30 years:

1. Significantly improved radio reliability by eliminating vacuum tubes and mechanical tuning mechanisms, while adding "SEEK" and "SCAN" search tune features and high-quality sound systems.
2. Improved powertrain performance, while significantly reducing emissions and improving fuel economy. It also enhanced powertrain system reliability and greatly reduced system maintenance by replacing numerous wear-prone electromechanical components with more durable electronically equivalent systems.
3. Added safety features such as anti-lock brakes and air bags.
4. Developed electronic displays to show driver information, readily such as: compass headings, in-vehicle and outside temperature, fuel consumption data and diagnostic information.
5. Improved driver convenience and comfort with automatic temperature control, electronic speed control, remote entry systems and improved lighting control.

During this 30-year period, the auto industry experienced a tremendous growth in automotive electronics to the extent of creating a whole new industry. This success story, however, was not without its failures.

Experience has shown that the success of past automotive electronic products, which directly interact with the driver, was very sensitive to user acceptance. In a number of cases, negative consumer reactions have caused affected products to be dropped from the market. Notable Chrysler Corporation examples are discussed below to exemplify what can be learned that might benefit the implementation of IVHS technologies:

1. *Citizens band (CB) radio*. This was very popular in North America during the mid-1970s and was introduced by Chrysler as an OEM (Original Equipment Manufacturer) product in 1978. In 4 years it was dropped due to reduced car buyer interest and resulting lack of market demand.

The lesson we learned is that an image product not consistently meeting car buyer expectations will cause them to drop the option for their next vehicle purchase.

2. *Search tuned radio*. Chrysler introduced an advance push button controlled electronically tuned radio in 1978. It was radically different from previous radio offerings, required the user to relearn how to operate a radio and did not offer any additional feature value other than search tune and digital display. It was replaced 4 years later, due to lack of market demand, with an electronically tuned radio containing the search tuned feature and user accepted rotary knob controls closely resembling previously mechanically tuned radio offerings.

The lesson we learned is that car buyers will reject radically different high-tech products, which require them to learn how to operate the new product while offering little additional consumer value.

3. *Electronic voice alert*. This feature is sometimes referred to as the "Chrysler Talking Car". It received a mixed reaction in the market-place. Many liked the feature. Others disliked it because they felt it was annoying, nagging and irritating. Others complained that it would come on unexpectedly and interrupt their radio programme. This feature was dropped in 1988.

The lesson we learned is that when applying voice synthesis technology there is a need for fine-tuning-related human factors and potential reactions. It is interesting to note, that Chrysler's negative experience surely caused other auto manufacturers not to include this feature in their vehicles.

4. *Trip computers*. Early Chrysler Corporation trip computers provided many features and information to the driver regarding his or her trip and vehicle performance. The more sophisticated systems included the following features:

- (i) estimated time of arrival (ETA);
- (ii) distance to go;
- (iii) distance to empty;
- (iv) instantaneous fuel efficiency;
- (v) average fuel efficiency;
- (vi) fuel consumption;
- (vii) time of day;
- (viii) day and month; and
- (ix) distance travelled.

Car buyers perceive the trip computer as a very accurate device. As a result, they would complain when the fuel consumption reading did not accurately correlate with the pump reading at their service station. Of course, sensor accuracy, fuel pump cut-off variations and individual influence on fuel tank filling made it impossible to achieve an accurate and repeatable fuel consumption correlation. In addition, drivers needed to learn how to use the complicated trip computer features and in some cases set-up was required before the feature could be used. As an example, for the ETA feature, the driver needed to input an estimate of the distance to his/her destination and needed to understand that accuracy of the ETA prediction was dependent on the accuracy of their estimate and ability to maintain their current average speed throughout the trip.

The average driver expected an accurate response and often was not willing to or capable of fully understanding the limitations of the system. To avoid these problems, more recent trip computer offerings were simplified by providing only time of day, instantaneous and average fuel consumption, distance travelled and distance to empty.

The lesson we learned is not to provide the feature unless it is convenient to use, easily understood by the average driver and the implied accuracy is readily achievable.

These examples illustrate that high-tech automotive electronic products will not survive in the market-place unless they are easy to use, meet consumer expectations and add consumer value, which the car buyer is willing to use and purchase.

In some cases, acceptance of new automotive features by the car buyer was delayed many years. Anti-lock brakes and air bags, as an example, were introduced more than 15 years ago with very little success. Cost and low perceived value caused the lack of market demand at that time. Today, with the recent car buyer interest in personal safety and lower cost, these options have become very popular in the market-place. To preclude this delay phenomenon for IVHS, it will be necessary to make consumers aware of IVHS technology, show them how it will satisfy their needs and be able to manufacture related products at a price they are willing to pay.

3.2.2 Present Trends

While electronics greatly enhanced vehicle performance and value, the resulting proliferation of electronic modules escalated vehicle packaging problems as available space was being decreased due to vehicle downsizing. Additional wiring complexity required to interconnect each of the modules compounded the problem. Vehicle wire counts rose from approximately 250 wires to more than 1000 wires today. The increased complexity has led to additional engineering effort, increased plant complexity and more difficult serviceability. These problems are the major reasons prompting the industry to shift into phase III of the automotive electronics evolution. Chrysler Corporation used an integrated electrical/electronic (E/E) systems strategy employing multiplex wiring to resolve the wiring complexity problems caused by the electronic module proliferation.[2] This design strategy enabled Chrysler to decrease overall vehicle complexity, while improving costs and reliability, providing better diagnostics and fault isolation, and increasing vehicle design flexibility. The remaining paragraphs for this section describe how the integrated E/E system strategy was employed at the Chrysler Corporation and the author's expectations on the strategy's future trends.

Simply stated, the integration strategy is a top down approach, wherein, the vehicle functions are partitioned into the fewest module assemblies. Recognizing that this will increase module complexity, ASIC (Application-Specific Integrated Circuit) technology is then employed to place as many functions on integrated circuits as possible to simplify the module assembly. Multiplex wiring is then used to accommodate the increased module input/output functions.

Similar processes have been used to develop digital watches, hand-held calculators, and now personal computers. The history of these products has demonstrated the capability of integrated electronics to provide more functional features, while at the same time simplifying product design, increasing reliability and reducing overall cost.

E/E systems integration represents a significant departure from the traditional subsystem approach in use by the automotive industry. For this reason, Chrysler decided to implement the integration strategy in an evolutionary manner. This was necessary to gain customer acceptance and provide the practical experience needed to measure the effectiveness of the integration strategy.

The advantages of E/E systems integration and multiplex wiring are generally well known; however, Chrysler's goals and objectives were to achieve these advantages in a cost-effective manner. It was recognized early on that E/E systems integration and multiplex wiring would be essentially transparent to their customers; they will benefit from the improved reliability, superior diagnostics and fault isolation. However, although these features would lead to increased customer satisfaction and value, in themselves, they do not necessarily justify increasing vehicle costs to the customer. As a result, E/E systems integration and its implementation were guided by quality, cost, performance and/or packaging considerations. Generally, E/E systems integration was employed where equal or better quality, cost or performance were achieved, and/or where packaging problems need to be resolved.

In model year 1988, Chrysler introduced the integrated E/E systems strategy in the Chrysler New Yorker and Dodge Dynasty vehicles. The affected systems and related multiplex wiring network is illustrated in Figure 3.2.

In the New Yorker vehicles, a 25% savings in cost was realized with eight fewer electronic modules and 111 fewer wiring harness circuits over the traditional subsystem approach. The integration was accomplished by combining the power control, logic, speed control and voltage regulator module functions into the engine control module; features such as the key lamp delay, illuminated entry, intermittent wipe, auto-door locks and chime into the body control module; and compass/temperature, electronic monitor and navigator module display functions into the electronic vehicle information centre (EVIC) display module. In the Dodge Dynasty vehicles the Traveler is used in place of the EVIC display and engine compartment node modules.

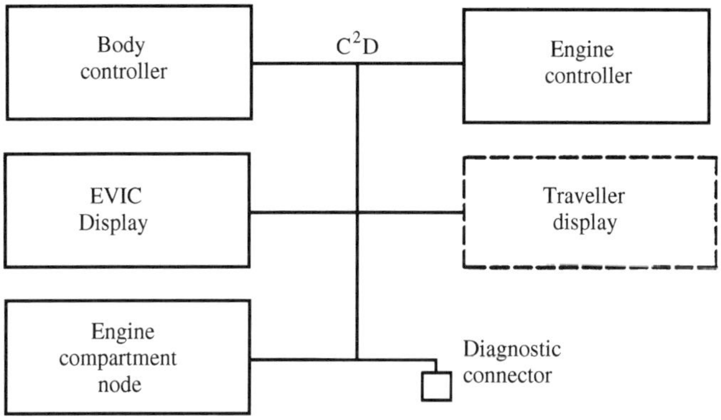

Figure 3.2. Model year 1988 multiplex system

The multiplex system, illustrated in Figure 3.2, is a six-node network using Chrysler's collision detection (C^2D) multiplex wiring system.[3,4] Key characteristics of the C^2D multiplex system are listed in Table 3.1.

Since its introduction, the strategy has been extended to more Chrysler vehicle lines and expanded to include more vehicle electronic modules.

In model year 1989, the New Yorker multiplex system was expanded by adding the electronic automatic transmission (EATX), electronic instrument cluster (EIC) and vehicle theft alarm (VTA) modules to the multiplex wiring system. The resulting system, shown in Figure 3.3 is an eight-node network. These modules were added to the bus primarily for signal sharing and system diagnostics. Opportunities existed to integrate the EATX into the engine controller and the VTA into the body controller. However, it was generally felt to be premature at this time as both modules are relatively new and their designs needed to mature before they can be considered for integration into other modules.

In model year 1990, the multiplex wiring system was extended to the Chrysler Imperial vehicle. In this case, the system was expanded by adding the air suspension system control module for diagnostics and signal sharing. The resulting system, shown in Figure 3.4, is a nine-node network.

Figure 3.5 illustrates the multiplex wiring systems in the new Chrysler 1993 LH series vehicles.

One of the immediate benefits realized was the ability to monitor and diagnose the electronic system operation. A hand-held diagnostic tool, called the Diagnostic

Table 3.1. C^2D key characteristics

Type	Class B
Protocol	Arbitration
Baud rate	7812.5 Hz
Encoding	10 bit NRZ
Input/output	Differential current mode

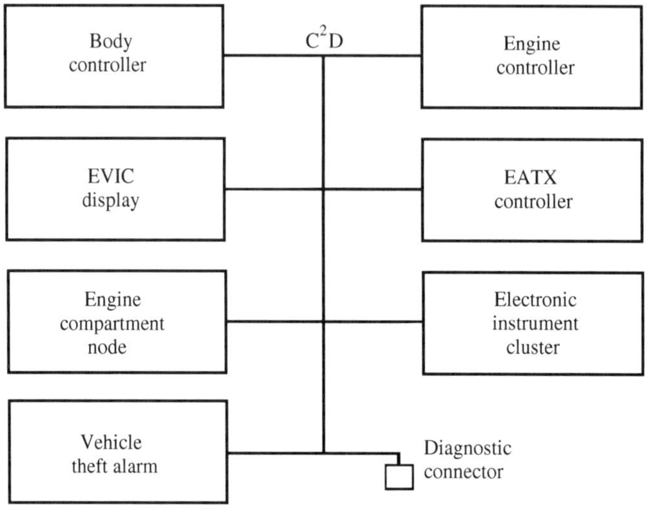

Figure 3.3. Model year 1989 multiplex system. EATX, electronic automatic transmission; EVIC, electronic vehicle information centre

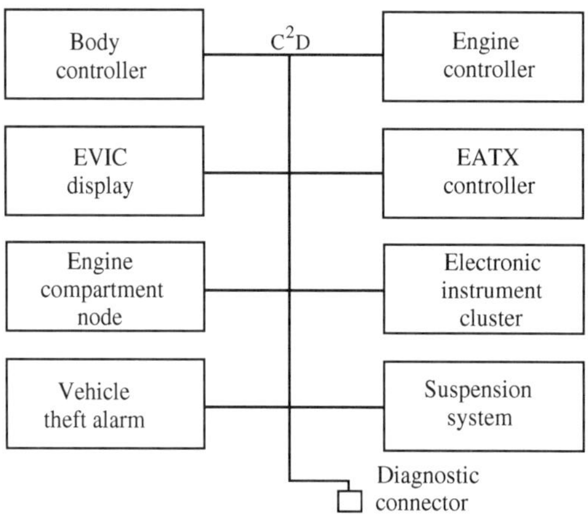

Figure 3.4. Model year 1990 multiplex system. EATX, electronic automatic transmission; EVIC, electronic vehicle information centre

Readout Box II (DRB-II), was released by Chrysler Corporation in model year 1988 for servicing the multiplex systems and related subsystem assemblies. The DRB-II was an improved version of the DRB-I hand-held tester introduced in 1984 to service EFI Engine Controllers. The DRB-II differs from the DRB-I in that it contains programmable intelligence and it utilizes plug-in memory cartridges to provide year-to-year test program updates. In addition, messages and data are presented in plain language text along with appropriate measurement units to provide an interactive and a more user friendly tester.

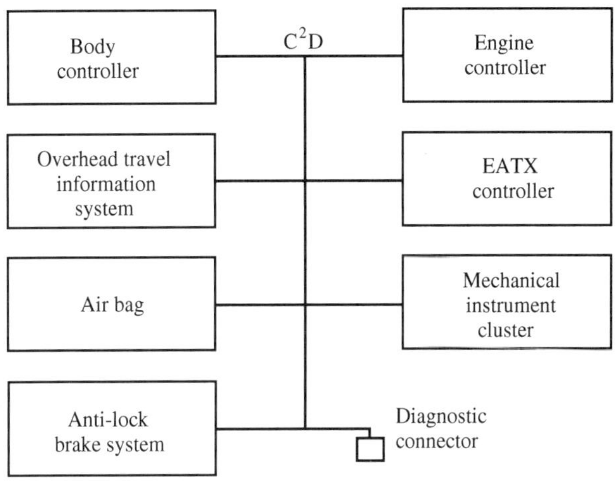

Figure 3.5. Model year LH 1993 multiplex system. EATX, electronic automatic transmission

The DRB-II can monitor sensor outputs, activate actuator and relay functions, and bring displays to life. These capabilities enable testing individual elements in the system, thereby enhancing service personnel's ability to diagnose and identify the fault before disassembly of the system. This is particularly helpful in trouble shooting wiring problems which are difficult to find and often leads to a misdiagnosis followed by removal of a properly functioning electronic module. Wiring problems represented about 25% of an E/E system warranty and are believed to be partially responsible for the estimated 20–30% removal of apparently good (no trouble found) electronic assemblies.

One surprise in the program was the demand for DRB-II hand-held testers by engineering and assembly plant personnel. Specially programmed versions were requested by many areas for real-time monitoring of vehicle multiplex systems and subsystem performance.

3.2.3 Future Trends

The E/E systems integration strategy implementation is continuing at Chrysler Corporation. It has led to lower costs, fewer modules, fewer circuits and provided a sorely needed diagnostic capability. However, although Chrysler accomplished a significant number of their goals, they have yet to realize the full capability of the strategy. The potential for more improvements does exist and can be illustrated by the following analysis of model year 1990 Imperial wiring assemblies.

A fully loaded Imperial has 15 wiring assemblies containing a total of 1020 circuits. In comparison, the base car Dynasty wiring assemblies contain a total of 562 circuits. A circuit is defined as a wire of any length requiring a termination at each end. As reliability and cost are generally a function of the number of circuit terminations, the system design objectives are to reduce them to a minimum.

Table 3.2 identifies where the circuit terminations occur on the fully loaded Imperial.

An examination of the connector pin terminal assignments, shown in Table 3.3, provides an insight on where the circuit terminations are used and the potential for reducing their number in this vehicle.

About 32% of the connector pin terminals were required to interconnect the 15 wiring assemblies used in the vehicle. A unified wiring assembly was a possible solution; however, vehicle assembly problems and the trend toward modular vehicle assembly precluded its consideration. As a first-order solution to resolve this type of problem, where it is possible, wiring assemblies should be combined to eliminate the need for interconnections. Otherwise, systems integration and multiplex wiring would be required to reduce the number of interconnection terminals. The interconnections cannot be totally eliminated as they are still required for power distribution, multiplex wiring networks and redundant circuits.

More opportunities exist to continue reducing the number of control modules by further systems integration. Where it makes sense to delete control modules the

Table 3.2. Circuit terminations

Connector pin terminals	1447
Double crimp connections	93
Harness splice connections	500
Total	2040

Table 3.3. Connector pin terminal assignments

Vehicle hardware connected		Connector pin count
1. Wiring assembly interconnections		468
2. Modules		323
Control	271	
Display	52	
3. Output devices		224
Relays	82	
Motors	81	
Actuators/solenoids	61	
4. Input devices		222
Control switches	102	
Sense switches	68	
Analogue sensors	32	
Pulse sensors	20	
5. Lamps		102
Exterior	70	
Interior	32	
6. Power distribution		72
7. Miscellaneous		36
Total		1447

related power, ground and redundant sensor circuits would be eliminated from the wiring assemblies.

There appears to be a trend developing suggesting that the system integration strategy could lead to three intelligent modules called powertrain, chassis and body controllers. If this proves to be the ideal system architecture, a number of modules seen today will simply disappear. As an example, in current systems Chrysler eliminated the voltage regulator and speed control modules. In time the engine and transmission modules could evolve into the powertrain module; the engine compartment node, VTA, memory and air bag modules could be merged into the body controller; and the anti-lock brake and suspension modules evolve into the chassis controller. In the future, for the more sophisticated IVHS systems, a navigation module may be added to accommodate the IVHS functions.

At this time the module integration progress is somewhat constrained by a related increase in module connector size and pin count. As more functions are added to each module, an increase in module input/output (I/O) connector pins occur due to the additional switch, relay, actuator and signal connections. In addition, a proliferation of body controller part numbers is occurring due to large variations in pin assignments caused by different options and option rates within and between car lines.

In most cases, the display modules, I/O devices and lamps listed in Table 3.3 cannot be eliminated by integration due to specific design and/or functional requirements. Therefore, multiplex wiring would be required to reduce the connector pin count, part number proliferation and wiring assembly interconnections. Most display modules will probably continue using the C^2D network. The I/O devices and lamps, with a few exceptions, are low response devices that could be accommodated by a class A type network. One solution appears to be a class A local area network for each module to interconnect related lamps and smart sensors, motors, actuators and/or solenoids.[5] This system approach could also eliminate relays now used to control other output devices. Cost savings related to the relays and associated circuit reductions would help offset the higher cost of these smart devices.

The class A network architecture would contain a combination of smart devices and nodes. The nodes would be used where the present I/O devices are clustered and are not sensitive to option selection. Examples include fuel injectors, transmission solenoids and switches, exterior lamp clusters, control switch clusters, etc. Other items sensitive to options and/or distributed in isolated areas throughout the vehicle would remain as they are or use smart switches, motors, actuators and solenoids. Examples include level sensors, security system sensors, fuel pumps, locally switched lamps, etc.

A typical body controller I/O configuration is simply illustrated in Figure 3.6, wherein a number of switches and relays are connected in parallel to the module. As module functions were added, more connecter pins were added and different pin assignments were needed as functions varied between vehicles. These conditions led to the larger pin count and proliferation of body controller part numbers. In addition,

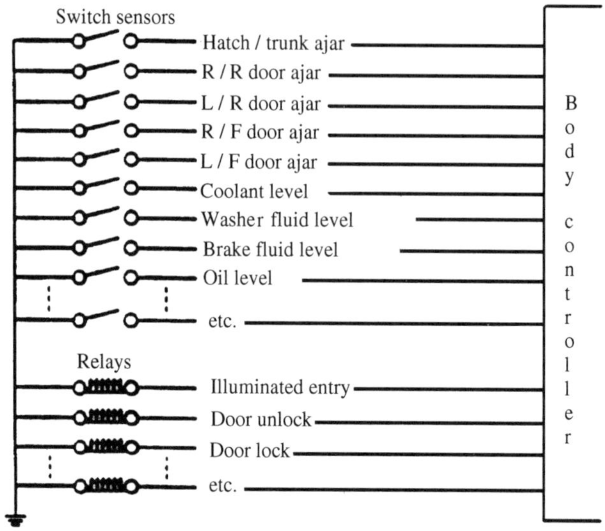

Figure 3.6. Typical input/output configuration

in its present configuration, operation of related subsystems would be required to verify the integrity of the wiring system. As an example, switches would have to be closed for circuit verification. In the case of relays, they have to be operated and some feedback provided to indicate closure. (Audible feedback or occurring events, e.g. light coming on, is used today to check relay operation.)

Ideally, a common body controller would be desired for all car lines, which can readily adapt to the various options. A body controller module configuration that provided connector pins for all the possible I/O functions needed for all car lines could be provided. However, this approach would be quickly discarded because of the large connector size, cost and hardware redesign required to accommodate new functions.

An ideal system would allow the addition and/or subtraction of switches, relays, actuators and signal functions without affecting the body controller hardware configuration — in other words, a software-configured system. The addition of a class A local area network (LAN) appears to be a solution. This led to the development of Chrysler's Sensor & Control (CSC) data bus.[5,6] Key attributes of the bus are listed in Table 3.4.

A typical I/O configuration possible with a CSC LAN is illustrated in Figure 3.7.

Table 3.4. CSC key attributes

Type	Class A
Protocol	Polling/command
Baud rate	1000 Hz
Encoding	Level detection/PWM

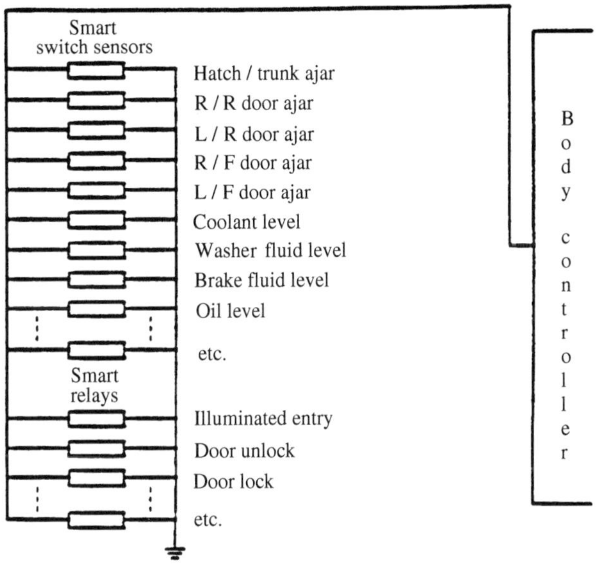

Figure 3.7. Chrysler's Sensor & Control input/output configuration

The CSC system currently in development can accommodate up to 30 input functions and 32 output functions. The I/O devices can be connected in any combination of series, star or ring network configurations. Any combination of switch, relay, actuator or signal functions can be added to the bus as long as software is provided to process the functions. What is needed is a series of CSC compatible smart switches, relays, actuators and signal sources.
Capabilities of the system are summarized as follows:

1. Significant reduction in number of module connections and connector size is realized.
2. Discrete functions can be added to the bus in any combination as long as supporting software is present in the module resident microcomputer and advised of the specific system configuration.
3. The system can be more responsive to differing requirements within and across car lines, and to model year changes as new features are added to the vehicles.
4. Less engineering support will be required due to the decrease in controller hardware configurations and related model year changes.
5. Standard hardware configurations lead to high-volume manufacturing, thereby improving efficiency, lowering costs, increasing reliability and quality.
6. Smart switches, relays and actuators can be polled to verify wiring connections, device operation and their state (on/off, full/empty, open/close, etc.) without operating vehicle mechanisms.

7. Fewer configurations to service in the field will improve dealer service response time leading to better customer satisfaction.

The availability of cost-effective technology will most likely dictate the class A network system architecture for some time. As an example, multiplex wiring of the exterior lighting and door wiring systems are heavily dependent upon the availability of low-cost power MOSFETs. In a fully loaded model year 1990 Imperial, these systems contain 38 lamp filaments and 15 motors which could require up to 98 power MOSFETs to implement a multiplex wiring system. To date, cost savings due to related low current switches, circuit deletions, warranty reductions and simpler vehicle assembly have not been sufficient to offset the MOSFET costs.

Cost issues are preventing broad application of the CSC bus concept at this time. The industry is faced with the usual high-tech-low-volume versus low-tech-high-volume hardware cost predicament. The finely tuned, automated manufacturing facilities used today to build commodity type automotive parts (i.e. switches, relays, motors, etc.) are difficult to compete against when replacing these parts with smart devices. This dilemma is expected to continue until an adequate market is developed for smart devices worthy of the investment needed to drive their costs down. Innovative component and system designs are needed to generate cost offsets required to justify the increased smart device component costs. In the meantime, targets of opportunity will develop as more creative designs evolve in the years ahead.

3.3 TECHNOLOGY PUSH VERSUS MARKET PULL

Early in the development of automotive electronic products, very few data were available on driver needs and how they would react to new products responding to perceived needs. As a result, the designs of these products were heavily influenced by what the technologists *believed* were needed by the driver. Although some car buyers liked them, products with low perceived value, often intimidating, and requiring frequent driver interaction were frequently referred to as electronic "gadgets". Most buyers showed no real interest in the products and generally avoided them due to potential repair costs. In time, they were dropped or radically changed to make them more acceptable to most buyers.

To avoid development of the "gadget" image for IVHS a careful evaluation and understanding of the car buyer's true needs and perceived value of related in-vehicle products are required. These considerations will be influenced by a number of factors including, but not necessarily limited to, the following:

1. Their primary needs are to provide a vehicle to transport themselves, other people and goods. Specific needs will dictate whether they buy a passenger car, mini van, large van or truck.

2. Once their primary needs are satisfied, their discretionary funds will be used to buy an array of in-vehicle options designed to increase their vehicle image, performance, safety, comfort and/or convenience. Their selection will be heavily influenced by past purchasing practices, good or bad experience with an option, perceived value of the option, its cost and ability to pay for the option.

3. Selection of new IVHS option offerings will be highly dependent upon satisfying a need, the frequency of the need, availability of alternative solutions and the cost/benefits of the option as compared with the alternative solutions.

The new in-vehicle IVHS products will be competing with many other in-vehicle options and alternative solutions available on the market. The perceived value of the IVHS product and available discretionary funds will heavily influence the buyer's selection of IVHS options. Its success in the market-place will depend highly on the value received for the investment, which in turn, is dependent upon availability of an IVHS infrastructure. The capability of the IVHS infrastructure to provide timely and credible traffic advisories will play a key role in forming and more importantly maintaining the buyer's interest in IVHS.

Until "smart highways" are in place the in-vehicle IVHS systems, if any, will most likely be limited to autonomous navigation and/or route guidance type systems. With limited value and relatively high cost, the market potential for these systems may be confined to niche applications and/or image enhancements for higher-priced vehicles. In the meantime, the ultimate benefits of IVHS will not be realized until the capabilities of all in-vehicle IVHS systems can be reinforced by dependable traffic advisory information.

Not all driver needs are the same and not all auto buyers will be able to afford every in-vehicle IVHS product offering. However, to justify the billions of dollars required for the "smart highways", affordable in-vehicle systems must be available to allow most, if not all, drivers to access the traffic advisory system. To resolve this issue, a variety of in-vehicle IVHS offerings will be required to satisfy the different driver needs and car buyer's budget. These offerings may span from simple highway advisory radios (HAR) to very sophisticated route guidance systems.

3.4 INTELLIGENT VEHICLE-HIGHWAY SYSTEMS DEPLOYMENT

Technology exists today to deploy a variety of IVHS systems, as exemplified by the worldwide development of in vehicle navigation systems employing digital map databases, directional sensors, wheel sensors, CRT displays and related computer processing systems. Some of these navigation systems also include global positioning locator systems.

The benefits of these systems have been advertised and demonstrated worldwide. Their high cost, however, will most likely limit their application to commercial interest where sufficient benefits and cost savings can be realized to offset the investment cost or to provide a high technology image for a limited number of high line vehicles.

What is not available, at this time, is an IVHS system architecture that defines the related infrastructure and how it interfaces with various in-vehicle navigation systems. Vehicle/infrastructure communication standards and their universal acceptance (at least within broad market regions such as North America, Europe and Japan) will be required before IVHS systems will be deployed as in-vehicle OEM installations. The auto industry cannot economically build vehicles to operate with a variety of different infrastructures uniquely developed for specific regions (e.g. New York, Chicago, Los Angeles) unless the in-vehicle equipment is compatible or can be readily made compatible with all infrastructure systems.

The OEM installations will not take place until sufficient interest is shown in the auto buying public to justify the IVHS in vehicle system development, vehicle modification and related production costs. In the meantime, it is expected that IVHS after-market installations will probably prevail which will help define and establish a firm IVHS market.

Primitive forms of IVHS exist today, which can be viewed as alternative solutions to the more sophisticated in-vehicle IVHS systems now being developed. In most cases the related IVHS services are essentially free or readily affordable for the average driver which will challenge the marketability of the new in-vehicle IVHS systems. The following is a list of the systems currently available in the USA.

1. *Highway advisory radio (HAR).* The HAR was introduced in the USA during the late 1970s. It has seen limited use primarily because of the lack of supporting infrastructures. Some of the applications now in use include:

(i) airport traffic and parking information;
(ii) athletic event stadium and traffic information;
(iii) local tourist information;
(iv) road construction and repair information; and
(v) real-time freeway system traffic information.

2. *Cellular phones.* The two-way communications capability of cellular phones has added a new dimension to traffic monitoring. In many cities, local broadcast stations use cellular equipped drivers as traffic probes, thereby, improving their capability to provide near real-time traffic information. In some cases traffic control centres use cellular equipped drivers as a source of traffic information and real-time reporting of road emergencies

3. *Highway exit road signs.* On many Interstate highway exits, road signs are provided to inform the driver of available service stations, restaurants and lodging at the marked highway exit.

4. *Welcome centres and rest stops.* Many US states are using welcome centres and rest stops to promote state tourism and readily provide information for the travelling public, such as:

(i) free state and city road maps;
(ii) brochures on hotels, bed and breakfast accommodations, restaurants, factory outlets, golf and tourist attractions; and
(iii) traveller guides for food, lodging, shopping and attractions including discounts.

5. *Computerized route guidance.* These systems are currently used by some car rental agencies. It provides a detailed route description to the location desired by the vehicle renter. It is reasonable to expect in the future, similar systems may be used in hotels, service stations, welcome centres and possibly on home computers throughout America Online Compuserve or Prodigy linked services.

At the State of Tennessee Welcome Center a computer terminal is available to provide route guidance to attractions within the state. It is called "Touch and Go Tennessee". It provides free information to the traveller on hotels, restaurants and tourist attractions available within the state of Tennessee. It provides a printout of driving directions and, in some cases discount coupons, for destinations selected by the traveller.

6. *Automatic toll collection booths.* Automatic toll tags are beginning to appear and their use is rapidly expanding throughout the USA. Toll tags appear to be one of the early winners for IVHS. As an example, in the case of the Mississippi River Bridge in New Orleans the toll for cars is US$1.00. The bridge has 12 toll gates. Three are unmanned and for toll tags only as marked with the large blue and white signs above the gate. After paying a nominal subscription fee, the automatic toll tag users are only charged $0.70, a 30% saving over cash payers. This technology provides a win–win situation for both the infrastructure and vehicle drivers. It also illustrates how rapidly technology can be deployed once its cost-effectiveness has been demonstrated and realized by the driving public.

Until the IVHS infrastructure is in place and functioning reliably, further IVHS in vehicle deployment will most likely be limited to vehicle enhancements, such as:

(1) route guidance/navigation systems;
(2) improved highway advisory radios; and
(3) radar sensors for station keeping speed control system, backup and adjacent warning systems.

The route guidance/navigation system would have the most significant impact on current vehicle designs. OEM in vehicle installations will require repackaging of the vehicle instrument panel to accommodate a CRT, related vehicle navigation controls and displays. This is necessary because space available in the vehicle instrument panel area is limited and human factors will dictate specific locations for related displays to enhance driver convenience and maintain safe vehicle operation. One solution implemented by General Motors deleted the vehicle radio and heater controls and integrated these functions into a CRT display which could be used as well for route guidance/navigation. For safety reasons, the CRT use for IVHS will most likely be limited to navigation system set-up and route evaluation. Additional displays will be required to provide specific driver instructions (e.g. turn left or right) and ideally should be located in the drivers line of sight. This suggests a heads up display or a simpler display mounted on top of the instrument panel. All of this will require a substantial redesign of the instrument panel.

Before the auto industry will undertake serious consideration to provide this product, some evidence that a reasonable market exists will be required to justify the investment. This of course will be heavily dependent upon driver acceptance, available infrastructure and sustained use of in-vehicle IVHS systems.

3.5 DRIVER ACCEPTANCE

In the final analysis, the car buyer, based on his/her experience as a driver, will judge the acceptability of future in-vehicle IVHS products by buying or not buying the product. Car buyers come from all walks of life, a variety of different cultures, with different levels of education and a wide range of resources. To date, they generally have been very conservative in what they buy and very sceptical about the value of new electronic "gadgets". We don't expect them to change buying habits in the future other than becoming more conservative if their discretionary funds are diverted to other needs.

To improve the probability of driver acceptance, the in-vehicle IVHS system attributes need to be user-friendly, pleasing, trustworthy, evolutionary and provide consumer value. In contrast, some proposed IVHS system attributes appear intimidating, have a potential to be irritating, may be undependable, are revolutionary and to some extent project an image/fad-type product. The task facing the industry, is to find the right compromise between these two extremes, as outlined in Figure 3.8, which will lead to products the driver will accept and use.

1. User friendly	⟷	Intimidating
2. Pleasing	⟷	Irritating
3. Trustworthy	⟷	Undependable
4. Evolutionary	⟷	Revolutionary
5. Consumer value	⟷	Image / fad

Figure 3.8. Driver acceptance compromise

The following are some of the author's viewpoints, based on past automotive electronic product experiences, on how these issues could be approached:

1. *User-friendly–intimidating*. Drivers are more likely to accept products that are simple, easy to operate and readily understood. They are less likely to accept a product requiring considerable time to learn, set-up and operate unless the benefits are well understood and believed to justify the extra effort.

2. *Pleasing–irritating*. IVHS could easily lead to a data overload for the driver and cause him/her to be irritated with the system. The challenge for IVHS is to provide the driver with *only* the information they need, in a way they would like to receive it and at the time they need or want the information.

3. *Trustworthy–undependable*. The need for timely and credible traffic information from the IVHS infrastructure cannot be overemphasized. Each time the driver is unnecessarily diverted from their primary route, due to erroneous traffic information or other cause, they will have another reason to rely on their natural instincts and no longer use their IVHS system.

4. *Evolutionary–revolutionary*. A revolutionary product should not be introduced unless the market-place has been properly prepared to accept the product. Negative market reactions to a new product introduction may have lasting affects on future marketability of it or related products. Introducing new products in an evolutionary manner would entail less risk.

5. *Consumer value–image/fad*. Car buyers, when they can afford it, will often buy a new product solely because of its image and perceived value. However, if the product does not live up to their expectations, they will most likely drop the feature when they purchase their next vehicle.

In summary, driver acceptance of IVHS will be highly dependent upon his or her perception of value received. As a result, for IVHS to be successful, it must satisfy a driver's need, in an acceptable manner, at a price he/she is willing to pay.

3.6 CONCLUSIONS

Automotive electronics has grown rapidly over the past 30 years. It is forecasted to continue this growth into the next century where car buyers perceive its value and are willing to pay the additional cost.

Integrated E/E systems and multiplex wiring will enable the industry to more cost-effectively deploy in vehicle IVHS functions than the current add-on equipment being offered today. Where applicable, the IVHS functions can be integrated into other vehicle systems and sensor signals shared to minimize the additional electronics required to complete the in-vehicle IVHS systems. These techniques and future electronic developments will assist vehicle manufacturers to bring in-vehicle IVHS systems closer to an economic reality.

To date, automotive electronic products have been generally well accepted in the market-place. However, products that directly interact with the driver have shown to be very sensitive to user acceptance; and in some cases, developed an electronic "gadget" image which limited their market growth.

Car buyers are generally conservative on what they buy and are looking for practical solutions for their needs. The perceived value of the IVHS product and available discretionary funds will heavily influence the buyer's selection of IVHS options. Their success in the market-place will depend highly on the value received for the investment, which in turn, is dependent upon availability of an IVHS infrastructure. The capability of the IVHS infrastructure to provide timely and credible traffic advisories will play a key role in forming and more importantly maintaining the buyer's interest in IVHS.

Not all driver needs are the same and not all auto buyers will be able to afford every in-vehicle IVHS product on offer. A variety of in-vehicle IVHS product offerings will be required to satisfy the different driver needs and car buyer's budget. These offerings may span from simple (HARs) to very sophisticated route guidance systems.

"Based on SAE Paper No 912780(c) 1991 Society of Automotive Engineers, Inc."

REFERENCES

1 Rivard JG. Automotive electronics in the year 2000. SAE Technical Paper 861027 *International Congress on Transportation Electronics, October 20–22, 1986.*

2 Heinrich BF. Electrical/electronic systems integration design strategy. S.I.A. Technical Paper 89055. *The 6th S.I.A. International Technical Congress, October 23–25, 1989.*

3 Collision detection serial data communications multiplex Bus. *SAE Information Report*, SAE J1567 August 1987.

4 Fassnacht RE, Madden WH. *Chrysler Collision detection (C^2D^{TM}) Bus Interface, Integrated circuit User Manual*. SAE Technical Paper Series 880586.

5 Meisterfeld, FOR. *Vehicle Multiplexing Strategies*, Automotive Technology International '90 Sterling Publications International Ltd., London.

6 Vig R, Wroblewski T. Addressable single wire magnetic sensor IC for multiplexed control. *3rd International Conference "Automobile Electronics — Interface and Environment," SIA, Toulouse, France, April 1988.*

PART 2

DYNAMICS OF INTELLIGENT VEHICLES

4

Intelligent Vehicle Systems and Control

4.1 INTRODUCTION

More than a century ago, the first workable four stroke internal combustion (IC) engine was invented in Cologne, Germany, by a young merchant's assistant named Nikolaus Otto. Subsequently, Karl Benz, Gottlieb Daimler and Henry Ford produced the world's first cars with IC engines. By the early twentieth century various practical car design were appearing internationally. This was followed by a great revolution in personal mobility.

Prior to the period in 1886 when Karl Benz patented the first vehicle, most people of the world including Europe and Japan were not able to travel more than 35 km from their homes except under duress to enlist in conquering armies or join groups of religious travellers.[1] In 1893, Henry Ford in Dearborn, Michigan, USA, placed his first car on the road. In 1901, the Daimler motor company predicted that worldwide car sales would never exceed 1 million because the population at large could not provide more than a million people capable of learning how to drive their vehicle.[2] As can be read from the subsequent paragraph, Daimler grossly underestimated. From this historical standpoint, it becomes interesting to see how the road vehicle has become the great liberator in terms of personal and mass mobility.

Early in the Second World War the world vehicle fleet was less than 50 million units, but by 1987 the total unit has grown to approximately 516 million, of which 395 million were passenger cars. By the year 2000, it is estimated that the total fleet would be approximately 536 million, with increases in the numbers of commercial vehicles, the total fleet unit of 720 million in the year 2000 is expected. Owing to

Advanced Vehicles and Infrastructure Systems, Edited by C.O. Nwagboso

increasing economic activities in Russia, China and fast developing countries, the total fleet would rise even further. Taking these increases down to a personal level, if every individuals eligible to drive has his or her own car the average fleet may reach 630 cars per thousand population (see Table 4.1). If some individuals kept two cars, which can be the case, the total might reach 700 per thousand.

In Europe alone more than 500 billion ECU are spent on road transport products and services every year. More than 10% of the average family budget is devoted to transport. Car ownership has been increasing by 4% per year and there are already more than 120 million cars in Europe. Within the European community, international traffic increases by 5% a year and it has been estimated that 1% growth of the GNP generates 1.5% growth in passenger transportation.

There can only be few people today who are not aware of the advances in road vehicles of today. Before now, the road vehicle for personal and mass mobility was still considered as under-developed, especially in terms of performance, safety, security, intelligence and the interaction between the vehicle and external infrastructures. The issue of technological under-development resulting in problems of traffic congestion and environmental pollution have been accelerated by the increase in the number of road vehicles.

Based on the foregoing trends, there is undoubtedly a big gap between the development of intelligent vehicle systems and control of today and that required for the future. The engineers of today are charged with the technical challenges of bridging this gap. The main challenges lie in achieving substantial advancement and automation of the vehicle and the infrastructure systems through the application of electronics, information technology and telecommunications. This chapter will highlight the essential technologies and automation for intelligent vehicle systems and control. It will describe the main technologies for the infrastructure systems required to support intelligent vehicles. It will concentrate on basic concepts and methods of control that have been proposed, developed and implemented.

Table 4.1. Forecast of UK car ownership per thousand people

Year	Low	High
1995*	382	404
2000	411	450
2005	438	490
2010	463	526
2015	487	557
2020	508	585
2025	529	608

*Mid-year rate of ownership.

4.2 VEHICLE ELECTRONICS AND CONTROL

Electronics and control technology is dramatically speeding up the automation and evolution of the road vehicle as the personal and mass mobility system. The electronics and control technology used in vehicles are influenced by in-vehicle devices and subsystems, such as, sensors and actuators, electronic control unit (ECU), software, vehicle electrical systems and multiplexing subsystems. Figure 4.1 shows an example of the application of electronics in the road vehicle. The cost associated with the three essential electronics functional blocks used in automated control

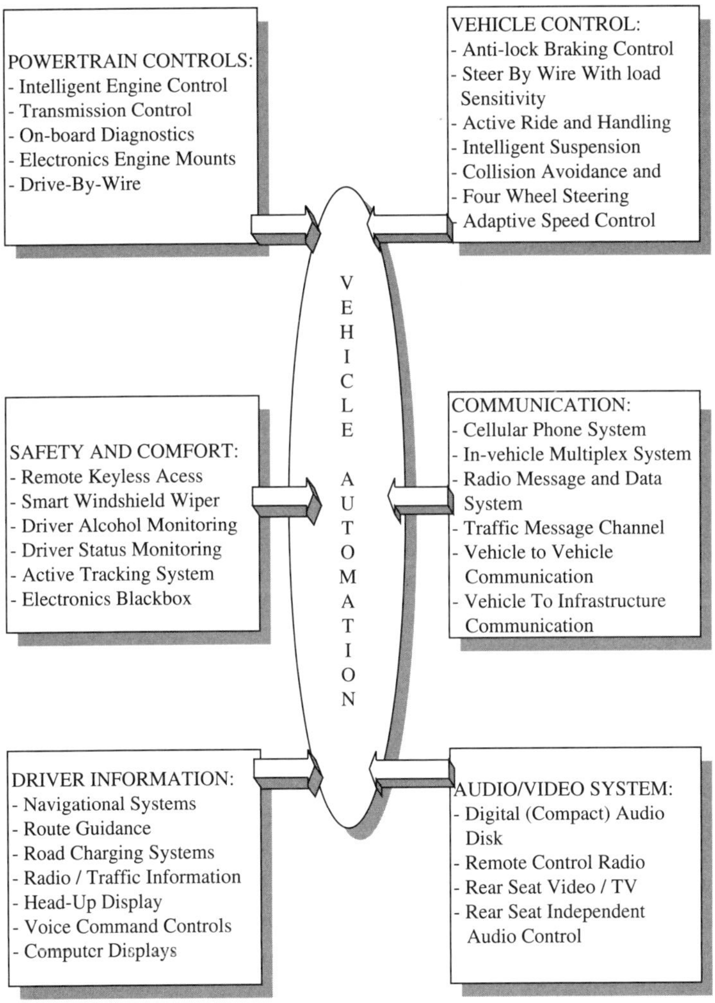

Figure 4.1. Electronics inclusions in road vehicle

systems have remarkably reduced over the last 10 years. For example, the price performance ratio of the electronic actuators used commonly in automated vehicle systems has improved by a factor of approximately 10:1. This has made it possible for the faster introduction of motorized seats, mirror adjustment and automatic doors as seen in most Japanese taxies. The development of the microcontrollers, signal transmitters and programmable miniature computers have been used for wide range of vehicle electronic applications. For example, the keyless access system incorporates a receiver which checks an incoming signal against a unique code in a memory, and then release the locking system if the code is correct. Other possible applications are based on the reverse procedure, whereby a coded signal from a vehicle is transmitted to a roadside toll booth or more commonly to a security gate. On a worldwide scale, the increasing developments in the enabling technologies and the decreasing cost of the electronic devices are providing engineers with opportunities to design advanced and intelligent transportation systems. Meanwhile, the user of the road vehicle is presently experiencing some of the outcomes of the new information technological inclusion in today's vehicles. This ranges from the modest improvement in vehicle performance, improved driver and occupant safety, emission control, convenience, comfort and vehicle traffic management systems not thought possible even 12 years ago. So the primary reasons for the development of electronics and control systems for automation of the road vehicle is to provide these improvements. One can therefore forecast that, by the year 2005, the electronic content in an advanced vehicle will account for more than 20% of the cost of the vehicle, with the cost of electronic management control alone at about £3.2 billion in Europe at that time. These electronic systems will be integrated, with shared data obtained via communication links (data highway network) rather than the current "add on" and "stand alone" compromises.

4.2.1 Vehicle Sensors

In the automation of vehicles, we are concerned with the necessary control of the dynamic system, such as the control and management of the ignition systems or the adaptive control of the advanced suspension system. The control of the behaviour of the dynamic system can be simplified to three functional blocks required in the directional control of a vehicle (Figure 4.2). The sensor is part of the vehicle electronics with the functional responsibility of detecting, acquiring and transmitting information from the given stimulus, in the case of vehicle steering, the condition of the road. This information is passed to the computer (processor), which determines the appropriate action to be taken, and finally the actuator is instructed to bring about the changes required to maintain the necessary vehicle dynamics control for the actual position on the road.

Vehicle sensors are essential components of advanced vehicle and infrastructure systems. A variety of sensors exist for automation of an advanced vehicle. Most of the present advanced vehicles contain more than 30 sensors on average; most of these

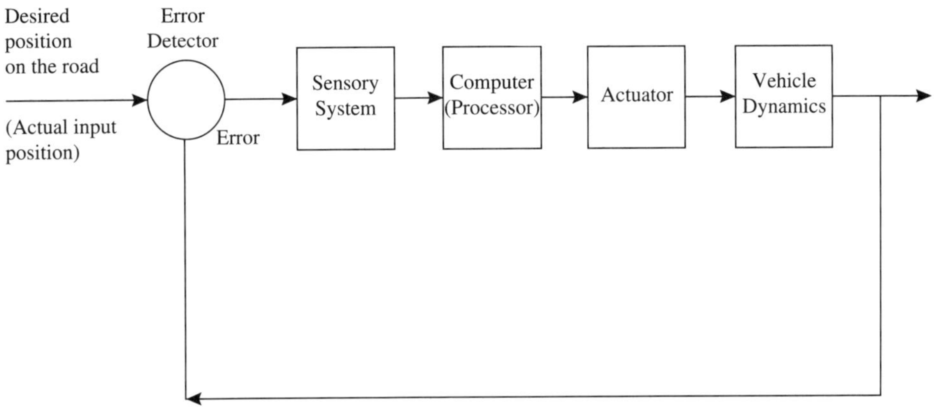

Figure 4.2. Three functional blocks of the control of vehicle steering

sensors are detailed in a relevant book[4]. For an in-vehicle system, the majority of the sensors are used for the monitoring, managing and control of the powertrain, vehicle ride and handling, driver behaviour, safety systems, distance sensing, navigational systems and driver/occupant comfort. Normally, each sensor is connected to its own local signal condition circuit and has its own wiring. The size and cost of the wiring used in vehicles is rapidly becoming larger. This problem will indeed be increasing in the future, especially as it has been predicted that the number of in-vehicle sensors will exceed 100 by the end of the century.[3] This problem will be alleviated by the use of intelligent sensors. Vehicle to infrastructural systems interaction requires sensors such as roadside sensors which are able to detect changing aspects of the road traffic environment. This will include the volume of vehicle traffic on the road, the prevailing weather conditions, exhaust pollution emission levels, integrated road payment system and law enforcement systems.

The future lies in the use of intelligent or smart sensors for advanced vehicles. In vehicle electronics and automation, there has always been a discussion about the suitability or otherwise of using the term "intelligent" or "smart" for sensors. None the less, the technical meaning of this now well established term has become quite clear during the course of time, and therefore may be defined within various texts.[4,5] This is developing sensor technology which is based on the concept of integrated sensory systems; instead of being a "stand-alone sensor" providing a voltage output to an electronic unit, the sensor itself becomes an integral part of the instrumentation system, with different systems requiring access to one and the same sensor signal. In the event of sensor failure or malfunction, a diagnostic, self-correcting or fault tolerance function is invoked. Intelligent sensors may be identified by means of the features shown in Figure 4.3 and typical operational factors shown in Figure 4.4.

The basic intelligent sensor can be described by using a conceptual silicon micromachined pressure sensor model shown in Figure 4.5. The first stage of the model is the initial functional element, which may be a silicon diaphragm. This

STANDARDIZED INTERFACE:
- Bidirectional
- Adaptable
- Communication with Digital/Analogue Systems (Bus Capability)

CHARACTERISTICS:
- Inherent Monitoring
- Multivariable Data Acquisition
- Self-recalibration Ability
- Additional Output of Influencing Variables
- Performance Diagnostics and Prognostics

LIMITATION OF THE TOLERANCE USING ELECTRONICS:
- Acquisition and Consideration of Disruptive and influencing Variables on the Spot (Static Correction)
- Improvement of the Dynamic Behaviour (Dynamic Correlation)
- Consideration of Specimen-specific Characteristics (Self-calibration)

Figure 4.3. Features of intelligent sensors

modifies the raw input, i.e. the applied pressure, to a more acceptable state, which is the mechanical deflection. The second stage of the model is the primary transduction element which is the piezoresistive bridge. This provides the change in the energy state from mechanical deflection to an electronic analogue signal. The next stage of the model is the signal conditioning element which may be provided to generate a suitable signal (e.g. an amplified temperature compensated voltage signal). The subsequent stage is the analogue to digital (AD) conversion element, with the AD converter which must, however, fully satisfy the corresponding measurement signal requirements regarding the converter speed, resolution and accuracy. The next stage in the model is the correction unit, which can be a microcontroller or a dedicated processor for the specific digital signal correction algorithm. The digital correction unit is interfaced with the programmable memory unit for model parameter variation. The final stage in the model is the sensor interface designed to accept the corrected digital signal for active or passive function within the intelligent vehicle.

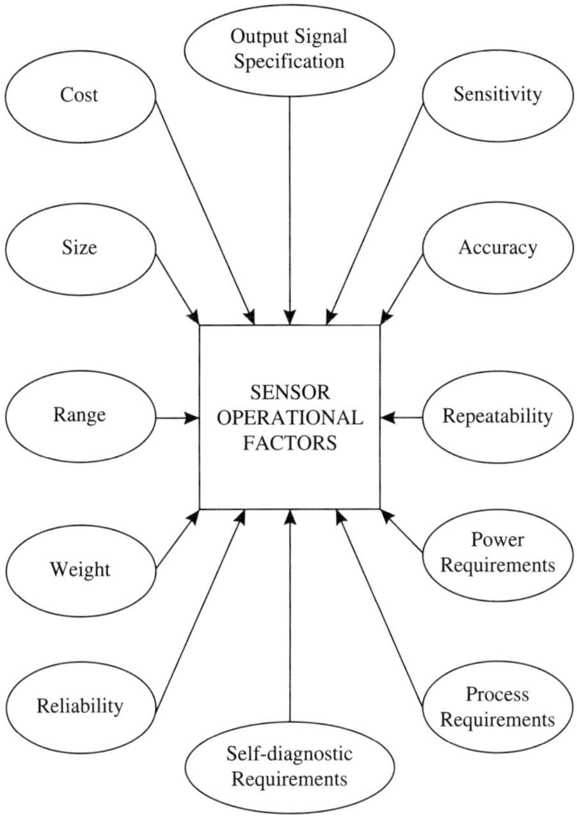

Figure 4.4. Typical intelligent sensor operational factors

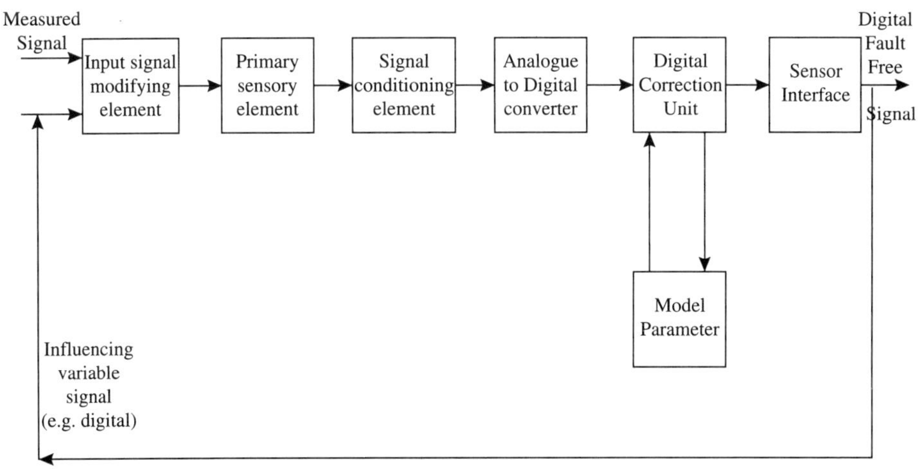

Figure 4.5. Functional intelligent sensor configuration

The model in Figure 4.5 is one version of an intelligent sensor. There are other types of intelligent sensors for specialist application and of simpler configuration.[4]

Sensors and components interconnections

As the number of sensors and electronic devices in the vehicle grows, so does the need for a standard network to ensure that the various devices, sensors and modules will not only communicate with each other, but also will be capable of sharing power resources, software and central computation power. A great deal of systems development is currently been done in this area involving the use of multiplex wiring system.

A multiplex wiring system architecture will be adopted in some vehicles with modular intelligent sensors and devices. The preferred objective is to get the individual, independent, separately developed vehicle electronic modules and sensors to be interconnected in a network protocol. The International Standard Organization (ISO) approved several network protocols which can be used for vehicle networking and automation. Of the three ISO accepted protocols (SAE-J1850, VAN, CAN), Controller Area Network (CAN) is the popular choice within the European automotive industry, it is also gaining acceptance with the American automotive industry. Figure 4.6 illustrates how the independent components can be linked to one another and to a master controller by a signal carrying a CAN linear bus network with different transmission speeds. This type of bus system is known to have a high data transmission rate (1 Mbit/s) and it is flexible with special facilities ensuring an interference suppressed operation.[6–8]

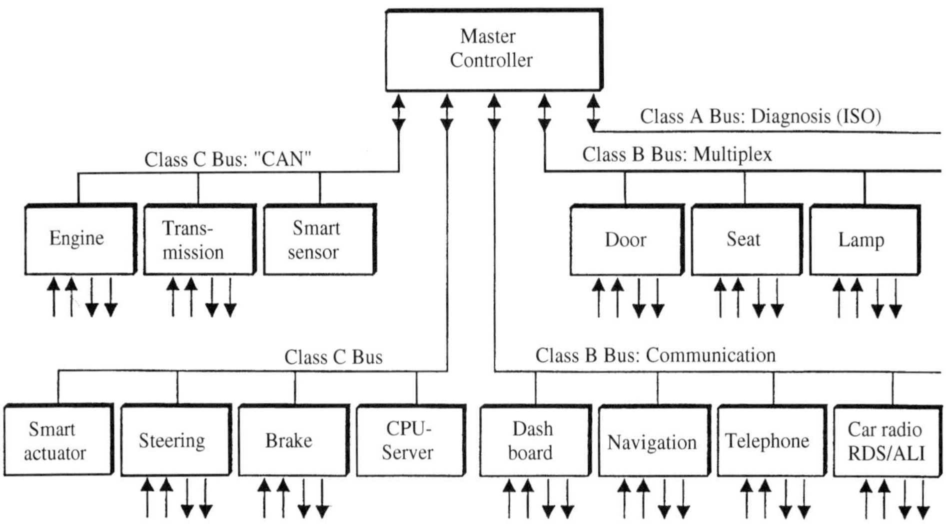

Figure 4.6. Decentralized sensors and electronics systems interconnection

CAN was developed by Bosch for automotive applications to reduce the complex wiring systems in road vehicles with multiple microcontroller-based control systems (e.g. engine management, suspension control, ABS, etc.). CAN is a multi-master bus topology that has a layered structure specific to systems designers, as shown in Figure 4.7. For data transmission, it uses carrier sense multiple access with collision detect (CSMA/CD), with non-destructive bit-wise arbitration for bus contention. To provide a faster transmission of real-time information, CAN utilizes broadcast-type functional addressing with a non-return to zero (NRZ) data format with bit stuffing at data rates in the range of 15 Kbit/s and 1 Mbit/s.

There are presently various microchip manufacturers that have entered into the CAN controller market, such as, Philips, Motorola, Siemens, NEC, Intel

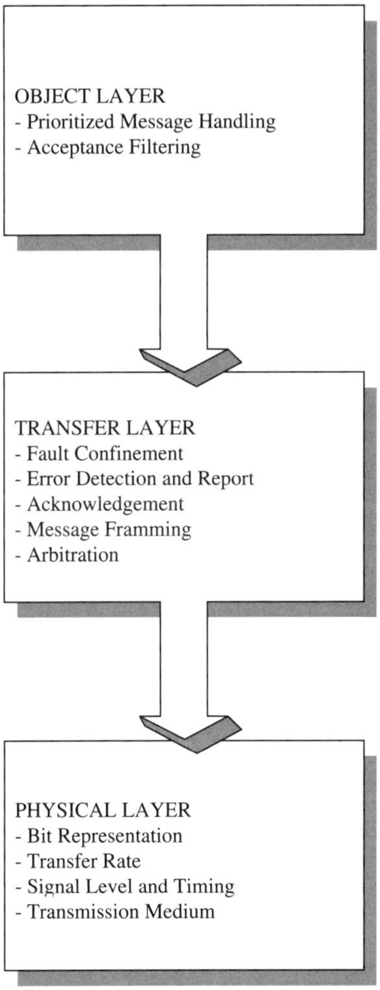

Figure 4.7. CAN layered architecture

and National Semiconductor. It is expected that with time, more automotive manufacturers that are presently reluctant to take up standard network structure will adopt the CAN system.

Electronic control unit

The ECU is the centre for the in-vehicle electronics systems management, and it interfaces with the most essential devices used for the monitoring and control of many vehicle functions. As shown in Figure 4.8, the ECU can receive a varying

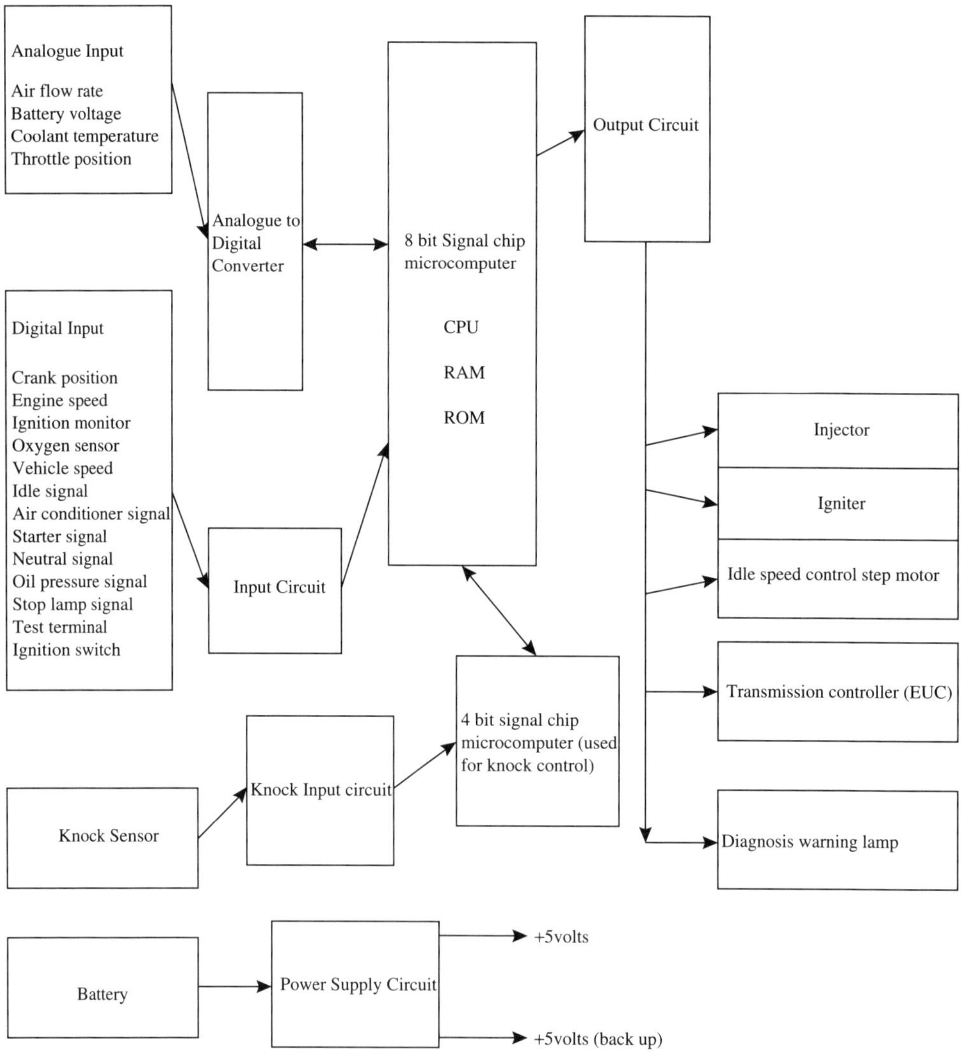

Figure 4.8. Basic systems of the electronic control unit

range of analogue or digital signals from sensors. The microprocessor uses these values to calculate the required control values based on preprogrammed settings. As a result of the calculation, signals are sent to drive actuators, motors, solenoids and other devices to adjust the required activities within the vehicle.

The ability of an ECU to perform a range of control function within the vehicle depends on the size of the ECU which is dependent on the number of signal inputs and outputs it has. The early ECUs only had five to eight inputs and outputs, some of the recent designs can have up to 70 inputs and outputs. Some examples of this type of systems are shown in Figure 4.9 typifying sensor usage in advanced vehicles, and in Figure 4.10 showing the Siemens engine management system via an ECU. However, as the systems and devices required to automate advanced vehicles grows the number of inputs and outputs in an ECU will also grow, but above all, the complexity of the unit will also grow.

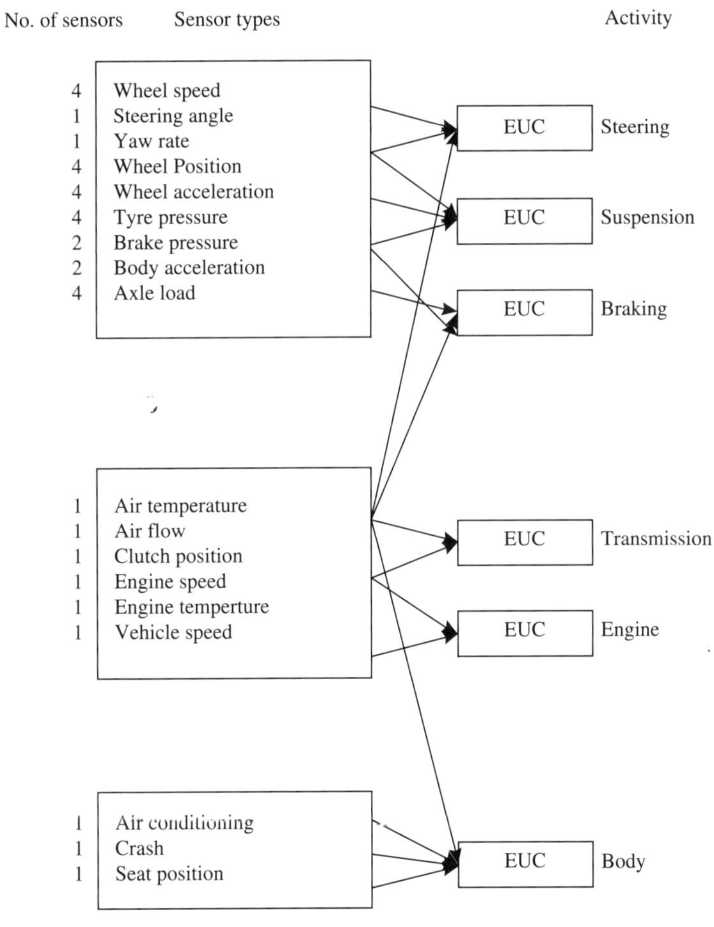

Figure 4.9. Example of sensor interconnection in an ECUs

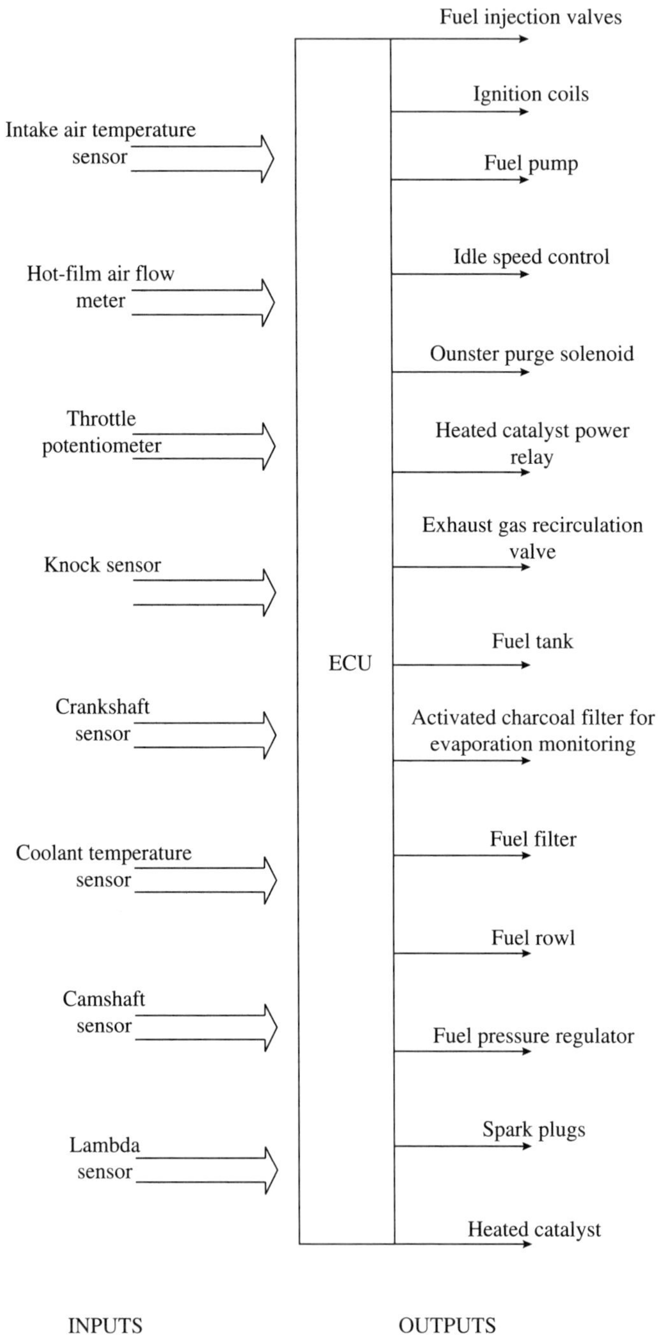

Figure 4.10. Digital engine management system (Siemens Automotive)

4.2.2 Basic Control Systems Technology

Control systems technology is now widely used in the automation of road vehicles thanks to electronics allowing the development of, for example, headway control, anti-lock braking systems (ABS), cruise control systems, controlled four-wheel-drive systems (4WD), controlled four-wheel-steering systems(4WS), controlled suspension systems, traction control systems, powertrain control, etc. The wider spread application of control technology will invariably continue to increase in modern vehicle and the relevant interfacing traffic infrastructure, but more importantly, as further intelligence is implemented into road vehicles, requiring reasoning and performance of a complex system's tasks, intelligent control technology will also be essential. Considering the general purpose of the control system's methodology in the road vehicle, it is more appropriate in this subsection to focus on the survey of basic control concepts and the most important fundamental control methodology that can be used in intelligent vehicles.

The fundamental requirement of developing control technology for the systems used in intelligent vehicles is to apply numerate treatment of the vehicle systems based on the facts that the behaviour of all dynamic systems can be governed by a series of differential equations of the form:

$$f\left(a1, \ldots a_n, x, x' \ldots \frac{\mathrm{d}^n x}{n}\right) = f(u, v, w, \ldots) \tag{4.1}$$

The coefficients in the equation may take into account the linear and non-linear relationship and $f(u, v, w)$ represents any physical disturbance in the system which include environmental effect and externally forced disturbances. In a simple mechanical system, the equation can be reduced to a relationship governed by a linear equation with the solution obtained easily in analytical form. To survey different types of control methodology available, the governing equation of the vehicle systems behaviour which form part of the control concept may be represented by:

$$M\frac{\mathrm{d}^2 x}{\mathrm{d}t^2} + f(x')\frac{\mathrm{d}x}{\mathrm{d}t} + f(x)x = F(t) \tag{4.2}$$

where x = linear displacement
M = System's mass
$f(x')$ = function governing the variation of resistance to motion with velocity component, such as, aerodynamic resistance
$f(x)$ = function governing variation of force with displacement, such as stiffness of the system, springs, etc.
$F(t)$ = function governing the external force or thrust applied to the system as a function of time

Taking equation (4.2) into consideration, two basic control system's concepts namely, open-loop and closed-loop control can now be surveyed.

Open-loop control

This type of control concept applies to the normal driver–vehicle interaction that exist during driving conditions shown in Figure 4.11. If the driver is driving on the road along the principal axis "x" where there are traffic constraints imposed by a series of speed-enforced limits displayed on the driver information system, the immediate reaction of the driver will be to control the speed of the vehicle to within the legal limit. Assuming the limit meant a reduction of the vehicle speed from 60 to 50 km/h, the driver will proceed to operate the accelerator, brake and clutch pedal accordingly in order to reduce the speed of the vehicle. From the driver's point of view the vehicle is an open-loop control system responding to the operation of the accelerator, brake, clutch pedal and possibly the gear box actuation. For sake of brevity, if we assume that the speed reduction was brought about by only the actuation of the brake pedal and the driver is experienced enough not to depress the accelerator pedal while braking, we can determine the system's response to the brake pedal pressure, where the brake pedal actuation is used as the input to the system. As the vehicle was going 60 km/h, to simplify the analysis further, let us assume that the effect of the engine output torque on the reducing vehicle speed is negligible over the comparatively small variation imposed by these conditions. The further depression of the brake pedal is regarded as an immediate reduction in torque from the prior braking value to the new value. Based on the known dynamics of vehicles,[9,10] this torque will act on the mass of the car via the road wheel so as to produce deceleration additional to the retarding forces imposed on the vehicle by aerodynamics, any gradient and rolling resistance. The deceleration force will also act on the mass of the powertrain, but for the purpose of this analysis, we can consider that this mass is lumped together with the mass of the vehicle (M). Similar

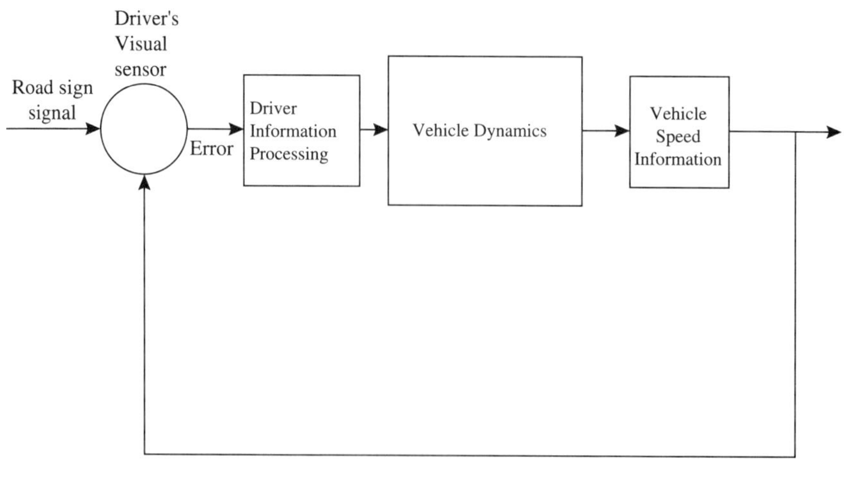

Figure 4.11. Vehicle driving using close-loop "feedback system

to equation (4.2), the systems equation becomes:

$$M\frac{d^2x}{dt^2} + \phi\frac{dx}{dt} = F \tag{4.3}$$

where F = the deceleration force acting on the wheel

ϕ = the coefficient governing the relationship of the aerodynamic, rolling and vehicle speed which is assumed linear.

If we use the "D" operator concept, equation (4.3) becomes

$$MDV + \phi V = F \tag{4.4}$$

Let $M/\phi = T$ dividing equation (1.4) by ϕ we get

$$V(TD + 1) = F/\phi \tag{4.5}$$

The "particular integral", $V = F/\phi$, is obtained when t approaches infinity and to describe the system's performance at finite time this result must be combined with the "complementary function", which describes the recovery rate of the system from any external forcing effect, therefore:

$$V(TD + 1) = 0 \tag{4.6}$$

If we assume a solution of the form $V = Xe^{rt}$ and substitute it in equation (4.6), then

$$Xe^{rt}(Tr + 1) = 0 \tag{4.7}$$

we can therefore say that:

$$r = -1/T \text{ and the solution becomes } V = Xe^{-t/T}$$

If we link equation (4.5) with the solution obtained above to get $V = F/\phi + Re^{-t/T}$ and assume an initial condition of $t = 0$, $V = 0$, therefore $R = -F/\phi$. Subsequently, the speed response of the vehicle as it accelerates from rest against a resistance varying linearly with speed may therefore be represented as:

$$v = F/\phi\,(1 - e^{-t/T}) \tag{4.8}$$

The expression enclosed in the brackets of equation (4.8) is known as the exponential lag due to the included exponential term and because it described the lag of the response of the system to a sudden change in condition. Equation (4.8) cannot provide an accurate description of the acceleration of the car from rest to its maximum speed because in real terms ϕ varies in a more complicated manner than was assumed and also the engine torque does not really remain constant with change in engine speed, even disallowing the effect of gear changes, which is not practicable. However, at a low speed of about 60–50 km/h it will be sufficiently accurate to use the basic expressions in equation (4.8) to study the response of the system in terms of the acceleration of the car at constant torque. Readers are therefore invited to obtain the response of the system by considering the initial conditions of time and relating it to the values of the vehicle speed.

The simple control system model we have used cannot be effective as a vehicle control system and it will fail in most vehicle driving conditions because no account is taken from other system's input variables and external effects. One method of enhancing the model as an open-loop control is by considering the accelerator pedal, the engine and transmission units, road wheels, etc. as an additional system relating the angle of displacement of the accelerator pedal with engine torque. We can further increase the flexibility of the model by depicting the region from the output of the engine crankshaft to the axle by an integrated system which enables such factors as the rotational inertia of the transmission system, the gear changing actuation and clutch slip to be incorporated into the modelling of the general vehicle dynamics and control system.

Closed-loop control

For an open-loop control system a single action on the vehicle controls will result in a definite response which is determined by the characteristics of the vehicle and the road environment. The driver who doesn't possess the skill of driving might provide either too much or too little input to the accelerator pedal to make any control adjustments and would indeed realize the mistake and make further adjustments. The alternative method of controlling the vehicle is somewhat simple and it is the technique used by all experienced drivers. The principle is to use the effects observed by the driver on the controlled variables of the vehicle to adjust the dynamics of the engine and transmission system and therefore influence the directional control and power of the vehicle. The means by which the driver becomes aware of the needed adjustment must incorporate some means of sensing the vehicle output, feeding an indicator of the output and a means of communicating the indication of the output to the control system. In the case shown in Figure 4.11, the speed display unit provides the indication of the speed, the eye of the driver is the vision sensor and the brain of the driver is the processing computer, that interprets the information which is transmitted to the foot for accelerator pedal actuation. The feedback of information required for a closed-loop control is obtained as the transmission of information about the speed and speed limits on the traffic information display (controlled variables) are fed to the driver (the controller). As the driver becomes aware of the speed limit, the difference between the instantaneous vehicle speed and the speed limit is in fact the error in the speed and this is what the driver will attempt to minimize. The actuation of the accelerator pedal then becomes the function of the error used to control the vehicle.

To replace the driver with an autonomous and intelligent driving system, a controller system will then be required so that the driver is replaced with a system element which among other requirements, would convert the speed error into an accelerator unit actuation. The controller system could follow several simple and complex control laws which will be discussed later, but the simplest of the laws is the proportional control in which we can assume that the variation in the engine torque is proportional to the error between the desired speed limit and actual speed. In brief, if the tractive force and the vehicle resistance are combined into a transfer operator $(TD + 1)$ the effect of a proportional closed-loop control can be evaluated using equation (4.5), such that $V(TD + 1) = F/\phi$, where F/ϕ is now a function of the error.

If the desired speed input is represented as $\emptyset_i$ and the actual speed output is represented as $\emptyset_o$, the error can be represented by $\emptyset_e = \emptyset_i - \emptyset_o$ then $F/\phi = K\emptyset_e$ and $V = \emptyset_o$, therefore:

$$\emptyset_o = \frac{K\emptyset_e}{TD + 1} \tag{4.10}$$

where $K\emptyset_e/TD + 1$ is the forward path transfer operator and $\emptyset_o/\emptyset_i$ is the closed-loop transfer operator. We can now define the output to error speed as $\emptyset_o/\emptyset_c = \check{S}$ then:

$$\emptyset_o = (\emptyset_i - \emptyset_o)\check{S},$$

$$I/\check{S} = (\emptyset_i - \emptyset_o)/\emptyset_o, \text{therefore} \emptyset_o/\emptyset_i = (1/(1/\check{S} + 1))$$

From these expressions the output speed can be obtained as:

$$\emptyset_o = \frac{K}{TD + 1 + K}\emptyset_i \tag{4.11}$$

The complementary function and particular integral can be obtained by specifying the conditions of the input and output speed, and by applying the same procedure as applied to arrive at equation (4.8), the speed response of the closed-loop control can be derived as:

$$\emptyset_o = \emptyset_i\{1 - e^{((-t/T)(1+k))}\}(k/(1 + k)) \tag{4.12}$$

As a proportional system the effect of the feedback would indeed be to increase the rate of response, with the effect of the amplifier having a gain of K in the system, the response would increase further but be limited by the tyre/road interfacial adhesion.

There are several other control laws which can normally be combined with the proportional controller system and the transfer functions of the vehicle system in order to enhance the performance of the controlled output of the vehicle system. Most of these laws which can range from proportional plus integral to fuzzy and neural network control laws are fully treated in modern control engineering texts[11–13] and are shown in Figure 4.12.

Control of the air/fuel (A/F) ratio

An area of system's automation which continues to receive automotive engineers attention is the implementation of control system for exhaust emission reduction.

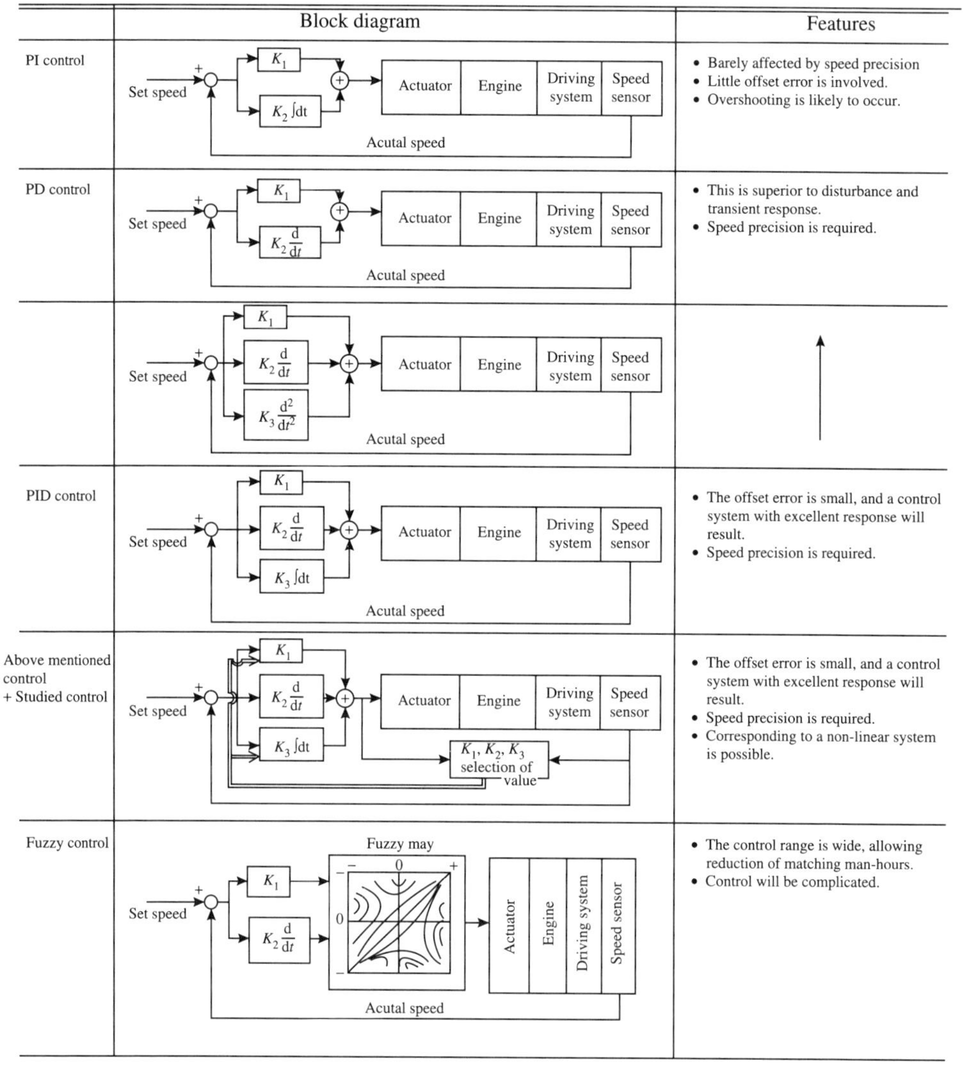

Figure 4.12. Methods of vehicle systems control laws

One method of doing this is to have a control scheme which monitors and adjusts the A/F ratio automatically in order to influence the pollution constituents of the exhaust gas. In a gasoline engine, such a control scheme can be used in conjunction with a "three-way" catalytic converter that will oxidize excess level of carbon monoxide (CO) and unburned hydrocarbon (HC) and reduces oxides of nitrogen (NOx). Generally, most catalytic converters can be ineffective when the A/F is more than 1% different from the stoichiometric threshold; therefore, a feedback control system is required to maintain the A/F within ± 1% of that threshold. This requires the use of an exhaust sensor downstream of the engine, just before the catalytic converter, and the capacity to adjust the fuel metering electronically using some form of an actuator.

Shown in Figure 4.13 is system's block diagram of a four cylinder engine with the control law shown as proportional plus integral. A sensor with a gain of K with normally a non-linear functional relationship has been linearized in terms of the relationship of the A/F ratio and the voltage output. The dynamics of the inlet manifold are represented by a transfer function with a time constant representing the characteristic fuel flow in the manifold. The time delay is the sum of the delay caused by the four piston stroke from the intake process until the exhaust process and then the time required for the exhaust to travel from the engine itself to the sensor, which may be located about 300 mm away from the engine. In order to account for the A/F mixture time, a sensor lag is also included in the process. The system's frequency response due to the A/F controller can be evaluated by analysing the magnitude and phase characteristics.

On-board control tasks

With the advent of the integrated electronics control strategies for road vehicles, one can safely say that a new era has arrived in the approach to providing intelligent control schemes for vehicle onboard tasks. This raises the issues on where further control schemes can be implemented to solve some of the onboard control tasks. For an intelligent vehicle, four essential categories of control tasks can be extracted from Figure 4.1 and related to road traffic infrastructures.

1. *Vehicle performance-enhancement tasks*: engine and transmission control; integrated chassis and suspension control; active vehicle speed control;

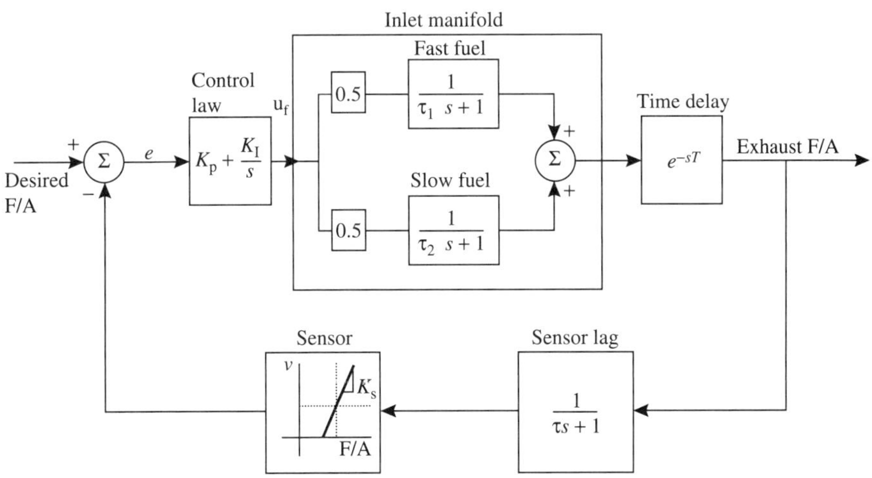

Figure 4.13. Simplified representation of a closed-loop A/F control system

electronic voltage and ignition control; active camber and castor angle control; cybernetic/active steering control; lateral control enhancement; and active torque management and control

2. *Safety and security tasks*: intelligent maneuvering and control; headway control; anti-lock braking control; driver fitness and engine prestart control; air bag control; active side-impact bars; seat belt supervision and control; and anti-theft and vehicle tracking systems control.

3. *Driver assistance tasks*: climatic control; power seat positioning control; voice command control; real-time intelligent driver support systems; autonomous driving control; and cooperative driving control.

4. *Traffic and infrastructure*: route guidance control, planning and displays; navigation control systems; automatic debiting and electronics payment systems; vehicle location antennas and control; and remote speed enforcement.

It is quite obvious that to deal with the control tasks necessary to implement most of these tasks, the control laws that will be applied will range from proportional control to more advanced control which requires fuzzy logic and neural network and possibly the combination of these two. For a normal vehicle speed range of 0–50 km/h, a human driver is able to handle the vehicle comfortably, especially after basic training. However, as the speed range is increased, human's visual sensing and information processing abilities with the necessary speed of his/her reaction to motion are considered unable to match sufficiently the high-speed driving. Although with extensive training, driving at the high-speed range is normally achievable, but with a risk. Therefore, during the next decade, a major contribution can be expected in the areas of intelligent control especially within the onboard systems interfacing with a four-level control hierarchy identified as advanced vehicle control, route guidance, navigation system's control and traffic flow control. The human–machine interface problem of such systems will form part of the development process.

4.2.3 Vehicle Following

In designing collision avoidance system for an intelligent vehicle control system, one of the vehicle's performance requirements will involve the understanding of the vehicle's interaction with other vehicles on the road. This will require the understanding of various vehicle following concepts and models. For example, under normal traffic conditions, vehicle following by a driver who has to drive the vehicle as part of a string of vehicles on a highway can indeed be a complicated control process. The driver may take account of the fact that the vehicle in front will require some time and distance to stop or accelerate. Therefore if the driver can brake or accelerate as quickly as the driver in the front or rear, collision can be avoided provided that the distance separating the vehicles is sufficient to compensate for the reaction time required by the second driver to apply the brake or accelerate after the first driver has done so. This example does not take into account the fact that obstacles can suddenly appear on the roadway and that different drivers can react differently to a given set of traffic circumstances. It also assumes that the vehicles travel in a single lane, and no overtaking occurs. If we take this somewhat simplified example it becomes possible to provide a one-dimensional quantitative model that can describe the driver's behaviour in a manner which may have significance in the understanding of a collision avoidance system.

As shown in Figure 4.14, the driver as part of the vehicle following scheme can be treated as a control system, where the input will be the position, speed and acceleration of the car which he/she is following and the output will be the behaviour of their own vehicle. This is in line with various vehicle following models,[14–16] where the behaviour of a vehicle, to some extent, is dependent on the behaviour of the vehicle in front and the modelling task is to find the relationship between each pair of vehicles.

The dynamics of the vehicle following theory can be modelled by the so-called microscopic traffic flow model. The basic but simplifying assumption in this model is that every driver in a single lane reacts to a stimulus from the immediate environment in accordance with the relationship which states that:

$$\text{Reaction of } n\text{th driver at time } t = \beta n\{\text{Stimulus at time } t - Tn\}$$

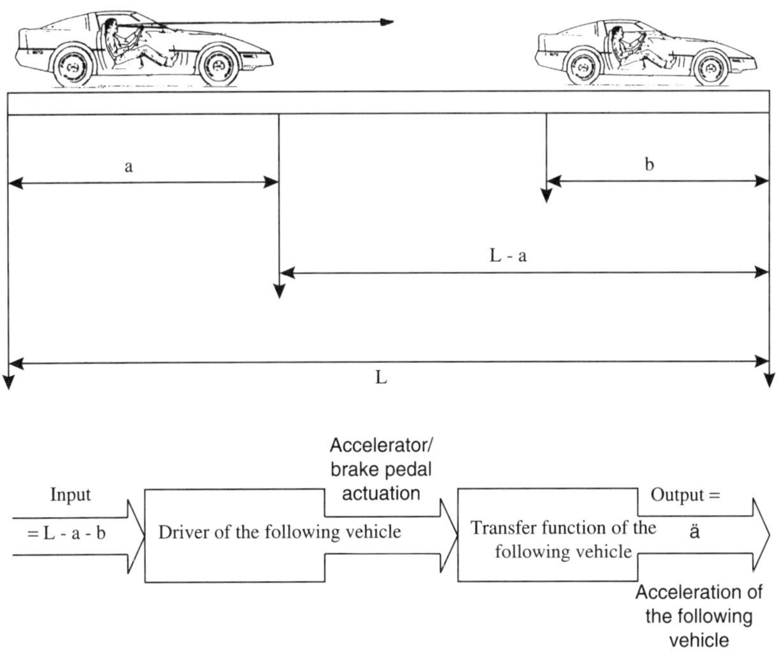

Figure 4.14. Vehicle following control

Now suppose that each vehicle travelling within a single lane of roadway follows or trails the one in front, thereby forming a platoon. A platoon in this case, is identified as a chain of N vehicles, preceded by a lead vehicle, which behaves independent of all others and followed by $N - 1$ other vehicles; each is strictly influenced by the behaviour of the vehicle which is directly in front. We can represent the position, velocity and acceleration of the nth vehicle in the platoon, at time t by $X_{\mathrm{n}}(t)$; $X_{\mathrm{n}}(t)$; $\ddot{X}_{\mathrm{n}}(t)$, respectively. In a platoon, if the driver finds that the vehicle is catching up with the preceding vehicle, the driver will slow down and if the vehicle falls behind in speed, the driver will speed up. Under this condition, relative to the position, and velocity of the vehicle ahead, each vehicle in the platoon will take a finite time to react to changes instigated by the driver. This is normally taken into consideration in the model by incorporating a time lag so that the response to a situation existing at time t does not occur until $(t + \tau)$. Established vehicle following models[15,16] involve a response in the form of an acceleration, $\ddot{X}_{\mathrm{n}}$, of vehicle n, at time t, to a stimulus proportional to the relative velocity, $\dot{X}_{\mathrm{n}-1} - \dot{X}_{\mathrm{n}}$, of vehicle n and $n - 1$, at earlier time $t - \tau$. When the vehicle attempts to match the speed of the vehicle ahead, this condition can be modelled by a standard linear vehicle following the model in equation (4.13), where the positive integers l and m are scaling exponents and a is a real constant. The function X_{n} describes the nth vehicle's sensitivity to changes in the relative velocity and X_n is a reaction time lag. The model does not take into account the effect of intervehicle spacing independently of the relative velocity.

$$\ddot{X}_n(t) = \beta_n[\dot{X}_{n-1}(t-\tau) - \dot{X}_n(t-\tau)] \tag{4.13}$$

where: $$\beta_n \equiv a\frac{[\dot{X}_n(t)]^m}{[X_{n-1}(t-\tau) - X_n(t-\tau)]^1}$$

If equation (4.13) is considered in terms of fundamental control theory, the speed transfer function can be derived as:

$$G_n(p) = \frac{\dot{X}_{\mathrm{n}}(t)}{\dot{X}_{n-1}(t)} = \frac{\beta_n \exp(-\phi\tau_n)}{\phi + \beta_n \exp(-\phi\tau_n)} \tag{4.14}$$

A system described by the transfer function in equation (4.14) will become unstable for $\beta_n\tau_n > \pi/2$, this will result in collision of a two-vehicle system. The vehicle following dynamics of a platoon of vehicles can be determined by equation (4.14) using varying values of β_n and τ_n, and additionally assuming special driving maneouvres of the leading vehicle. Figure 4.15 shows such an example, where $\pi/2 > \beta_n\tau_n > 0.5$. In this case, the two-car system remains stable and collision-free, but collision may occur at the end of the platoon (see position of vehicles 7 and 8 in Figure 4.15). Instability and collision will occur if a driver reacts too slowly (large τ_n) or too sharply (larger β_n) to speed changes of the leading car. A small variation in speed of the leading vehicle will in this case be magnified, resulting, for a platoon of vehicles, in collision between the cars at the end of the platoon.

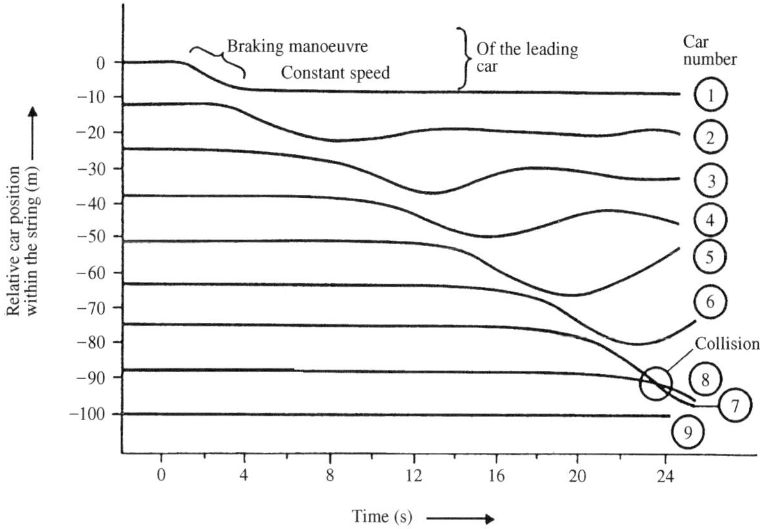

Figure 4.15. Examples of simulation of collision of vehicles in a single lane platoon

4.3 COLLISION AVOIDANCE SYSTEMS

We have already established the basic concept of vehicle collision using the basic vehicle following model. The model so reviewed is based on vehicle to vehicle interaction, but collision can arise from other situations such as visual blindspots, other road users emerging from junctions, driver falling asleep, driver not paying attention or distracted, incidental objects on the road, etc. These situations cannot be easily predicted and are very difficult to model, but one way of taking account of these variables is to use the collision avoidance system.

The collision avoidance system is an active driver support and vehicle safety system aimed at preventing or mitigating accidents by enabling obstacles and developing critical situations to be detected in a sufficiently safe time for the driver or the system to take appropriate avoidance action. The fundamental methodology of obstacle avoidance is based on perceiving the vehicle outside environment and traffic situations using appropriate sensory systems and subsequently providing the most important driver assistance under the hazardous conditions. The driver assistance can be through driver information system or through active and just-in-time intervention-based vehicle control. As shown in Figure 4.16, there are three essential subsystems of a collision avoidance system: the multisensory subsystem acquires the environmental information on the world around the vehicle; the processing subsystem interprets and evaluates the environmental or traffic situation, and subsequently assesses and prioritizes the threat potential of the obstacle; the output subsystem provides prioritized data to the driver through the driver interface system and also issues fast and just-in-time intervention data to the actuators for vehicle

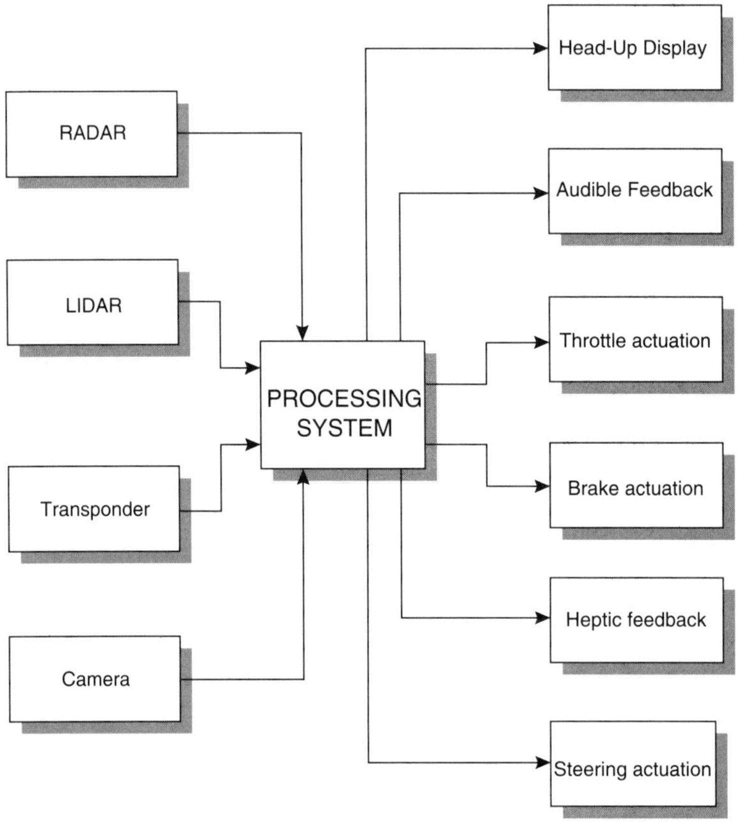

Figure 4.16. Potential subsystems representation of an obstacle avoidance system

longitudinal or lateral control in a situation where there is no chance for the driver to react in time.

There are several sensor technologies available for use in an obstacle avoidance system, some of these are shown in Figure 4.16, but the most common of the sensors used are the RADAR (microwave device), laser type (LIDAR) and camera (imaging systems), with the RADAR presently most commonly used for intelligent vehicle application, especially when considering environmental conditions[17]. The RADAR antenna mounted on a vehicle can provide obstacle detection in excess of 100 m from the vehicle. The frequencies greater than 100 GHz can be used with acceptable antenna size, in Europe and the USA the specific range of 76, 77, 94 (see Figure 4.17) and 144 GHz band are used. A typical system uses 38.5 GHz VCO (voltage-controlled oscillator) with frequency doubling to obtain about 40 mW of power at 77 GHz. A frequency modulated continuous wave or pulse-modulated system can be used. The system uses GaAs devices to meet the frequency and power requirements. A typical systems specification is shown in Table 4.2, this is similar to one of the Philips models.

Figure 4.17. RADAR antenna (Courtesy Phillips Research Laboratories, UK)

Table 4.2. A typical specification for an automotive RADAR

Maximum indicated range	127 m
Operating range	76.5 GHz
Data range	50 Hz
Range resolution	10 cm
Angular resolution	2° (azimuth) by 6° (elevation)
Azimuth coverage	2° single beam, 6° 3-beam sensor (can be extended)
Power supply	12 V nominal, <4 A
Output data	Range, range rate, confidence
Data interface	Serial bus

To utilize the full potential of an obstacle avoidance system, an integrated system that is capable of having some basic intelligence which allows it to sense possible threat on the road and continuously transmits a signal which is acquired by the central processor (onboard computer system). In the event of any obstacle, the processor then calculates the distance of the threat from the time taken for the

RADAR signal to return from the obstacle. The obstacle size may also be deduced from the change in the signal frequency. When the signal is fully processed, the resultant data warns the driver and the support vehicle manoeuvres through the actuation of the brake, throttle, steering, haptic feedback or any of the relevant vehicle dynamics and control systems.

4.4 CRUISE CONTROL SYSTEM

To alleviate driver stress, avoid accidents and uneconomical driving, a road vehicle speed control system can be arranged as a simple throttle latching device to a sophisticated computer control system that constantly maintains a set vehicle speed under varying driving conditions. The main function of a basic cruise control system shown in Figure 4.18 is to control and maintain a certain level of engine speed such that

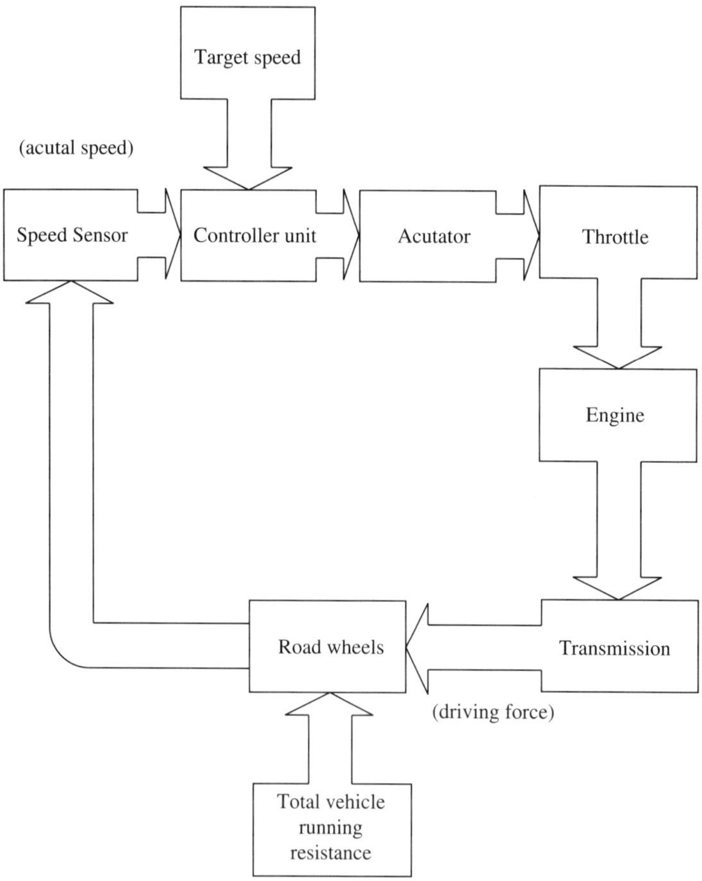

Figure 4.18. Block diagram for a simple cruise control system

the vehicle travel can be maintained at a certain speed without the driver needing to depress the accelerator pedal once the speed is set.

The cruise control system is a closed-loop control system (Figure 4.18) and it consists of an actuator that opens and closes the engine throttle valve in lieu of the driver's depression of the accelerator pedal. The actuator is linked to the controller that controls the actuator motion. The input signals to the controller are the driver's speed set-point, the vehicle's actual speed and the signal transmitted to the controller to enable the relinquishment of the cruise control. The output signals are the ones that activate the throttle servo control actuator, the cruise ON indicator, data for diagnostics plus information to the engine and transmission system.

When the cruise control is switched on, the control unit, using any of the control laws discussed previously in Figure 4.12, initiates control by means of the set signal, memorizes the speed at the set signal input time as the set speed, and controls the actuator so that the actual speed will coincide with the set speed. In addition to this basic functional requirement, cruise control system has other functions, such as:

1. *Acceleration and coasting function* which allows the setting of the travel speed to be changed during cruising.

2. *Resuming function* where the cruise travel is released when the brake is applied due to difficulty in maintaining the necessary distance from the car ahead or when the clutch is used to change gears.

3. *Overdrive release function* allows the cruise control to operate with a vehicle equipped with automatic transmission, such that fuel consumption and improvement in transmission quietness can be obtained, especially during repeated kickdown and overdrive return.

4.4.1 Intelligent Cruise Control System

An intelligent cruise control system, which may be regarded as autonomous or adaptive intelligent cruise control, performs the standard function of a cruise control and additionally acts as a driver assistance system for intelligent control of the relative speed and distance between two adjacent vehicles in the same lane. By adapting vehicle speed automatically to the demands of traffic flow, driving at a safe distance, increasing road capacity as a result of speed harmonization, enabling smoother driving, reducing fuel consumption and subsequently exhaust emission, adaptive cruise control will become part of intelligent vehicle systems.

The use of appropriate sensors capable of measuring up to a few hundred metres away from the vehicle with a tight focal point in all weather conditions is fundamental to the use and development of an intelligent cruise control system. Intelligent cruise control operates in a similar fashion as the collision avoidance system, but with the ability to track the vehicle in front. As shown in Figure 4.19, several types of sensors have been used by different European automotive manufacturers. Similar

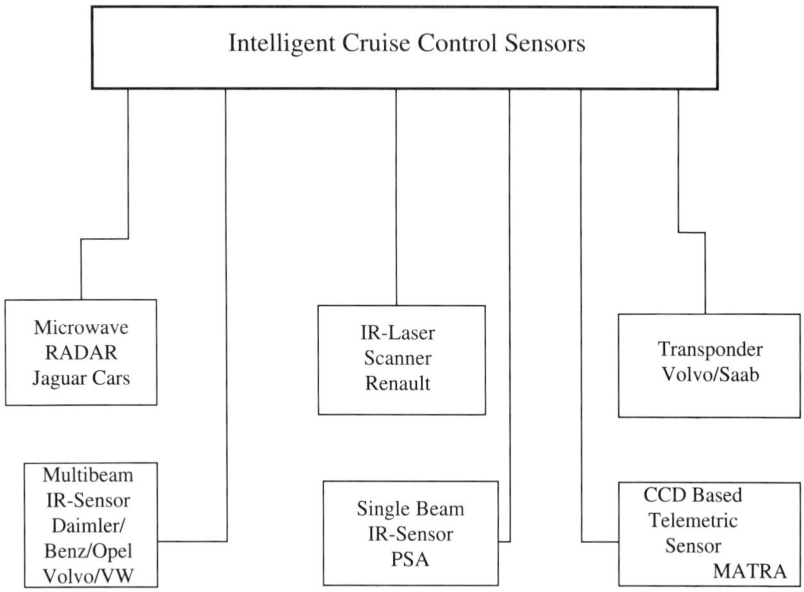

Figure 4.19. Types of sensors for intelligent cruise control

to a standard cruise control systems, actuators for longitudinal and lateral control of the vehicle is an important unit of the intelligent system. Software for applying cruise control strategies and driver information is used. The cruise control software uses any of the control laws described in the early part of this chapter. A communication system which allows roadside to vehicle interaction for traffic regulation may be used.

The performance of the systems is supported by a configuration of several subsystems shown as block diagram in Figure 4.20. The six blocks of the subsystem's units are functionally affected by the output information from the data processor with exchange of information through a communication network.

Pulse encoders are installed at the wheels, this functions as the wheel speed sensors. They detect the number of wheel revolutions and the information is fed into the data processor (main computer) and used to compute the wheel speed. The intelligent cruise control sensor detects the presence and the positional coordinates of the vehicle in front, the information is fed to the data processor, and this is combined with the speed sensor signal. Based on this information, the strategy selection subsystem decides whether the cruise control should be effected through the actuator control unit or the driver informed of the safety distance to the preceding vehicle, whereby the system switches to distance control and the alarm is activated in case of a large speed difference with the preceding vehicle. The intelligent cruise control system can be activated or deactivated by the driver when the environment and the driving conditions are adequate. The system can also be overridden at any moment by the action on the accelerator or brake pedal.

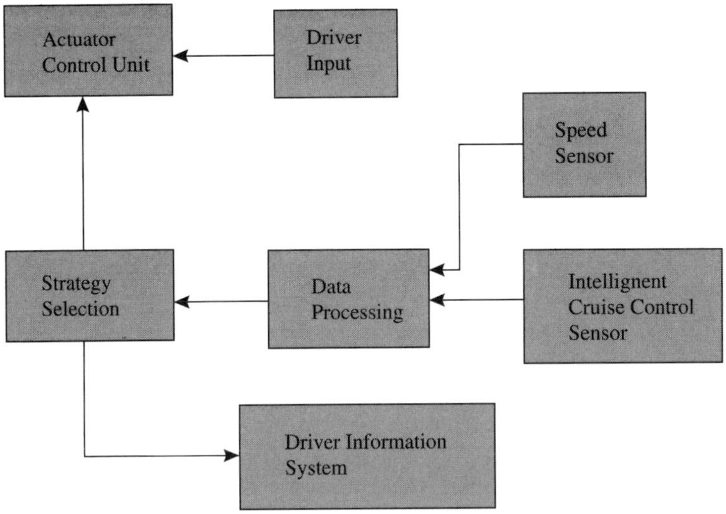

Figure 4.20. Configuration of an intelligent cruise control system

With the data processor receiving input from the appropriate sensors, the intelligence of the system is developed mainly through the control software. For example, the response time and gain of the cruise control system can be adaptively adjusted to match individual drivers. In this case, some class of drivers may prefer to allow the vehicle to slow down slightly when climbing a gradient and then respond quickly to maintain a set speed. Some other drivers may prefer a constant speed at all times, while other drivers may prefer very slow responding cruise conditions to minimize fuel usage and pollution. At a human-machine interface level, the cruise system can be adapted by either driver selection switch or by prior analysis of the classes of driver's acceleration/deceleration habits during non-cruise conditions. At later stages of the development of the intelligent cruise control system, the adaptive software can be improved by using a self-learning neural network technique for classifying the different types of drivers that will use the system.

4.5 NAVIGATION AND ROUTE GUIDANCE SYSTEMS

One of the fundamental requirements of an intelligent road vehicle is to assist the driver in finding their anticipated destination in the most economical, reliable and safe way. There are several questions usually resident in a driver's mind, for example, "Where am I?", "What route do I have to take to reach my destination?", "How can I best react on different traffic situations?" and "Where can I get particular services?". As part of road vehicle automation, these questions can be answered by using an in-vehicle system capable of providing relevant driver assistance that includes accurately identifying the actual location of the desired destination, determining the optimum and high-quality route and assisting the driver to follow the route in order to avoid traffic congestion. Generally, vehicle navigation and route guidance systems provide the necessary assistance required to answer all of these questions.

4.5.1 Vehicle Navigation Systems

Navigation is the process of directing the movements of a user from one point to another. Successful navigation is dependent on establishing where the user is, where the user wants to be and how to get there. Therefore, a vehicle navigation system allow the driver to specify his/her current location and desired destination and it provides some variety of computer-based support to assist the driver in getting to that destination. The type and level of support given to the driver depends on the type of system used. From the early years of vehicle navigation technology (e.g. the south-pointing chariot — China, c.a. 1000 BC)[18] to when mechanical vehicle navigation systems were introduced in the USA at about 1910, the system has always incorporated the information of the route maps in various forms, which include sequential instructions printed on turntable, punched on a rotating disk, and printed on a moving tape, all driven by odometer shaft in synchronization with the distance travelled along the route. Some elements of this fundamental technology have formed the basis on which the first and second generation of vehicle navigation system is developed.

Several systems with a variety of functions can be grouped as "navigation system". These can be broken down into the following general categories:

1. The system that only display map of road network. This only serve as a substitute for paper map and it is a passive system.
2. Those systems which only display current position of the vehicle or local area in which the vehicle navigates, these include:
   (i) systems which display names of the place and coordinate number;
   (ii) systems which display the names and patterns of intersection along the route;
   (iii) systems which display the vehicle's current position on a map with the direction and route travelled.
3. Those systems which give the guidance to the driver's destination:
   (i) these are the systems which display the vehicle's direction and the distance from the commencement of the journey to the end destination;
   (ii) systems which display a map showing the optimum route from the start point to the destination with added information on the current location of the vehicle;

(iii) systems which display intersections and indicate right or left turns along the optimum route selected manually or automatically.

4. Those systems which use road infrastructures as an added information source and subsequently provide comprehensive information for navigation.

(i) systems which display maps indicating traffic congestion and provide traffic control information advisories;

(ii) systems with computerized means of receiving and processing traffic congestion and general conditions information, and use that information to optimize the vehicle route.

There are in general various types of navigation systems which fall into the categories described, for example:

1. *Pre-trip information systems.* These are quasi- and static-based navigational aids (e.g. Auto-route PC-based route planner.

2. *Traffic information systems.* These are systems which give event-based information on the traffic conditions in the road network. Development using the radio data system which is a digital sideband information channel (RDS), and Traffic Message Channel (TMC) for driver information is central to this type of navigational system. TrafficMaster is another type of a traffic information system in this class.

3. *Autonomous navigation systems.* In these systems, information is presented to the driver to help guide the driver to a predetermined destination either by showing an appropriate segment of a map or by providing a turn by turn icon and voice information to the driver (e.g. Philips CARIN or Bosch Berlin systems).

4. *Dynamic route guidance systems.* Navigational systems of this type provide route guidance for an optimum route after taking account of prevailing traffic conditions. This may be achieved by an in-vehicle communication with an infrastructure of beacons providing a centrally calculated route (e.g. Siemens EUROSCOUT) or by integrating traffic information from, e.g. RDS-TMC or the System of Cellular Radio for Traffic Efficiency and Safety (SOCRATES), into the route calculation of an autonomous navigation system.

The most important requirement of a vehicle navigation system is its ability to position the vehicle and the driver at the required location using the necessary technologies. Most of the navigation systems use dead reckoning, map matching and global positioning systems (GPS) as their main positioning technologies. For the system to function within the vehicle, a means of sensing the vehicle speed and the orientation along its principal direction is required. This is usually achieved by using an electronic equivalent to the odometer, such as, speed encoders/optical sensors on the vehicle wheels to sense the speed and heading sensors such as geomagnetic sensors, gas-rate gyro sensors or fibre-optic gyroscope.

Dead reckoning

The term "dead reckoning" comes from the more accurate description "deduced reckoning". The dead reckoning process is a vehicle locating method used as part of the vehicle navigation system for identifying vehicle position within given world coordinates. It is a process involving an analytical deduction of the vehicle present location by integrating the measured increments of the distance and direction of travel of the vehicle relative to its reference location. The process identifies the vehicle's present $X_n{:}Y_n$, coordinates on the road relative to its previous $X_0{:}Y_0$ coordinates. This can be represented by a basic mathematical expression in equations (4.15) and (4.16.)

$$X_n = X_o + \sum_{i=l}^{n} dX_i = X_o + \sum_{i=l}^{n} dl_i \sin \emptyset_i \tag{4.15}$$

$$Y_n = Y_n + \sum_{i=l}^{n} dY_i = Y_o + \sum_{n}^{i=l} dl_i \cos \emptyset_i \tag{4.16}$$

where $\emptyset_i$ is the vehicle heading coordinate angle associated with the vehicle principal direction l_i at the ith measured increment of vehicle travel.

Dead reckoning is the basis for most navigation systems, but over a long period of time, it can accumulate position errors which can be minimized by the use of position fixing sensors, such as roadside beacons or through receiver hardware and software optimization. Immediate vehicle position input to the dead reckoning process is provided by various types of sensors, such as, geomagnetic sensors, gyroscopic sensors, differential wheel speed sensors, GPS, beacons, etc.

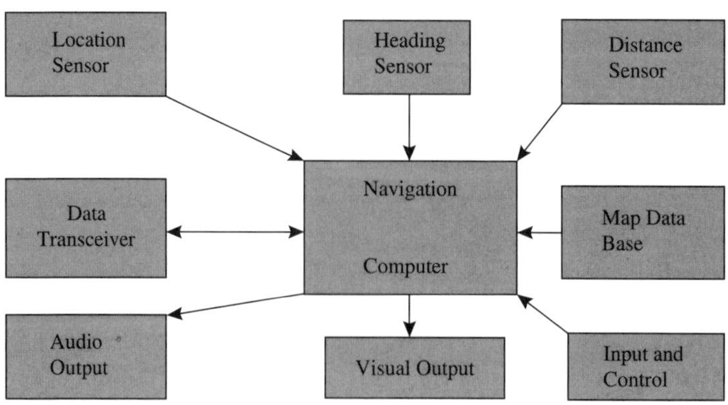

Figure 4.21. System's block diagram of a vehicle navigation system

Digital road map matching

For the navigation system to make sense of the vehicle's environment, it needs to be aware of the road map from which the driver's destination can be identified. The navigable road map can be a digitized map in a database centrally located in control centre as a traffic infrastructure system, but the most common is the digitized road map resident in-vehicle within the navigation system's computer. The database provides a complete and accurate detailed-level information about the roadway infrastructure in a manner that facilitates computer processing. Within urban areas, the map should include precise geometry, full characteristics, names, address ranges for all navigable roads at all levels, travel speed variation within road segments, directionality, traffic signals, etc. In normal cases some of the roads can be easily more identified by the landmarks within a recognizable vicinity (e.g. shopping centres, government/major office buildings, air/rail/bus terminals, etc.). These landmarks must therefore be included. The database should also contain towns and cities within the surrounding urban areas.

These road maps are digitized for computer processing using matrix encoding and vector encoding. A matrix-encoded map is a digitized image of the map in which each pixel on the computer screen with its $X-Y$ coordinates represents profile shades and colour of the road. Matrix-encoded maps require large computer memory storage and are ill-suited for fast analytical treatment required for dynamic map matching and route guidance. The vector-encoded map uses mathematical modelling to represent the geometrical features of the road, boundaries and surroundings with minimum data abstraction. Most of the navigation systems use the vector encoding method where each road is represented as a series of straight lines and each intersection of the road as a node.

The digitized image is stored in the memory of the computer, such as CD Rom, but for the vehicle to navigate through the road network, map matching of the road network on which the vehicle is driven to the digitized road map stored in the memory is essential. Map matching is therefore a knowledge-based process which may be said to use artificial intelligence to recognize the present vehicle location by matching the patterns of vehicle path, as approximated by the dead reckoning and/or any external roadside beacons, with the digitized road map stored in the computer. There are various methods of software-based map matching, but they can be classified into two types on the basis of the matching concept used. The two types which have been named in the past are called "semideterministic" and "probabilistic".[20]

The semideterministic approach is to search for the most likely road that the vehicle is travelling by comparing driving directions and angles at each intersection assuming that the vehicle is essentially confined to a predefined route or road network. In this approach all roads that the vehicle is likely to travel are normally stored in a computer memory or computer disk. This technique may be illustrated with the aid of Figure 4.22 which shows a simplified route in a basic road network;

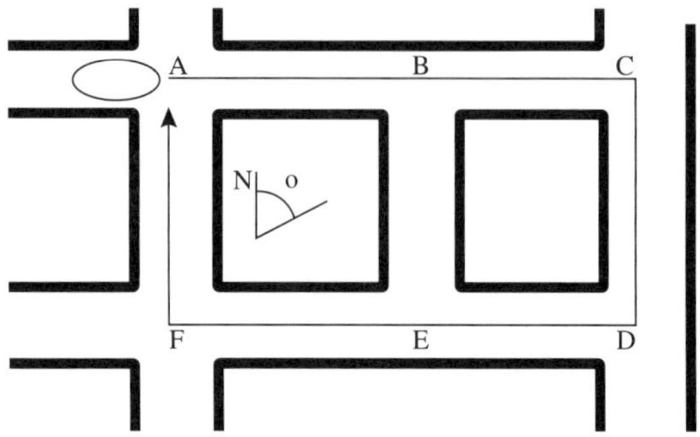

Figure 4.22. Example of a simplified route in a basic road network

the vehicle shown in the network approaches the network at node A and travels through nodes B, C, D, E, F and back to A in the direction of the accumulative distance L. Taking into consideration the alternative route in the network, the chosen route is represented by vectors and a map matching routine is used to ascertain the optimum vehicle path.

A simplified map matching algorithm for the semideterministic technique is shown as a flow chart in Figure 4.23. At the starting point after initialization, when 90° and $L = 0$ the algorithm repeatedly asks, "Is the vehicle still on the route?" and "What is the present location along the route?" The vehicle is then confirmed as being on route if certain tests are satisfied. The location along the route is estimated by the wheel sensors, any error in the estimate is automatically deleted at each node where it is ascertained that an expected change in the vehicle heading actually occurs.

The wheel sensors normally installed at the left and right wheels to sense the vehicle speed transmit the signal interrupt which drives the algorithm. The distance L from the beginning of a route segment is updated by adding an increment ΔL for each left wheel sensor signal interrupt, also the vehicle heading ϕ towards its principal coordinate axis, is updated by adding an increment $\Delta\phi$ derived from the difference in travel of the left and right wheels since the count N was last set to zero.

The N counter controls the monitoring for the unexpected changes in the vehicle heading over a relatively short distance. The turn flag can be set to denote that the vehicle is approaching a distance L, where the vehicle heading change should occur, otherwise, count N is checked after each sensor signal interrupt to determine if it has reached a limit C corresponding to an arbitrary amount of travel on the order of say 75 m. When the count limit C is reached, a test is made to determine if ϕ is within a limit of say $\pm 5°$. If that is the case then ϕ is reset to ϕ_0 in the direction of the vector, and N is set to zero to recommence another cycle of monitoring for the unexpected changes in the vehicle heading. If any of the preceding tests are outside the limits, the vehicle is presumed to have turned off the route in the encoded map, hence other vehicle path recovery or safety routines can be invoked.

Within the road network, when the vehicle approaches within a given distance of say 70 m of a node where a change in vehicle heading should occur, the turn flag

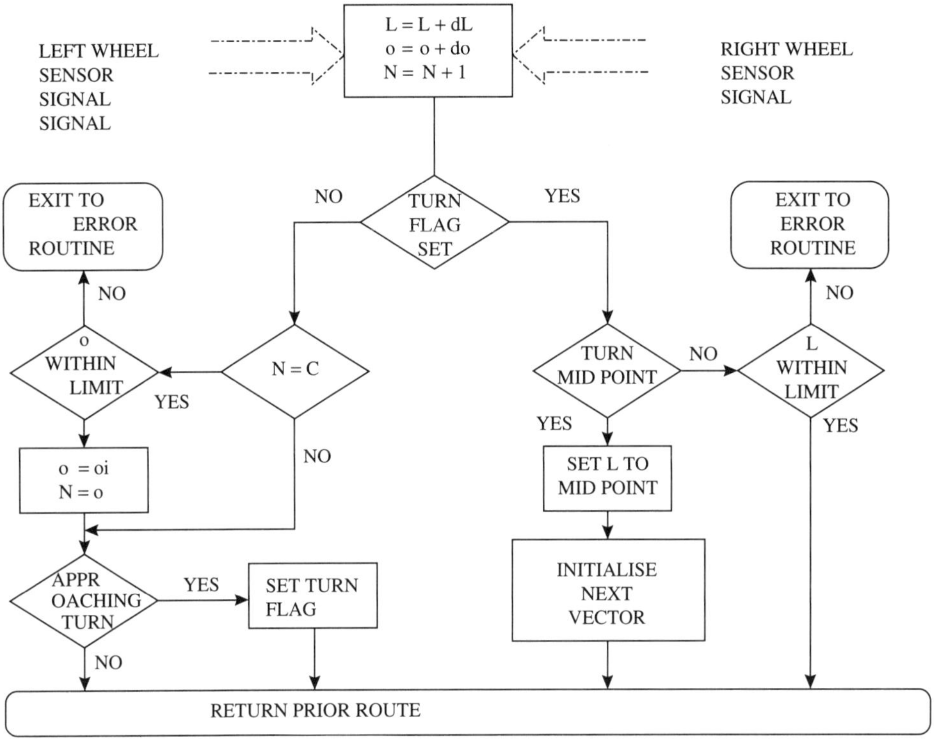

Figure 4.23. Example of a simple map matching algorithm[20]

is set and a route guidance instruction is issued. This will give the direction of the vehicle turn and possibly a pointer to the subsequent road to be approached. At this juncture, the algorithm monitors for the changes in to confirm that the mid-point of the expected turn is reached within a given limit of say 10 m of the value of *L* specified for that node, and subsequently, should confirm that the turn is successfully completed.

The probabilistic approach is used for vehicle navigation not constrained to a predefined road network. In this case, when the vehicle commences a journey from a known route within a road network, the vehicle's present correct location is deduced by collating a travelling locus with the pattern of the road shape without assuming that the vehicle is always travelling on a road. This means that the vehicle's current location can be determined even when the vehicle is travelling on a road not encoded in the digitized map (e.g. parking space or departure as a result of a dead reckoning error).

The concept may be explained by tracking the location of the vehicle travelling in a straight line within a simplified road network shown in Figure 4.24. When the vehicle travels from the left side of the road network with the navigational aim of road (B), but instead turns right, possibly due to a dead reckoning error,

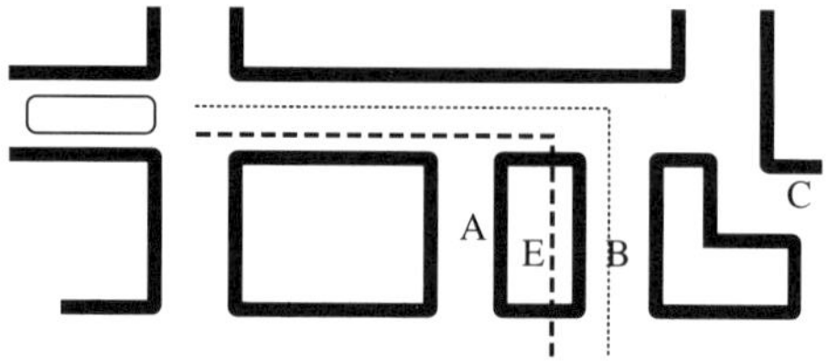

Figure 4.24. Example of simplified route

along an unrecognized route (E), the probabilistic technique will compare the road situation and the vehicle dead reckoned coordinates with the existing roads (A, B, C) in the neighbourhood and repositions the course to road (B) with the highest probability. A navigational system using this technique depends on the input from the dead reckoning sensors, such as, geomagnetic sensors to detect the vehicle directional angles, differential wheel speed to detect the vehicle speed and GPS to detect the global coordinates of the vehicle. Unlike while the vehicle is on a define road network, map matching adjustments do not prevent accumulation of the dead reckoning error. Therefore, depending on the distance travelled off the road network and the accuracy of the dead reckoning sensors, there may be considerable uncertainty in the vehicle coordinates, which could produce misleading conclusions when tested against the surrounding and possible routes in the road network. This problem can normally be reduced by the use of accurate sensors, reinitialization of vehicle position using roadside beacons and the use of map matching algorithms of high precision.

A map matching algorithm that can be used is shown as a flow chart in Figure 4.25. The main task of an algorithm of this type is to repeatedly ask the navigational system, "Where is the vehicle?", with no prior presumption that the vehicle is on an acceptable road. At the commencement of the algorithm, several alternative roads near the current vehicle location are selected from the road network by searching as the vehicle travels on the current route. Thereafter, both alternative roads and loci are approximated by a series of vectors of equal length, to find the matching value in equations 4.17 and 4.18. This procedure is achieved using equations (4.17) and (4.18).

$$\theta = \frac{\sum_{i=1}^{n} \left[w(i) \cdot \cos^{-1}\left(\frac{I_i r_i}{I_i r_i} \right) \right]}{\sum_{i=1}^{n} w(i)} \tag{4.17}$$

$$S = \frac{\sum_{i=1}^{n} \left| \sum_{j=1}^{i} I_j - \sum_{j=1}^{i} r_j \right|}{\delta(n)} \tag{4.18}$$

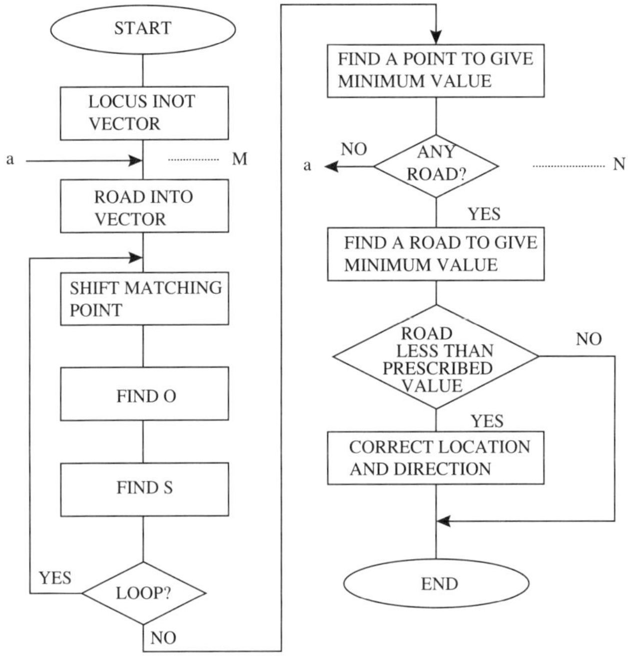

Figure 4.25. Probabilistic map-matching[21]

where $I_{i,j}$ = vectors of locus
$r_{i,j}$ = vectors of road
$I_j = \mathrm{rot}(I_i\theta)$
$w(i)$ = weight function
$\delta(n)$ = normalized function

In the map matching procedure, the angular difference between a road and locus is derived using equation (4.17), while the area bounded by the road and the locus when the angular difference between the road and the locus is minimized can be derived from equation (4.18). The value of the area calculated from equation (4.18) is obtained by shifting the vector of equal length to find a point which gives the minimum value and this is used as a matching evaluation in the algorithm. A series of recursive operation is carried out between "M" and "N" in Figure 4.25, for all candidate roads selected by the route searching routine in order to find the road of the smallest matching value. At any point during this operation, if the matching value is found to be smaller than the preset threshold value, then the present location of the vehicle is corrected to this newly determined location.

Generally, probabilistic map matching algorithms seek to minimize the potential of off-road errors by maintaining a running estimate of uncertainty in the dead reckoned location. The estimate of location uncertainty is minimized each time in the search when it is ascertained that the vehicle location is on an acceptable road.

Global positioning systems

GPS is the term used to define the system of navigation satellites with the capability of providing accurate position, velocity and time data to earth-based users. There are presently two operational satellite systems, one Russian and the other American. The Russian GPS system is known as GLONASS (GLObal Navigation Satellite System) and the American system is known as NAVSTAR, and they operate on different orbits.

The first GLONASS satellite was launched on 12 October 1982 and the system has not been as successful as was hoped. There were 67 GLONASS satellites launched in 1982, but 48 of the initial number are no longer useable because of worse than expected reliability. The Russians are presently launching three satellites at a time to try and provide adequate coverage.

The American NAVSTAR GPS provides 24 operational satellites and three spare ones' which together continually orbit the earth every 12 h. NAVSTAR GPS provides position–, velocity–time (PVT) information to anywhere in the world. A previous French system had operated on similar principles but did not provide worldwide coverage. Therefore, most of the Western world application of GPS uses the NAVSTAR system, which is owned by the US Department of Defence (DoD).

The operation of a GPS system is made of three segments: space segment, ground based control system segment and user segment. The space segment consists of the 24 satellites which orbit the earth every 12 h at a height of 20 200 km. The ground-based control system consists of the ground-based monitoring and control stations situated in five locations around the world. These track the satellites and update the messages sent by the satellites to reflect information such as satellite clock errors, orbit path and satellite health condition monitoring. The user segment is basically the receiver mounted in the vehicle. In order to calculate the position on earth of a receiver, it is necessary for the receiver to acquire signal information from three to four satellites at one time.

GPS signals are digital coded signals which contain the satellites number and clock information, and they are transmitted with a carrier frequency of 1575.42 Mhz modulated by a spread spectrum technique. Figure 4.26 shows the GPS systems for road vehicles with the three Cartesian coordinates. When the three Cartesian coordinates of the satellite vehicles (sv) are given as (U_i, V_i, W_i), the four basic non-linear navigation equations that are used to derive the pseudo-range between the satellites and the road vehicle are given as:

$$R_1 = \sqrt{(X - U_1)^2 + (Y - V_1)^2 + (Z - W_1)^2 + T} \tag{4.19}$$

$$R_2 = \sqrt{(X - U_2)^2 + (Y - V_2)^2 + (Z - W_2)^2 + T} \tag{4.20}$$

$$R_3 = \sqrt{(X - U_3)^2 + (Y - V_3)^2 + (Z - W_3)^2 + T} \tag{4.21}$$

$$R_4 = \sqrt{(X - U_4)^2 + (Y - V_4)^2 + (Z - W_4)^2 + T} \tag{4.22}$$

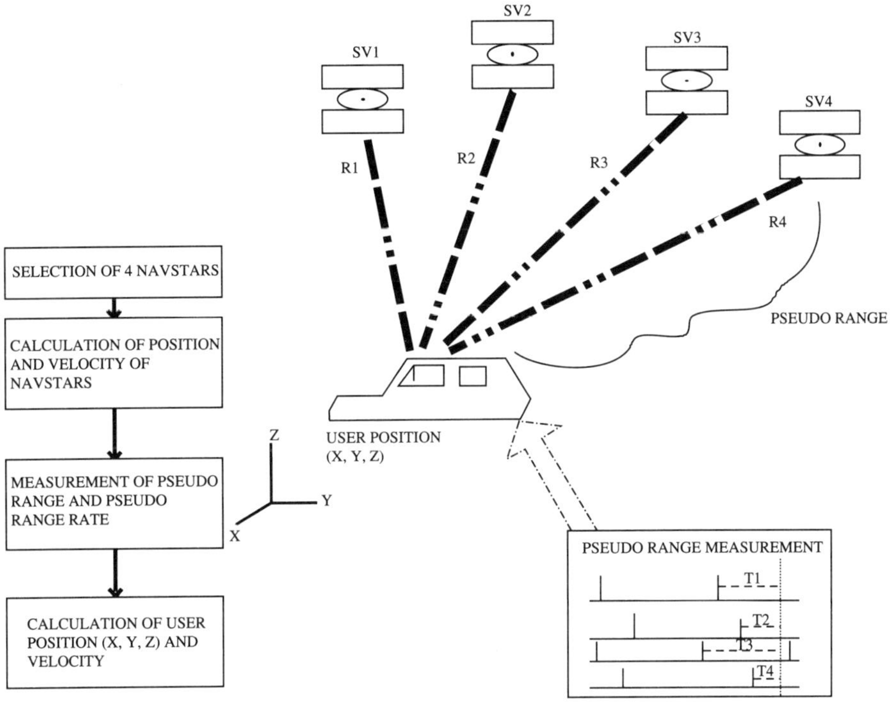

Figure 4.26. Global positioning system for road vehicles

where:

$$X, Y, Z, T = \text{vehicle position and clock bias}$$

$$X_i, Y_i, Z_i = SV_i \text{ position}$$

$$R_i = \text{Measured pseudorange to } SV_i$$

Having determined the position of the vehicle using equations (4.19)–(4.21), it becomes relatively easy to determine the velocity component from the GPS position calculation as these are time tagged. This can be achieved by dividing the distance travelled between GPS readings by the difference in time between these readings. The accuracy of this procedure is limited by the errors from the uncorrected coordinates of the GPS coordinates. The amount of this error can be influenced by the factors affecting the geometric dilution of precision (GDOP), infrastructural obstructions of the GPS signal and the actions of the US DoD. The DoD is known to have deliberately distored the GPS signal for non-military users, and this is known as *selective availability,* where a minimum accuracy level of $\pm$ 50 m for 95% of the time for civil use has been specified. This error can be minimized by using a GPS receiver that outputs the speed directly by integrating into the receiver a Doppler effect to measure the rate of change of the GPS signal and hence the speed.[22]

Most GPS sets use a digital signal processing technique which employs a Kalman filter to provide the navigation capability. Compared with the four equations and four unknowns in equations (4.19)–(4.22), the Kalman filter improves position estimates, without dramatic improvement in the velocity estimates. It provides dead reckoning capability, with a good estimate of the navigation solution quality. As the Kalman filter is a linear recursive least mean squares estimator, to use the filter the navigation equations (4.19)–(4.22) must be linearized such that:
Let ΔX_n, ΔY_n, ΔZ_n, ΔT_n be the nominal values of X, Y, Z, T (a priori best estimate) and ΔX, ΔY, ΔZ, ΔT be the corrections to the nominal values, then:

$$\begin{aligned} X &= X_n + \Delta X \\ Y &= Y_n + \Delta Y \\ Z &= Z_n + \Delta Z \\ T &= T_n + \Delta T \end{aligned} \tag{4.23}$$

We can also define:

$$R_{ni} = \sqrt{(X_n - X_i)^2 + (Y_n - Y_i)^2 + (Z_n - Z_i)^2 + T_n} \tag{4.24}$$

Then:

$$R_i = R_{ni} + \Delta R_i = \text{Nominal pseudorange to } SV_i$$

Using binomial expansion and ignoring Δ squared terms, the linearised equation can be derived as:

$$\Delta R_i = \frac{1}{(R_{ni} - T_n)}\{(X_n - X_i)\Delta X + (Y_n - Y_i)\Delta Y + (Z_n - Z_i)\Delta Z\} + \Delta T \quad (i = 1, 4) \tag{4.25}$$

This can be put into matrix form as follows:

$$\begin{bmatrix} \Delta R_1 \\ \Delta R_2 \\ \Delta R_3 \\ \Delta R_4 \end{bmatrix} = \begin{bmatrix} \alpha_{11} & \alpha_{12} & \alpha_{13} & 1 \\ \alpha_{21} & \alpha_{22} & \alpha_{23} & 1 \\ \alpha_{31} & \alpha_{23} & \alpha_{33} & 1 \\ \alpha_{41} & \alpha_{24} & \alpha_{43} & 1 \end{bmatrix} \begin{bmatrix} \Delta X \\ \Delta Y \\ \Delta Z \\ \Delta T \end{bmatrix} \tag{4.26}$$

One main receiver's environmental-based disadvantage of the GPS is its poor performance when its signal is obstructed. When used in the cities, town and villages, it may not cope very well with features like tall/tower buildings or heavily densed tree lines that obstruct or reflect away its clear line of communication with a sufficient number of satellites. When the GPS experiences these communication problems, in some cases, it takes no reading until the vehicle/receiver has cleared the obstruction, then the receiver will regain its lock to the available satellites, in other cases, the receiver will attempt to make the best guess of where the car is until its signal contact with the satellites is re-established. However, as stated earlier, the use of Kalman filtering should be able to reduce these problems. The future of GPS will centre around the minimization of these inherent errors using advanced technology such as the differential GPS (DGPS) that has been designed with the potential of providing accuracies of less than a metre. The basic principles of DGPS is based on the concept that the apparent errors is determined at a known fixed point that has been previously surveyed and this is then applied to the receivers mounted in moving vehicles. The receiver at the known location must be able to track all the satellites in view in order to compute their differential pseudorange corrections. These corrections are obtained by comparing the measurement taken by the receiver at the given location with those based in the vehicle which can be termed as the instantaneous receiver position prior to corrections. After the comparison, the resultant valid corrections for each satellite in view are formatted in a standardized protocol and modulated in the appropriate signal which is subsequently sent to the vehicle. The vehicle then uses this information to compute a more accurate position which can be in the region of less than 10 m.

4.5.2 Dynamic Route Guidance Systems

Dynamic route guidance systems are vehicle navigation systems with an added in-vehicle facility which allows it to self-determine the optimal route using the technologies already described in this chapter with the road map stored as a database in a CD-ROM device. This type of system may also have dual mode route guidance

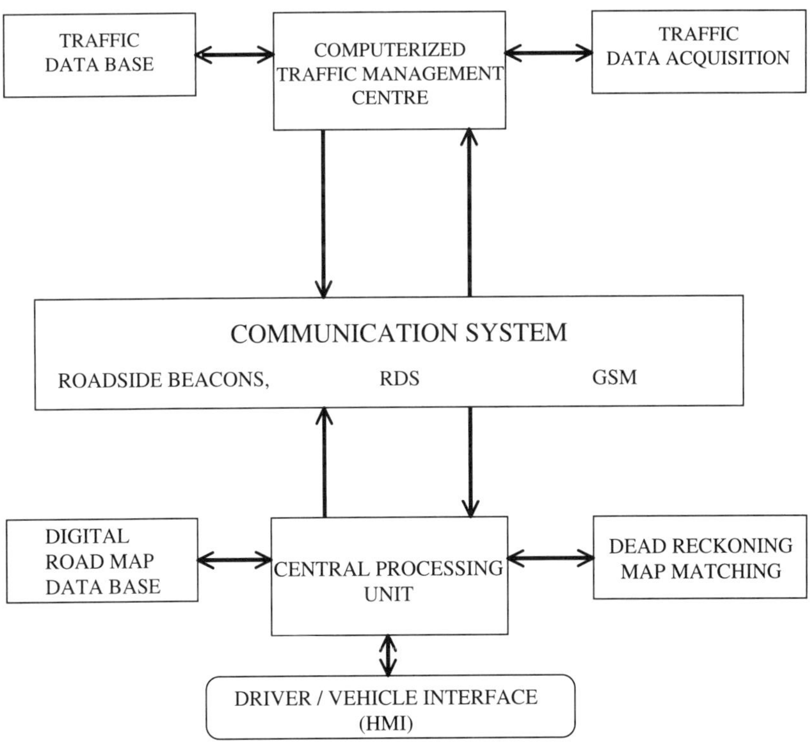

Figure 4.27. Dual mode route guidance system

Table 4.3. Carin system's specification

Navigation specification	
Average elapsed time for planning of a route	<8 s
Distance allowable for driving "off map" without requiring a relocation	>20 km
Average distance between vehicle relocation on digitized roads (without any GPS unit)	>10 000 km
Average distance between vehicle relocations on digitized roads with GPS option	>100 000 km
General specification	
CD format	CD-i bridge
CD data capacity	600 Mbytes
Microprocessor	68EC030 running at 24 MHz
Dynamic memory	2 Mbytes
Non-volatile memory	1 Mbyte
Interface to human–machine interface	Arcnet, data rate of 2.5 Mbit/s
Interface with ABS wheel sensor	Minimum of 40 pulses per metre travelled (detectable signal at a speed <0.3 km/h)
GPS and RDS-TMC interface	Full duplex asynchronous RS422 serial line
Vehicle speed range (forwards)	0–250 km/h; 0–50 km/h
Time required to load software from CD	45 s
Power up time (software loaded)	<5 s
Power requirements	12 V DC, negative grounds
Operating current	<2 A @ 12 V (fuse rate at 3 A)
Operating temperature range	−20° C −60° C
Dimensions (W × H × D)	180 × 51 × 160 mm (DIN ISO 7736)
Weight	1.8 kg
Vehicle type	Coupe, Sedan, Limousine and Hatchback (vehicles towing a trailer or fitted with roof rack are excluded)

capacity which allows it to communicate with the hardware provided by the traffic infrastructure such as roadside beacons (e.g. ALI-SCOUT systems such as Siemens EUROSCOUT* and Autoguide) for dedicated short-range communications via infrared sensors inside a receiver mounted in the vehicle or by incorporating traffic information and management systems from, for example, RDS-TMC† or SOCRATES‡ The switchable modes provide a system's capability which allows it to perceive, ahead of the vehicle, the dynamics of traffic conditions such as road link times, traffic incident reports and traffic flow, and use the results of that perception to guide the vehicle dynamically through the optimal route. To enhance the communications of several vehicles with the traffic infrastructure without information capacity restraints, specific communication channels from the digital mobile phone network which allows a multiple-access protocol, such as the pan-European GSM§ system, is necessary.

The fundamental framework for the dual mode system is shown in Figure 4.26. The driver/vehicle interface is normally provided by a human–machine Interface (HMI) control station. In addition to other facilities provided by the HMI, a display unit which shows the driver a simplified picture of the road and a direction arrow is provided. Normally, the navigation function superimposes vehicle location on the map display screen, highlights on the map then suggests the optimal route, and route guidance instructions are issued. Voice and LED can be used to provide the driver with a route guidance which may include direction information before junctions and other route characteristics. This system is similar to the TravTek system which was used in 100 General Motors vehicles, where prototype multifunctional architecture provided navigation, route selection and guidance, real-time traffic information, local yellow pages and a cellular phone service.[23]

An example of a system which uses some of the technology described in this section is the Carin system (CAR Information and Navigation system). The CARIN system is based on CD-i (interactive) and digital map technology, and it consists of basic components, such as, magnetic compass, connection to ABS, GPS systems, HMI, CD-i mapping and RDS-TMC that allows digital traffic information to be received. A basic system's specification for the Carin system is shown in Table 4.3.

4.6 COMMUNICATION SYSTEMS

The information age has altered completely the whole nature of time, space and distance relating to personal mobility. The telecommunication revolution encapsulated in the "information superhighway" is progressing at a galloping pace, and it

* Euroscout = Infrastructure base route guidance system-part of ite Siemens Plessey "Universal Vehicle Information system"

† Radio Data System-Traffic Message Channel

‡ System of Cellular Radio for Traffic Efficiency Safety

§ GSM = Group Special Mode-digital Pan European cellular radio system will underpin the most profound development and changes in the social and work pattern over the next 10–20 years. The effect of this on personal mobility is as yet not fully known. What can be said is that the role of vehicles are shifting from a pure "personal mobility" oriented system to what can be safely termed a "personal-Info-mobility"-oriented system. We now live in a wired society, where internet, E-mail, satellite communication, mobile phone, voice recognition and digital network is part of the communication system. In addition to the traditional requirements of vehicles, it is apparent that vehicles of today are becoming an extension of a living space and a means of conducting business which would have otherwise been carried out in a normal office, it is also part of an integral traffic system.

We have discussed in the previous sections the communication system which allows the vehicle to navigate through an optimum route taking into consideration the traffic conditions. The communication system which is deemed at present very important and popular is that which allows the vehicle, driver and the vehicle passive occupants to communicate with the outside world while the vehicle is still in motion or stationary, as the case may be. There are several sources of communicable information which can be interfaced with an intelligent vehicle, but the basic practical sources are shown in Figure 4.28. These are essentially, radio frequency (RF) communication systems capable of operating within a short and wide-area range through high, medium and low data rate communication channels.

As most of these communication systems have been discussed in previous sections. Other communication systems such as radio massaging and cellular network systems have been covered by many other authors.[24–26] It is important to note that due to the limited frequency allocation for cellular networks, the new trend on communication

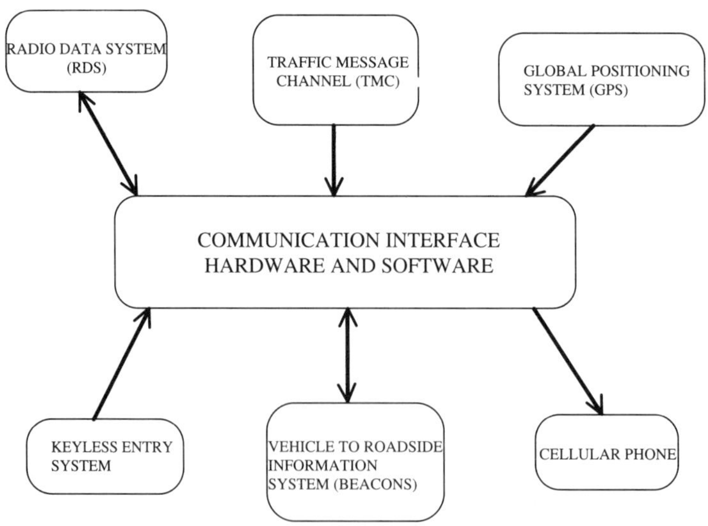

Figure 4.28. Vehicle radio frequency communication systems

systems for intelligent vehicles will be on the use of digital network to increase capacity, provide clearer communication, improve security and provide applications for greater data communication use.

The overall systems requirement is therefore to provide the driver of an advanced vehicle with a communication system-based framework which converts him/her to become an "informed driver" driving a safe and secured vehicle, capable of communicating with other vehicles, traffic infrastructural systems and business/social units. The "informed driver" is the driver that has the in-vehicle communication and information system that enhances the driver's capability to be linked into the wired society, possibly through the information superhighway. The communication systems within the advanced vehicle should provide the "informed driver" with the communication channel linked within the global concept of the "informed traveller", where an automatic intermodal travel interface is provided so that he/she is able to use an alternative transport system (e.g. airline, train, bus/coach, etc.) when

Table 4.4. European radio frequency allocation for vehicle systems

Type	Frequency	Application
Vehicle to vehicle communication	66 GHz	Lane control, collision avoidance
Roadside to vehicle communication	5.8 GHz	Road pricing, car park access, etc.
Onboard radar	76-77 GHz	Intelligent cruise control, collision avoidance

Table 4.5. USA allocated frequencies for application in intelligent transport systems

Function and application	Frequency or band allocated
Real-time traffic data	FM/TV best subcarrier
Safety and warning	220–222 MHz, Highway Advisory Radio (AM best band)
Vehicle probe	220–222 MHz, VHF (150–450 Hz land mobile band)
AVI/AVL/LMS (tagging)	902–928 MHz, 5.8 GHz
In-vehicle signaling	220–222 MHz, VHF land mobile, Special mobile radio (800/900 MHz)
Differential GPS best "value added" services (e.g. yellow page, parking, etc.)	FM best subcarrier cellular
Collision avoidance	24, 60, 77 or 94 GHz

required. The "informed traveller" concept is in the recent Technology Foresight report[27] published by the UK Office of Science and Technology.

As increasing demand is made of radio frequency communication, it is a normal practice to allocate a different frequency range to different users for intelligent vehicles. Table 4.4 shows examples of the agreed European frequency allocations given to different devices in vehicles. In the USA, as shown in Table 4.5, different frequency range are also allocated to different functions of systems used in intelligent transport systems.

4.7 DRIVER VISION ENHANCEMENT

A high percentage of accidents occur because the driver's visibility is poor or the driver finds the visual conditions of the traffic in front of the vehicle very poor. The ability of the driver to visualize objects and traffic in front of the vehicle is dependent on several factors. These include, among other factors, the driver's age, weather conditions, the time of the day and the health of the driver. So any system capable of providing enhanced visual capability of the driver should form part of an intelligent vehicle.

Driver vision enhancement is a vehicle safety system which provides an enhanced driver vision under varying weather and time of day conditions. This is achieved by perceiving the traffic environment through sensory systems that monitors the road scenery condition and subsequently processes the visual data in order to ascertain the traffic and road conditions in front of the vehicle and provide relevant visual information to the driver. The basic systems identifying the main subsystems are shown in Figure 4.29. Figure 4.28(a) shows the block diagram and the relationship between the individual subsystem, while in Figure 4.29(b) a type of in-vehicle system with the near infra-red illumination and camera is shown.

The type of sensors used in a driver vision enhancement are typically imaging sensors such as a charged couple device (CCD) camera, thermal imaging system with far infra-red camera and gated intensified camera with pulsed illumination. For these cameras to work, appropriate illumination of the road scene, which may be a dark or fog road scene, is required. Examples of the illumination systems used include ultraviolet illuminators and near infra-red illuminators. The illumination systems are normally mounted in the headlamps or adjacent to it (Figure 4.29(b)). In order to make sense of the road scenery, image data acquired from the sensor is processed by an image and data processing unit. Irrespective of the type of camera used, the main task of the processing system is in fact to enhance the image by using different types of image enhancement routines in the data processing system. Once the image is processed, the resultant information is relayed to the driver. This can be done by displaying the information on a head-up display system shown in Figure 4.30.

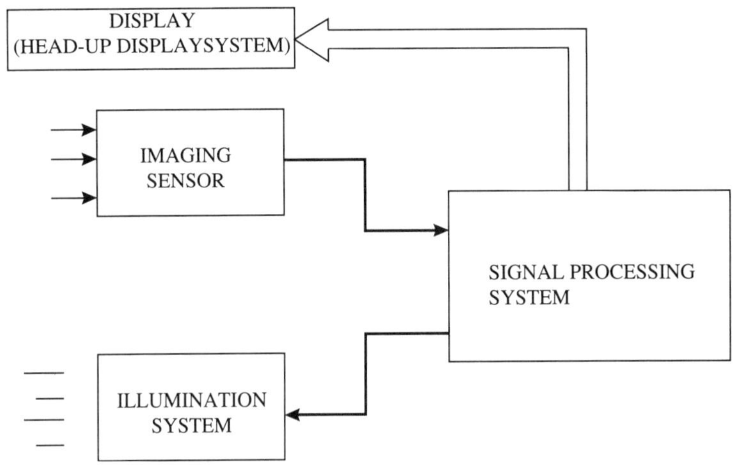

(a)

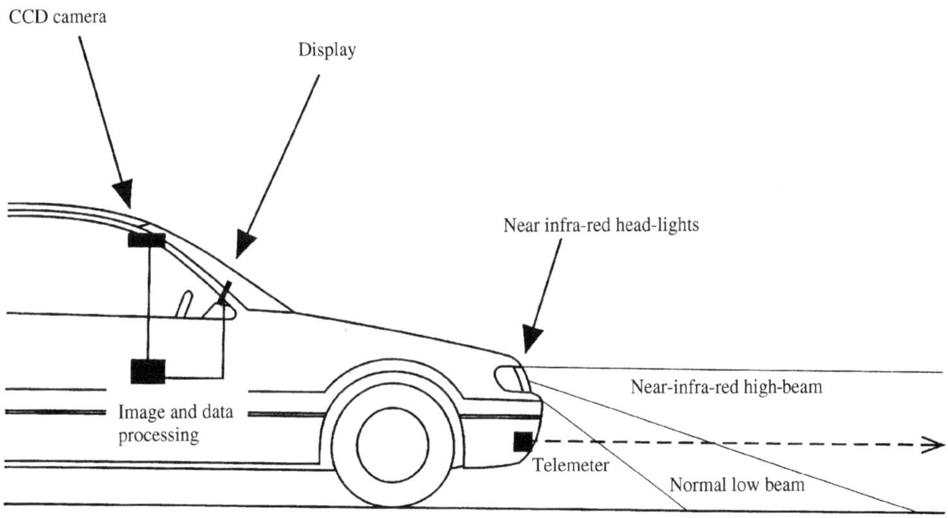

(b)

Figure 4.29. a) System's block diagram b) *In-vehicle systems layout* System's diagram of driver vision enhancement

The head-up display system allows the overlapping of the scene the driver sees with the image acquired by the camera sensor. With the type of head-up display shown in Figure 4.30, the road images acquired by the camera are projected on to the holographic optical element (combiner) through a spatial light modulator and the zoom lens.

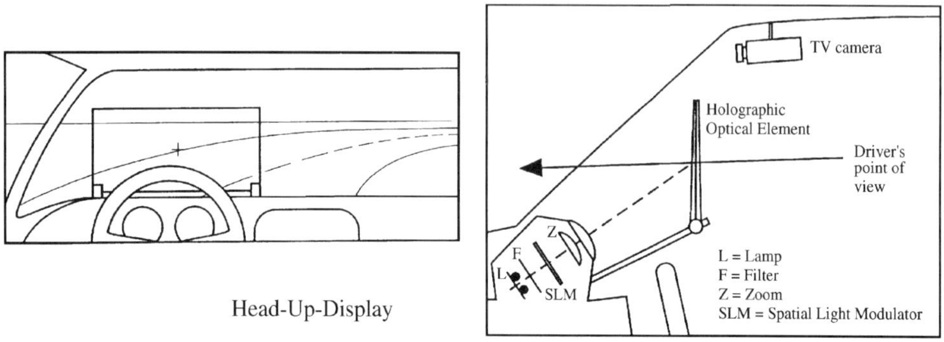

Figure 4.30. Head-Up display for vision enhancement system

4.8 CONCLUSIONS

The vehicle as it is known to date is currently going through several developments that undoubtedly will bring about some intelligence in the vehicle. In the USA they have been several million dollars invested into the intelligent transportation system, the European countries, under the cooperative frame work of the European Union, has invested several million ECUs in PROMETHEUS, DRIVE and the 4th Framework Programme, and in Japan an extremely large amount of money is spent on similar projects. Lately, several independent countries are also starting to realize that a separate national programme on intelligent vehicles is needed to support their respective automotive industries This chapter has therefore examined some of the systems developed so far to slot into the technological jigsaw necessary for intelligent or smart vehicle. The development of these systems should take onboard user requirements, HMI problems, cost/benefits issues and systems reliability.

4.9 REFERENCES

1 Anon, The motor car as liberator. *The Boston Globe*, p. 31 1986; February.
2 Renner M. Rethinking the role of the automobile. *Worldwatch* paper 84, June 1988.
3 Hominger B. *et al.* Integrated electronics for automotive application in the EUREKA programme PROMETHEUS. *Proceedings, Essire '91, Grenoble, France*, 1990.
4 Nwagboso CO. *Automotive Sensory Systems*. Chapman & Hall London 1993.
5 Trankler HR. Sensorspezifissche MeBsignalverarbeitung. NTG Technical Report No. 93, *3rd Professional Conference*, Bad Nauheim, Germany, March, 1989.
6 Kiencke U, Dais S, Litschel M. *Automotive Serial Controller Area Network*. SAE Technical Paper No. 860391, 1986.
7 Kiencke U, and Dais S. *Application Specific Microcontroller for Multiplex Wiring*. SAE Technical Paper No. 870515, 1987.

8 McLaughlin RT. *EMC Susceptibility Testing of a CAN Car.*, SAE Technical Paper No. 932866, 1993.
9 Nwagboso CO. Vehicle dynamics, Lecture Note (unpublished). *Beng Automobile Engineering Course.* Bolton Institute, UK, 1991.
10 Gillespie D. *Fundamentals of Vehicle Dynamics.* SAE, 1992.
11 Kuo BC. *Automatic Control Systems.* Prentice-Hall, Boston USA, 1991.
12 Eveleigh VW. *Adaptive Control and Optimization Technique.* McGraw-Hill, New York, 1967.
13 Franklin GF, Powell JD, Emami-Naeini A. *Feedback Control of Dynamics Systems.* Addison-Wesley, Massachusets 1986.
14 May AD, and Keller HEM. Non-Integer Car Following Models. Washington HRB, **199**, 1967; 19–32.
15 Gipps PG. A behaviour car following model for computer simulation. *Transportation Research*, 1981; **15**, (Part B), 105–111.
16 Leutzbach W. *Introduction to the Theory of Traffic Flow.* Springer-Verlag, The Netherlands 1988.
17 Raffaelli L, Stewart E, Borelli J, Quimby R. Monolithic components for 77 GHz automotive collision avoidance Radars., *Proceedings Sensors Expo*, 1993; 261–268.
18 French RL. Historical overview of automobile navigation technology. *Proceedings, 36th IEEE Vehicular Technology Conference*, Dallas, USA, May 20–22, 1986; 350–358.
19 Sasayama T, *et al.* Sensing technology for autonomous navigation system. *Proceedings 25th ISATA conference*, 1–5 June, Florence, Italy, 1992; 521–528.
20 French RL. Map matching origins, approaches, and applications. *Proceedings 2nd International Symposium on Land Vehicle Navigation*, Munster, Germany, July 4–7, 93–116 1989
21 Iihoshi A *et al.* Honda navigation system. *25th ISATA conference*, 1–5 June, Florence, Italy, 1992; 109–116.
22 May MB. Measuring velocity using GPS. *GPS World*, 1992; **September**: 58–65.
23 Krage MK. The TravTek driver information system. *Proceedings Vehicle Navigation and Information Systems Conference*, Vol. 2, October 20–23, Michigan, USA, 1991; 739–748.
24 Walker J. *Mobile information systems.* Boston: Artech' 1990.
25 Davies P and Kleein G. RDS-ALERT — Advisce and problem location for European road traffic. *Proceedings of the Advanced Telematics in Road Transport, DRIVE Conference, Brussels.* Elsevier Amsterdam, 1991.
26 Underwood G. *et al. Information Technology on the Move*: John Wiley, Chichester 1994.
27 Office of Science and Technology (OST). *Technology Foresight (Transport): Progress Through Partnership.* 1995.

5

Intelligent Automotive Systems: Development in Full-time Chassis Motion Spheres for Intelligent Vehicles

5.1 INTRODUCTION

Intelligent automotive systems is an esoteric engineering science. There are only about 50 hard core practitioners in the world. As with all human endeavours they must have a leader and one of them has been Jerzy ("George") Zulawski (1874–1915). He is referred to by Professor Dr M. Gregory ("Greg") Bekker [best known as the "Father of Terrain-Vehicle Systems" and the inventor of a metal open-worked-caterpillar track for tracked vehicles and pioneering construction of a wire-mesh wheel. Others remember him for contemporary lunar roving vehicles (LRV) for the National Aeronautics and Space Administration (NASA)], and many regard him as the "Father of Intelligent Automotive Systems", still others see him as one of the science-fiction writers, the true creators of the term "extraterrestrial vehicle", who first saw road vehicles (RV), including the first extraterrestrial LRV, conceived in his science-fiction novel[1] as an intelligent road vehicle (IRV). All of these references are accurate reflections of what he has meant to the individuals and the science of intelligent automotive systems.

The vision of intelligence has been lost in the real world of automotive vehicles (AV), and it does not, or rather does not yet, exist in actual RVs. The demands on the performance of RVs increases, however, as they tend to be used for

Advanced Vehicles and Infrastructure Systems, Edited by C.O. Nwagboso

more sophisticated duties than simple transportation functions. For IRVs, too, the number of functions should be greater, for instance, for electrically energized and mechatronically controlled axleless full-time chassis motion [all-wheel-steerable (AWS) conversion × all-wheel-absorbable (AWA) suspension × all-wheel-driveable (AWD) propulsion × all-wheel-brakeable (AWB) dispulsion] as well as vehicle route guidance navigation and collision avoidance functions.

A simple one-lane RV, for example, a motorcycle, scooter or moped enlarges the velocity and the range of the rider in a very adaptable way, but without the human driver (HD), it will be without any autonomy. In each case, it would be desirable to enlarge the number of functions and add the missing capabilities. For a two-lane RV, this leads to IRVs that can be made more adaptable for extending the spectrum of duties. IRVs, however, will obtain more and more autonomy using in-vehicle sensors (IVS) in order to relieve human beings from heavy physical or mental work. The intelligence of a RV for such work depends on its ability to acquire information about its environment, and to process it in such way that it reacts according to the situation. This desired extension of adaptability and autonomy sooner or later will always require an interaction with a human being and an intelligent behaviour of the full-time AWS × AWA × AWD × AWB IRV.

Today, global environmental issues and their related energy problems have begun a new trend toward RVs. Following this, an trying to conceive a novel full-time AWS × AWA × AWD × AWB IRV giving ideas and expectations for the present and the twenty-first century as a nucleus of R&D proposition for IRVs world-wide, which will achieve high driving performance using automotive high technologies (high-tech).

5.2 PROOF-OF-CONCEPT FULL-TIME 4WS × 4WA × 4WD × 4WB INTELLIGENT ROAD VEHICLES

In the RV field an example of study is represented by the proof-of-concept electrically powered and mechatronically controlled full-time four-wheel-steering × four-wheel-absorber × four-wheel-drive × four-wheel-brake (4WS × 4WA × 4WD × 4WB) IRV with extremely high mobility and steerability, which was originated and designed by myself.[2,3] The proof-of-concept axleless full-time chassis motion (4WS conversion × 4WA suspension × 4WD propulsion × 4WB dispulsion) spheres are designed for installation in a full-time 4WS × 4WA × 4WD × 4WB smart tri-mode supercar.

Figure 5.1 shows an overall view of the full-time 4WS × 4WA × 4WD × 4WB smart tri-mode supercar, that is the *Poly*technic all-round energy-efficient electrically energized and mechatronically controlled smart tri-mode *supercar* called the Poly-Supercar. The barycentre is located at the centre of the body. The two independent-sprung steer-, autodrive- and autoabsorbable wheels (SA^2W) or steered

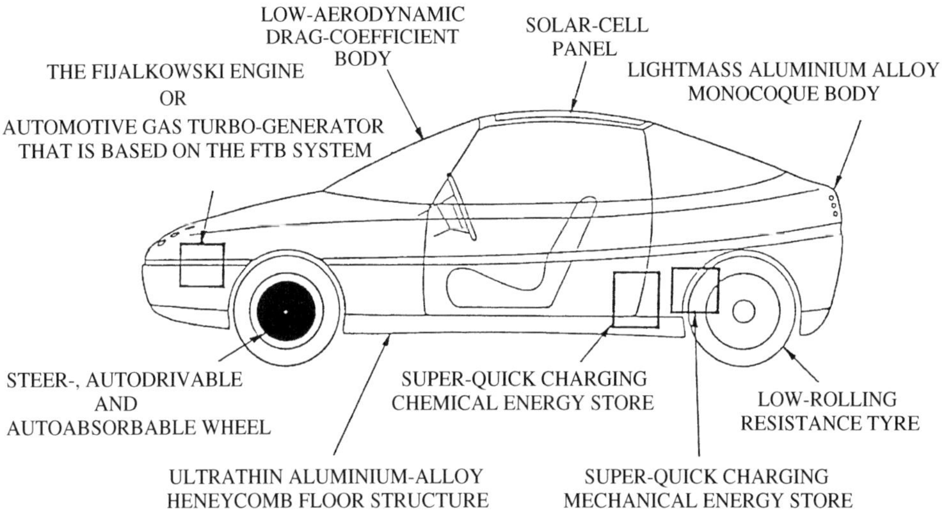

Figure 5.1. Principle layout of a high-performance all-round energy-efficient smart tri-mode supercar called the Poly-Supercar

and motorized wheels (S&MW) on each side of the body have independent electrorheological fluid (ERF) variable-damping ERF shock-absorber suspensions. The four (two front and two rear) independent-sprung SA^2Ws or S&MWs are driven individually by four direct current (DC) to alternating current (AC) and/or AC-to-DC specific application integrated matrixer (ASIM), macroelectronic commutator (ASIM MCommutator) electromechanical (E-M) and/or mechanoelectrical (M-E) wheel-hub motors/generators, and the rotating speed of each front-wheel drive (FWD) or rear-wheel drive (RWD) and each SA^2W or S&MW can be arbitrarily controlled by driver-vehicle and terrain-vehicle real-time expert hypersphere, incorporating neuro-fuzzy (NF) reasoning-based motion control (MC) of a full-time 4WS × 4WA × 4WD × 4WB IRV.

The acceleration and jerk (A&J) sensor[4] may be attached at the geometric centre of the IRV. The required steering power of each FWD or RWD can be measured by the wheel torque gauges, which are specially designed to eliminate the bending effects of FWD or RWD on SA^2W or S&MW bearings.

In the Poly-Supercar, the piston-type internal combustion engine (ICE) or automotive gas turbine (AGT) that is based on the Fijalkowski turbine boosting (FTB) system[5] is replaced by the Fijalkowski engine with constant output[6] and the rear or front thermomechanical (T-M), two-wheel-driveable (2WD) propulsion sphere is replaced by the proof-of-concept series hybrid thermomechanoelectromechanical (T-M-E-M) and/or E-M 4WD propulsion sphere.

The full-time 4WS × 4WA × 4WD × 4WB smart tri-mode supercar called the Poly-Supercar (Figure 5.1) has unique type of the dual-mode hybrid 4WS conversion sphere, which incorporates adjustable torque and/or speed front-wheel steering

(FWS) and rear-wheel steering (RWS) gears. When the steering wheel (SW) is turned the power steering mechanisms are activated and push/pull the IRV until the desired turn is made.

This dual-mode hybrid 4WS conversion sphere has proven to be a relatively effective mobility feature because there is no power loss in braking the FWD and RWD for steering as on a conventional FWS and/or RWS RV. When steering the modern full-time 4WS × 4WA × 4WD × 4WB IRVs both FWD and RWD are still powered, providing smooth and continual power to them, which allows it to achieve maximum tractive effort.

The components of axleless chassis motion spheres are interconnected through all these heterogeneous discrete automotive functional spheres, and the converting and suspending as well as propelling and dispelling power flows are controlled by an on-board artificial intelligence (AI) hypersphere electronic control instrument (ECI) including AI sphere electronic control units (ECU) based on the application specific integrated circuit (ASIC) NF microcomputer or transputer PID (Proportional Integral Derivative) microcontrollers (ASIC NF PID μControllers), and AI hyposphere electronic control bits (ECB), which receive commands from the human- and/or telerobotic-driver (H&TD).

The full-time 4WS × 4WA × 4WD × 4WB smart tri-mode supercar called the Poly-Supercar will have extremely high mobility and steerability, and be driven by the same four independent-sprung SA^2Ws or S&MWs. All four wheels are consequently all-electric drive-by-wire (DBW) steered, driven (motorized) and absorbed (generatorized) ones.

5.3 PURPOSE OF INTEGRATED IRV MOTION CONTROL BETWEEN INDIVIDUAL 4WS, 4WA, 4WD AND 4WB CONTROLS

The mechatronically controlled predictive and adaptive 4WA suspension sphere will make it possible to control attitude, ride comfort, stability and vehicle height, and partially control the longitudinal and lateral manoeuvrability.[7] Also dual-mode hybrid 4WS conversion sphere will make it possible to improve stability and controllability, which will be the lateral manoeuvrability.

Each of these heterogeneous discrete automotive functional spheres will greatly improve performances and greatly influence performances other than the intended ones. This will be because, as shown in Figure 5.2, the basic functions of IRV's chassis-motion dynamics, such as turn (lateral motion) and bounce/rebound (vertical motion) as well as drive/brake (longitudinal motion) will be closely related to each other through the road-contact force of wheel tyres which will be the sole contact points between an IRV and ground surface.

In Figure 5.2, the heavy line shows the basic reaction, the fine line the actions against other functions, and the broken line the restrictions. As the bettering effect

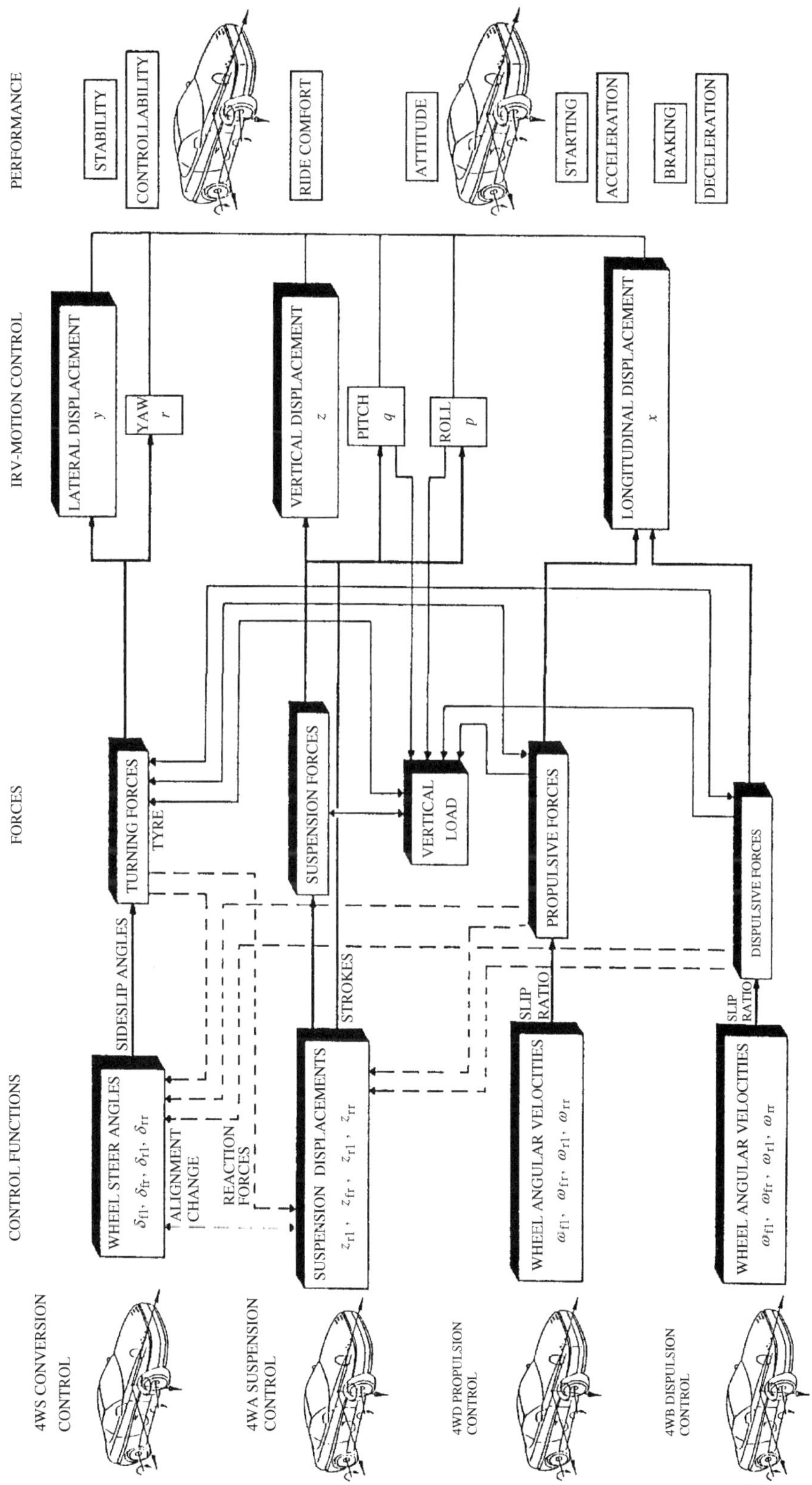

Figure 5.2. Cooperative intelligent road-vehicle chassis-motion control

of more performances will be expected by increasing each automotive-functional-sphere control gain, the effect other than the heavy line or the actions against other automotive functional spheres cannot be ignored. For instance, if roll is decreased by the mechatronically controlled predictive and adaptive 4WA suspension sphere, the yaw response will be often weaken. Meanwhile, if the yaw responses of the dual-mode hybrid 4WS conversion sphere can be improved, roll convergence may also weaken. Under steady state, the roll–understeer function cannot be used and roll–oversteer disposition will appear by eliminating the roll. Therefore, individual control of each dynamical automotive sphere may often affect performances other than the intended ones. It is important for a better integrated performance of IRVs to eliminate unwelcome actions of different forces. Cooperative control of each dynamical automotive sphere will allow the whole IRV, that is the dynamical automotive hypersphere to provide a balanced performance. Moreover, performances will be improved by aggregation of individual controls of each dynamical automotive spheres.

The proposed on-board AI hypersphere ECIs have a concentrical configuration in which all AI sphere ECUs are centred in one place. This tends to make the electrical connections (cabling) connected to IRVs' IVSs, as well as reciprocating E-M actuators and rotating E-M motors, long. It stands to reason that a hybrid (concentrical/dispersal) configuration, shown in Figure 5.3, ought to be used. Exclusive AI hyposphere ECB placed at each SA^2W or S&MW station will be combined together with a bisensual-direction fibre-optic (FO) in-vehicle data link for information exchange. Thus, the on-board AI hypersphere ECI will be placed at the centre, which will do the calculations for the whole IRV motion control. It will also exchange data between 4WS conversion, 4WA suspension, 4WD propulsion and 4WB dispulsion spheres' AI sphere ECUs, which will be translated to the AI hyposphere ECBs at the SA^2Ws or S&MWs. In this way, it will be possible to fit together AI reciprocating E-M actuators and AI rotating E-M/M-E motors/generators with unified ones and their AI hyposphere ECBs, which will seem to be a promising IRV motion control for the future.

The distinction between randomness and imprecisions can perhaps be made clearer by a consideration of crisp or fuzzy sets. The universal set U may be an on-board AI hypersphere ECI for integrated IRV motion control and the four sets A, B, C and D may be the AI sphere ECUs for 4WS, 4WA, 4WD and 4WB controls, respectively. The union is the crisp or fuzzy set of all the AI sphere ECUs for 4WS, 4WA, 4WD or 4WB controls or all, i.e. integrated IRV motion control. The intersection is the crisp or fuzzy set of all the AI sphere ECUs for all individual 4WS, 4WA, 4WD and 4WB controls. The primary purpose of the integrated IRV motion control between individual dynamic automotive spheres can adequately be explained by the Venn diagram shown in Figure 5.3.

"4WS + 4WA + 4WD + 4WB" in Figure 5.3 represent simple aggregation of individual 4WS, 4WA, 4WD and 4WB controls, while "4WS × 4WA × 4WD × 4WB" represent integrated IRV motion control between 4WS, 4WA, 4WD and

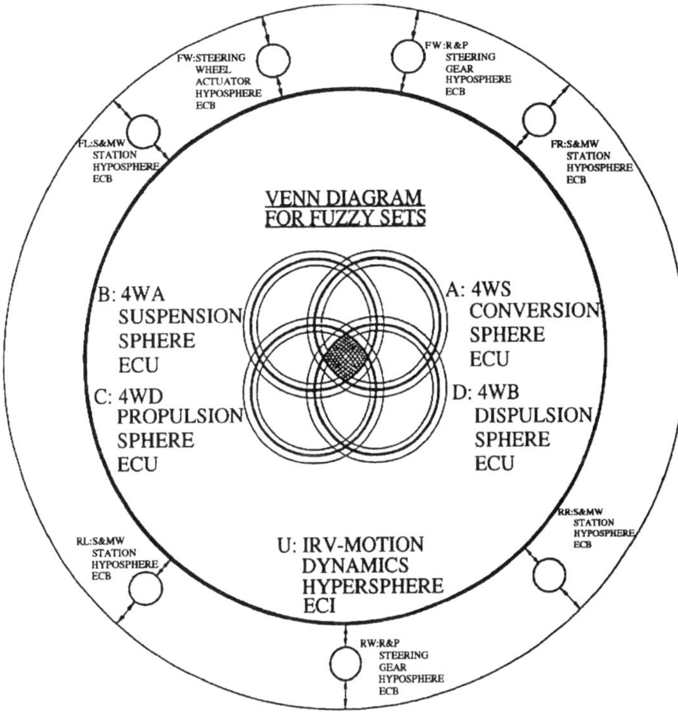

Figure 5.3. Concept of integrated intelligent road-vehicle chassis-motion control between 4WS, 4WA, 4WD and 4WB controls

4WB controls, which shows "4WS × 4WA × 4WD × 4WB" improves performance more than "4WS + 4WA + 4WD + 4WB" does. This integrated IRV motion control between dedicated automotive functional spheres is sophisticated because it consists of many components.

5.3.1 Dual-Mode Hybrid 4WS Conversion Sphere

As a base a conventional rack-and-pinion (R&P) steering gear, shown in Figure 5.4, will be used.[8] The tubular linear DC-to-DC ASIM MCommutator conversion-actuator motor is placed concentric to the rack, and drives it, which produces the necessary rack force (12 kN) for a full-time dual-mode hybrid 4WS conversion sphere. Hardware for the all-electric DBW steering gear is also shown in Figure 5.4.

The steering converts the SW rotary motion into a turn motion of the SA^2Ws or S&MWs of the full-time 4WS × 4WA × 4WD × 4WB IRVs. A perfect dual-mode hybrid 4WS conversion sphere should utilize not only a feedback control for IRV stability, but also a feedforward control signal for better steering response.[9,10]

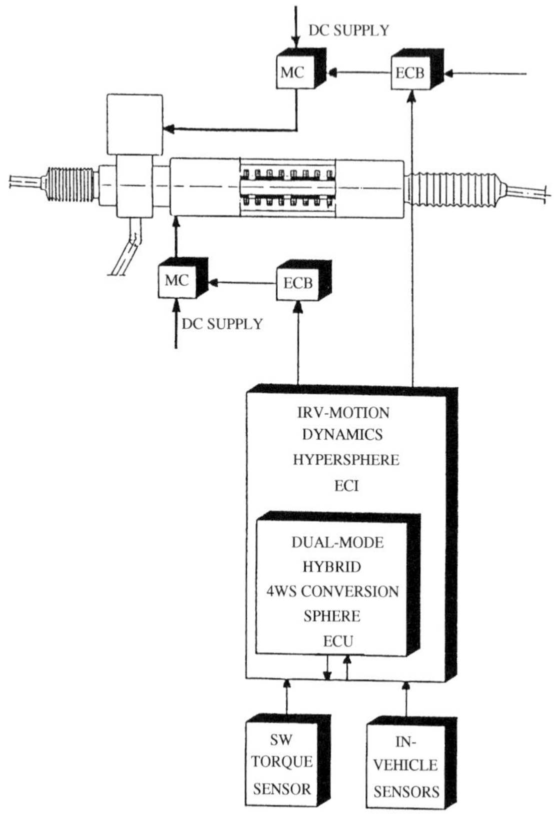

Figure 5.4. Principle layout of an electromechanical rack-and-pinion steering gear with the tubular linear DC-to-DC ASIM macrocommutator electromechanical steering-actuator motor and its hardware (MC — ASIM macrocommutator; ECI — electronic control instrument; ECU — electronic control unit; ECB — electronic control bit)

The dual-mode hybrid 4WS conversion sphere will be vehicle-speed controlled by an AI sphere ECU based on the ASIC NF PID μController, and AI hyposphere ECBs for each of the SA^2W or S&MW stations. Tighter turning and reduced scrub will be the benefits.

The series hybrid 4WD propulsion sphere mentioned will allow the omission of the conventional front- and rear-axle as well as inter-axle M differentials. Therefore a superimposed IRV motion control will be necessary to co-ordinate both FWD and RWD units. Both front and rear independent-sprung SA^2Ws will be controlled individually. Obviously, the distribution of wheel torques can influence driving behaviour very much. If both front and rear wheel torques will be equal, the 4WD propulsion sphere will behave like a conventional one with M differentials lock. In practice, each distribution of wheel torques will be possible within the technical limits. This means, for instance, maximum positive wheel torques at right-hand front and rear SA^2Ws and maximum negative wheel torques at the left-hand ones at the same time. Such inner-front- and/or inner-rear-wheel torques distribution will lead to yaw torques around the vertical axis, that is a turn of the IRV. Consequently, the influence of the 4WD propulsion sphere on to the driving behaviour of the IRV can be compared with a torque-steering mode of the dual-mode hybrid 4WS conversion sphere.

Thus, the torque steering can be used to control the yaw motion of the IRV simultaneously with the conventional front-wheel steering (FWS) and rear-wheel steering (RWS) gears. If a yaw angular velocity sensor can be applied, it may be

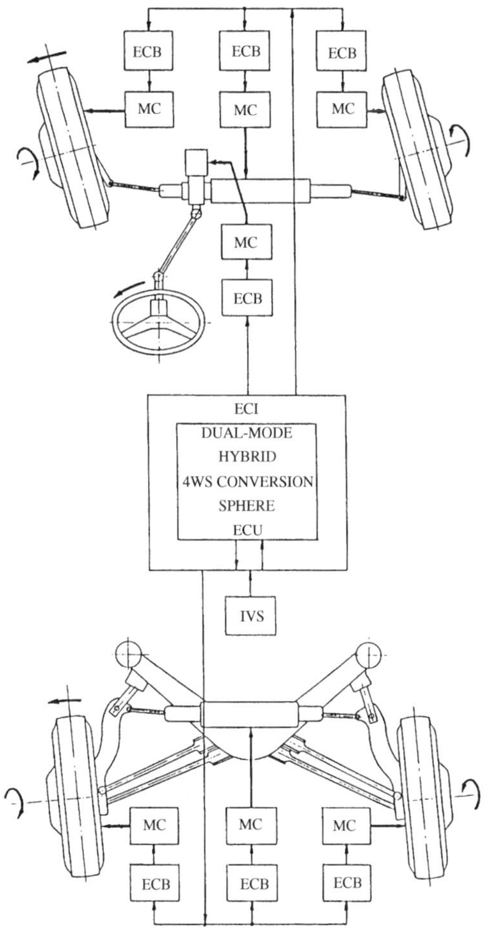

Figure 5.5. Principle layout of a full-time dual-mode hybrid 4WS conversion sphere with the front and rear electromechanical rack-and-pinion steering gears, two right-side and two left-side electromechanical/mechanoelectrical steer-, autodrive- and autoabsorbable wheels, the steering wheel and its hardware (MC — ASIM macrocommutator; ECI — electronic control instrument; ECU — electronic control unit; ECB — electronic control bit)

useful to combine an open-loop control with superimposed closed-loop control of the yaw angular velocity as described by Inoe.[11]

One possibility to control the yaw angular velocity is the application of an AI sphere ECU based on the ASIC conventional PID μController. Also an AI sphere ECU based on the ASIC NF PID μController can already give satisfactory results with little knowledge of lateral-motion dynamics of the dual-mode hybrid 4WS conversion sphere, which has to be controlled. NF reasoning based lateral IRV motion control may be easier to adapt. Thus, using torque and/or speed controls of 4WD propulsion sphere's inner SA^2Ws can increase this effect especially at recuperative braking with the inner SA^2Ws acting as the AC-to-DC ASIM MCommutator M-E wheel hub generators, because the front gravitational forces on to the IRV become greater than the respective rear ones. At the same creep this leads to greater horizontal (longitudinal and lateral) forces.

Figure 5.5 shows the dual-mode hybrid 4WS conversion sphere configuration. The rear independent-sprung SA^2Ws will be steered according to the front-wheel steer angle by installing the steering mechanism to steer the rear SA^2Ws at the rear suspension and electrically connecting it with the front steering mechanism by electrical connections (cabling) only. The relation of steer angles between the front and rear SA^2Ws will be mechatronically controlled according to the front or backward movement of the IRV and the vehicle speed.

5.3.2 Predictive and Adaptive 4WA Suspension Sphere

In vibration control, the AI sphere ECU based on the ASIC NF PID μController first will obtain five system-state variables from the sensors. Then, the road-surface condition, whether smooth or rough, will be judged by statistics operations of relative vertical velocities to select control gains. From these data, the control amounts will be calculated, and the control DC voltages of the damping force ER fluidical valves will be decided. Furthermore, floor vibration due to other SA^2W or S&MW manoeuvres will be compensated based on the gains selected by its mode and the SA^2W or S&MW system-state variables of all the SA^2Ws or S&MWs. Vibration control will be realized with the ER fluidically suspended IRV. Position US sonars or linear potentiometer or electromagnetical sensors alongside the variable-damping ERF shock absorbers will read the on- and/or off-road surfaces. The cornerstone of ERF automotive high-tech has been the ER fluidical valve and the resultant ERF shock absorber. In the ER fluidical valve the ERF is pumped through a channel capacitor wherein the DC voltage difference between the walls may be varied to regulate the flow resistance up to the point where the fluid solidifies and flow ceases.[12] A schematic diagram of a force-control-based predictive and adaptive 4WA suspension sphere with the ERF shock absorber and ER fluidical valve is shown in Figure 5.6. Variable-damping ERF shock absorbers exploit the property of electric-field-dependent stiffness for a virtually limitless range of damping response. When

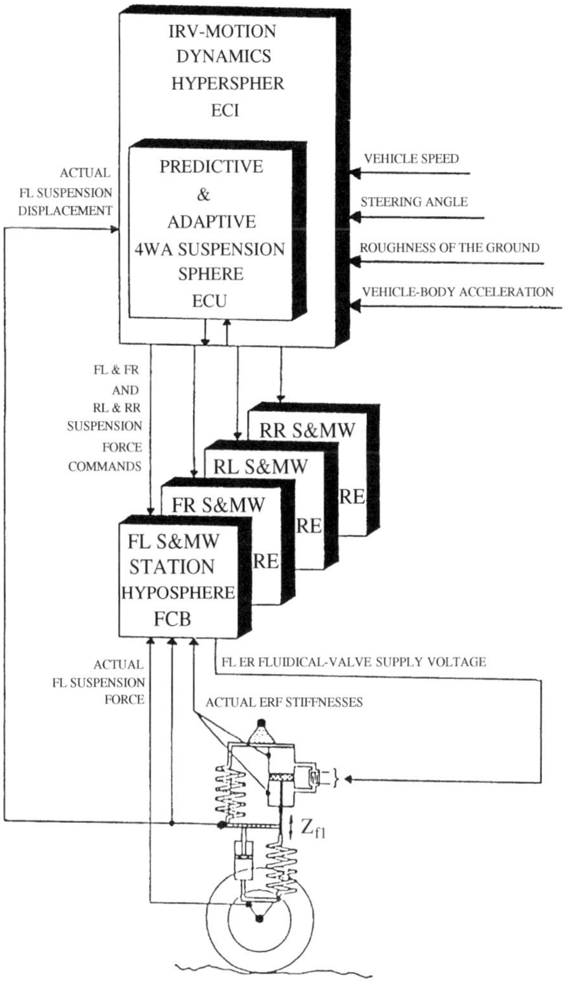

Figure 5.6. Principle layout of a full-time force-control based predictive and adaptive 4WA suspension sphere with the ERF shock absorber, ER fluidical valve and its hardware (ECI — electronic control instrument; ECU — electronic control unit; ECB — electronic control bit)

one considers the milliseconds response time involved, and exciting automotive high-tech in position US sonars for anticipating on- and/or off-road surface roughness, it is easy to understand what motivates competitors in the automotive industry, to be the first to offer an ultrasmooth, AI ASIC microcomputer- or transputer-controlled predictive and adaptive 4WA suspension sphere for controlling stability, ride comfort, attitude and vehicle height.

An AI sphere ECU based on the ASIC NF PID μController will be responsible for achieving commanded forces despite disturbances and non-linearities. This force — servo approach may facilitate the design of the on-board AI hypersphere ECI, but there will also be systems where AI sphere ECUs for the SA^{2}Ws or S&MWs will only be used for some local stabilization, but the on-board AI hypersphere ECI will directly inject control signals into the ER fluidical valves.

In general, a large number of sensors, four ER fluidical valves and a large number of mechanical degrees of freedom of an IRV will create a demanding multivariable control problem. All motion interact. Control designs must also take the non-linearities of fluidics into consideration.

In the attitude control, the reactive real (measured) displacements between the top and the bottom of the suspensions will coincide with the aimed displacements by feedback controls. In addition, for the immediate reduction in roll angles caused by steering manoeuvres, feedforward controls based on the estimation of roll angles from steer angles and compensation of roll angles based on changes in 4WA suspension sphere's electric-field-dependent stiffnesses due to lateral acceleration will be performed. Attitude control for roll and pitch directions decreases low-frequency vibration. Acceleration in the horizontal (longitudinal and lateral) directions of a full-time 4WS $\times$ 4WA $\times$ 4WD $\times$ 4WB IRV during turning, acceleration and deceleration will be detected by the A&J sensor[4] or the other acceleration sensor, the so-called G sensor[7] and the control signal will be given to the linear control ER fluidical valve of each wheel to control the shock-absorber–cylinder ERF solidification. As a result, attitude change including roll and pitch will be eliminated and the IRV attitude will be kept constant. To compensate for the delay of control response, roll will be controlled by foreseeing the start of the roll with the steer angles and vehicle speed sensors at the beginning of turning.

The predictive and adaptive 4WA suspension sphere detects the abruptness of a on- and/or off-road bump almost quickly enough to react before the shock of the impact reaches the IRV spring. In practice, the predictive and adaptive 4WA suspension sphere swallows high-speed undulations, prevents bottoming and retains a smooth ride on harsh on- and/or off-road surfaces.

5.3.3 Series Hybrid 4WD Propulsion Sphere

An advantage of the series hybrid 4WD propulsion sphere is that a novel form of ICE can be used, designed specifically for high efficiency constant reciprocating speed and power operation and minimum emission. Other longer-term options include future hydrogen-fuelled ICEs and fuel cells, with the prospect of the virtually zero emission vehicles (ZEV). The 4WD propulsion sphere using E-M/M-E SA^{2}Ws with the DC-to-AC/AC-to-DC ASIM MCommutator E-M/M-E wheel-hub motors/generators will eliminate the need for axles and drive shafts, which will allow a separate suspension of each independent-sprung SA^{2}W (Figure 5.7). The rotating housing is made in the form of a wheel hub, designed to fit standard rim sizes on the market. The proof-of-concept DC-to-AC ASIM MCommutator E-M wheel-hub–motor transmission offers complete freedom in the design of the IRV.

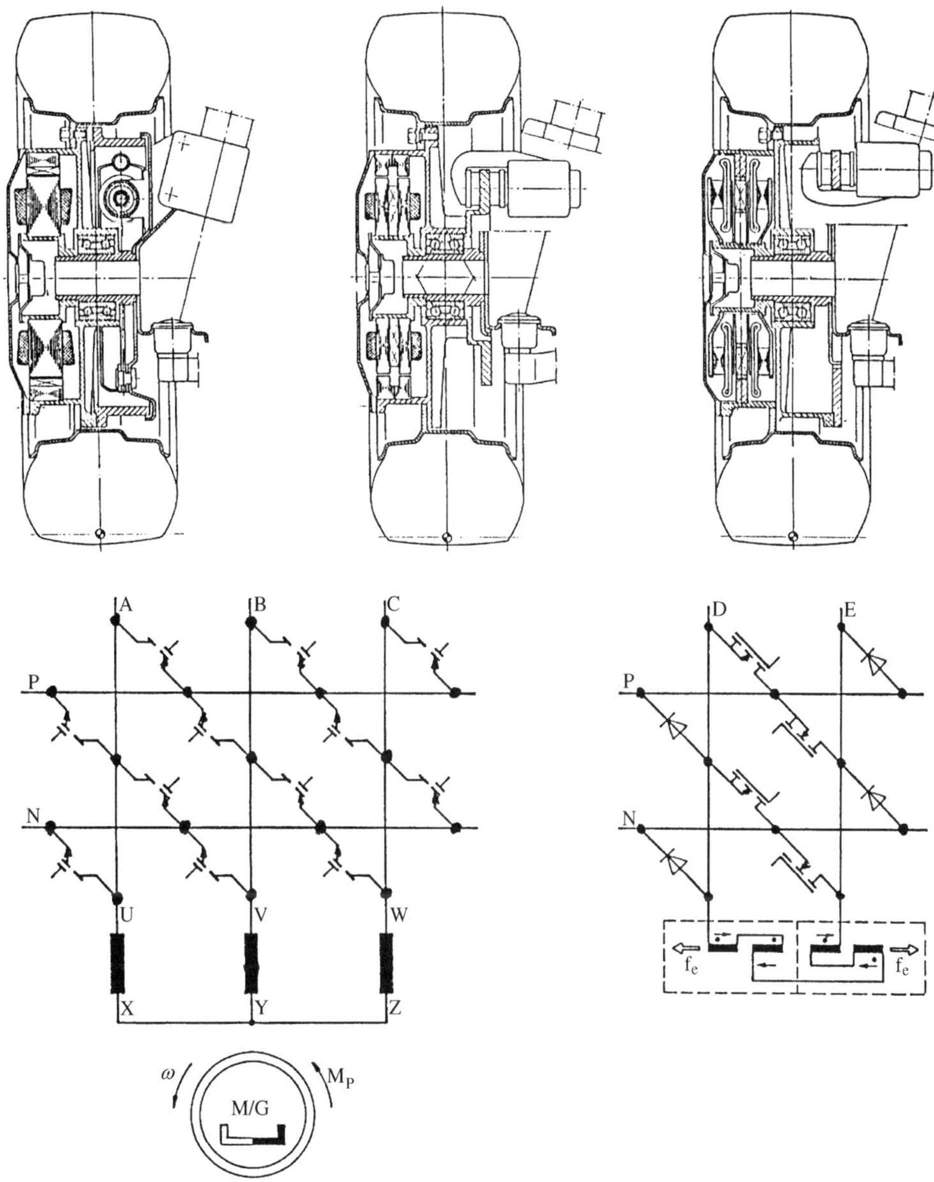

Figure 5.7. Principle layouts of three different independent-sprung, planetary gearless, electromechanical/mechanoelectrical steer-, autodrive- and autoabsorbable wheels — MAGNET ROTA^{TM}: RX-V, RX-VI and RX-VII with rotational DC-to-AC/AC-to-DC ASIM macrocommutator (unwound outer rotor and wound inner stator) wheel-hub motors/generators and tubular linear DC-to-DC ASIM macrocommutator wheel-drum, -disc and -ring brake-actuator motors with their DC-to-AC/AC-to-DC and DC-to-DC ASIM macrocommutators

While the basic principles of all-electric DBW IRVs remain the same for virtually all classes of IRVs, the actual arrangements vary — for instance, some may have 2WD, that is either FWD or RWD, and others 4WD.

Another requirement for the all-electric DBW 4WD propulsion stems from the fact that, when the vehicle is cornering, the outer SA^2Ws must roll faster than the inner ones which will be traversing circles of smaller radii, yet their mean speed, and therefore both the reciprocating speed of the Fijalkowski engine and the translational speed of the vehicle, may be required to remain constant. So far as effectiveness of traction is concerned, FWD is better than RWD, especially on difficult terrain, including ice or snow. This is partly because the mass of the ICE on the front SA^2Ws enables them to grip the surface better, which also applies to rear-engine RWD IRVs. Principally, however, the advantage is gained by virtue of the fact the the propulsive force, i.e. tractive effort is always delivered along the line in which the front SA^2Ws are steered. Another factor is that front SA^2Ws tend to climb out of holes or ruts, whereas rear SA^2Ws tend to thrust the front undriven wheels deeper down and, in any case, not necessarily in the direction in which they are being steered.

The use for full-time 4WS × 4WA × 4WD × 4WB IRVs will grow rapidly over the next few years. This is due to the increasing demand for IRVs with higher performance and power. It is well known that the distribution of gross tractive effort (thrust) and slip between the front and rear S&MWs of full-time 4WD and/or 4WS IRVs has considerable effect on the efficiency of operation. Differences in theoretical speeds (wheel speed when no slip or skid occurs) and the concept of cooperation between the front and rear S&MWs generally affect the gross tractive effort and slip distribution.

Analysis and experiments indicate that the development of new concept axleless FWD and RWD as well as inter-wheel-drive (IWD) E-M differentials, which can provide the optimum distribution of gross tractive effort and slip between front and rear S&MWs is of practical importance.

Typically in FWD and RWD as well as the IWD E-M differential layout (Figure 5.8) a series electrical connection (cabling) is interposed between the electrical power source, that is the Fijalkowski engine and/or DC CH-E storage battery power-outputs as well as FWD and RWD units. The function of a series electrical connection is to transfer the drive electrical power from the electrical power source to both the front and/or rear E-M/M-E SA^2Ws or S&MWs. The IWD E-M differential, in the series electrical connection, is also necessary to distribute the drive equally between the front and rear E-M/ME SA^2Ws or S&MWs and to allow for the fact that, when the IRV is driven in a circle, the mean speeds of the front SA^2Ws are different from those of the rear SA^2Ws and therefore the speeds of the two FWD and RWD units must differ too. Other factors include different rolling radii of tyres due to, for example manufacturing tolerances, different degrees of wear and, perhaps, different tyre pressures. Provision is usually made for locking this IWD E-M differential out of operation to improve the performance and reliability of traction when the smart automotive vehicle is driven on slippery ground. For IRVs intended

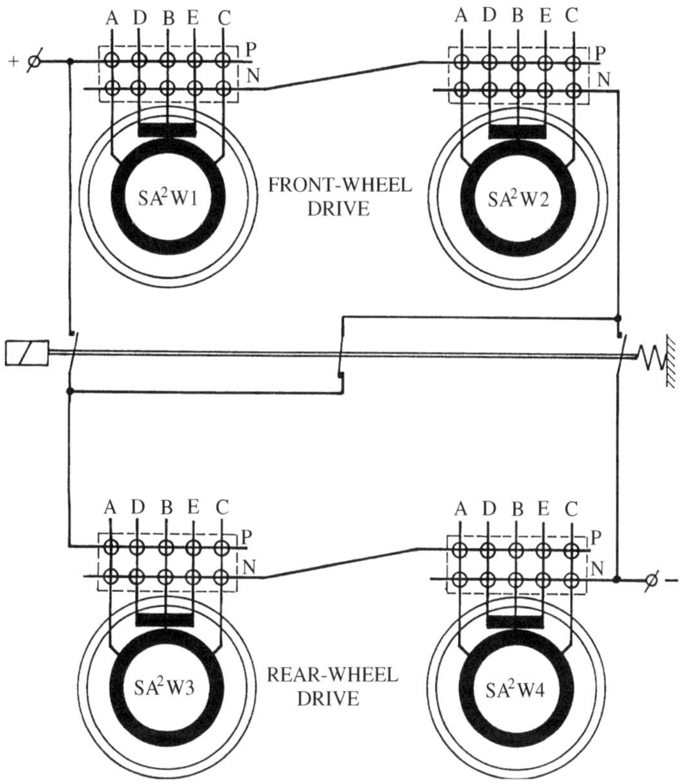

Figure 5.8. Elementary wiring connections for axleless full-time 4WD propulsion and 4WB dispulsion spheres with the torque-proportioning front-wheel-drive and rear-wheel-drive as well as inter-wheel-drive electromechanical differentials

mainly for operation on soft ground, the IWD E-M differential may be omitted from the drive E-M powertrain line, but some means of disengaging 4WD, leaving only single pair of SA^{2}Ws to do the driving, is generally provided for use if the IRV is required to operate on metalled roads.

As is well-known from the principle of Ackermann's 2WS conversion sphere, the front E-M/M-E SA^{2}Ws always tend to roll further than the fixed-geometry rear E-M/M-E SA^{2}Ws, because their radius of turn is always larger, a parallel electrical connection (cabling) can be interposed between the electrical power source as well as FWD and RWD units and the IWD E-M differential may be omitted from the drive E-M powertrain line (see Figure 5.8). In practice, this usually takes the form of two separate FWD and RWD units, single on each front or rear E-M/M-E SA^{2}Ws, on which there are rotary controls which can be locked by the H&TD, but they have to stop and get out to do so. As soon as the H&TD again drives the IRV on firm ground, however, the H&TD must remember to unlock the FWD and RWD units. Should the rear E-M/M-E SA^{2}Ws lose traction, on the other hand, and therefore tend to rotate further than the front ones, the drive will automatically be transferred to the front E-M SA^{2}Ws even if they are in the freewheeling mode.

5.3.4 Anti-wheel-lock and Anti-wheel-spin 4WB Dispulsion Sphere

During unexpected conditions, e.g., without warning braking on rain-slick or icy road surfaces, the wheels may lock and the IRV may slide in a direction completely unrelated to H&TD's SW operation. Under these conditions, skilled and experienced HDs release the left-foot brake pedal to recover wheel locking, and depress the left-foot brake pedal when they obtain wheel recovery. Such braking operations are repeated to stop the IRV stably and efficiently. These countermeasures are acted automatically by an AI sphere ECU based on the ASIC NF PID μController in the anti-wheel-lock and anti-wheel-spin 4WB dispulsion sphere.

The relationship of the wheel–slip ratio and the friction coefficient between the on-road and/or off-road surfaces and tyre differs according to the on- and/or off-road surface conditions and vehicle speed. The friction coefficient may peak at a slip ratio of 0.2 on a particular on-road and/or off-road surface, but at 0.05 on another one.

Thus, how to determine the desired command signal, under the moment-to-moment on-road and/or off-road surface conditions and vehicle speed, becomes the major aspect in stably and efficiently stopping the IRV. Besides, in an anti-wheel-lock and anti-wheel-spin 4WB dispulsion sphere based on optimal ASIC conventional PID μController, feedback gain must be determined within certain limits so as to eliminate wheel speed fluctuations. With this type of gain, a problem of slow responsiveness results when the on-road and/or off-road surface conditions change unexpectedly causing rapid variation in the disturbance to the ASIC conventional PID μController.

Realizing that on-road and/or off-road surface conditions (on-road and/or off-road friction coefficient) are unintelligible or uncertain factors which cannot be accurately measured, Matsumoto[13] began to apply fuzzy logic in order to separate the two problems discussed above.

An HD perceives on-road and/or off-road surface conditions by visual sensation and senses wheel slippage by bodily sensation. By comparison with past experience, the HD decides on a given amount of the braking force to be applied, that is, left-foot brake pedal pressure. While fuzzy logic is applicable to qualitative approaches, it can replace perception and sense. Thus, a common on-board AI hypersphere ECI including AI sphere ECUs based on the ASIC NF PID μControllers ought to be used. The drum-, disc- or ring-brake actuator voltages will be measured and supplied to the AI sphere ECU based on the ASIC NF PID μController as analogue voltage signals, which are then converted by high-speed high-resolution analogue-to-digital (A/D) converters. The ASIC conventional PID μController based on the optimal one may be applied to the part that can be modelled mathematically. On the other hand, a ASIC NF PID μController ought to be applied to the part that cannot be modelled mathematically, containing non-linear elements. It will be described by IF-THAN-ELSE rules.

Fundamentally, in case of braking the IRV, the front SA^2Ws must lock earlier than the rear ones; otherwise, the IRV's stability is dangerously lost and the vehicle would skid, because the locked tyre cannot contribute to side-stability. Thus, yaw torques cannot be balanced. Consequently, in IRVs without the anti-lock and anti-spin 4WB dispulsion sphere, the braking force between the front and rear SA^2Ws is distributed from approximately 0.7 to 0.3.[14] If an anti-wheel-lock and anti-wheel-spin 2WB dispulsion sphere is built in, it acts always on the RWD, never only on to the FWD.

The co-ordination of a recuperative braking with the SA^2Ws acting as the AC-to-DC ASIM MCommutator M-E wheel-hub generators in order to optimize recuperation of kinetic mechanical energy is very difficult for FWD or RWD propulsion spheres, especially for RWD ones.

In addition, in IRVs with a FWD or RWD propulsion sphere, the kinetic mechanical energy contribution of the specific kinetic mechanical energy (SKME) can only partly be recuperated into the CH-ES, that is the DC CH-E SB and/or the M-ES, that is the high-density mechanical energy-storing high-speed autodrive- and autoabsorbable flywheel (A^2F) with the DC-to-AC/AC-to-DC ASIM MCommutator E twin-flywheel-hub motor/generator.

Series hybrid 4WD propulsion spheres offer the possibility of recuperative braking, that is recuperation and conversion of kinetic mechanical energy at each SA^2W until the maximum torque and/or limit of adhesion is reached. Then, the activity of E-M drum, disc or ring brakes must also be controlled by an anti-wheel-lock and anti-wheel-spin 4WB dispulsion sphere to avoid locking of a single or all SA^2Ws. The optimization of recuperation is only useful if many changes in the vehicle speed occur and/or distance between two following stops of the IRV is short. Otherwise, the SKME which can be used for recuperation, is negligible compared with the specific basic mechanical energy (SBME), which is needed to overcome the rolling resistances of S&MWs. This means, a 4WD propulsion sphere that is optimized towards recuperative braking is especially interesting for IRVs, which are used in urban and rural areas.

5.4 CONCLUSIONS

In the IRV of the future, axleless full-time chassis motion (AWS conversion × AWA suspension × AWD propulsion × AWB dispulsion) and in-vehicle route guidance navigation and collision avoidance spheres might not only do much of the work, they might also monitor themselves, respond adaptability to changing demands and emergencies, and have the intelligence to keep the heterogeneous discrete automotive functional spheres operating smoothly and efficiently. Such is my hope as I work towards this goal using several AI techniques. Telerobotic-driven unmanned IRVs will never become a reality in on-road and/or off-road transport, as there will always be situations in which decisions must be left to the HD. Because of this, automotive high-tech will continue to give the HD final responsibility for their decisions.

This chapter was a brief and very personal selection of some aspects of intelligent automotive systems tackled using a systems approach. A full and complete point of view can be achieved from a study of the increasing amount of literature on this subject. I have little doubt that much of the material presented herein will have been subsumed by more general ones, or perhaps by an alternative of more evident utility, even before this book is published.

REFERENCES

1 Zulawski J. *Na Srebrnym Globie*, Tom I, (*On the Silver Globe*), Vol. 1. 1903. (in Polish).

2 Fijalkowski BT, Krosnicki JW. Concepts of electronically-controlled electromechanical/ mechanoelectrical steer-, autodrive- and autoabsorbable wheels for environmentally-friendly tri-mode supercars. *Journal of Circuits, Systems and Computers* 1994; **4**(4): 501–516.

3 Fijalkowski BT. The concept of a high performance all-round energy efficient mechatronically-controlled tri-mode supercar. *Journal of Circuits, Systems and Computers*, 1995; 5(1): 93–107.

4 Yamakada M, Kadomukai Y. A jerk sensor and its application to vehicle motion control system. *Proceedings for the Dedicated Conferences on Mechatronics & Supercomputing Applications in the Transportation Industries in conjunction with the 27th ISATA*, Aachen, Germany, 1994.

5 Fijalkowski BT. Very advanced propulsion automotive gas turbines power boosting with low specific fuel consumption and low emission capability. *Proceedings of AUTOTECHNOLOGIES '85: International Forum on New Autotechnologies*, Monte Carlo, Monaco, 1985, GT.1–GT.21.

6 Fijalkowski BT. Future hybrid electromechanical very advanced propulsion systems for civilian wheeled and tracked all-terrain vehicles with extremely high mobility. *Proceedings of EVS 8: The 8th International Electric Vehicle Symposium*, Washington, DC, 1986, 428–443.

7 Yokoya Y, Kizu R, Kawaguchi H, Ohashi K, Ohno K. Integrated control system between active control suspension and four wheel steering for the 1989 CELICA. *SAE Technical Paper Series*, 901748, 87–102.

8 Fijalkowski BT, Mechatronically-controlled and electrically-powered four-wheel-steering conversion spheres. *Proceedings for the Dedicated Conference on Mechatronics in conjunction with the 26th ISATA*, Aachen Germany, 1993, Paper 93ME010, 297–304.

9 Irie N, Shibahata Y, Ito H, Uno T. HICAS — Improvement of vehicle stability and controllability by rear suspension steering characteristics. *Proceedings of the FISITA 21st Congress*, 1986; Paper 865114, 2.81–2.88.

10 Hosaka A, Taniguchi M, Ueki S, Kurami K, Hatori A, Yamada K. Steering control of an autonomous vehicle using a fuzzy logic controller. *Proc IMechE* 1989; C391/005, 291–296.

11 Inoue H. Allradlenksystem im Toyota Soarer. *HDT-Tagung Allradlenksysteme bei Personnenwagen*. Essen, Germany, 1991.

12 Jordan TC, Shaw MT. Electrorheology. *IEEE Transactions on Electrical Insulation* 1989; 24(5): 849–878.

13 Matsumoto N, Kuraoka H, Ohka N, Ohba M, Tabe T. Expert antiskid system. *Proceedings of SPIE, Vol 857, IECON'87: Automated Design and Manufacturing*, Cambridge: MA, 1987, 810–816.

14 Newton K, Steeds W, Garret TK. *The Motor Vehicle*. London: Butterworths, 1987.

6

Trends in Traction Control for Improved Vehicle Dynamics

6.1 INTRODUCTION

Traction control made an appearance as one of the vehicle control systems in Formula 1 car racing during the 1992 season. The development and application of new technologies such as this is also flourishing in mass-production vehicles. As a result of the multiplicity of driving stages, it has been said that the heuristic necessity of such a control system and its application occurred in this area earlier than in the racing scene. In particular, there are now countless varieties of control technologies that aim at improving vehicle dynamic performance, involving the brake steering traction and suspension systems, or even just the traction system by itself. On top of this, the increase in the concerns of society in recent years regarding vehicle safety demands a high degree of integration with the performance for vehicle dynamics to give the degree of "driving pleasure" that drivers have previously been looking for.

This Chapter deals, first, with the current history of the development of traction control technology and examines the issues left unsolved by this technology. It then examines the latest traction control systems based on actual implementations of these systems. Following this, it gives an overview of the way in which the fundamental form of ideas for control have been introduced into traction control, and how they have been used to test the role and patterns of current traction control systems in order to uncover their meaning. The hierarchical structure which is inherent in the control logic of this fundamental form is also analysed to give an insight into the relationship between the control system on onc hand and the driver and society on the other. Lastly, it discusses how traction control systems have been incorporated

Advanced Vehicles and Infrastructure Systems, Edited by C.O. Nwagboso

into overall vehicle dynamics control technology, while also giving an overview of future developments and trends in control technology and the main themes which should be dealt with.

6.2 HISTORY OF TRACTION CONTROL SYSTEMS

If the accelerator pedal is used excessively when driving on slippery road surfaces, such as snowy roads, the driven wheels start to spin and the vehicle does not travel as intended by the driver, or the steering wheel jerks to one side and the vehicle becomes unstable, something that is often experienced by people driving in snowy areas. It was largely because of such road surfaces which provided only a small frictional force between the tyres and the road surface that four-wheel-drive (4WD) systems came to be used, in order to compensate for this insufficient traction force.

Incidentally, insufficient frictional force on snowy roads is an obstacle to driving safety more during braking than during normal driving. As four-wheel-brake systems are now implemented on virtually all vehicles, preventing the wheels from locking in order to absorb efficiently the energy of movement of the vehicle when braking, while also maintaining handling performance, became an important implementation feature. To achieve this, first, anti-lock brake system (ABS) technology was taken from technology used in the aircraft industry and adapted, with actual application in vehicles beginning in the last half of the 1970s in Europe, and in Japan from the beginning of the 1980s. A short time after the introduction of ABS, traction control systems (TCS) that prevented the driven wheels from spinning were designed and publicized[1] in the early part of the 1970s as a technology that was as important as ABS. However, this technology was finally implemented in a real way only after the application of the ABS electronic control technology in road vehicles, and was commercialized[2] during the mid-1980s in Sweden, which was one of the major centres of research into vehicle safety. During this period, four-wheel-steering (4WS) technology independently developed in Japan to improve accident-avoidance capabilities, and electronically controlled 4WS was also being inplemented. In addition, electronically controlled technology for vehicles that began with ABS has thrust itself into the age of fully fledged control technology using digital computerization during the latter part of the 1980s.

TCS began with engine control, but has been expanded and developed over time to cover a variety of types of systems utilizing the drive transmission system and the brake system, and was implemented in Europe and Japan at almost exactly the same time. On the other hand, 4WS research was one of the triggers that revived research into vehicle dynamics. This did not only mean linear handling performance, but also covered research into driving performance and braking performance in order to investigate the full range of vehicle performance in detail. The end of the 1980s saw the birth of TCS and electronically-controlled 4WDs that took into account handling performance using traction control technology, and also active suspension that controlled steering characteristics using four-wheel load control technology.

Specifically, this was used as a means to control forces acting on the tyres. Indeed, the 1980s were the dawn of the age of control of vehicle dynamics by computer.

However, because controls in each field covering the whole range of vehicle dynamics performance were developed independently, almost no consideration was given to the framework of an overall system, and some important effects were overlooked. Also, as these control systems were expanded while this state of affairs still existed, the relationship between the driver and the vehicle became more and more complicated, resulting in the problem that vehicles with fewer advantages were created. Thus, it is evident that there is a need to investigate an overall control theory that considers comprehensive vehicle dynamics. This theory includes people and vehicle systems that are called total control or cooperative control.

As we entered the nineties, two unique systems for traction control appeared from what seem to be opposite directions. On one hand, there was the "lateral acceleration regulation" type TCS that aimed for a brand new improvement in driving stability, and then there was the "driving pleasure" type automatic transmission that skilfully attempted to make the driver's task easier. These two systems hint at the future of traction control and an overall theory for vehicle dynamics.

In addition, in line with the pursuit of higher levels of vehicle performance, it has also become evident that TCS does not fully correspond to the pattern of the originally intended "foolproof" application control, which could take steps to recover from mistakes made by the driver. An example of this is where the reduction in the load on the inside wheels during lateral acceleration turning is relatively quicker than the turning movement of the vehicle. In this case, the wheels soon start to spin and the transmission of drive power in order to come smoothly out of the turn would not be possible, or the TCS control operates too quickly and the vehicle movement characteristics desired by the driver could not be obtained. Furthermore, there are also cases of impetus being applied by transient movement with complex time lags over a non-linear range in suspension systems, an area which is always seeing continual improvement, and in the reductions in surplus camber angle due to the recent increased flatness of tyres. Moreover, if tyre performance under dry conditions is targeted, then performance under wet conditions or on snow-covered surfaces will be affected relatively easily. This has not been done for the reason that TCS will be effective only on snow-covered roads.

Thus, in line with improvements in vehicle performance, TCS could not provide complete control logic within the range of a single physical quantity such as wheel rotating speed. If this is termed "micro-control", then it has become necessary to estimate the "macro" driving view, i.e. vehicle movement, and apply this to the control logic. This can also be referred to as progression to a stage where control logic should be extended from a mechanism to prevent wheel slippage to a vehicle TCS. As a part of the man–vehicle–environment relationship and as one of the control systems implemented in vehicles, TCS can be said to be moving from the field of traction performance to provide assistance in the harmonization between the whole of this relationship. From this point of view also, we believe that it is necessary to question anew the role of TCS in the context in which it has been placed.

6.3 RECENT TRENDS IN TRACTION CONTROL

Implementation of a typical "foolproof" type TCS to keep the slip ratio of the driven wheels within a constant range has been proceeding. Therefore, let us take an overview of the technological trends in this field since 1989, when systems which incorporated a variety of new features[3] were introduced. First, the way in which current vehicle control systems are handled in today's transportation society is shown in Figure 6.1.

This figure shows the roles and relationships of the vehicle controller in a mobile society, and this society seems to consists of mainly five elements, i.e., "man" "vehicle" "controller" "environment" and "traffic society". Each of the elements seems to have the principal and subordinate relationships indicated by the size of the elements in the figure.

For example, the smallest element in this system, i.e. the controller, is presently almost completely dependent on the vehicle, with the amounts of signals received from the person and feedback signals being very small. As a result, even if the controller compensates for a mistake in the car arising from a human, environmental or mechanical factor, the control barely exceeds the ability of the vehicle itself to exert a direct influence on the performance of the whole system.

People are the main factor behind traffic on roads. Even so, for a large number of people when they are driving a car, what can be termed the self-awareness and subjectivity of their consciousness as a driver and the unconsciousness and habituality of their consciousness as a passenger are not clearly separated. Thus, the data signal from the controller is usually barely considered, and while the steering and inertial

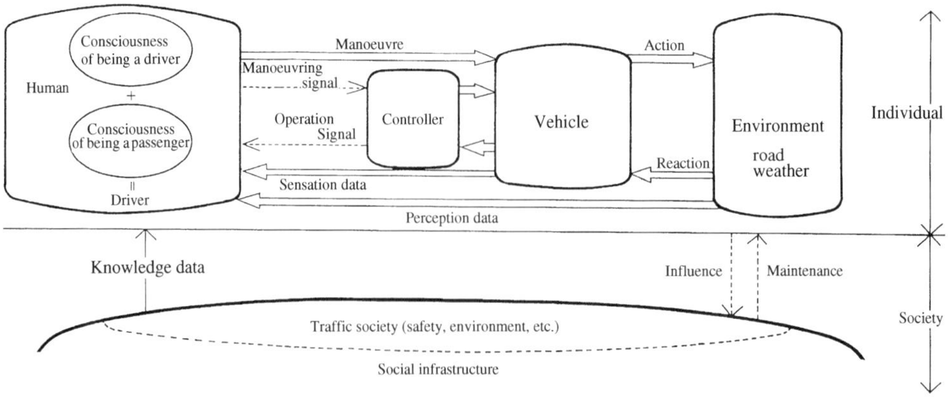

Figure 6.1. Conceptual view of vehicle control system in traffic society

force from the vehicle are felt tactually to a slight degree, the change in behaviour of the vehicle is visually appreciated mainly from the road and the view from the vehicle. The driver does not objectively understand the conditions of the road and the weather, and compares them with traffic information and consciousness gained from past experience. When the driver's consciousness is strongly that of a passenger, he perform his programmed operation and if a mistake or an unexpected condition occurs, a correction is made based on the feeling of responsibility as a driver.

Based on the state of these mutual activities, let us consider how current TCSs function and for what purposes. Systems A through D which are detailed hereunder are new types of traction control systems developed since 1989.

6.3.1 Nissan ETS System (System A)

This is a 4WD system, with variable torque distribution to the front and rear wheels.[4]

The features of this system can be broadly stated as follows:

(1) adoption of a 4WD system based on rear-wheel drive;

(2) traction force distribution control based on the difference in rotation speed between the front and rear wheels;

(3) optimized control under a variety of conditions by means of lateral acceleration detection;

(4) all-round control in combination with a four-wheel anti-lock brake system.

Of these, the features (1), (2) and (3) related to the subject of this chapter are explained in Figure 6.2.

First, the difference in rotation speed between the front and rear wheels (ΔV_H) occurring as a result of the side slip angle of the vehicle only is calculated from the lateral acceleration and the vehicle speed signal. Next, the difference in rotation speed between the front and rear wheels $\Delta V_W (= V_{WR} - V_{WF} - \Delta V_H)$ occurring as a result of braking and traction torque is calculated. Then the coefficients K_1 and K_2 for the lateral acceleration (K_1 is the gain when the rotation speed of the rear wheels is greater, and K_2 is for when it is smaller) are included to calculate the basic control command value (T_v: torque distribution signal to the front wheels). As shown in Figure 6.2, K_1 and K_2 decrease as the lateral acceleration increases. Because of this, when the vehicle is turning on roads with a high frictional surface, the torque distribution to the front wheels is reduced, so that the lateral force of the front wheels is approximately maintained, while the rear wheel torque can be increased or decreased to allow the lateral force of the rear wheels to be controlled by the drive power. As a result, the driver can actively control the vehicle's behaviour by their accelerator operation.

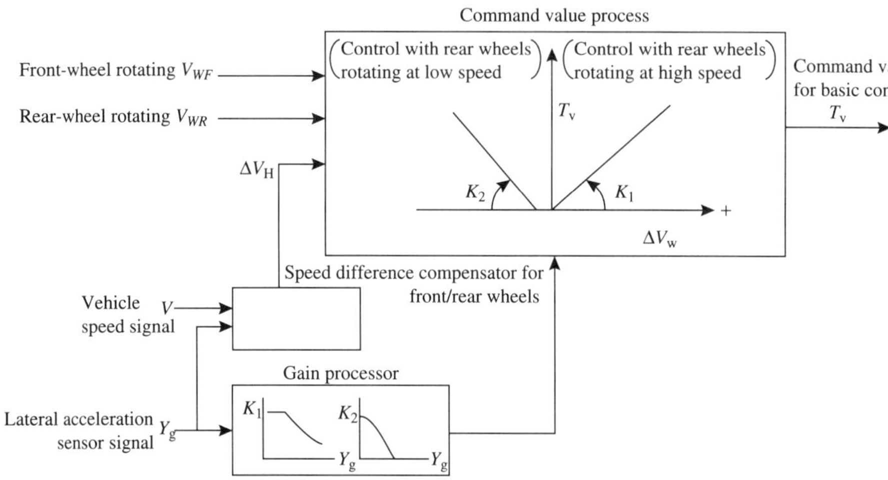

Figure 6.2. Basic control block diagram of Nissan E-TS (System A)

6.3.2 Porcshe Tiptronic System (System B)

This system is basically the same as the automatic transmissions used up to the present.[5]

A point to note is the AT shift control logic. Data processing by the control unit with regard to this is shown in the block diagram in Figure 6.3.

Firstly there are the gear shifting characteristics. In this system, the mode selection is automatic within the range from SK1 (economy) to SK5 (sporty) by means of the opening and closing speed and the position of the throttle valve. The result of this is that if the opening and closing speed of the throttle valve is high, the SK5 (sporty) shifting pattern is selected and on entering a corner upshifting is prevented. Then, during cornering, shifting changes are prevented for a uniform period of time, depending on the amount of lateral acceleration. Using this control, the driver's requirements can be reflected in the timing of the gear shifting.

Additionally, in decelerating on roads with low frictional coefficients, if the rear wheels tend to become locked, the transmission changes up to reduce deceleration torque in order to prevent a reduction in the lateral force of the rear wheels.

6.3.3 Mitsubishi TCL System (System C)

As shown in Figure 6.4, this system consists of two main controls.[6] The first is a normal traction control function that keeps the slip ratio within a constant range (slip control). The second function improves the trace performance by controlling excessive traction caused when the accelerator pedal is depressed too far during turning (trace control). The nature of this trace control is as follows.

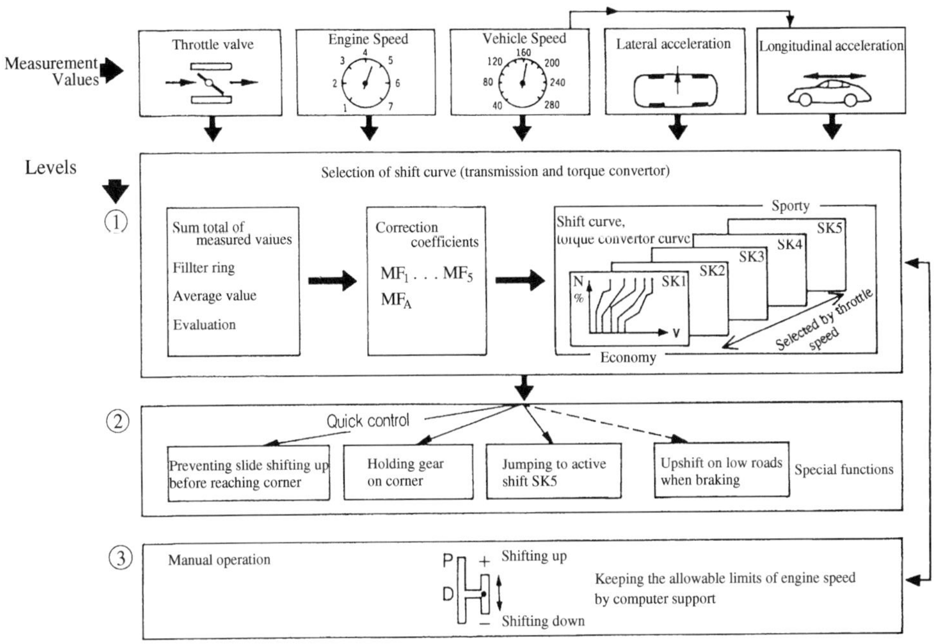

Figure 6.3. Control block diagram of Porsche Tiptronic (System B)

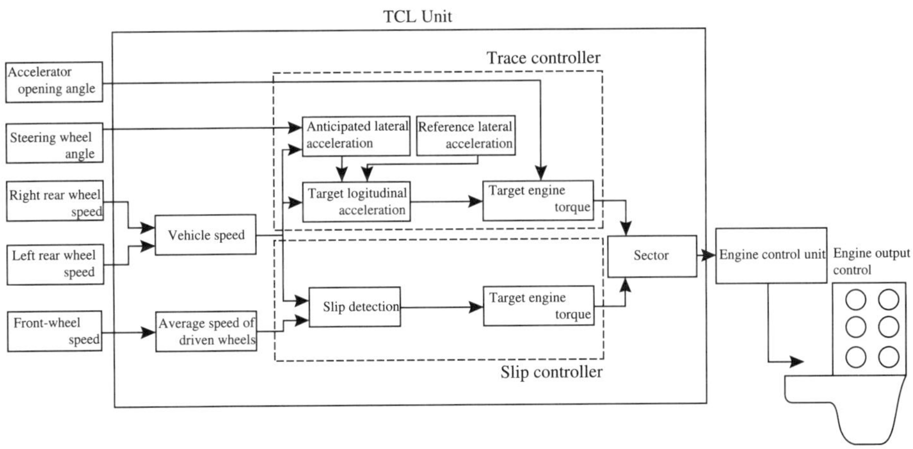

Figure 6.4. Control block diagram of Mitsubishi TCL (System C)

1. The anticipated lateral acceleration is obtained from the steering wheel angle and vehicle speed.

2. The level of lateral acceleration at which the driver can drive with confidence in relation to the vehicle speed (within the tyre gripping range) is initially set as the reference degree of lateral acceleration. The target engine torque is then

reduced when the anticipated lateral acceleration exceeds the reference lateral acceleration.

3. A revision factor resulting from the amount of acceleration intended by the driver, obtained from the throttle opening angle and the engine speed, is added to this target engine torque to obtain the target engine torque for trace control.

With this, acceleration during turning is maintained at the level of lateral acceleration with which the driver can drive with confidence, and the stable trace control during cornering is made possible.

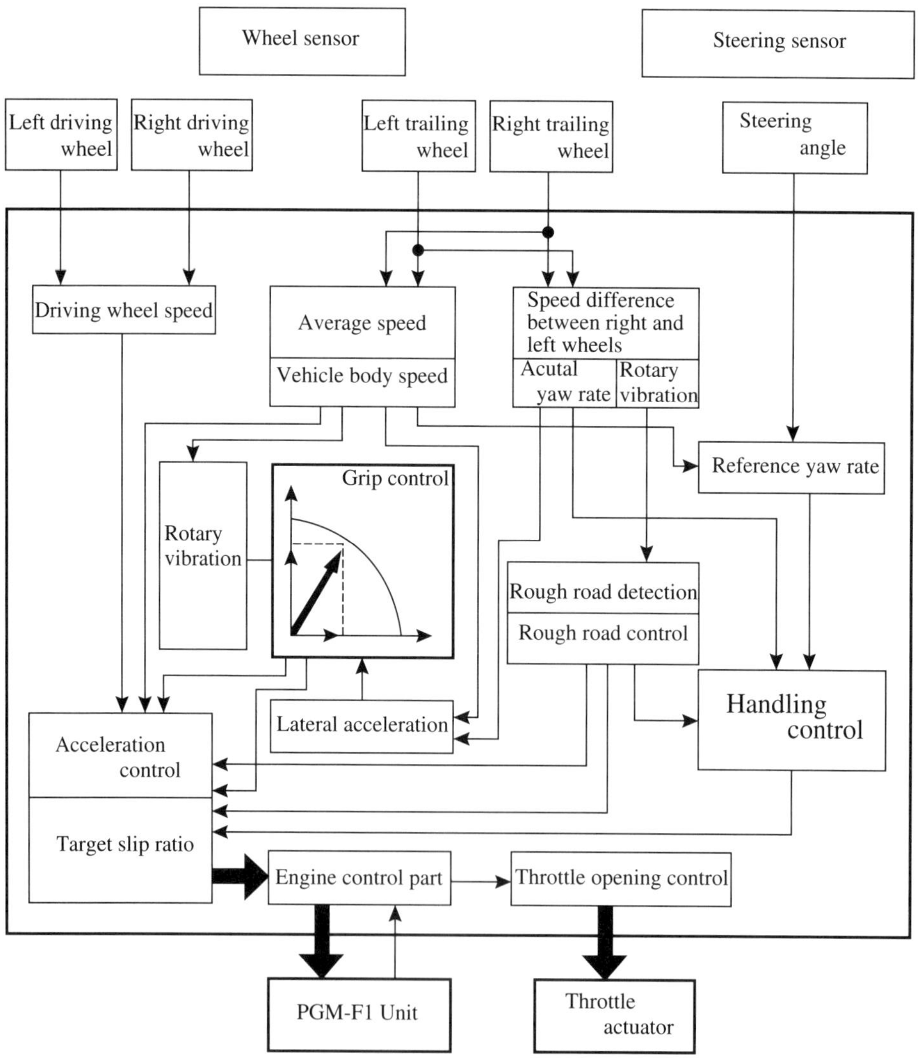

Figure 6.5. Control block diagram of Honda TCS (System D)

6.3.4 Honda TCS System (System D)

As shown in Figure 6.5, this system consists of four control sections.[7] Those related to the topic of this chapter are the acceleration control, the handling control and the grip control sections. The main section is that of acceleration control, which performs the function of typical slip control.

The handling control section, which provides one of the subcontrols, section which calculates the amount of transient understeer/oversteer characteristics and sends the result to the acceleration control section. This value is used to change the target value and the control gain for the slip control to provide slip control that takes into account the changes in steering characteristics.

Another subcontrol is the grip control section. This control section makes an estimate of the vehicle's longitudinal acceleration and lateral acceleration in order to monitor the force of the rear wheel tyres. Then, after the lateral acceleration is multiplied by the weighted coefficient for the rear wheel load, the result is squared and added to the square of the longitudinal acceleration. The square root of the resulting total gives an estimate of the value of the frictional force, which further gives an estimate of the friction coefficient between the tyres and the road surface. These data are used to minimize the target slip ratio when driving on slippery road surfaces such as icy roads to get the maximum degree of performance.

6.4 FUNDAMENTAL FORM OF TRACTION CONTROL

The above four types of TCS consist of a variety of internal control sections, with mutual principal and subordinate, or equal relationships between these control sections. The way in which each control section can be broken down into basic patterns is explained below. The assumptions behind the pattern of these controls are shown in Figure 6.6.

The vertical axis in this figure represents the control variables. The controlled dynamics of the vehicle become more stable in from the bottom to the top. Incidentally, if the vehicle movement becomes more stable and movement changes are reduced, the driver's focusing image states of the vehicle motion (operation selection range) dissipate in the background of the driver's consciousness. As a result, the driver's subjective manoeuvring is reduced. Therefore, it can be said that stabilization of the vehicle motion is equivalent to a decrease in the driving information which cause disparities in the actions of the driver. Accordingly, this vertical axis corresponds to (the reciprocal value of) the driving information in the human–vehicle system. The horizontal axis represents the dynamic state variables. From left to right, the vehicle characteristics change progressively from linear to non-linear until the limit is reached. The meaning attached to this axis is that from the left to right, the driving mentality changes from safety to comfort, to pleasure and then to danger; the driving mode changes from economy to sporty. These

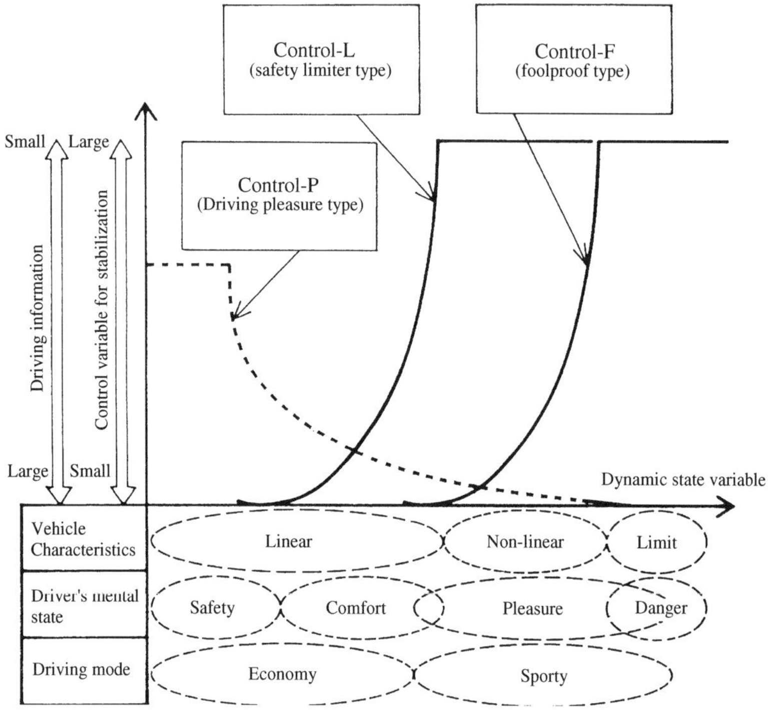

Figure 6.6. Three fundamental forms of control

curves are shown in Figure 6.6. Each is given its own name below (P = pleasure, F = foolproof, L = limiter), and these form the basic pattern of control.

(1) *P* control (driving pleasure type);

(2) *F* control (robust foolproof type);

(3) *L* control (safety limiter type).

A solid line indicates that control and information transfer occurs as indicated by the line in proportion to the state variable. A dotted line shows that control is performed with conditions attached, or that either control or information transfer fails to occur.

Control *P* reduces the control variable for stabilizing the vehicle behaviour, in proportion to the dynamic state variable by which the vehicle characteristics approach the non-linear range while moving towards the limit. However, the driving information is increased. For instance, the driver can predict that the limit is being reached when the lateral *G* force moving the vehicle becomes non-linear, because the driving information is increasing. Therefore, before the controller requests *F* control, i.e., foolproof control, it is the driver's skill which is responsible for controlling the vehicle. Thus, *P* control is regarded as a kind of man–machine system in which the driver also actively participates in a part of the controller's action.

The shape of *L* control in the figure is almost symmetrical that of *P* control. This means that *L* control stabilizes the vehicle motion far earlier than *F* control; for example, at the stage where the vehicle dynamics exceed the linear region, in order to achieve a reliable degree of driving safety that does not rely on the driver. This is therefore a safety limiter type of control that gives priority to social responsibility over the context of the individual inclinations of the driver. Thus, *L* control gives insufficient information for driving from the point of view of the driver's consciousness, and *P* control gives insufficient control for stabilizing.

6.5 CLASSIFICATION OF TYPES OF TRACTION CONTROL

The control features from system A to system D are discussed using three basic types of traction control. Figure 6.7 shows how the basic type of control is applied to the main control and subcontrol of each system, and classifies the results.

The main control is something that can be brought into effect independently. When the main control is not functioning, the subcontrol will not be brought into effect. In Figure 6.7, each system is characterized by symbols to link the basic patterns of control (*P*, *F*, *L*). The symbol *x* links the main control on the left and subcontrol on the right in the manner of a logical AND expression. The symbol + means the sum of the essential elements of control for which the type of control variable is in the same level, but the point at which functioning occurs is different.

According to this method of classification, it is possible to say that the composite pattern of system A is similar to that of system D. In system B, the main control and subcontrol types are the reverse of those in system A. However, in system C, the main control is the same as that of system A and it has the special feature of being equipped with *L* control which can be operated independently of the main control.

In this way, the four representative TCSs from A to D can be narrowed down to systems A, B and C, which possess various unique features.

As shown in Figure 6.8, as system A does not normally control to stabilize the vehicle behaviour, it is able to transmit satisfactory information to the driver through the basic chassis performance. When the vehicle motion enters the non-linear domain and the changes in the vehicle behaviour become non-uniform, the system begins to operate. Once control begins, it is implemented so that the driver can continue their driving pleasure, and in an emergency it operates as a foolproof system. In short, system A seeks to strike a balance and rhythm between safety and pleasure, while taking the relationship between human and vehicle into consideration. Issues which still remain to be resolved include such things as whether foolproofing will be brought into effect to its maximum extent after changing to 4WD, and whether the system will be able to appropriately select the points of convergence noted

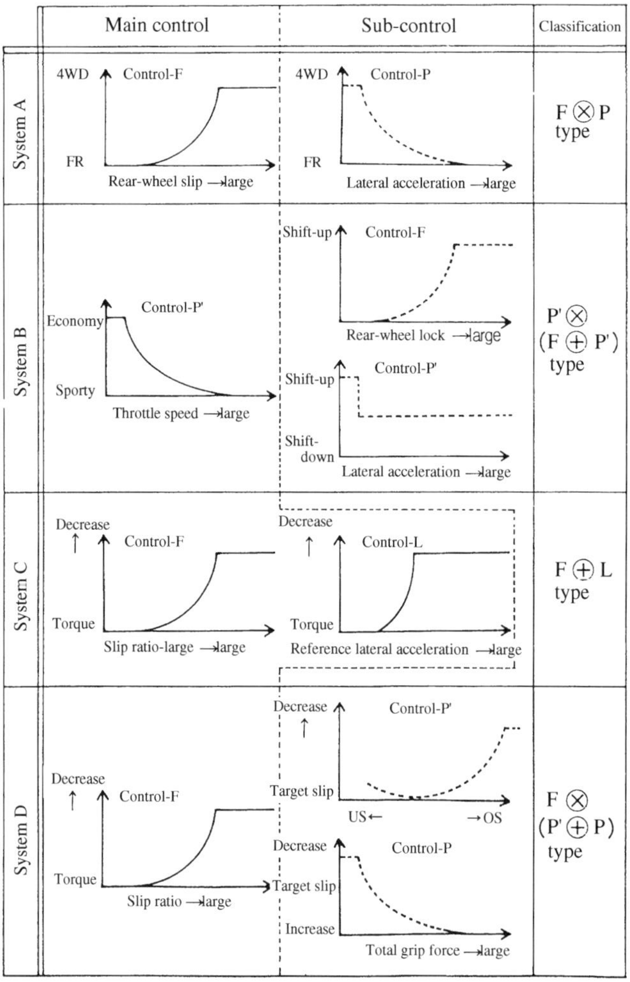

Figure 6.7. Classification of traction control systems

above during driving on a variety of road surfaces and in a variety of driving situations.

System B is a type of control which is the reverse of that of system A. This is because system B is not intended to be a dynamic control system but a manoeuvring operation system. Figure 6.8 also shows the control action of system B. This system continues the open-loop control for normal driving conditions and strives to bring pleasure to the driver. However, when vehicle movement enters a certain domain, control is stopped so that the control function itself does not disturb the vehicle dynamics. If the movement exceeds this domain, a foolproof operation will be attempted even though the control might not be perfect. In short, the manoeuvring operation system also tries to ensure that the balance between safety and pleasure will not be lost. As

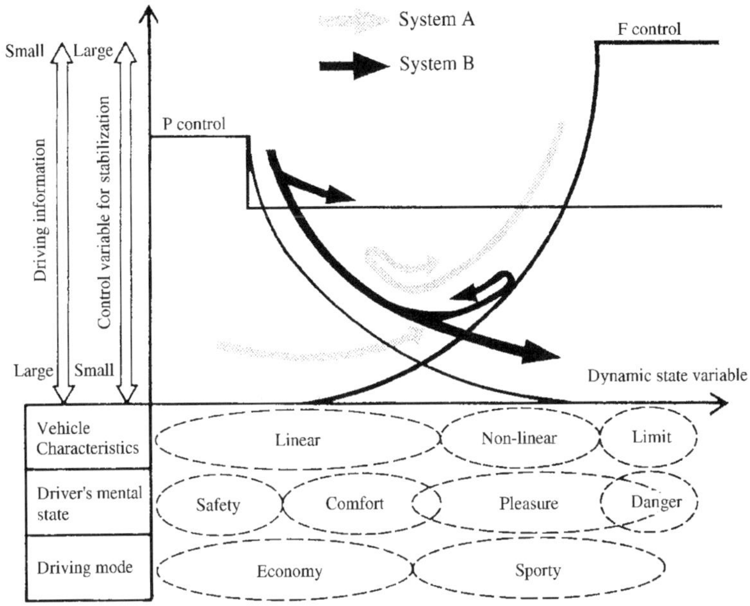

Figure 6.8. Theory of balance between safety and pleasure

noted previously, system C has the same foolproof control as system A but has one more main control. This system is thus a type of safety limiter which gives preference to the transportation society rather than to individual driving pleasure, and thus operates in a different control context. As mentioned before. system D resembles system A in as much as the main control is a foolproof control. However, the subcontrol is a pleasure type by effectively augmenting the performance of the tyres, and it is a foolproof type by limiting vehicle movement. So in these respects it can be differentiated from system A. The structure of system D will be explained in more detail in the next section.

6.6 HIERARCHICAL RELATIONSHIPS IN TRACTION CONTROL SYSTEMS

As explained in the previous section, by using a classification system for the fundamental forms of traction control, one aspect of the meaning behind the operation of each type of traction control system can be expressed. Nevertheless, the meaning of individual operations under a variety of driving situations is supported by larger meanings which encompass the whole domain of smaller factors. These meanings can then be considered to consist of even greater and deeper meanings which are linked in the manner of a hierarchy. Accordingly, the fundamental forms of each type of control element will be treated as a hierarchical construction, and the meaning and role of each will be re-examined.

By this means, we believe that the position of and relationship between control systems which are currently being used in actual vehicles can be understood clearly.

First, there are three logical layers in the hierarchy which can be considered according to the limits of the extent of the world as perceived by the driver, or in other words, the extended processes of the limbs, body and the outside world, from the centre and extending toward the periphery. They are;

(1) the sliding movement of the tyres, which the driver is relying upon and manipulating;

(2) the movement of the vehicle which the driver is driving; and

(3) the natural environment and the transportation society in the midst of which the driver finds him/herself.

Following this, the provision of control elements is carried out, with system D[7] given as the example (refer to Figure 6.9). In the control logic for the TCS, the initial layer of the fundamental form of control consists of:

(1-A) Oversuppression control of slip ratio limit.

Then, the control element referred to as "handling control" is also utilized, but this is a form of control for determining the target value for the slip ratio of the tyres while control is being performed. It has two functions as follows:

(1-B) variable control of target slip ratio which is within the first layer of the hierarchy; and

(2-A) reference vehicle response following system which is in the second layer, and has the role of maintenance of order in this layer.

Next, the control element which is known by the name of "grip control" is also utilized. The role of this element is in the rationalization of elements such as the target slip ratio in response to the grip conditions between the tyres and the road surface detected by the overall acceleration (lateral acceleration in particular) of the vehicle. If this is looked at in a different way, it can also be called a control which varies the reference vehicle response in accordance with the vehicle movement conditions such as the lateral acceleration of the vehicle. Accordingly, this control can also termed:

(2-B) variable control for reference vehicle response, which is also in the second layer of the hierarchy.

There is a certain range within which controls listed up until now are technically possible as controls inside the vehicle. If the levels are exceeded, a solution would no longer be possible with purely technical arguments and methods.

Below, we examine the third layer of the hierarchy. If driving safety in the transportation society is taken from the point of view of vehicle driving conditions, then the existence of:

(3-A) oversuppression control of vehicle driving conditions could be established like an *L* control; but (3-B) variable control of vehicle driving conditions (driver's freedom with responsibility) also exists and the two must be balanced. These should be determined primarily on various levels such as the customs, governments, religions and culture of each respective country, and it is considered that they could not be represented on a conventional scientific and technical level except by way of concrete methods and procedures.

Incidentally, if these relationships are shown graphically (Figure 6.9), issues that we as engineers are currently dealing with and the issues that need to be dealt with in the future, or in other words the borders of technology, seem to become clearer. Before examining these issues, an explanation will be given of Figure 6.9. In this

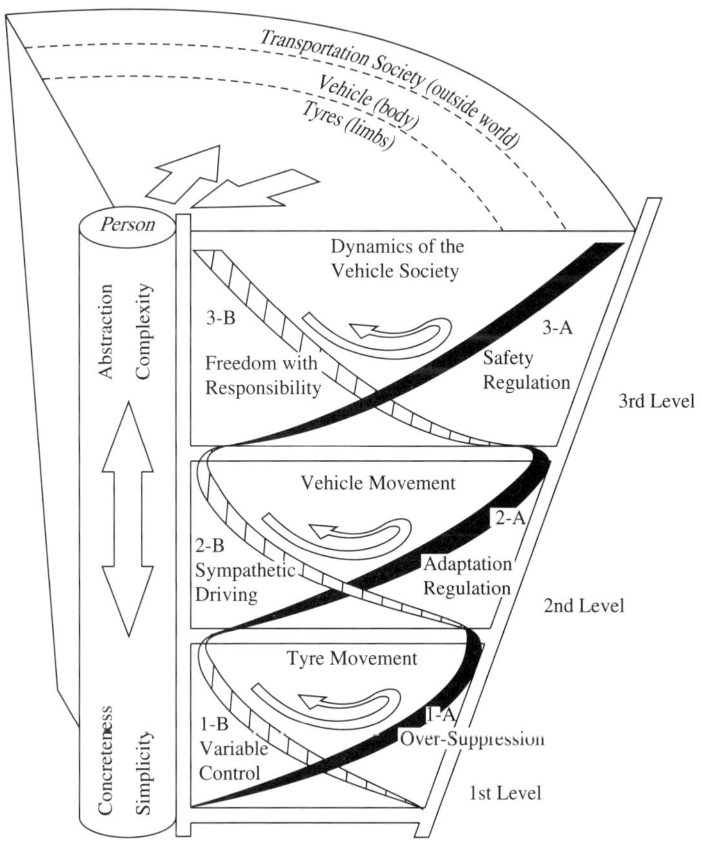

Figure 6.9. Hierarchical relationships and rhythms in control logics

figure, three logic levels are shown, and are numbered in the order 1, 2, 3 starting from the bottom. Each level consists of a pair of control elements, referred to as A and B, respectively. As explained in the previous section, each of the type A controls can be called adaptive controls in the level in which they are found. Type B controls are not only variable controls, which cause changes in the response characteristics for type A controls. They also harmonize within the driver based on higher-level logic, by letting the driver use the freedom of variable control to the level desired, even if they do not provide explanations for higher levels.

That is to say, the first logic level consists of *F* control in position 1-A, and *P* control in position 1-B, and the second logic level consists of *F* control in position 2-A and *P* control in position 2-B. Then, the third logic level consists of the society-oriented *L* control (3-A), which goes beyond the level of the individual. In the current situation where vehicles are operated by people, there is no control existing to which 3-B can be applied.

In this way, traction control operates at multiple levels within a hierarchical structure defined by vehicle movement and social dynamics (and is therefore not just concerned with tyre performance). The respective levels are considered to be mutually connected. Because the range becomes greater and the degree of complexity increases as we progress to a higher level, the control information that can be used becomes increasingly abstract, and the setting of control variables, central values and threshold values becomes more and more complex. Accordingly, in control which crosses over the boundary of a level, the restrictions of a high degree of abstractness ignore the context (meaning and process) of an intermediate level, and control may move strongly into a lower level. Because of this, a double-binded contradiction can easily occur. Moreover, in the real world these levels are stacked up in a way that surrounds the driver and they can only be perceived from the inside, so that identifying the type and numbers of levels could be considered to be very difficult.

6.7 HARMONIZING TRACTION CONTROL WITH OTHER SYSTEMS

Up to the previous section, traction control has been divided into different types based on fundamental forms, and these fundamental forms have been used to search for the meaning behind traction control from several different angles. In this section, these basic forms will be applied to the harmonization relationships between traction control and other systems, and the future of overall control systems centering on traction control will be examined.

Vehicle control systems adjust the "relationship" among the man–vehicle–society/environment system, and thus can be said to assist in the total harmonization of this relationship. Let us then position the operation of the fundamental forms of the control elements mentioned previously into the system operation process of the man–vehicle–environment relationship (Figure 6.10). Thus *P* control can

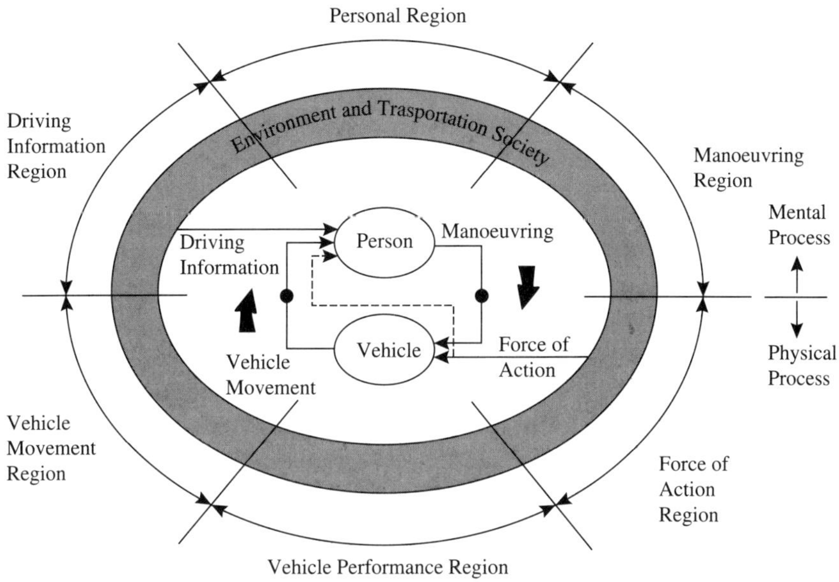

Figure 6.10. Mutual interaction in the human–vehicle–environment system

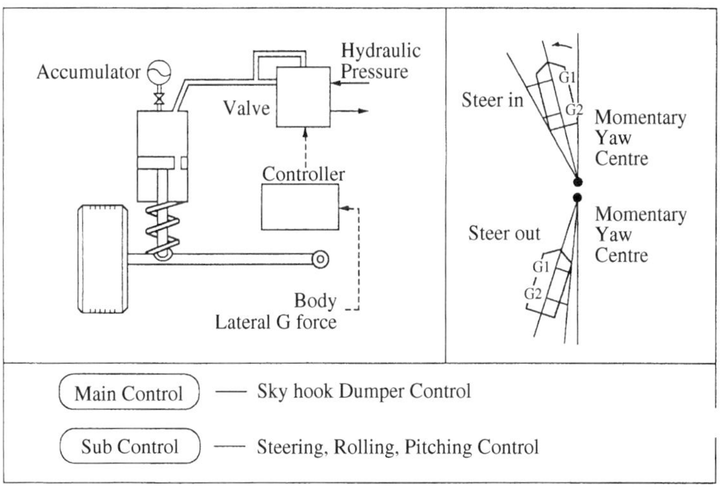

Figure 6.11. Nissan hydraulic active suspension

be considered to be the element which operates through the personal process of receiving information from the vehicle and environment resulting in manoeuvring the vehicle. *F* control is the motion process where the vehicle receives the manoeuvring force from the driver and the disturbance from the environment, and then moves. *L* control is principally the surrounding process where the environment and the transportation society feed information to the driver and external forces act upon the vehicle.

These will be considered in terms of three examples of vehicle control systems based on the three fundamental forms *P*, *F* and *L* in this changed point of view (Figures 6.11, 6.12, 6.13).

1. *Active suspension.*[8] In this system, the vehicle uses the *F* control system (as described above) to compensate for and suppress vehicle vibration caused by uncomfortable variable forces from the environment. This system carries out the main control using a skyhook damper driven by hydraulics, and added to this is the transient control of steering characteristics corresponding to *P* control, and also vehicle movement control such as roll and pitch control. This control system utilizes micro-movement control, i.e. the vertical changes to the position of the wheels, in order to try spontaneously to create a macro-ordering, such as the exacted steering characteristics and the desired vehicle movement.

2. *Yaw rate feedback-type 4WS.*[9] In 4WS also, yaw rate feedback-type 4WS has been implemented not only in order to improve vehicle steering characteristics, but also to compensate for and suppress vehicle movement tendencies arising from disturbances, such as crosswinds. In mechanical-, electrical- and hydraulic-type 4WS systems, not only is the transient control of steering characteristics corresponding to *P* control carried out, but also the effects from unexpected disturbances from the environment such as crosswinds are also compensated for as a part of *F* control. This system always attempts to realize and maintain a macro-ordering of a constant reference of steering characteristics.

3. *ABS.*[10] This system is the oldest example of *F* control, and it has become widely used. Recently, a type of ABS which senses the lateral acceleration of the vehicle has been implemented. The braking force applied to all four wheels during a turn is controlled and balanced using sensors. Accordingly, the handling performance and braking performance are both said to be maintained, and this can also be said to be the case of *P* control. Nevertheless, the essential characteristic which defines this control is that the ABS self-adjusts the micro-control, such as wheel slip control, in response to the macro-driving conditions such as the lateral acceleration of the vehicle.

As can thus be seen, *P*, *F* and *L* controls are all present in the many vehicle control systems, including TCSs, in current usage: *P* control is a control which applies desirable qualities such as a tool from the person to the vehicle; *F* control is a control which returns compensation from the vehicle to the person and the environment; and *L* control can be called a control which safely protects the person and the vehicle from the environment, in particular the transportation society. These three forms of control adjust the mutual actions between the person, vehicle and environment, respectively, and provide a total ordering and harmonization.

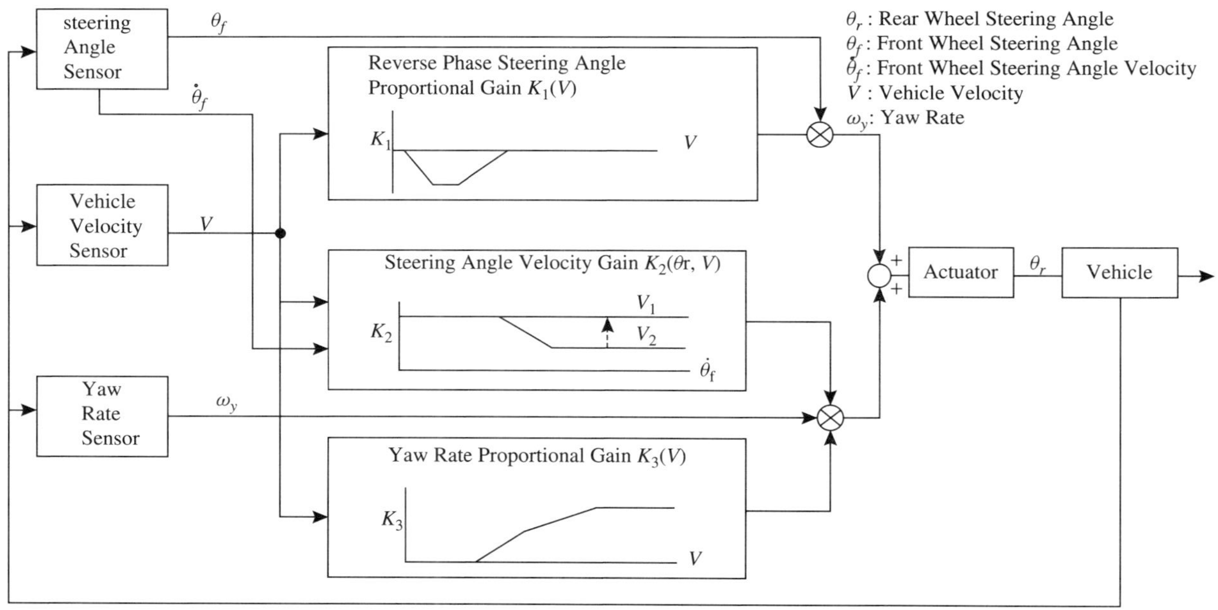

Rear Wheel Steering Angle Control

$$\theta_r = \underbrace{K_1(V).\theta_f}_{\text{Steering Angle Proportional Term}} + \underbrace{K_2(\theta_{f1}V).K_2(V).\omega_y}_{\text{Yaw Rate Proportional Term}}$$

Figure 6.12. Toyota 4WS control block diagram

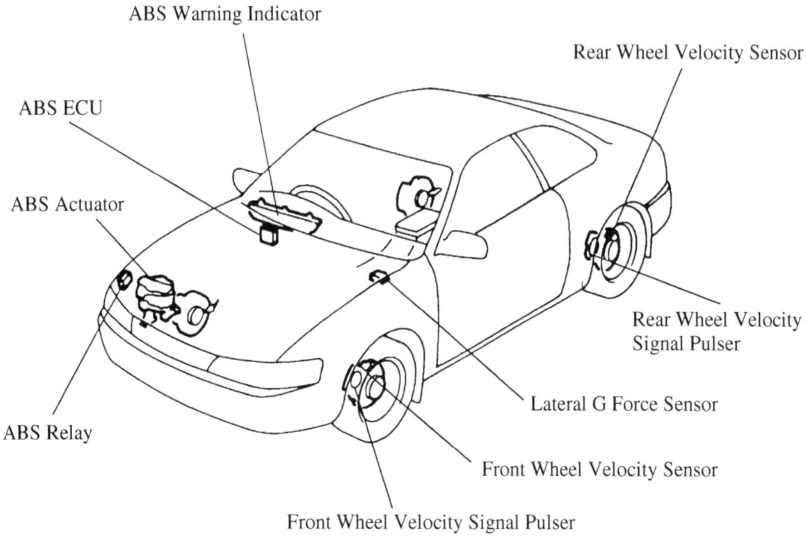

Figure 6.13. Toyota 4WS ABS

In addition to corresponding to all of the principal orthogonal components of the forces acting on the tyres, the three control systems mentioned above and the TCS (a total of four systems) are also independent in terms of their movement functions. Therefore, if these four systems are integrated to maintain the ordering in the area of vehicle dynamics in the second layer as elaborated in Section 6.6, they would be expected to form a single comprehensive harmonization control.

These four control systems have been given the following names.

(a) four-wheel anti-lock brake system (ABS);

(b) rear wheel steering system (RWS);

(c) traction control system (TCS);

(d) active suspension system (SUS).

Regarding these control systems, the four-wheel steering system (4WS) which has been implemented in actual vehicles is here included as a RWS system. That is to say, the front wheel steering system is a part of the driver's manoeuvring operation system in a wide sense, and the manoeuvring operation system in this chapter includes mechanical systems such as the steering wheel, accelerator, brakes and transmission system, and other electronic assistant systems.

The features of comprehensive control for vehicle dynamics are different from conventional total control which depends on signal transmissions between cooperating systems. The result of the elemental systems at the micro-level spontaneously generating an ordering at the macro-level is that a field is created where all elemental systems exert an effect, and through the application of this field, other elemental system groups carry out self-adjustment, so that the application starts to become apparent not only as elements but as a whole system. The ordering then becomes stronger, and a new ordering is continually being generated. That is to say, information is transmitted through this ordering, and the information is received by the union of the fields. This situation is illustrated in Figure 6.14.

The advantages of a universally comprehensive control are that signal transmission is simply rendered unnecessary; elemental systems have universal interactions, and the mechanism is thus made independent. This means that, if an element is suddenly taken out of the system, or if it is replaced with a more advanced element, or even if the number of elements is increased, only the macro-control performance is affected, and the total control will automatically continue to function as always. This is a very important quality, and can be applied to vehicle dynamics control in the following way.

As is shown in Figure 6.14, a target operation such as cornering is considered. When the full force of the brakes is applied and the vehicle enters the ABS operation range, the universal ABS first selects the optimum transient ordering of the steering characteristics for the braking action from several alternatives in line with the driving rhythm of the driver, and then commences the wheel anti-lock operation in order to maintain this ordering. At this time, the conditions of load transfer caused by braking are formulated, the resulting vehicle movement is sensed

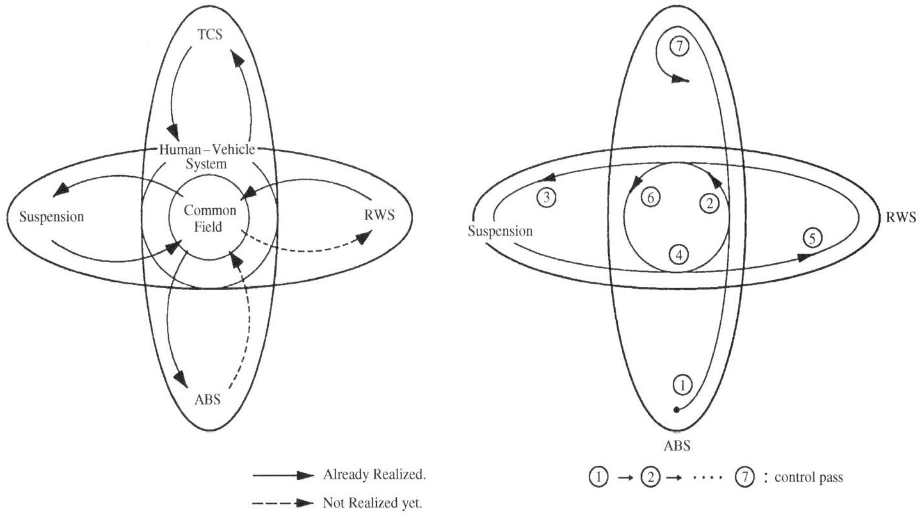

Figure 6.14. Circular-type cooperative control system

and the auto-adjusted universal rear wheel steering system and active suspension system controls are activated and the control is adjusted in response to the driving. Because this causes the optimum steering characteristics to be formulated during the cornering operation, the universal TCS which senses this in the field selects the optimum steering characteristics when straightening up after the turn in response to this field and carries out auto-adjustment control. The result not only for the TCS but also for the driver contributes in to producing a typical control according to a pattern of control; or in other words, there is a greater possibility of predicting those vehicle movements which will occur next, with no losses in the flow of overall driving during cornering. This is the reason why it is possible to achieve a high level of both handling performance and traction performance in the human–vehicle–environment system. Moreover, direct electrical signal transmissions are no longer needed in any case.

Figure 6.15 makes an assumption of what will play a part in vehicle control systems in the near future. The size of the controller in this figure is much greater than that shown in Figure 6.1. The meaning of the controller as another cultural system shows an increase in response to the requirements of individuals and society from the relationship of the existence of the vehicle as a machine. With respect to *L* control, which did not feature so prominently in the examples given earlier in this discussion, the fields where this control should be applied will become more clearly defined within the context of the maturing of society; and *L* control should make a renewed appearance in the transportation society. It is in a society in which the road environment and the traffic flow are monitored by the social computer as shown in the figure that the validity of this will be demonstrated. And what is more, as we progress to higher levels as explained in Section 6.6, the control which is the optimum target control for a level becomes more concrete, more localized and more limited in its time frame. The reason for this is not that the simple actions of people or the movements of objects themselves are good or bad, but that within the relationships and differences between these actions and movements,

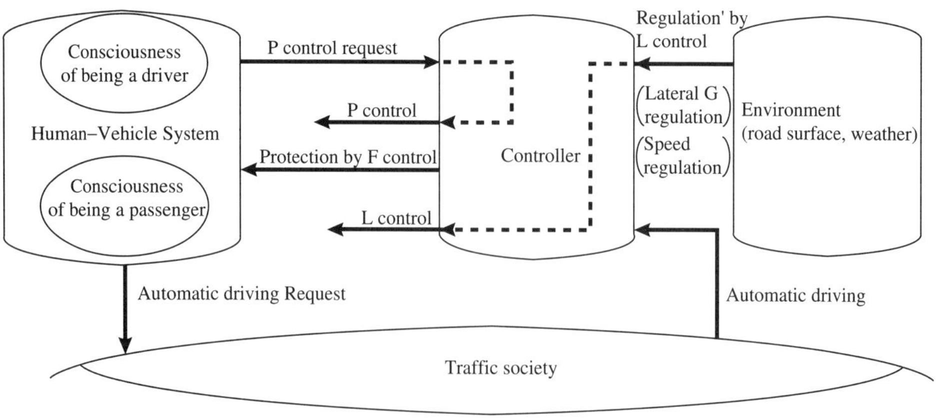

Figure 6.15. Role of the vehicle control system in the near future

the variety of other activities that support them, the background which specifies the time and place, and the context. Judgements should therefore be based on this overall perspective.

Accordingly, controllers for vehicles must progress and develop in both the vertical direction in terms of logic levels, and in the horizontal direction in terms of harmonization control; the fields of movement must become more concrete and the moving action must become more subtle and skilful.

6.8 CONCLUSIONS

Vehicles as objects and information as events are being spread throughout the world by means of large-volume reproduction and high-speed transportation systems. Therefore, our frontiers are gradually becoming lost. Environmental problems are not only at saturation point from the sheer number of problems, but from the characteristics of these systems themselves. The various problems associated with driving safety nowadays could be said to be the same. The age of horizontal development of control technology could also be said to be over. Nevertheless, the vertical depth of the control logical hierarchy is limitless, and even more so the concrete field in which this logic activates is a situation which will never be repeated. Accordingly, it is the people themselves who create these ideal hierarchies and concrete fields and recognizing and actions of people as groups which will become the focus of the frontier of technology in the future.[11] There are several huge international projects for providing harmony in the human–vehicle–environment system, which have already been started. The technology which will be necessary here should be quite a lot more than a level of simple objects and events. There is no guarantee that these projects will succeed, but in order to obtain a single foundation for transportation that will support the daily life-styles of the people from a global point of view for the first time in history, harmony between nations is no doubt necessary. Moreover, it is not only the scientific research of peoples' recognizing and actions, but the search for the meaning of human existence which is indispensable as a viewpoint for the manufacturing of vehicles. This is because the selection of each subsequent process is based upon decisions about what is pleasant or unpleasant, good or bad, which exist in the values and meanings that people perceive in between processes. Vehicles that offer support for our daily life-styles and control systems which adjust the "relationship" between people, vehicles and the environment are topics for which labours and improvements will probably never end.

REFERENCE

1 Moran JH, Grimm RA. *Max Trac-Wheel Spin Control-by Computer*. SAE 710612.

2 Lind L. The Volvo electronic traction control-concept for safe driving. *9th International Technical Conference on ESV, 1982, Kyoto.*

3 Shiraishi S, Yamamoto O, Kiryu H, Nishihara T, Sato M. Control concept for traction control system (TCS) and its performance. *12th International Technical Conference on ESV, 1989, Goteborg.*
4 Matsuda S. *et al.* Nissan ETS: a new electronic torque sprit system for improving vehicle dynamics, Pre-print of JSAE (in Japanese), No. 891074, 1989.
5 Maier U *et al.* Porsche Tiptronic. *ATZ Automobiltechnische Zeitschrft* 1991; **92**.
6 Isoda K. Traction control system for improved driving safety. Symposium of JSAE (in Japanese), '90 No. 3. 1990.
7 HONDA NSX Press Information 1990; 9, 13.
8 *Hydraulic Active Suspension* Nissan Press Information (in Japanese) 1989–9.
9 Development of four wheel steering system using yaw rate feedback control. *Toyota Technical Review* 41–1. (in Japanese)
10 *Corolla, Sprinter*. Toyota Press Information, 1991. (in Japanese)
11 Shiraishi S, Yamamoto O, Kin K, Akuta Y. Traction control for improved vehicle dynamics. *JSAE Review* 1992; **13**(1).

PART 3

AUTONOMOUS VEHICLES SYSTEMS

7

A Closed-loop Four-wheel-steering Control System: Toward an Autonomous Vehicle

7.1 INTRODUCTION

Four-wheel-steering (4WS), in which not only the front wheels but also the rear wheels of vehicle are steered actively, is a control technology developed recently for automobiles to enhance their manoeuvrability and control performance. Many Japanese car manufacturers such as Nissan, Honda, Toyota, Mitsubishi and Mazda have already introduced their 4WS products into the market to demonstrate the superior performance of 4WS vehicles in handling, stability, obstacle avoidance and easiness for parking. Among many studies on 4WS, the concept of closed-loop control has been proposed to provide the active control strategy for the vehicle, such that the automobile can be designed in a way similar to the control-configured vehicle (CCV) approach, which is currently applied to advanced aircraft designs.

As more advanced control technology being used, the vehicle may become more intelligent such that the autonomous vehicle on the highway can be realized quite soon in the near future. In fact, many research programmes such as the Intelligent Vehicle Highway System (IVHS) in the USA, Super Smart Vehicle in Japan, and PROMETHEUS (Program for a European Traffic with Highest Efficiency and Unprecedented Safety) in Europe are actively in progress. Also many conferences on intelligent vehicles have been held to discuss its future trend.

In this chapter, the 4WS technology with closed-loop feedback control is examined and studied to investigate its development as one way toward the autonomous vehicle

Advanced Vehicles and Infrastructure Systems, Edited by C.O. Nwagboso

of the future. First, 4WS is studied and compared with front wheel steering (2WS) to illustrate its design philosophy using examples. Then the closed-loop control is introduced to provide some advanced performance for the present 4WS systems based on the feedback signals measured from the vehicle motions. For developments of autonomous vehicles, a description is made on how to incorporate the closed-loop 4WS control technology with sensed information from the environment or highway to make autonomous vehicles more intelligent. Finally, conclusions are given to show the developing trends for the future.

7.2 4WS CONTROL SYSTEM

7.2.1 An Overview of the 4WS Control System

Since the first automobile appeared in the nineteenth century, 2WS has been the standard design for steering until recent years. As the improvement of vehicle performance and the accumulation of knowledge in vehicle design, the concept of active 4WS has been introduced to achieve a quicker response in steering and handling performance for course changing and safer driving.

4WS is a steering control scheme of which both front and rear wheels are steered. In this way, the steer angle of the rear wheels can be directly controlled to generate the cornering forces effectively from the rear wheels. Whereas in 2WS vehicles, the cornering force of the rear wheels can only be generated by the yaw or sideslip motions of the vehicle. Therefore, quicker response can be achieved in 4WS vehicles to provide better handling performance. Furthermore, the stability of the 4WS vehicles during braking may also be augmented by properly steering the rear wheels. The merits of the 4WS system as compared with the 2WS vehicles are listed below:

1. *Quicker response or shorter phase lag.*

4WS is commonly considered a better choice than 2WS when quicker steering responses for lane changes are needed, because 4WS can effectively generate the cornering force from its rear wheels. Sano *et al.*[1] found that the phase lag of the 4WS vehicle during lateral acceleration is less than that of the 2WS vehicle under test conditions of course changes at high speeds (80 and 100 km/h) in their static functional type of 4WS. It is also proved that the 4WS vehicle has less phase lag in the frequency response of the yaw rate to steering wheel angle by a steering force feedback-type 4WS (HICAS-High Capacity Actively Controlled Suspension).[2]

2. *Less off-tracking.*

In the cornering process, sideslip motions may have some undesirable effects and should be minimized. For example, when the vehicle is moving around a circular path, the driver would like to have a zero sideslip motion, such that the vehicle can head toward the tangent direction of the curve to reduce off-tracking. This can be easily achieved by 4WS[3], and more details will be explained in Section 7.2.2 using a computer simulation.

3. *Augmented stability.*

An important requirement for designing a steering control system is the safety or stability of the vehicle at high speeds. One key factor which causes system instability is the cornering force saturation due to the large slip angle of tyres. For 4WS vehicles, the slip angles of front and rear tyres can be controlled in a balanced way to avoid large slip angles causing the saturation of tyre force in either the front or rear wheels. 4WS vehicles can therefore augment the stability compared with the 2WS vehicle especially at high lateral acceleration manoeuvres or at high speeds.[4] Furthermore, an inherent unstable vehicle can be made stable by the feedback control.[5] In fact, the concept of so-called computer-aided steering has been suggested by Lugner *et al.*[6–8] which, independently and autonomously, steers each wheel by an on-board computer. In this way, it is possible to balance the tyre forces under different environments to avoid the saturation of tyre adhesion on any tyre, such that the stability of the vehicle can be augmented.

4. *Improvement of manoeuvrability.*

One feature of the 4WS system is that the turning radius of the vehicle is greatly reduced. Figure 7.1 makes a comparison of the turning radii of 2WS and 4WS vehicle with the same steer angle of front wheels (15.87°) at a speed of 40 km/h. In this figure, the trajectories of a 4WS vehicle and a 2WS vehicle are shown by dashed

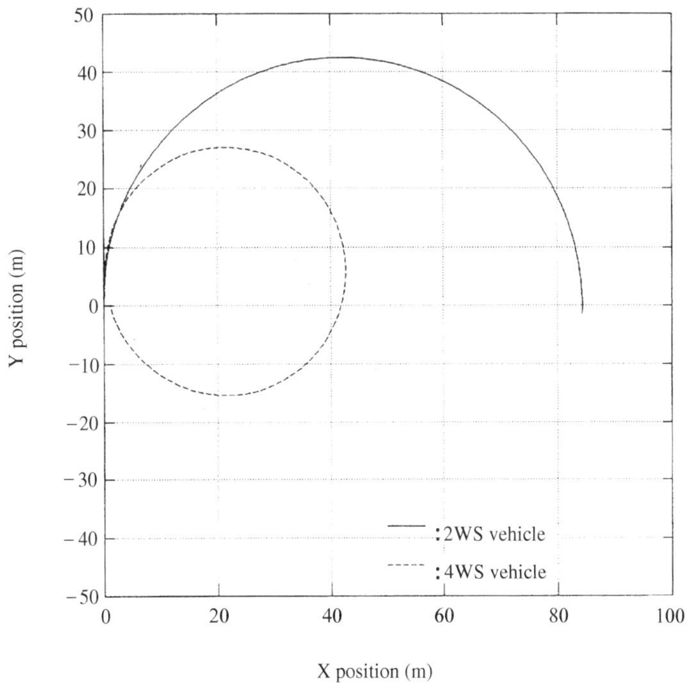

Figure 7.1. Trajectories of the 2WS and the 4WS vehicles for a circular turn

line and solid line, respectively. With steerable rear wheels, the turning radius of the 4WS vehicle is much smaller than that of the 2WS vehicle to provide better manoeuvrability.

5. *Pure sideslip motion.*
It can be a great help if the vehicle can perform a sideslip motion for some tasks. For example, parking a car would be simpler if the car could move laterally without changing the heading direction, that is, moving like a crab. On the other hand, for a lane change manoeuvre, people are accustomed to first turn the steering wheel to change the vehicle heading direction for a lateral motion, then turn the steering wheel back to make the vehicle face forward again. Turning the steering wheel and then turning it back is the only way to make lane changes for 2WS vehicles. The task can be easier if the lane change can be done by a pure sideslip motion. In fact, this can be easily attained by using 4WS as demonstrated by the simulation results in Section 7.2.2.

After introducing the general properties of 4WS and its merits above, a special design of the 4WS system, a multi-input 4WS vehicle, is introduced below to provide some advanced performances made possible by the 4WS system as illustrated by computer simulation results.

7.2.2 A Multi-input 4WS System

Independent control of sideslip and yaw motion are generally not available in the conventional 2WS vehicle or current commercial 4WS vehicles. However, in some situations, such as parking and lane change manoeuvres, the sideslip motion is more desirable than the yaw motion. To make a vehicle keep heading forward in the lane change manoeuvre or move around a circular path without off-tracking, pure sideslip or pure yaw motions are desired. However, if the driver uses only one input (i.e. turning the steering wheel) to drive the 4WS vehicle, the independent control of these two motions at a time is not possible. Therefore, a multi-input 4WS system is proposed[5,9] to provide the driver with the ability to control these two motions independently and simultaneously. As shown, conceptually, in Figure 7.2, the steering wheel of this multi-input 4WS system can be shifted laterally for the sideslip command and rotated for the yaw command. With this multi-input steering wheel, the driver can give the yaw command and the sideslip command independently and simultaneously.

To achieve the independent control for the sideslip motion and the yaw motion under these two steering commands, a 4WS vehicle is conceived as a multi-input and multi-output (MIMO) control system, as shown in Figure 7.3. Here, the desired sideslip command β_d or the yaw command r_d is given by the driver through the special designed steering wheel shown in Figure 7.2. Then the controller converts steering commands into the steer angles of the front and rear wheels through actuators. The vehicle can then perform sideslip and/or yaw motions using the

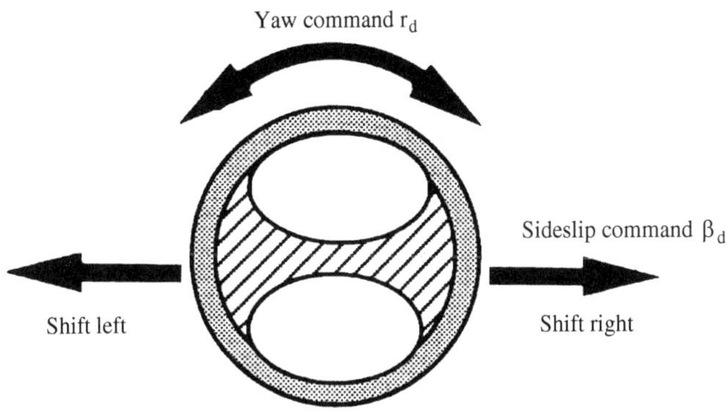

Figure 7.2. A multi-input steering wheel for a 4WS vehicle

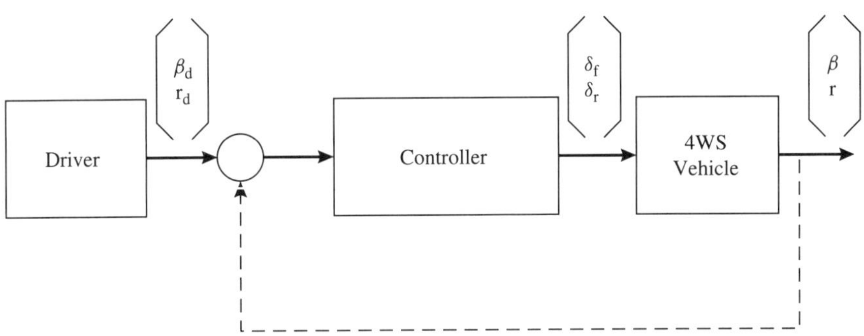

Figure 7.3. The 4WS vehicle as a multi-input and multi-output control system

proper steer angles of four wheels. This concept is similar to the flight-by-wire control technology used in the advanced fighter aircraft.

The controller design of the decoupling 4WS system

It can be seen from Figure 7.3 that the core of this 4WS system is the steering controller. This controller receives yaw or sideslip (β_d or r_d) commands from the driver and steers the front and the rear wheels with proper steer angles (δ_f and δ_r) in order to achieve the desired vehicle response. In addition, this controller has to make a proper distribution of the steer angles on the front and the rear wheels to achieve the pure sideslip or the pure yaw motion. That is, this controller enables the driver to control the sideslip motion and the yaw motion simultaneously and independently.

For a stable vehicle, one way to design such a steering controller is to find the inverse matrix of the 4WS vehicle. Consider the open-loop control system shown in Figure 7.4 as an example, where P represents the transfer matrix of the 4WS vehicle from steer angles of front and rear wheels (δ_f and δ_r) to yaw rate r and

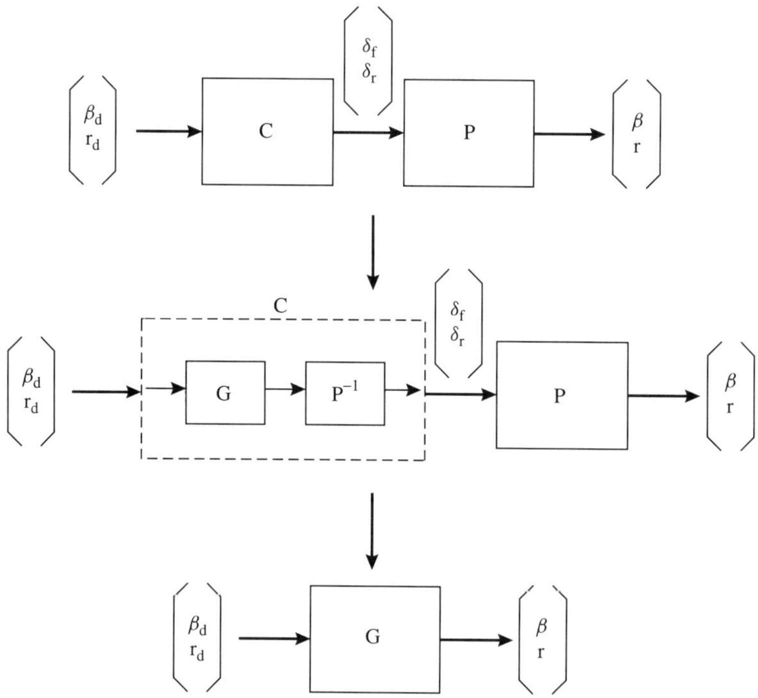

Figure 7.4. Design concept for decoupling controller of 4WS vehicle

sideslip angle β, C is the controller, and G is the diagonal matrix for the desired performance. Because the overall transfer matrix of the system is:

$$G = PC, \tag{7.1}$$

the controller C for the open-loop control system can be calculated by:

$$C = P^{-1}G, \tag{7.2}$$

once the diagonal transfer matrix G is specified. It can be seen that: if the controller C is composed of the product of a diagonal matrix G and the inverse matrix of the vehicle P^{-1}, the overall transfer matrix G is diagonal and the system is said to be decoupled.

Similarly, for the closed-loop system, the desired overall transfer matrix G is:

$$G = PC(I + PC)^{-1} \tag{7.3}$$

To design the closed-loop controller C so that G is a diagonal matrix with desired quality, a matrix D is defined as:

$$D = PC \tag{7.4}$$

Equation (7.3) becomes:

$$G = D(I + D)^{-1} \tag{7.5}$$

From equation (7.5), the relation between D and G is:

$$D = G(I - G)^{-1} \tag{7.6}$$

Thus, as the desired diagonal transfer matrix G is selected, the matrix D can then be calculated by equation (7.6). As G is a diagonal matrix, the calculation of D is straightforward. With this diagonal matrix D, the controller C for the closed-loop control system is:

$$C = P^{-1}D = P^{-1}G(I - G)^{-1} \tag{7.7}$$

To achieve the decoupling control with handling characteristics invariant to vehicle speeds, the symbolic inverse matrix of the vehicle plant P^{-1} must be found so that the controller can be modified for varying vehicle speed. A method[9] was presented to obtain the symbolic inverse matrix of the vehicle without the computation of matrix inversion. With this symbolic inverse matrix, the speed-adaptive controller can be designed in a very straightforward way.

Thus, with available symbolic P^{-1}, the design task of the speed-adaptive steering controller can be easily accomplished for both the open- and closed-loop systems. Details of design procedures and computer simulation examples can be found in Yeh and Wu.[5,9]

Simulations for the multi-input 4WS system

To examine the performance of the multi-input 4WS system, some computer simulation results are presented. Figure 7.5 shows the simulated trajectories of lane change manoeuvre at a speed of 90 km/h, in which a comparison is demonstrated among three steering strategies including 4WS with yaw command, 4WS with sideslip command, and 2WS. Their trajectories are obtained by the following steps.

First, some important variables of the vehicle are defined as shown in Figure 7.6 with the relations shown below:

$$U_x = U \sin \psi + U \tan \beta \cos \psi \tag{7.8}$$

$$U_y = U \cos \psi - U \tan \beta \sin \psi \tag{7.9}$$

where:

U = the longitudinal speed of the vehicle

β = the sideslip angle of the vehicle body

r = the yaw rate of the vehicle

$\psi = \int r \, dt$: the yaw angle of the vehicle

U_x = the velocity in X direction

U_y = the velocity in Y direction

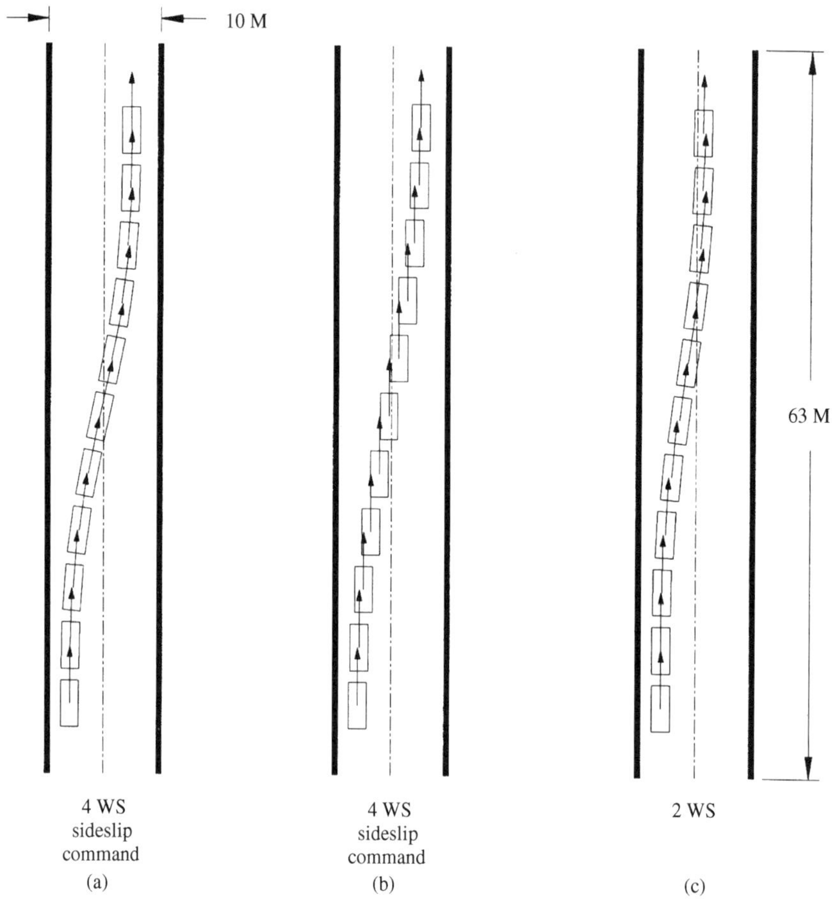

Figure 7.5. Trajectories of three steering strategies for lane change manoeuvre: (a) 4WS with yaw command; (b) 4WS with sideslip command (c) 2WS

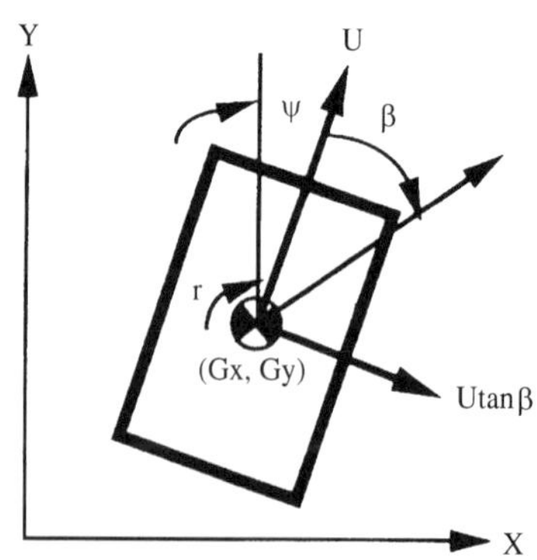

Figure 7.6. Coordinates for vehicle motion

and the position of the vehicle are obtained by

$$G_x = \int U_x \, \mathrm{d}t : \text{position of } X \text{ axis, the forward direction, and}$$

$$G_y = \int U_y \, \mathrm{d}t : \text{position of } Y \text{axis, the lateral direction.}$$

Then a closed-loop desired diagonal transfer matrix for the decoupling 4WS system is selected as:

$$G(s) = \begin{bmatrix} \dfrac{k_1 \cdot \omega_{n1}^2}{s^2 + 2\zeta_1 \omega_{n1} s + \omega_{n1}^2} & 0 \\ 0 & \dfrac{k_2 \cdot \omega_{n2}^2}{s^2 + 2\zeta_2 \omega_{n2} s + \omega_{n2}^2} \end{bmatrix} \tag{7.10}$$

with $\zeta_1 = \zeta_2 = 0.8, \omega_1 = \omega_2 = 10$ Hz, and $k_1 = k_2 = 1$. According to equation (7.10), yaw rate r and sideslip angle β of the vehicle are obtained by the given steering commands. Then using equations (7.8) and (7.9), the trajectories of the 4WS vehicle as shown in Figure 7.5(a) and (b) can be obtained under pure yaw or pure sideslip commands. Figures 7.7 and 7.8 shows the steering response of the 4WS vehicles for this lane change manoeuvre. In these two figures, not only the steering commands, the sideslip angle, yaw rate output, yaw angle and the steer angles of the front and the rear wheels are also illustrated.

The trajectory of the 2WS vehicle is obtained in Figure 7.5(c) with the same front wheel steer angle as that 4WS vehicle under pure yaw command, but with zero steer angle for rear wheels. Figure 7.9 shows the simulation results of the 2WS vehicle and the history of steer angles.

From these results, some conclusions about the 4WS vehicle can be made. In Figure 7.9, the front wheel steer angles of the 2WS vehicle are the same as that of the 4WS vehicle. But without steering the rear wheels, the yaw rate output of the 2WS vehicle reaches only about 70% of that of the 4WS vehicle. More steer angles are needed for front wheels if the same lateral displacement is desired. In addition, the yaw motion of the 2WS vehicle is accompanied by a sideslip motion. This is the reason why the off-tracking is worse for the 2WS vehicle. Figure 7.5 shows that the heading direction of the 4WS vehicle by a pure yaw motion is always tangent to the turning curve. This results in a smooth trajectory with very small off-tracking. On the other hand, a pure sideslip motion keeps the heading direction of the vehicle unchanged while moving laterally.

The lane change trajectories made by the three steering strategies are plotted together in Figure 7.10, where the 2WS vehicle can only move about 3 m laterally, while 4WS vehicles can move laterally close to 6 m. This means that obstacle-avoidance ability of the 4WS vehicle is better than that of the 2WS vehicle. No much difference on the lateral movement is found between the 4WS vehicles by the pure yaw and by the pure sideslip commands.

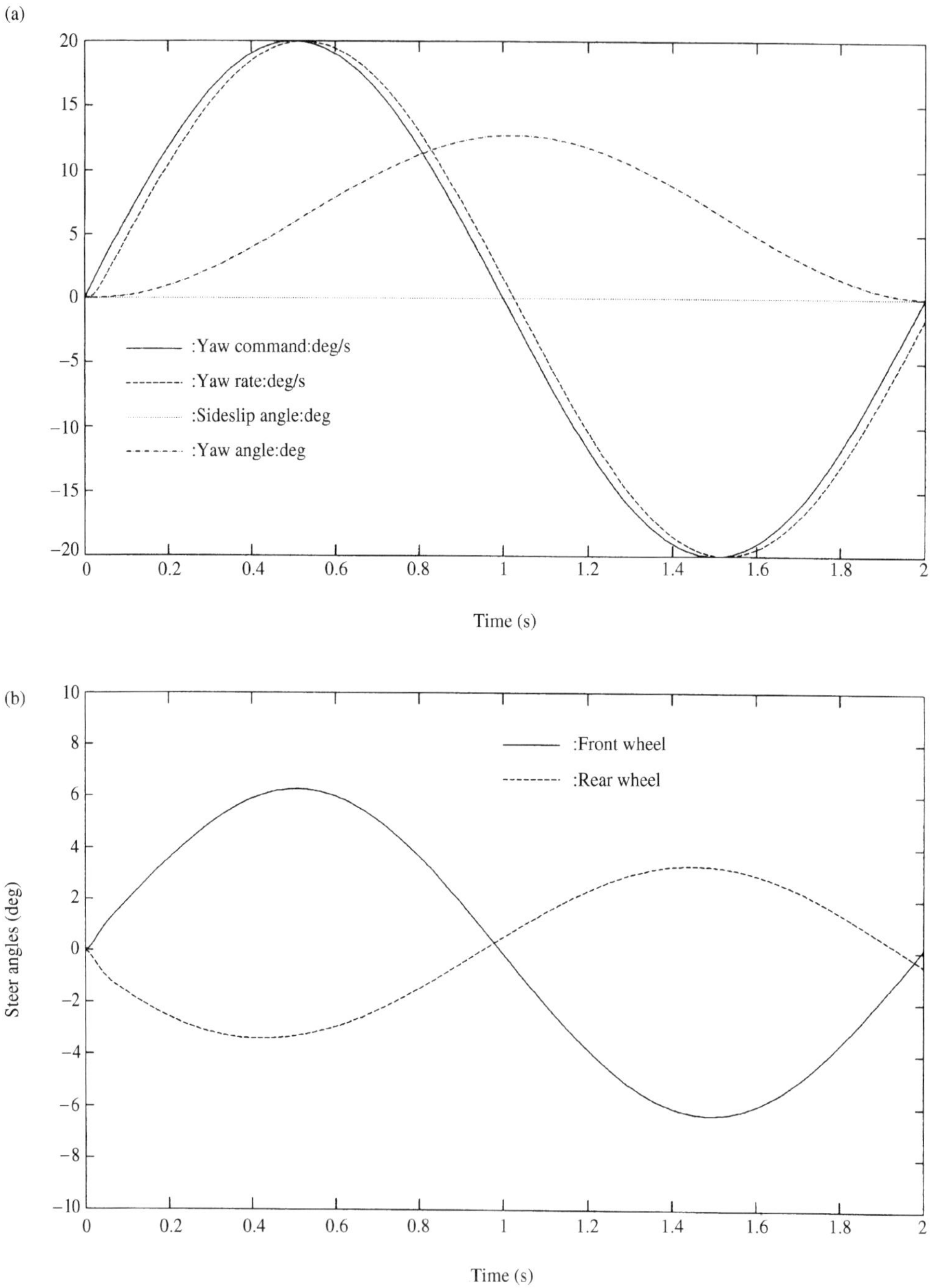

Figure 7.7. Simulation results of 4WS vehicles by pure yaw command: (a) outputs of the 4WS vehicle; (b) steer angles of front and rear wheels

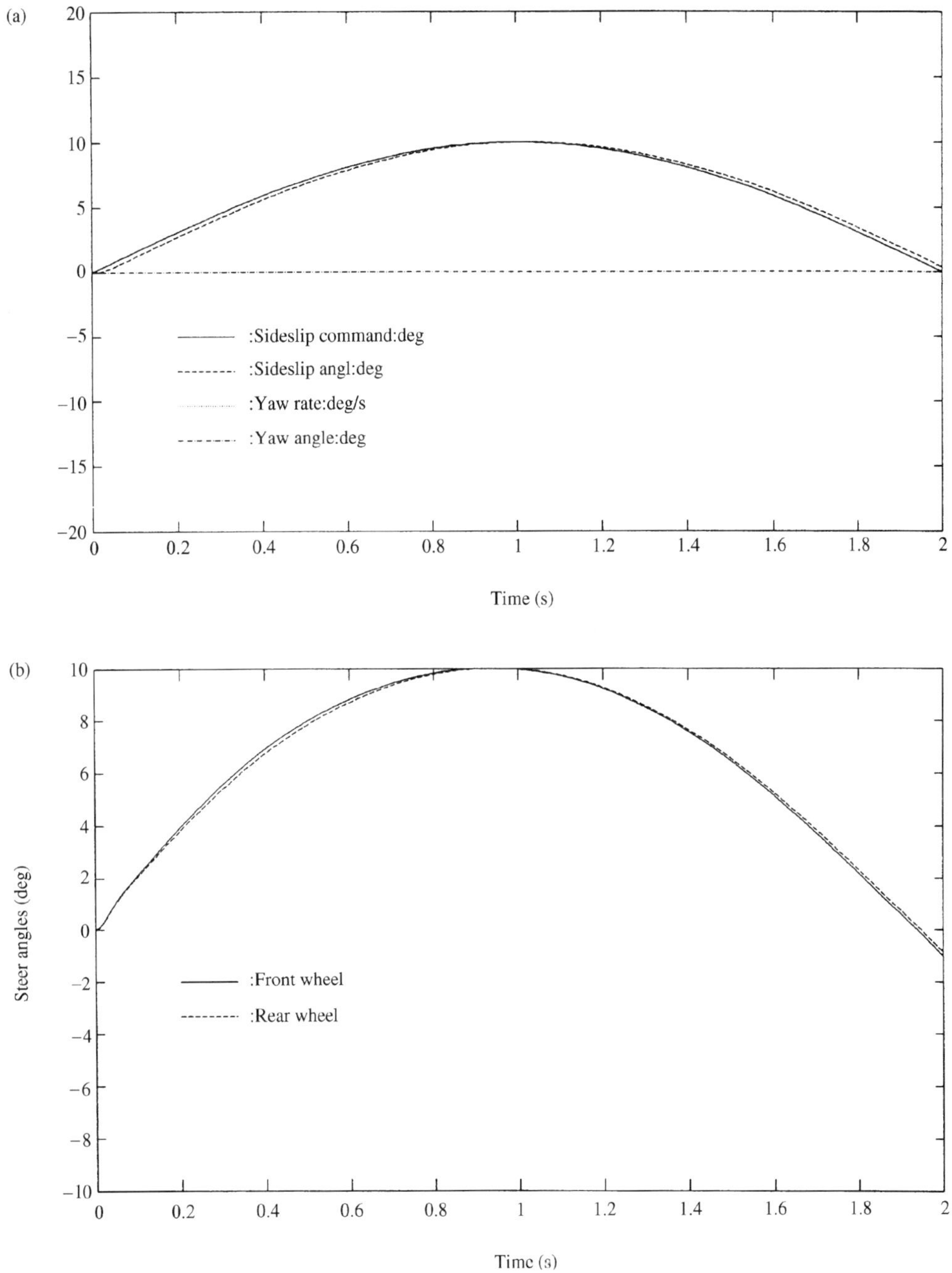

Figure 7.8. Simulation results of 4WS vehicles by pure sideslip command: (a) outputs of the 4WS vehicle; (b) steer angles of front and rear wheels

(a)

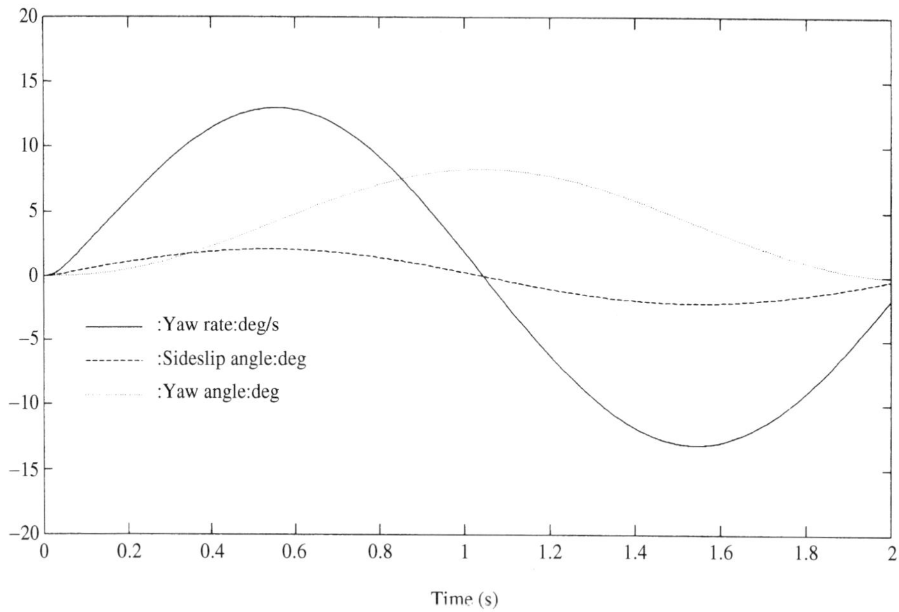

(b)

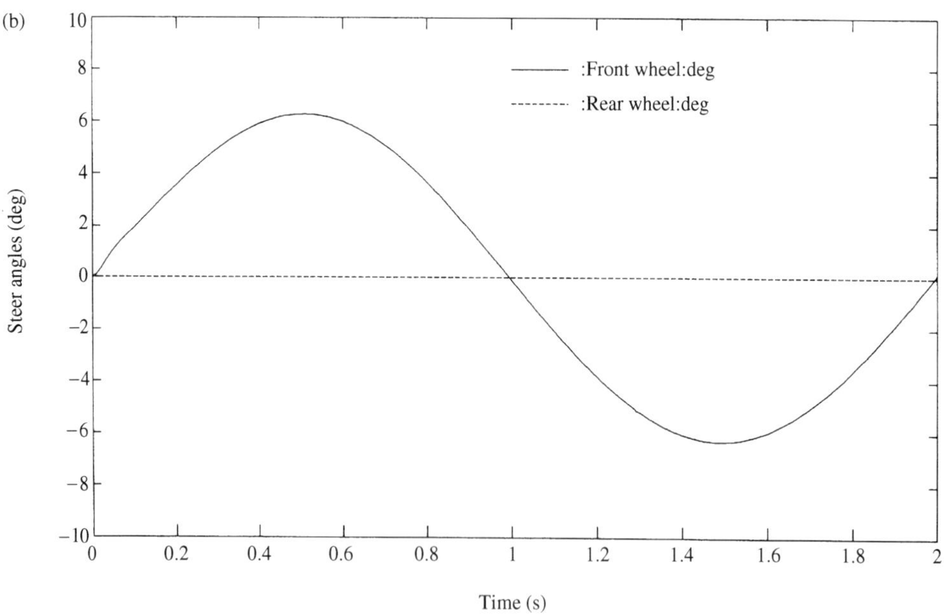

Figure 7.9. Simulation results of a 2WS vehicle: (a) outputs of the 2WS vehicle; (b) steer angles of front and rear wheels

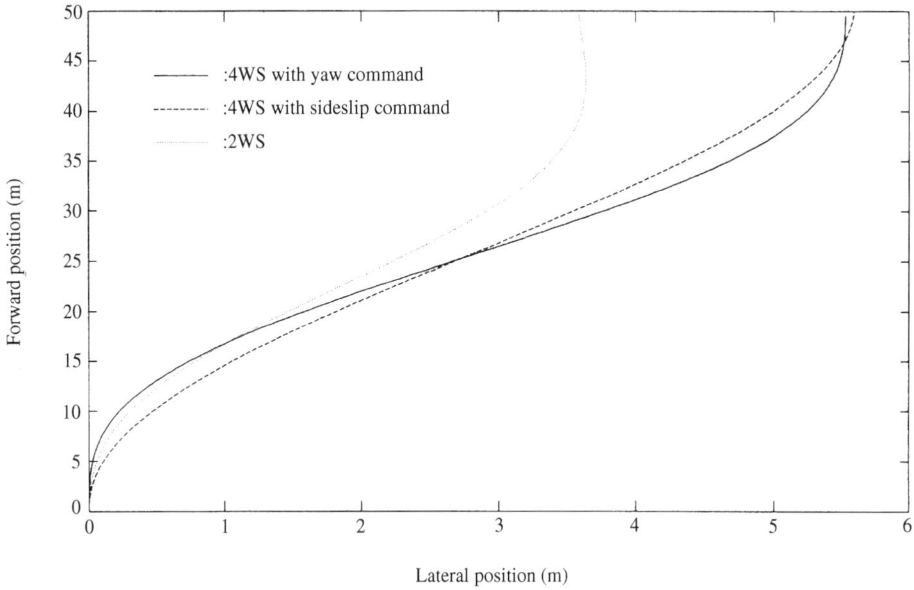

Figure 7.10. Comparisons of trajectories among three steering strategies for a lane change manoeuvre

However, from the steering commands shown in Figures 7.7 and 7.8, it can be found that for the same lateral movement, the amplitude and frequency of the yaw command is twice that of the sideslip command. The double cycle of the yaw command means that the driver has to make a faster steering effort while driving the vehicle. In fact, 50% of the steering command is used to bring the vehicle back to face forward again under the yaw command, i.e. to reduce the yaw angle of the vehicle back to zero. While by pure sideslip motion, the lane change manoeuvre can be made very easily with a smooth trajectory and the vehicle always keeps heading forward during the lane change manoeuvre, more attention can thus be focused on the front view by the driver.

Therefore, from the point of view of driving skill, it seems easier to use a pure sideslip command in a lane change manoeuvre instead of a pure yaw command, which is widely adopted in the present 4WS vehicles. It may be even better to use a pure sideslip manoeuvre instead of yaw to avoid an unexpected obstacle. Thus, it might be useful to design the autonomous vehicle to choose the proper motion command to achieve the desired manoeuvre.

7.3 THE CLOSED-LOOP 4WS CONTROL SYSTEM TOWARDS AN AUTONOMOUS VEHICLE

Closed-loop feedback compensation is a widespread control technology in control system design. It brings many advantages to the control system, including the improvement of output precision, reducing the system's sensitivity to disturbances or plant's parameters variations, and stability (or controllability) augmentation by rearranging the pole location of the system's transfer function by feedback compensation.

In fact, closed-loop feedback is commonly utilized in design. For example, some suspension mechanism are carefully designed to provide the "roll–steer" effect, which is a closed-loop mechanical feedback to adjust the steer angle of the wheel from vehicle's roll angle to improve vehicle stability. This is a traditional way of using mechanical mechanisms in a steering system to stabilize the vehicle passively. For lack of flexibility by the passive mechanical mechanism, a more advanced performance is difficult to obtain. With strict requirements in vehicle performance, advanced control concepts can be introduced into the vehicle steering system using closed-loop control technology to augment stability or controllability, similar to the CCV design concept for flight control. Because of the distinct advantage and flexibility, the 4WS system can be integrated with closed-loop control technology to achieve some advanced control objectives for vehicle steering. An example of this design is to stabilize an unstable oversteer car using an inner feedback loop with the measured sideslip angle by a pair of castors and yaw rate by a gyro sensor.[5]

The natural extension of closed-loop 4WS system in the future is to realize a more intelligent autonomous vehicle system. Actually, a similar concept was proposed in 1979,[10] which was an automated guideway transit (AGT) system using the closed-loop 4WS concept to reduce the turning radius in the AGT system. The control strategy of this kind of the 4WS vehicle is based on the feedback of position error measured by sensors, then proper steer angles can be determined for front and rear wheels to achieve the required motion to follow a desired path in a rapid and stable manner.

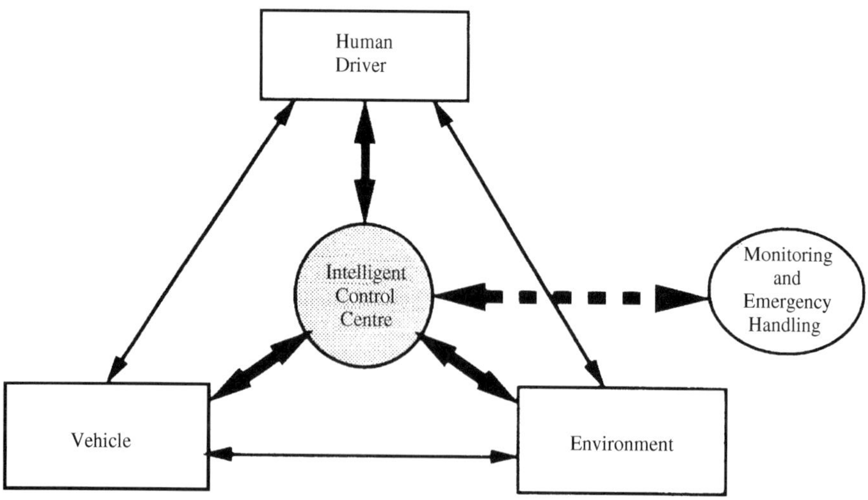

Figure 7.11. Architecture of an intelligent autonomous vehicle system

As rapid progress is being made in automotive electronics, some advanced sensors or sensing methods, such as the electrical cable guidance using electromagnetic wave[11] and the optical guidance using laser, or TV camera[12] can be utilized to measure the vehicle position for the automatic steering of a vehicle, in addition to the conventional mechanical guidance.[13] These advanced technologies can be integrated with 4WS vehicles by closed-loop position feedback to develop an autonomous vehicle system with some intelligence. A proposed design concept for such a closed-loop 4WS vehicle is described below.

This intelligent autonomous vehicle system consists of an intelligent control centre and the its interfaces between three main factors of the vehicle system: driver, environment and vehicle (see Figure 7.11).[14] For safety and reliability, on-board monitoring and diagnosis are needed for failure detections to handle emergency situations. From Figure 7.11, the function diagram of the whole system can be constructed as shown in Figure 7.12, where the subsystems in the intelligent autonomous vehicle are described in detail below.

7.3.1 Intelligent Control Centre

This is the heart of the autonomous system, which consists of three components:

1. *Vehicle commander.* The commander can automatically make a decision to execute a desired motion based on sensed information from either driver or fed back information from environment. For example, a desired yaw or sideslip motion can be executed by the closed loop 4WS subsystem as a result of the given commands (desired sideslip β_d and yaw rate r_d). The vehicle commander can also be used to change the vehicle's handling characteristics by selecting a desired reference model for the vehicle from the data bank when the driver intends to do. Several ideal vehicle reference models with different handling characteristics are stored for selection; for example, a fast sport car or a rather sluggish luxury sedan.

2. *Expert system for safety.* This can be used to sense the steering wheel angle, its time derivative and environment information, such as obstacles and vehicle positions to analyse the driver's steering pattern and to find out the driver's intention. Some driver's features such as driving skill, desired vehicle motion, and driver's preference can be estimated from sensors on board. The expert system can be developed in a self-organizing manner, for example, possibly using the neural network learning approach, or other machine learning approaches, in which the vehicular performance could be used as an index for learning.

3. *Expert system for steering.* This can be used to develop a series of steering strategies to satisfy the driver's desired objectives to accomplish some vehicle motions such as lane change manoeuvres, parking, curve tracking or even acrobatic motions (e.g. wheel-lift). These steering strategies can be extracted from experiences of the

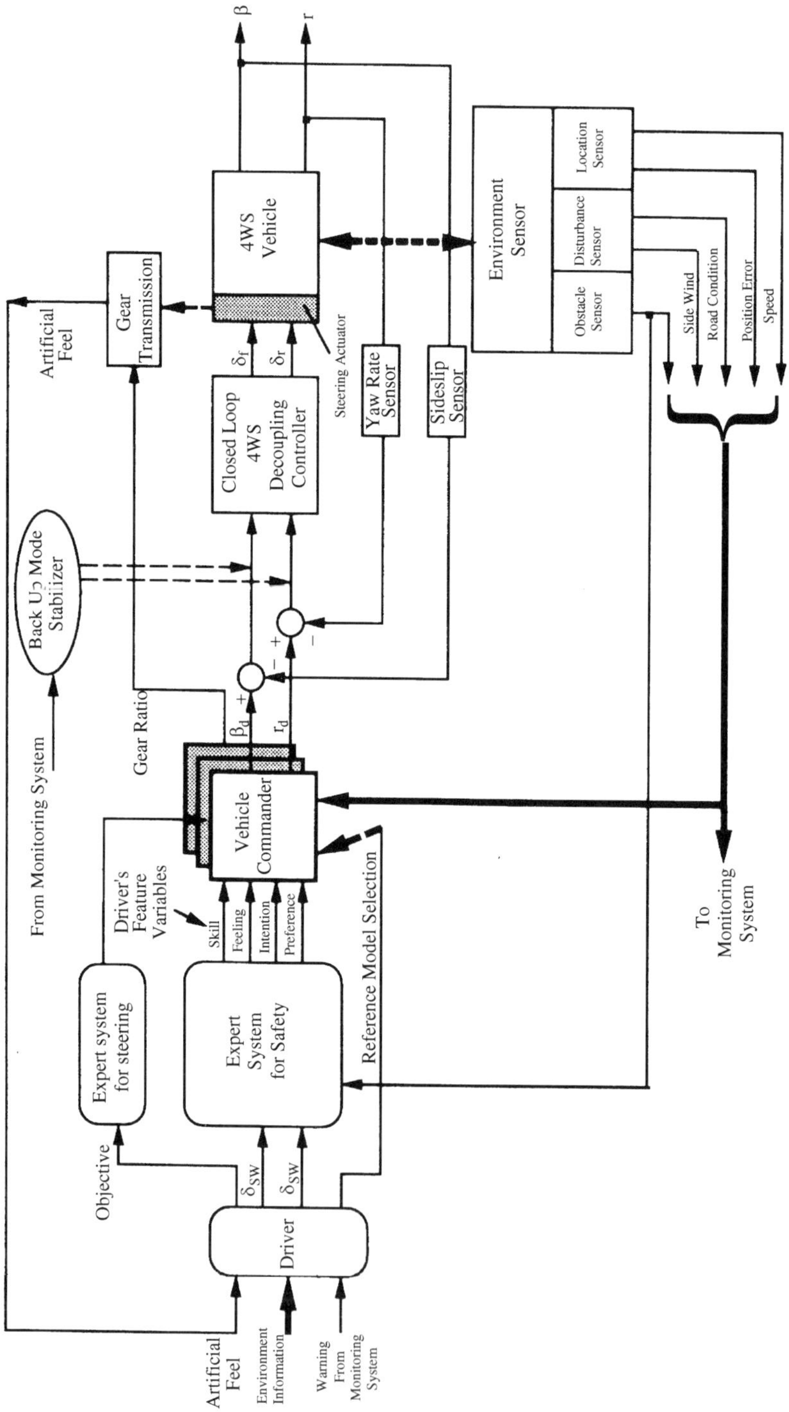

Figure 7.12. Function diagram of the intelligent autonomous vehicle system

experts, for example, using the fuzzy approach[14] to obtain linguistic rules of the expert system. Initially, the developed strategies may not be satisfactory. Therefore, a machine learning phase could be executed using neural network methods to improve performance. Further advances are expected.

7.3.2 Closed-loop 4WS Vehicle System

The vehicle control system may be a MIMO 4WS system as described in Section 7.2. It consists of the steering controller, 4WS vehicle and vehicle motion sensors, including yaw rate sensors and sideslip sensors. Recently, some low-cost yaw rate sensor have been introduced into the commercial market to detect the Coriolis force generated by vehicular yaw motion.[15] Besides the sideslip motion can be measured by a pair of caster wheels mounted on the front and rear axles.[9] Therefore, by these feedback information, the decoupling controller can determine steer angles of the front and rear wheels (δ_f and δ_r) for desired vehicle motions. These steer commands are then sent to the steering actuator which could be a DC motor or DC brushless motor to steer the wheels to the desired steer angles.

On the other hand, if electrical power steering (EPS) of the 4WS vehicle is implemented, it could provide a suitable artificial feel to the driver through the steering wheel, such that some road feel corresponding to road conditions can be artificially generated on to the driver. Furthermore, the gear ratio between EPS and the steering wheel could be controlled by the vehicle commander to provide a proper steering assistant torque based on the driver's feature variables: feeling, preference, skill, and intention. This human–vehicle interface should be studied carefully to avoid conflicts, and it may become an important topic of study in the future.

7.3.3 Environment Sensing System

The environment sensors are used for measuring various pertinent information. These information include vehicular position deviating from the desired path, vehicle speed, traffic condition, road condition, sidewind disturbance and obstacle. Recently, image processing technology using TV cameras to get environment information has been actively studied.[12] Other measuring approaches such as electromagnetic wave radars, optical lasers, and ultrasonic microphones to detect road markers or obstacles are also under study. In fact, the environment information can sometimes be provided by the highway control centre or even from satellites by the navigation or guidance function of global position systems (GPS).

Disturbance sensors are used to measure the magnitude of different disturbances on the vehicle. One of the disturbances is the sidewind which may be measured by a small wind vane on the surface of vehicle body. Road conditions such as wet and bumpy roads can possibly be sensed by TV cameras or even from the response of a suspension system.

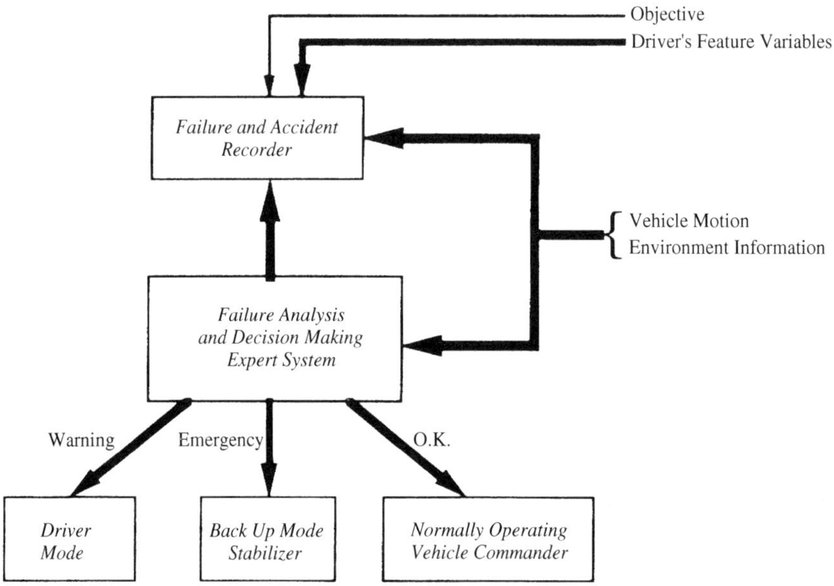

Figure 7.13. On-board monitoring and emergency handling system of autonomous vehicle

7.3.4 On-board Monitoring and Emergency Handling

This system (see Figure 7.13) is constructed for reliability and safety to avoid accidents caused by sudden breakdown of this intelligent autonomous control system. One of the components in this system, the failure analysis and decision-making expert system, can analyse various information from the environment in real time and check the possible system's breakdown from the vehicle performance. A proper decision can then be made to let the system operate continuously, or to start the back-up mode automatically to stabilize the vehicle in an emergency situation rapidly, or even to shift the control back to the driver when the autonomous system fails completely.

Besides, a driving recorder, which stores the measured signals about vehicle motions, can be used for accident reconstruction and analysis. In addition, it can provide the driver with useful information for diagnosis and repair.

7.4 FUTURE PROSPECTS FOR THE STEERING CONTROL SYSTEM

According to the current state of the art and the trend in steering control technologies, the following can be expected about the future development.

1. The rapid evolution of automotive electronics is obvious. The electrical technology, that is, electronically controlled EPS, may become popular for the

steering system. The so-called control-by-wire technique, developed from the fly-by-wire in fighter aircraft, will become widespread to achieve more flexible driving functions. Besides, the on-board computer equipped in the vehicle will support the steering system and assist the driver to control the vehicle properly and safely in an assistant mode, or it can be the control centre of an autonomous vehicle to handle the full driving work. Also, many smart schemes such as adaptive control or machine learning may be easily implemented with the aid of EPS.

2. "Integrated control" of the vehicle has gradually become a trend in automotive electronics. Therefore, the steering control system will become a subsystem of the integrated chassis control system in the future. The current advanced chassis control technologies, including anti-lock brake system (ABS), traction control system (TCS), active suspension, EPS or 4WS etc., will be parts of an integrated control system such that the whole vehicle behaviour and motion can be controlled more flexibly and effectively. Actually, a Japanese car[15] has already been designed with this kind of integrated chassis control system, and it will be a developing trend worldwide.

3. Environment sensing and information analysis will be necessary for cars in the future. With the aid of precise environment information, including the estimation of the disturbance, road conditions and traffic conditions for vehicle control, the vehicle can hopefully become more safe and intelligent to satisfy the driver's need.

7.5 CONCLUSIONS

In this chapter, 4WS technology with position feedback control is examined and studied, which is an attempt to develop an integrated system toward the intelligent autonomous vehicle of the future. An overview of 4WS vehicles is presented and some advantages of the 4WS system over 2WS are described. A special 4WS vehicle with multi-input steering wheel is proposed, and some computer simulations are made to compare the handling characteristics of 2WS and 4WS vehicles. With the aid of closed-loop position feedback control, the 4WS vehicle can be extended to an autonomous vehicle system, and a scheme to achieve this target is also proposed. Finally, developing trends in steering control technology are discussed by pointing out the important concepts in automotive electronics, integrated control and information analysis.

REFERENCES

1 Sano S, Furukawa Y, Shiraishi S. *Four Wheel Steering System with Rear Wheel Steer Angle Controlled as a Function of Steering Wheel Angle*. SAE Paper 860625 1986.

2 Mouri H, Kuroki J, Sugasawa F, Irie N. Handling and stability improvement achieved with four-wheel steering. *IMech E Proceedings of the Institution of Mechanical Engineers* 1988; C441/88: 191–197.

3 Takiguchi T, Yasuda N, Furutani S, Kanazawa H, Inoue H. *Improvement of Vehicle Dynamics by Vehicle-Speed-Sensing Four-Wheel Steering System*. SAE Paper 860624 1986.
4 Ohnuma A, Metz LD. *Controllability and Stability Aspects of Actively Controlled 4WS Vehicle*. SAE Paper 891977 1989.
5 Yeh EC, Wu RH. Closed-loop design for decoupling control of a four-wheel steering vehicle. *International Journal of Vehicle Design* 1989; **10**(5): 703–727.
6 Lugner P, Mittermayr P, Endlicher KO. Theoretical investigations on the behaviour of a car with additional four-wheel steering at μ slip conditions. *IMech E Proceedings of the Institution of Mechanical Engineers* 1988; C440/88: 181–190.
7 Lugner P. Some investigations on computer aided steering. *Proceedings of 9th IAVSD — Symposium*, 1985, 353–366.
8 Lugner P, Mittermayr P. Controlled additional 4-wheel steering at critical driving conditions. *Proceedings of the International Symposium on Advanced Vehicle Control 1992*, Yokohama, Japan, 1992, 245–251.
9 Yeh EC, Wu RH. Open-loop design for decoupling control of a four-wheel steering vehicle. *International Journal of Vehicle Design* 1989; **10**(5): 531–552.
10 Nisonger RL, Wormley DN. Dynamic performance of automated guideway transit vehicles with dual-axel steering. *IEEE Transactions on Vehicular Technology* 1979; **VT-28**(1): 88–94.
11 Fenton RE, Olson KW, Bender JG. Advances toward the automatic highway. *Highway Research Board 344, 50th Annual Meeting of Committee on Vehicle Characteristics*, 1971, 1–14.
12 Masaki I, (ed.) *Vision-based vehicle guidance*. New York: Springer-Verlag, 1992.
13 Hallmark WC. Dallas-Ft. worth airport airtrans control System. *Proceedings of 1974 Joint Automatic Control Conference*, 1974, 335.
14 Uyeda Y. Improvement of chassis control by fuzzy controller. *Proceedings of IEEE Roundtable Discussion on Fuzzy and Neural Systems, and Vehicle Applications*, Tokyo, Japan, 1991, Paper No. 2.
15 *Toyota and Automotive Electronics*, Press Information of Toyota Motor Co., 1991.

8

Collision-free Manoeuvring and Control for an Autonomous Vehicle

8.1 INTRODUCTION

Diverse types of planning problems share a number of attributes. They often have a goal or set of goal states, a current state, some measure of success and limits on how the system can progress from the starting state to any other state. The objective is to provide an optimal path to the nearest goal while avoiding the limits, which can be hard obstacles, movement based on rules of the game or even kinematic constraints. This chapter describes a framework that can be used to represent such problems, and then gives examples where it has been applied.

8.2 FRAMEWORK

There are several elements that are required in the framework: a problem state–space, a neighbourhood, a cost metric on the neighbourhood, a transformation of the obstacles/forbidden regions and a goal set. Each of these elements is described in further detail below along with a simple example. The general idea is that by employing these elements it is possible to transform the problem into an easier space in which to solve the problem. The solution is then transformed back to the original "task" space where it is carried out.

The state of the system must be characterized by one or more parameters, which can uniquely describe that state. For example, the two joint angles of a two-link robot arm uniquely describe the position of the robot. A test for the chosen parameters is to ask "Do these parameters give a unique state of the system?"

Advanced Vehicles and Infrastructure Systems, Edited by C.O. Nwagboso

The span of legal ranges for the parameters defines the problem state–space. The reformulation of the problem in this space is called *configuration space*[1] in robotics and is sometimes abbreviated CS. This is an easier space in which to solve the problem than the task space because the status of the entire system can be characterized by a single state within the configuration space. Continuing with the robot example, a robot that has shoulder and elbow joints that are fully revolute, would have a configuration space as in Figure 8.1. A single state describes a specific pose of the robot. Each state is specified by the values assigned to the parameters. Figure 8.1 also shows the mapping between a state and the corresponding pose in task space.

The *neighbourhood* in the configuration space is a set of possible motions for moving "legally" from one state to another assuming no obstacles. It describes the fundamental motions of the system, and is depicted by an arrow showing the direction of a particular motion. Therefore, if the example robot has one motor that is switched between the shoulder and elbow joint so that only one can move at a time, then the neighbourhood of possible motions is as in Figure 8.2.

This neighbourhood represents one unit of motion in the horizontal direction corresponding to a unit change in the shoulder angle, and one unit of motion in the vertical direction corresponding to a unit change in the elbow angle. This is a simplified example of the neighbourhood. If the joints could be moved simultaneously then four diagonal arrows would be seen as well.

A cost can be associated with each motion, resulting in a cost-weighted neighbourhood relative to a given state. The cost metric is usually expressed as a function of the "current" state and the neighbour state, or equivalently, the current state and the difference to achieve the neighbour state. (The word "metric" is used here more loosely than the strict mathematical definition. The cost must be >= 0, but the

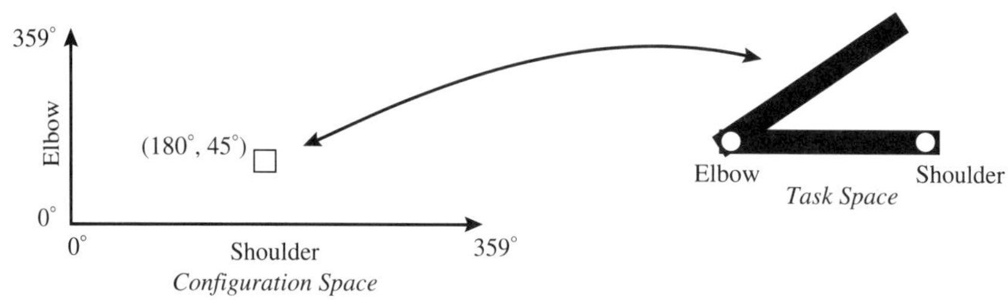

Figure 8.1. Transformation from task space to configuration (state) space

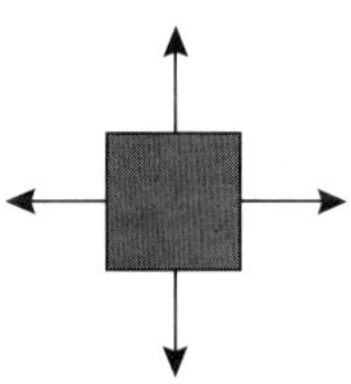

Figure 8.2. Simple neighbourhood of possible motions in configuration space

triangle inequality and commutative properties need not be enforced.) The metric need not be memoryless, that is, it may have cost factors that are a function of prior motions. Admissible heuristics also may be employed as part of the metric. The metric can be used to implement an optimality criterion, for example, a metric M that will minimize the distance travelled by the two-link robot's end effector (giving the straightest possible motion in the task space) can be expressed as in Figure 8.3.

The *transformation* of obstacles from task space to configuration space can be very difficult, particularly for high-dimensional spaces with complex shaped obstacles. Others[2,3] have examined such cases, but no fast, general transform methods currently exist. Fortunately, in problems such as point-to-point routing, rendezvous and strategic planning, 2 d.o.f. (degrees of freedom) robot arm path planning and vehicle manoeuvring, the transformation can be performed in real time.

Goal states are a specification of acceptable target locations. These may be supplied by a higher-level task planner,[4] or by manual specification (pointing to a grid on a map for example).

The *starting state* must be known in order to issue commands to the controllable entities. It is not required for the planning process unless a heuristic (using the starting state) is employed.

Once the elements of the framework are in place, the cost-weighted neighbourhood is simultaneously propagated from all goal states through the configuration space in an A*,[5] least-cost-first manner until the space is filled (or at least until the start state has been reached). In most machine-based problems it is preferable to propagate waves from the goal rather than the start, although wave propagation can clearly be performed in the reverse direction (with appropriate reverse calls to the metric function). Details for an efficient cost-wave propagation (CWP) technique are given in a previous paper.[6] The CWP marks each state with the cost to the goal and a pointer toward the adjacent state leading to the nearest goal.

The result of the CWP is a globally optimal path from every reachable state to the nearest goal. The total solution is a field of arrows as is shown in Figure 8.4. This is different from potential field techniques,[7] in that it is globally optimal, always finds a solution if there is one, and never gets stuck in local minima. In addition, if there is no solution to get to the goal, then this is immediately known, and a higher control authority can be summoned. Unlike the potential field technique, however, there

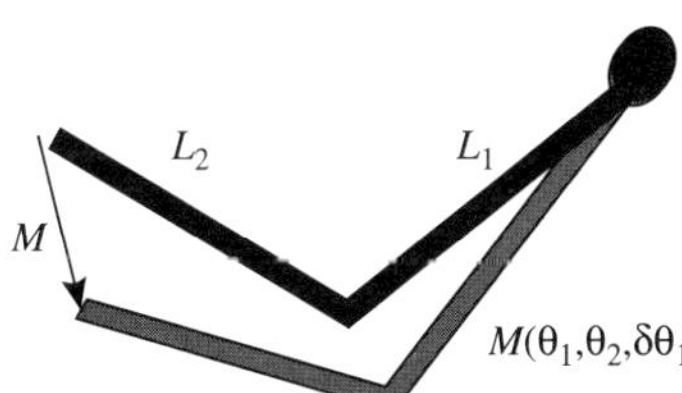

$$M(\theta_1,\theta_2,\delta\theta_1,\delta\theta_2)=\sqrt{(L_1\delta\theta_1)^2+(L_2\delta\theta_2)^2+2L_1\delta\theta_1 L_2\delta\theta_2\cos(\theta_1-\theta_2)}$$

Figure 8.3. Minimize distance of the end effector

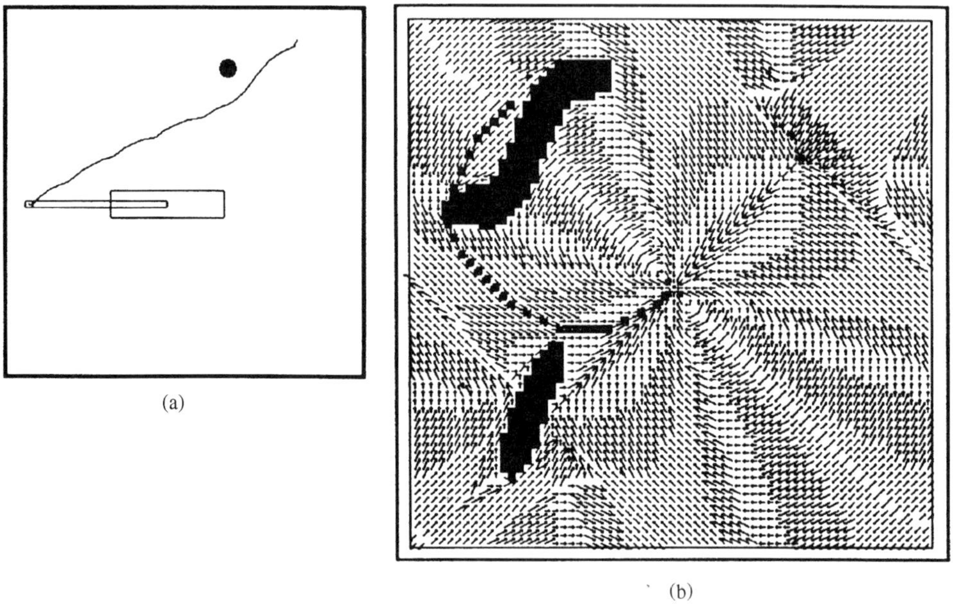

Figure 8.4. Path planning. (a) Task space path corresponding to configuration space solution. (b) resulting field of arrows from cost wave propagation in configuration space using minimum distance criterion

are optimal-direction arrows at all reachable states. This is useful when controlling a device or issuing commands that are not carried out perfectly. For machines in the field, unexpected slippery terrain or failing gear mechanisms could cause such problems. In this event, if the controlled device falls off the optimal path, then a new path can be found from any state *without* further computation.

An example of a CWP solution for the robot arm example is shown in Figure 8.4. The black region in the configuration space represents all of the configurations where the pose of the robot arm results in a collision with the round obstacle. For each state in the configuration space there is at least one arrow that shows the next pose (set-point) that the robot should achieve. By following these set-points the robot will move optimally and purposefully to the goal. Note that the imperfections in the end-effector travel shown in the task space of Figure 8.4 are due in part to the motion required for obstacle avoidance, but also to discretization error.

In live problems, the environment is not static but ever changing. Obstacles and goal opportunities arrive and depart over time. To keep pace with the changing environment, the planner must adapt the plan as quickly as possible. In the current implementation, a 10,000 (100×100) state configuration space with an eight-connected neighbourhood can be computed from scratch in 595 ms ($\sim$1/2 s) on a Sun-4 (SPARCstation IPX). A similarly connected 90,000 (300×300) state configuration space can be computed 6177 ms. Higher connected neighbourhoods require more time, as will large numbers of states. Even though these times seem fast enough, more complex problems are easy to create.

As global environments and the resulting plans often change piecewise it is more efficient to update only the changed areas and the subsequently affected areas. In these situations, a method called differential A*[8,9] can be used. This has two advantages. Not only does the computation take less time, but the unaffected regions can continue to be used for control. Thus, if an incident takes place at the opposite end of the area of interest from a controlled entity, then the controlled entity can continue its work while the incident region is recomputed. Another application is when information about the environment is being assimilated while the machine is operating. In this case the machine can, in a simple sense, learn and adapt to the environment quickly. For example, an intelligent vacuum cleaner might detect the layout of furniture while it is cleaning so as to create a better vacuuming pattern the next time.

Another improvement on the basic framework is to create a more concise representation of the freespace by grouping adjacent free configurations into hyper-parallelepipeds of various sizes.[10] Alternatively, the "swept bubble" concept[11] can divide recursively smaller spaces according to obstacle occupancy. Swept bubbles also provide a uniform geometric representation of the machine and the obstacle environment which provides quick intersection tests required for high-dimensional planning.

The framework that has been described allows the computation of quite diverse problems simply by modifying the relevant elements in the framework. This has proven quite powerful, especially when computing higher dimensional spaces.

A variety of machine examples follow that describe what modifications were necessary to each element that result in the desired plans. Finally, the framework is applied to the rendezvous problem which can be used to predict worst-case enemy strategies and efficient co-ordinated responses.

8.3 ALGORITHM ANALYSIS AND ACCURACY

Discretizing configuration space causes some error to be introduced. The amount of error for Euclidean distances in eight-connected and 16-connected neighbourhoods has been studied.[12,13] The eight-connected neighbourhoods are the horizontal, vertical and diagonal transitions in configuration space. A 16-connected neighbourhood additionally includes the eight knights moves. By using eight-connected neighbours in a Euclidean metric there is a 4% error from the optimal solution. By using 16-connected neighbourhoods there is a 1.4% error. Therefore a trade-off between increasing discretization and increasing neighbourhood size can be made.

The worst case timing for this algorithm is $O(N \log N)$ for N states in the configuration space. Wave propagation is a one pass method with no computation for obstacle states. Therefore, obstacle strewn environments give faster results than those with no obstacles. For robot examples, where the number of states is determined by the resolution R, and by the number of degrees of freedom D, $N = R^{D}$. For i neighbours, time $\propto iR^{D} \log R$. Even though this is exponential, many practical problems have still been solved in acceptable time by accommodating the most essential factors of each specific problem. Some key factors that improve computation time are described next.

The nature of the solution method makes it more efficient for spaces with more obstacles, which are also typically problems that are the most difficult to perform by inspection. The resolution, R, is a tunable parameter that can be adjusted for regions of greater sensitivity, and has even been adjusted dynamically.[10,11] For robotic applications, D may be adjusted by observing that most machines have the greatest volume of motion in only 3 or 4 d.o.f., whereby the remaining degrees of freedom may be treated in a fixed configuration until the final manipulation is required. Finally, the number of neighbours, i, may be adjusted for a slight gain in computing time, although this may not be desirable because of the loss in accuracy. The trade-off between the time, space and accuracy must be made separately for each implementation.

8.4 THE FIRE EXIT PROBLEM

The objective is to provide exit directions for inhabitants of an unfamiliar building (conference centre, hotel, etc.). Observe that currently in a fire, we are all trained to blithely follow the red exit signs even if they lead directly into the blaze of a fire. Oddly enough, the fire/emergency systems today can specifically pin-point the problem areas, but cannot convey that information to the inhabitants.

Directing people out can be performed using the framework described. The directions to inhabitants are envisioned as lighted arrows at about knee height distributed about the building at regular intervals and at logical junctions. The directions should lead the inhabitants out of the building by the most direct route, but by avoiding the problem areas where possible.

The building in the example shown in Figure 8.5 has two floors, two exits and one set of stairs. The first floor is on the left, and the second floor is on the right. The exits are in the upper left and lower left corners of the first floor. The stairs are in the lower right corner and lead to the second floor.

The configuration space is therefore three-dimensional, and has a "solid" floor between the two floors at all states not representing stairs. There is no difficulty in performing obstacle transformations. Walls, and the solid floor are the only forbidden regions. The exits are the goals. The cost metric given in the equation below is the minimum distance approximated by a straight line between six adjacent neighbours (up, down, right, left, forward, back). Additional penalties can be imposed by adding extra cost at regions where smoke (or other) alarms have been activated. In this example, there are three types of alarm. The darker colour

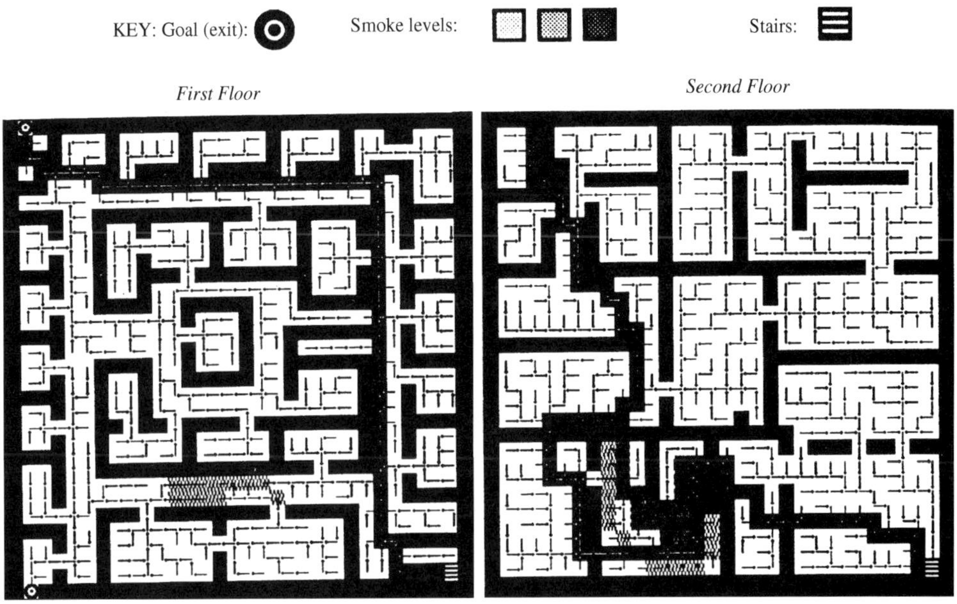

Figure 8.5. Intelligent emergency exits

represents a more severe alarm. From all locations in the building, a path is found to the nearest exit. If there is no exit from a particular location (perhaps because of being blocked in by an impassable alarm) causing people to be trapped, then this would also be apparent to the emergency personnel monitoring the system.

$$M(x_1, y_1, x_2, y_2) = \sqrt{(x_2 - x_1)^2 + (y_2 - y_1)^2}$$

While this problem initially seems specific to fire exits, it is also applicable to other planar point-to-point routing problems. Some examples are routing over varied terrain, selecting optimal roads for personal or emergency use, and routing automated guided vehicles on a factory floor.

8.5 THE ROBOT PATH PLANNING PROBLEM

The objective is to plan a path for the robot to follow from the current position to the goal position without hitting anything. It should move optimally in terms of the minimum energy, where the energy is a function of the mass of each link. The goal is for the arm to be stretched out to the left (180°, 180°), where the angle is measured absolute from a 0° reference pointing right.

Figure 8.6 contains the path computed from a starting state of about (45°, 100°). In this case, the robot can move both the shoulder and elbow simultaneously using a 16-connected neighbourhood.

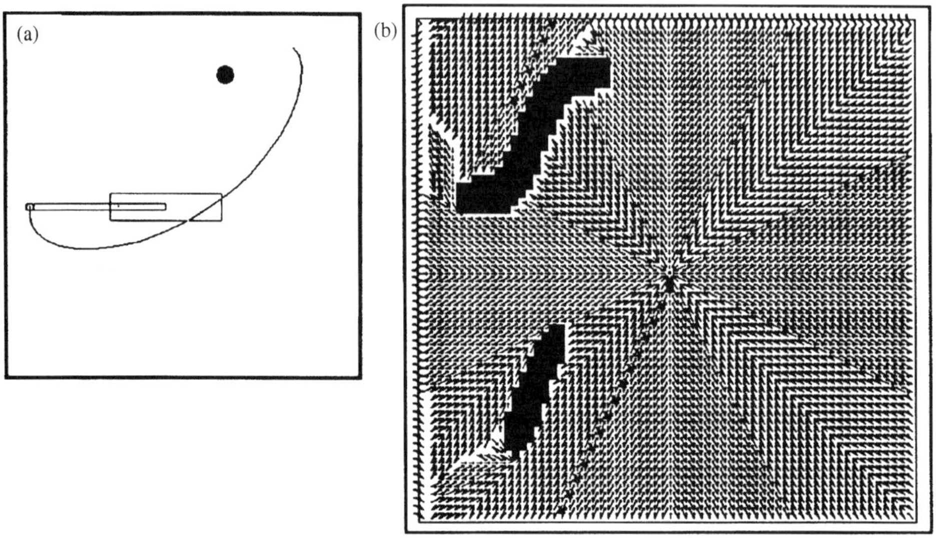

(c)

$$m_c(\delta\theta_1, \delta\theta_2) = \sqrt{\left(\frac{m_2\delta\theta_1}{m_1}\right)^2 + \left(\frac{m_2\delta\theta_2}{m_1}\right)^2}$$

Figure 8.6. Path planned in configuration space and executed in task space. (a) Task space minimum path corresponding to (b) configuration space. (c) Metric cost for change in joint angle where m_1 and m_2 are mass of link 1 and link 2

The configuration space for this problem is two-dimensional as described previously. The span of angles for the shoulder and elbow joints gives the axes and limits of the configuration space. The cost metric imposed on each of these motions is governed by the equation in Figure 8.6. The transform gives the two black areas in the figure which represent forbidden poses. They are computed in real-time based on a function.[14] Based on only this information, a path can be found from any reachable starting state to the goal by reading out joint states from the configuration space and sending them as set-point commands to the robot.

8.6 THE VEHICLE MANOEUVRING PROBLEM

The objective is to plan a path for a vehicle, such as a car, to follow from the current position to the goal position without hitting anything. The configuration space is best characterized in three dimensions by the x, y location of the vehicle and its θ orientation. It should move optimally in terms of the minimum distance that a specific point on the car travels. The goal in this example is to be parallel parked between two other cars. The vehicle's body dimensions and steering capability is known in advance. The control parameters are clearly the direction of the steering wheels and the forward/backward motion of the drive wheels. As the planning parameters and the control parameters are different, this problem is non-holonomic.

The neighbourhood is interesting and different from the previous examples. The vehicle is kinematically constrained. It must follow a particular pattern for each forward/backward motion given a fixed steering wheel orientation. By using the centre of the rear wheels as the reference point on the car, the hard-right, hard-left and straight motions (for forward and reverse) result in a bow-tie-shaped trace if drawn on the ground as the car drives. It is depicted in Figure 8.7 on the left. In configuration space, however, the bow-tie has three parameters, the change in x and y, and the change in orientation of the vehicle. This "twisted bowtie" is shown in Figure 8.7 on the right. The bowtie describes only the partial motion of the vehicle which results in a fixed circular arc based on the turning radius. Therefore, if distance travelled is the measure along an arc between specific neighbours, the metric is easily computed. For clarity, Figure 8.8 uses the planar version of the bow-tie to specify the neighbours and their respective costs.

Each neighbour in the neighbourhood has a dx, dy, $d\theta$ associated with it. In an implementation, improved performance and accuracy is achieved by computing and then storing a separate neighbourhood template for each angle θ.

The transformation of obstacles for vehicle manoeuvring can be simplified if it is assumed that the downward projection of any obstacle from the maximum height of the vehicle determines the illegal zone. In the vehicle manoeuvring examples, each obstacle is first enclosed by a rectangular area and then transformed. The controlled vehicle itself is also approximated by a rectangle to simplify the intersection test.

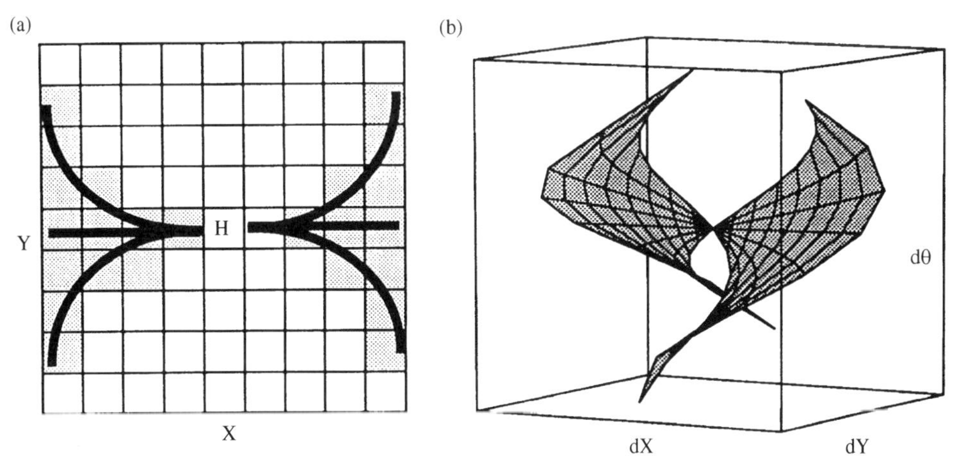

Figure 8.7. Neighbourhood for a vehicle. (a) Trace of possible motion in two dimensions. (b) Trace of possible in three-dimensional configuration space

Neighbour no.	Cost
1, 11	0.25
2, 12	0.50
3, 13	0.75
4, 14	1.0
5, 8, 15, 18	0.502
6, 9, 16, 19	0.757
7, 10, 17, 20	1.017

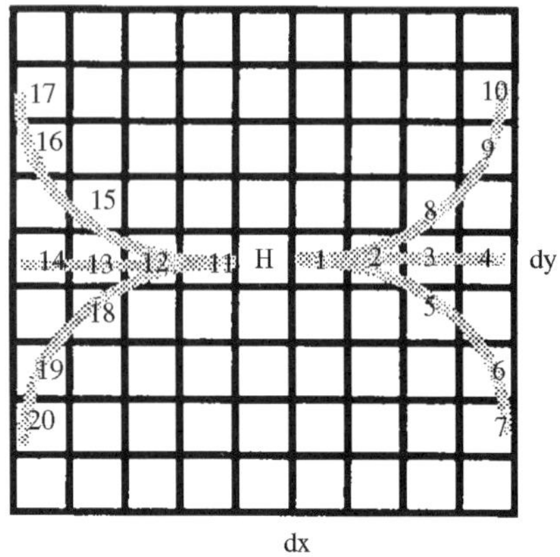

Figure 8.8. Example neighbourhood and associated costs

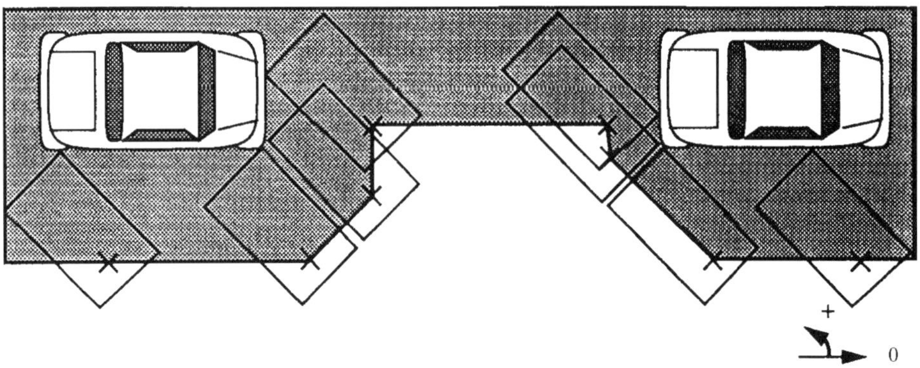

Figure 8.9. One slice (135°) of the illegal region. Position measured at centre of the rear axle

In Figure 8.9, two parked cars are transformed based on the dimensions of the controlled vehicle. This is achieved by determining the outer region (as measured by the centre of the rear axle) where the body of the controlled vehicle would intersect the parked cars or the curb. This can be done by simple geometry, and is most efficiently implemented by separately computing the illegal region of x, y for each possible angle. The transformation of one angle is a "slice" of the full transformation along the x, y plane. For each angle θ of the controlled vehicle, the illegal region in configuration space is determined by computing the transformation at each of the convex and concave corners and filling in the forbidden region. Figure 8.10 shows the full transformation of the configuration space.

The *goal position* is simple to transform. If the goal is to be parallel parked between the two parked cars, in a specific orientation (e.g. parked pointing left) then the transform of the goal x, y coordinate of the car and orientation can be given directly. Any number of goals can be accommodated using the CWP technique. If there are

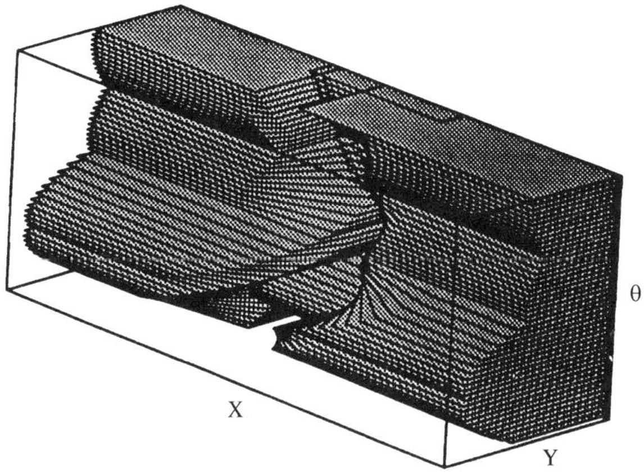

Figure 8.10. Full transformation of parked cars in three-dimensional configuration space

several parking spots to choose from, then each could contain at least one goal. In the current example, only one goal is used.

Once the framework is in place, the CWP method can be used with one modification. The original CWP method propagated waves by repeatedly evaluating the neighbourhood of the least cost state in the wavefront. No particular ordering of the neighbours was imposed. To propagate waves in the vehicle manoeuvring example, however, the neighbourhood must be evaluated so that the nearer neighbours along a particular direction are evaluated first. If a near neighbour is blocked by an obstacle state, then the remainder of the neighbours along that direction are not explored (see Figure 8.11). This feature is necessary to avoid collisions with convex

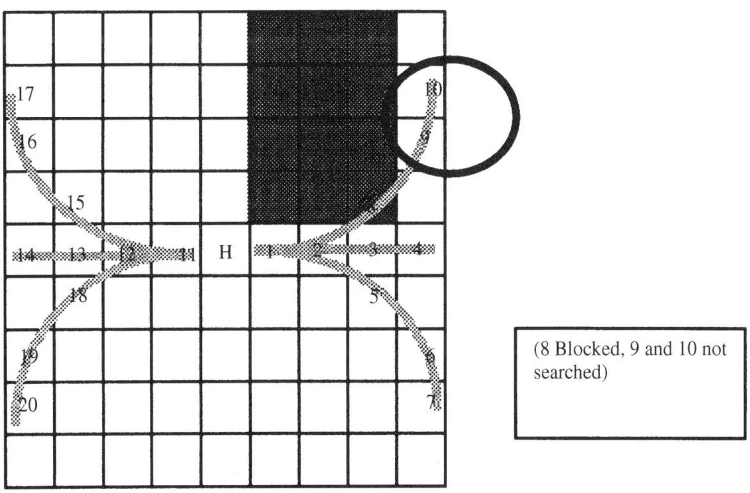

Figure 8.11. Later neighbours not searched if prior neighbour obstructed

corners because otherwise it is possible to step over obstacles when evaluating distant neighbours.

The result from the CWP is a direction arrow in each reachable configuration state indicating the steering wheel orientation and transmission direction (i.e. forward or reverse) for the vehicle. By following the arrows from any starting state, the minimum distance path will be travelled to the goal. Figure 8.12 shows the path resulting from a minimum distance criterion starting from a position parallel to the rear parked car. Figure 8.13 shows the manoeuvre necessary if the controlled vehicle is started in the parking spot but facing backwards. Of course, parallel parking is not the only manoeuvre that can be computed.

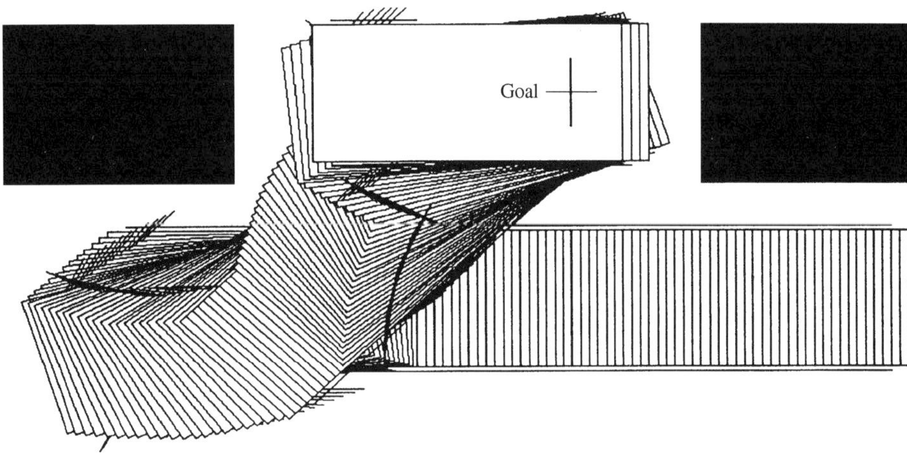

Figure 8.12. Parallel parking in minimum distance

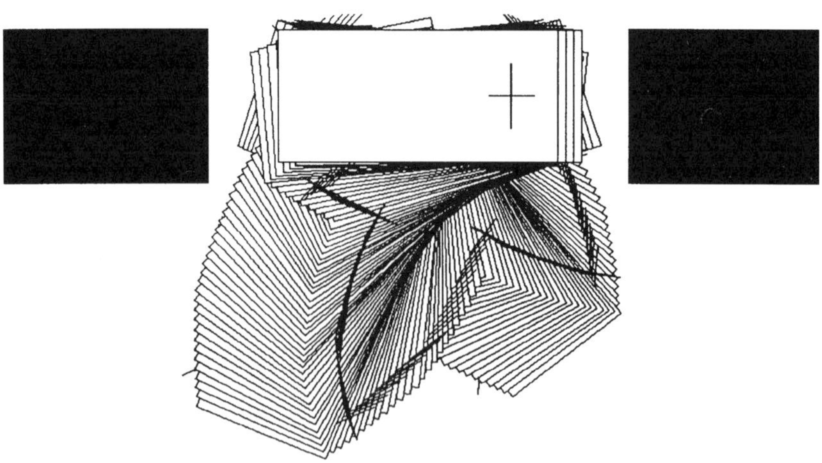

Figure 8.13. Reversal

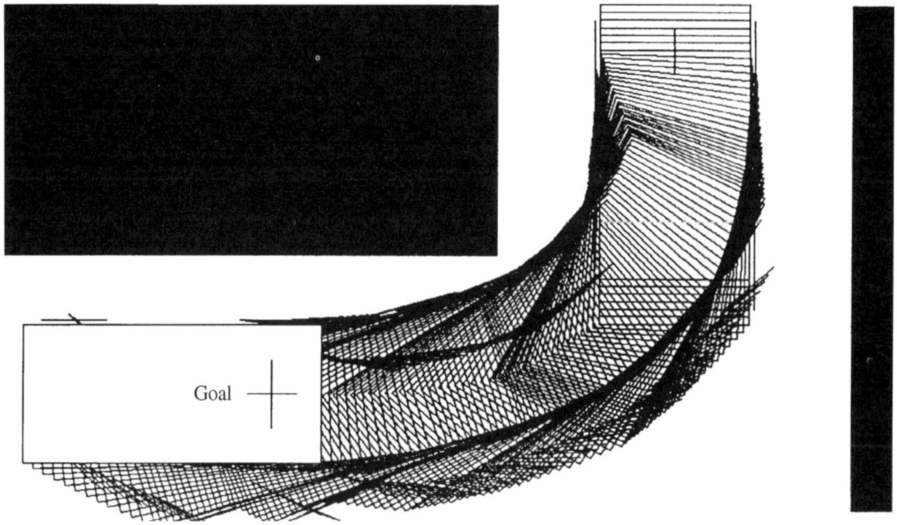

Figure 8.14. A right turn with forward-only constraint

8.7 A FORWARD-ONLY CONSTRAINT

If the above method was used to make a right turn, it would possibly lead to a manoeuvre with several transmission changes, because it is the actual minimum distance solution. This is obviously undesirable in ordinary traffic. This could be solved by greatly penalizing changes of transmission, although in this case it is simpler to eliminate the neighbours corresponding to the reverse direction entirely. This has the additional benefit of speeding the wave propagation considerably. All other aspects of the set-up and control are the same. Figure 8.14 shows a simple right-hand turn constrained to allow only forward motions.

8.8 USE OF THE SYSTEM FOR COMPLEX MANOEUVRES AND LARGER AREAS

Complex manoeuvres are required when there are many perhaps odd-shaped obstacles or a restricted space. The problem is set-up and computed in the same way as for simpler arrangements; however, the sensing problem is more difficult. In the case where the vehicle is performing anything much more than parallel parking, inexpensive sensors will not suffice. This is particularly true if complex obstacles must be sensed, or a large area is used for manoeuvring. Even if the environment is known, the position and orientation of the vehicle must be reported in real-time. This is perhaps the most difficult problem, as open-loop control is only accurate for a few motions. Only if the vehicle is performing in a restricted location, such as a warehouse, airport, ship or commercial establishment, can fixed sensors be placed efficiently to track or locate the vehicle. Global positioning with differential correction may be of use in some outdoor applications, although this is an area for future research. Other than the sensing problem, there is no difficulty in computing an optimal path around numerous obstacles.

The set-up and computation follow directly from the previous example. Given the vehicle's dimensions and steering capability, the neighbourhood can be computed automatically. The cost used is the distance the cross-hair marker travels. If the obstacles and vehicle are approximated by rectangles, then the transform resulting from a single angle of the vehicle is as shown in Figure 8.15. The region around an obstacle represents the closest that the cross-hair of the vehicle can get to the obstacle at the current orientation of the vehicle. The illegal region in configuration space is determined by computing the transformation at each of the convex and concave corners and filling in the forbidden region. More complex shapes are certainly possible. For each possible angle of the car, a transform is computed. The transforms for all angles of the car can be performed in real-time. The optimal path is generated by performing CWP from the goal (or goals). The waves avoid the transformed obstacle regions, and eventually fill the reachable three-dimensional configuration space. As mentioned before, the more obstacles that are present, the faster the solution is computed. By following the steering and transmission commands from location to location the path in Figure 8.16 is found.

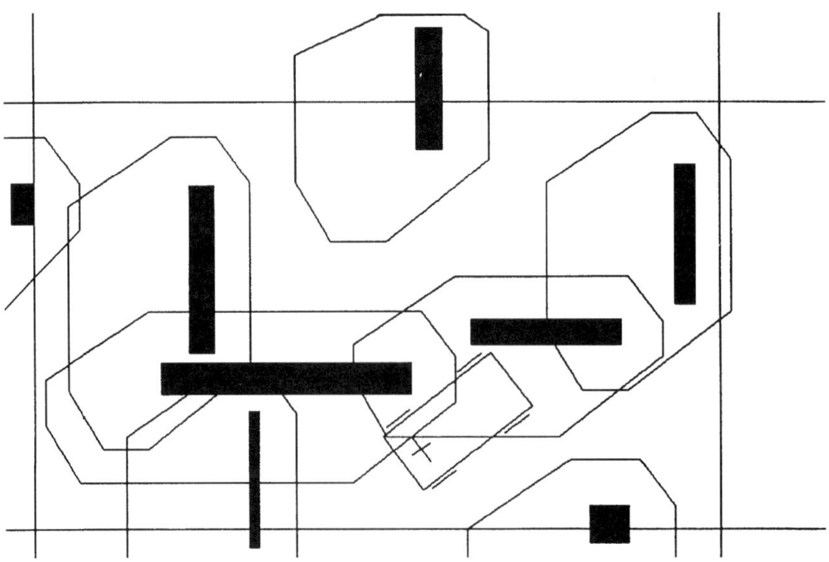

Figure 8.15. Transform of eight obstacles and four containing walls for a car at +45°

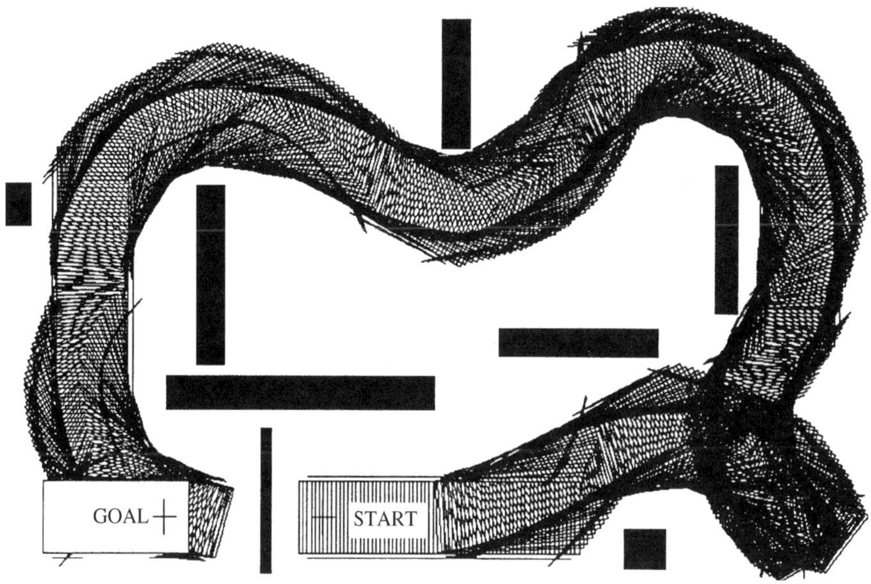

Figure 8.16. Minimizing distance travelled while avoiding obstacles

8.9 A RADIO-CONTROLLED EXAMPLE

The simulated path of the vehicle shown in Figures 8.12–8.14 have been executed in a test-bed environment shown in Figure 8.17. In addition to these, many more complex manoeuvres constrained by the four walls were accomplished. A Motorola 68020 based multiprocessor system called SPINE (structured processor interactive networked environment), developed at Philips Laboratories for real-time experiments, is used to co-ordinate and control the activities of the car. A stock 1/10 scale RC-10 radio-controlled car is used as the controlled vehicle. The body dimensions and steering ability are used as input for planning the motions. The test-bed also contains two car bodies as obstacles.

The position and orientation of the RC-10 are determined by an infra-red camera, mounted above, viewing three infra-red LEDs mounted on the top of the car body. The camera transfers the image to a Philips PAPS (picture acquisition and processing system) where the image is thresholded, the points are refined, and the position and orientation are reported at a nominal rate of 20 Hz to the M68020 processor controlling the car.

The RC-10 is controlled from a stock FM radio-control transmitter, fitted with D/A taps for computer control. One tap controls steering angle while the other controls transmission (forward/reverse). The drive-train of the RC-10 was altered so that it has a geared down rear axle and smaller motor. This was done both to reduce the speed of the vehicle and to extend the life of the on-board battery to about 4 hours.

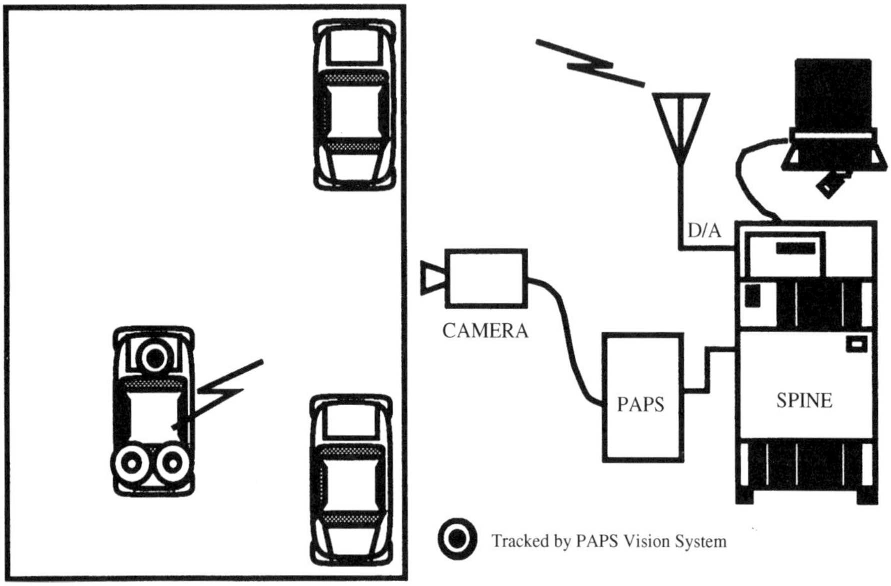

Figure 8.17. The test-bed environment

The set-up of the parked cars is input via a graphical user interface (GUI) on a Sun 3/160. In an expanded implementation, the location of the parked cars could be determined by sensors. The goal and selection of optimization criterion are also given in the GUI. Once the configuration space is computed, it is downloaded to the 4 Megabytes of common memory in SPINE.

From the common memory area, the controller can read the direction arrows indicating the proper control of the vehicle from any currently sensed position. In practice, a few enhancements to the previous control schemes (i.e. for the robot) are needed. These are needed because the direction arrows in the configuration space tend to point in wildly different directions for even minor variations in state, thus a smooth gradient field is not available for control.

The first enhancement is, as neighbours are arcs of limited length, any sequence of similar neighbours are concatenated to determine the next set-point for control. For example, if the path contains two directives that both result in forward motion with wheels at 45°, then the set-point should be at the end of the farthest 45° neighbour. This allows some smoothing of the discretization errors that necessarily occur when planning a continuous problem in a discrete space.

The second is that an arc is struck from the current active position through the next x, y, θ set-point and the vehicle is PID (Proportional Integral Derivative) controlled through that point. This helps correct for an imperfect mechanical system or minor sliding on the surface.

The third enhancement is that a control stopping criterion must be provided. The vehicle is considered to have reached the set-point when the vehicle's position crosses the line that runs through the set-point perpendicular to the set-point orientation. This line allows for some mechanical or frictional error, while still ensuring a fairly robust hand off between set-points.

Fourth, once the vehicle has crossed the line defining the stopping criterion, an evaluation is made. If the current location is within reasonable range of the set-point, then control resumes with the next neighbour in the path, otherwise the path is determined anew from the current location. As an alternative to this, an acceptability test can be made while control is underway. If the current position is wildly out of track with the set-point, such as if we had reached in and flipped the car around, the controller can abort the motion and continue from a newly determined starting position.

To use the test-bed once the system is loaded, the RC-10 is placed anywhere within the physical boundaries, and a "start button" is pushed. The car then drives to the goal.

Anyone interested in a free videotape (specify PAL or NTSC) showing the vehicle manoeuvring testbed may contact the author (see contributors list on pp xvii and xviii or electronically at kit@philabs.research.philips.com.

8.10 HIGH-SPEED VEHICLE MANOEUVRING

Initially, the planning for a scenario any larger than that of the complex region in the previous section would seem impractical, not only because of the real-time sensing issues, but also because the planning would be too slow and require too much memory. As manoeuvre planning determines the proper reaction in the sensing-reaction control loop, it is important to be able to perform this quickly to get adequate "reaction time". Fortunately, quick planning in a relatively small space can be performed for this problem.

In the situation where a vehicle is moving along a highway, the configuration space should be considered as the space moving *relative* to some other object or marker. Previously, all configuration spaces were considered to be relative to a fixed origin.

Looking at the "neighbourhood of permissible motions" for a high-speed vehicle, it is discovered that the vehicle can in fact move sideways, relative to other vehicles, simply by steering. Also, forward and backward motion can be achieved by accelerating or decelerating. This is a delightfully simple neighbourhood for planning purposes, as shown in Figure 8.18(a). Figure 8.18(b) gives the results of straightforward path planning in a relative space for vehicles moving at higher speeds. The goal is to pass the other cars. At each state there are directives such as "accelerate and move one lane to the left". Speed limits can be incorporated naturally by changing the neighbourhood as a function of the current speed. As the computation is very quick, any change in the relative positions of the other cars can be accommodated in a short time.

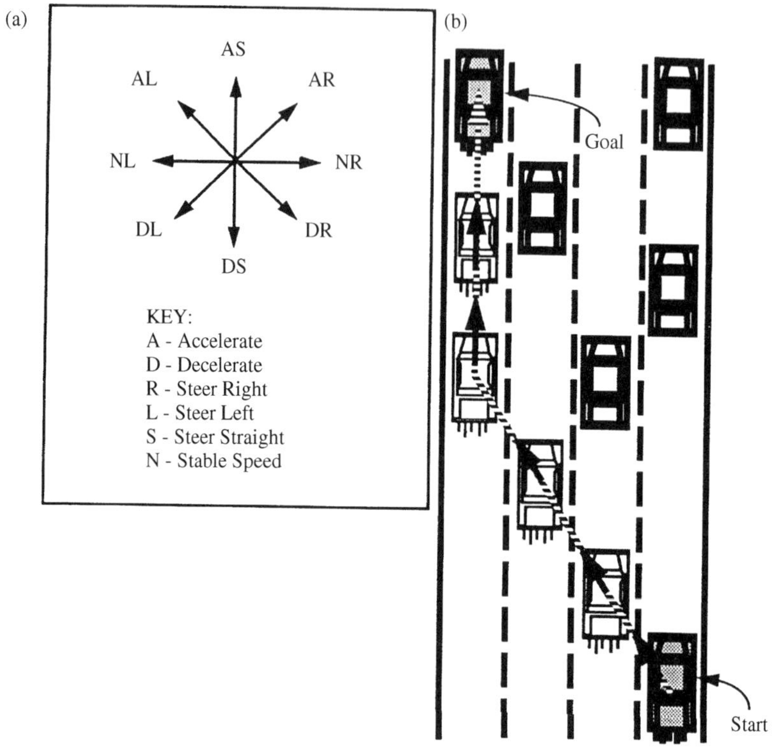

Figure 8.18. High-speed vehicle Manoeuvring. (a) Neighbourhood for possible moves for a relative frame of reference. (b) Overtaking manoeuvre planned automatically

8.11 INVITATION

Most of the information given in this chapter is covered by issued and pending patents. Anyone interested in collaborating or directly licensing these techniques is encouraged to contact the author at the given address. For example, a recent challenge to provide rendezvous co-ordination of multiple machines has already been addressed.[14]

8.12 CONCLUSIONS

A framework has been presented which can encapsulate a machine or agent's fundamental behaviour and automatically compute intelligent manoeuvres. By transforming the problem into a configuration space, path planning for a point can be used to yield optimal collision-free manoeuvres. It is general and efficient for many types of problems. A nice property is that the nature of the method makes it more efficient for spaces with more obstacles, which are also typically problems that are the most difficult to perform by inspection.

Many examples are given in this chapter, including a solution for a safer fire exit system, which provides the best route out of a building for all inhabitants based on the sensed alarms. This can be implemented by installing intelligent arrow-lights at regular intervals throughout the building which point toward the safest exit. This example is important because it can also be used for other planar problems such as optimal point-to-point highway selection and rough terrain routing.

Another example is a robot that moves optimally from one place to another without hitting obstacles. The specific machine movement can be incorporated into a neighbourhood which ensures that the resulting optimal path is feasible for a specific machine. Because the path is computed automatically, only higher-level directives or goals are needed for the machine to complete complex tasks.

The manoeuvre for a vehicle can be computed in much the same way as for the robot. The main difference is that the machine movement is more complex for a non-holonomic problem. The result is a car that can derive complex manoeuvres (such as parallel parking) based on the obstacle layout. A test-bed implementation has shown that the computed path is indeed controllable.

High-speed vehicle manoeuvring can be performed by using a relative frame of reference and by using the neighbourhood that maps acceleration and deceleration to forward and backward motion, and higher speed left and right steering to left and right motion. This makes the problem simple to solve in two dimensions, which in turn gives a relatively fast reaction time.

By continuing to challenge this framework with diverse problems, ever-evolving facets help shape its domain. New challenges are therefore always welcome.

REFERENCES

1 Lozano-Pérez T, Wesley MA. "An algorithm for planning collision-free paths among polyhedral obstacles." *Communications of the ACM* 1979; **10**: 560–570.

2 Kavraki L. "Computation of configuration-space obstacles using the fast Fourier." *1993 International Conference on Robotics and Automation* 1993; **3**: 255–261.

3 Lyons D, Hendriks A. "Planning for reactive robot behavior." *1992 IEEE International Conference on Robotics and Automation*, Nice, France, 1992.

4 Pearl J. *Heuristics: Intelligent Search Strategies for Computer Problem Solving*, Reading, MA: Addison Wesley, 1984.

5 Dorst L, Trovato K. "Optimal path planning by cost wave propagation in metric configuration space." *SPIE Conference on Advances in Intelligent Robotics Systems*, The International Society for Optical Engineering Nov., 1988.

6 Khatib O. "Real time obstacle avoidance for manipulators and mobile robots." *International Journal of Robotics Research* 1986; **5**: 90–98.

7 Trovato K. "Differential A*: an adaptive search method illustrated with robot path planning for moving obstacles and goals, and an uncertain environment." *International Journal of Pattern Recognition and Artificial Intelligence* 1990; **4**: 245–268.

8 Trovato K, Dorst L. US Patent Serial #4,949,277. *Differential Budding: Method and Apparatus for Path Planning with Moving Obstacles and Goals*. Issued 14, August 1990.

9 Brooks RA, Lozano-Perez T. "A subdivision algorithm in configuration space for findpath with rotation." *IEEE Transactions on Systems, Man and Cybernetics* 1985; **15**: 224–233.

10 Featherstone R. *Swept Bubbles: A Method of Representing Swept Volume and Space Occupancy*. Philips Public Document, MS-90-069, 1988.

11 Borgefors G. Distance transformations in arbitrary dimensions. *Computer Vision Graphics and Image Processing* 1984; **27**: 321–345.

12 Dorst L, Smeulders AWM. "Length estimators for digitized contours." *Computer Vision Graphics and Image Processing* 1987, 1990; **40**: 311–333.

13 Newman W. *High-speed robot control in complex environments*. PhD Thesis, MIT, October, 1987.

14 Trovato K. "General method for strategic assessment and planning problems." *Sixth Joint Service Data Fusion Symposium 1993*, 1993; **1** (Part 1): 305–335.

9

Optically-based Underground Mine Vehicle Guidance

9.1 INTRODUCTION

Underground hard rock mines are highly mechanized today and most of the operations involved in the extraction of ore, such as drilling, rock bolting and ore transportation, are performed by various machines. The transportation of ore, for example, is accomplished either by trains, conveyors or trackless vehicles like the load haul dump (LHD) vehicle or haulage trucks. Trains and conveyors are usually controlled automatically and are used for large-size exploitations where the infrastructures needed for these systems are economically justified. In small- and medium-sized exploitations, the use of man-controlled vehicles offers a maximum flexibility to mine operators but is labour-intensive. It also presents non-negligible risks to the safety and the health of operators. Automation of these vehicles according to the concept developed for the automated guide vehicle (AGV) system used in manufacturing industries is an important objective for the mining industry. Automated mining vehicles would offer good cost-saving potential while being adaptable to the evolving needs of mining exploitation. Moreover, the operator can be positioned at a location remote from the hazardous environment of underground haulage drifts.

The feasibility of applying an AGV system was demonstrated by researchers in 1972 in Amdahl and Lundstrom.[1] A classical electromagnetic guidance system using an embedded wire was used to control a 6-ton truck successfully over a drift. However, the hostile and unstructured nature of the mining environment, and the great variability in the exploitation of underground mines have made this classical system unsuited to most underground mine operations.

Advanced Vehicles and Infrastructure Systems, Edited by C.O. Nwagboso

9.1.1 Mine Context

To understand the requirements for an automated mining vehicle, the structure of an underground mine is briefly described. As illustrated in Figure 9.1, the general configuration of a mine is closely related to the structure of the ore body. As shown in the figure, the ore body is broken by explosives on its upper part, so that the resulting fractured boulders can be extracted on the production level at the bottom, from the draw points or the open stope. At this level, a haulage drift connects the draw points to the dumping point and the ore pass. In a typical exploitation cycle, the ore is transported by LHD from the draw points through the haulage drift to the dumping point over a distance of 200–800 m and at a speed of about 15 km/h. For larger distances, haulage trucks are used. These vehicles are both articulated and powered by diesel engine or by electric motor. Figure 9.2 shows an LHD vehicle.

The drifts are designed with the minimum width required to allow the safe operation of the largest vehicle travelling within it. Depending on local regulations, the lateral clearance between the vehicle and the wall of the drift varies between 0.9 and 1.2 m; the back of the drift is at least 0.5 m above the highest point of the vehicle. In haulage areas, the grade is less than 10% and the turning radius is designed to accommodate the vehicle.

The surfaces of the haulage drifts are rough and irregular as shown in Figure 9.3. The floor is usually covered with crushed stones and can be flooded with water in

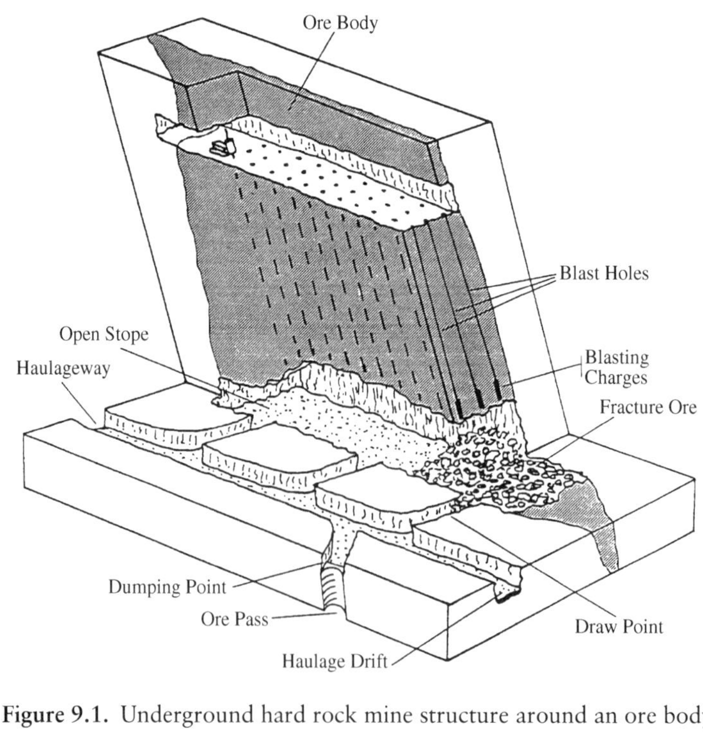

Figure 9.1. Underground hard rock mine structure around an ore body

Figure 9.2. An LHD vehicle

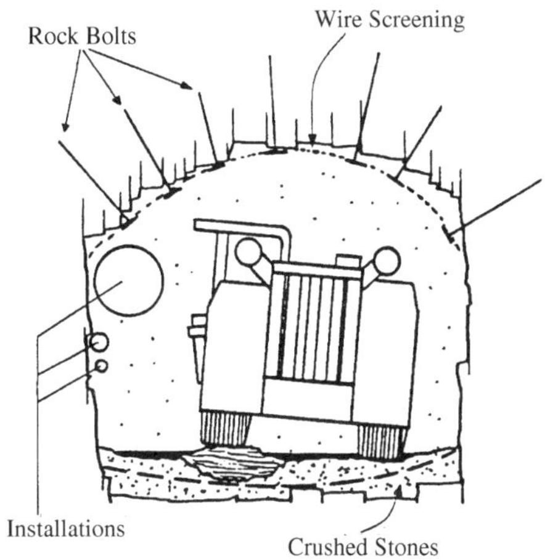

Figure 9.3. Cross-cut view of a typical haulage drift

some areas. The walls and the ceiling of the drift can be reinforced with rock bolts and, in some areas, wire mesh screening can be added. The atmosphere is rather dusty, humid and can be corrosive. During a rock-blasting operation, which is done on a daily basis between two work-shifts, a strong wind with debris and gases can flow through the drift. The immediate suroundings of the draw point and the open stope are continuously modified by this blasting and by the extraction of ore. For this reason, ground conditions in these areas can be qualified as dynamic. However, in the haulage drift and around the dumping point, ground conditions are stable and permanent installations are possible.

9.1.2 Automation of Ore Transportation

A guidance system properly adapted for underground mine applications must have the following requirements: be sturdy, able to resist shocks and vibrations, and be simple to understand and to maintain. The system must provide the capability to guide a vehicle in a network of drifts, typically in a point-to-point task. As mines are constraint areas where powerful machines and humans are present, the system must be predictable in any circumstance and must be fail-safe.

Over the last 5 years, many guidance principles have been proposed and discussed in the literature[2–4] e.g. the use of an electromagnetic field from an electrical guide wire as in most industrial AGVs,[1] the use of a trolley-type sensor following an overhead trolley wire,[5] the use of an ultrasonic or optical range finder to detect and to map the drift, the use of a two- or three-dimensional vision system in an unstructured environment, the use of optical point-like landmarks and the use of an overhead optical guideline.[6,7] A detailed evaluation of guidance principles[6] has concluded that the overhead optical guideline is the solution best adapted for flexible ore transportation in underground mines.

As illustrated in Figure 9.4, the optically based AGV uses an optical guideline installed on the back of the drift and two optical detectors mounted on the vehicle (one for forward motion, one for reverse motion). The guideline is made of a high-efficiency retro-reflector. The detector is composed of a camera and a light source. The detector has the capability to detect the deviation of the vehicle from the guideline, and a motion controller maintains the vehicle on the desired trajectory. This solution meets the requirement in terms of simplicity, low cost, sturdiness, performance and safety. The intrinsic safety is guaranteed by the fact that the vehicle moves only if a direct optical contact is maintained with the guideline.

The optical AGV system can be used to automate the displacement of the vehicle in the haulage zone, where the drift configuration remains relatively unchanged and where the optical guide can be installed permanently. The positioning of the vehicle in draw points and in open stopes as well as the loading of the bucket can be executed in a remote-control mode. This scenario will allow the operator to be located in a

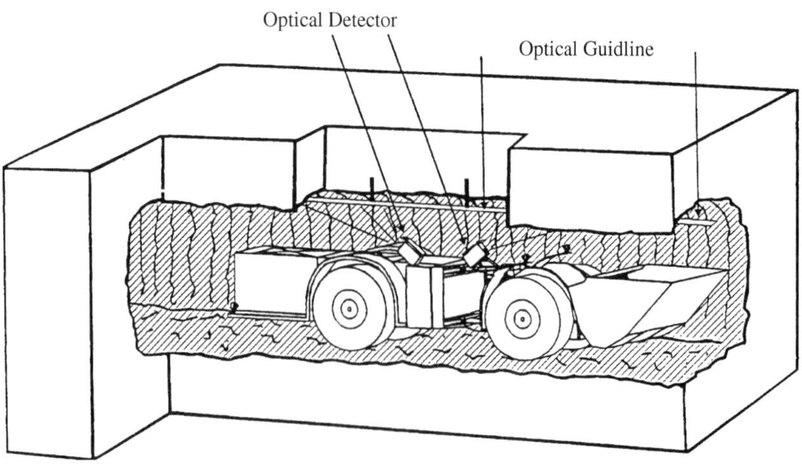

Figure 9.4. Optical guidance system

control room at a convenient distance from the vehicle and hazardous areas. The safety of the vehicle operator will be improved and productivity will be increased by having this operator supervising two or more vehicles at the same time and from the same protected location.

This chapter presents the research carried out and the technological developments achieved over the last 3 years to demonstrate the feasibility of an optical guidance system for an underground mine vehicle.[4,8] A general description of the automated vehicle is given and the operating principle for the optical guidance system is explained: the optical detector and the vehicle motion controller are described. Their co-ordinate performance is analysed and illustrated by simulation as well as on an experimental small-scale electric vehicle. Finally, a full-size demonstration prototype that has been tested at the Norita Mine in Matagami, Québec, is presented.

9.2 DESCRIPTION OF THE SYSTEM

The optical guidance system will be installed on a haulage vehicle for the underground transportation of ore, waste rock boulders or backfill material. Under normal operation, the miner would remotely program the vehicles by specifying a task sequence, as for example, repeating eight times: go to draw point A, load, go to dumping point D, unload, etc. Some of the actions in this sequence will be executed automatically; others can be executed by the operator in a remote-control mode, depending on the complexity of the task (loading for example). In this context, a modular architecture is needed to accommodate the various situations that will be faced by the vehicle and the automatic system in the mine. Figure 9.5 shows an architecture consisting of five object modules: the operator, the haulageway, the vehicle, the obstacle and the load. The various functions needed to automate the vehicle are embedded in these object-oriented modules, as they are perceived by the operator.

This architecture represents a complete system for an automated underground mine vehicle, which can be adapted, depending on the local context, to different levels of automation. Some modules can be manually operated, or partially or totally automated. This flexiblity is needed to accommodate the variety of situations existing in mines.

Each object module is composed of functions that are described by the following:

1. *Vehicle operator*. This is the supervisory module of the system. It contains functions related to operator responsibilities, in particular the production and safety functions. The production function contains a list of useful high-level sequences of operations and calculates the progress in the load flow. The safety function handles all the alarms and verifies the presence of humans in the drift or around the vehicle. The communication function provides the link between the operator and the machine, including the human–machine interface. Remote operation provides

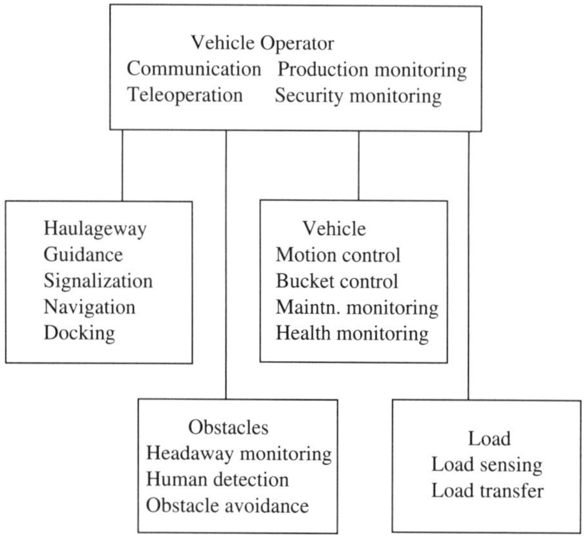

Figure 9.5. LHD automation system modules

the capability to operate the vehicle in a remote mode in any situation, introducing a functional redundancy to the other modules of the system. In particular events, the operator can bring the vehicle to an underground garage for immediate repair or operate the vehicle at the expense of production efficiency.

2. *Vehicle*. All the functions related to the operation of the vehicle are regrouped in this module: motion control (steering, engine and gear train), bucket control, and health and maintenance monitoring.

3. *Load*. The functions that are responsible for locating, transferring and identifying the load are present in this module. The identification of the load in terms of percentage of ore and waste is carried out at this level. This is presently done visually by the operator.

4. *Haulageway*. This module contains all the functions related to the location and operation of the vehicle in the haulageway. These are guidance, docking, signalization and navigation. The guidance and docking functions refer to the capability of the vehicle to travel in the drift and to stop at precise locations such as dumping points. The two latter ones, signalization and navigation, involve the possibility of identifying landmarks in the mine, situating the vehicle in a network of galleries and determining the path to follow in order to reach the final destination.

5. *Obstacle*. This module contains mostly event-type functions which detect any unpredictable conditions causing potential damage to material or to humans. Headway monitoring verifies the presence of any obstacles that can cause damage

to the vehicle while the human detection module is specialized in monitoring the presence of humans. The obstacle avoidance function provides a tactical plan in order to prevent collision when obstacles are present.

Obviously, guidance and the vehicle motion controller are the essential functions of the optically based guidance system. The fundamental concepts for the guidance of an LHD are defined by their geometric representation in Figure 9.6. The guidepath defines the desired trajectory of the vehicule reference point. On the LHD, the reference point is located between the leading wheels of the vehicle and depends on the direction of motion. Physically, the guidepath is achieved by an optical guideline which can be situated parallel to the desired trajectory.

The guideline gives only partial information to the guidance system. The presence of a branching point, a change of curvature of the guideline, the start or the end of a slope, the beginning of a rough surface and a garage entrance are information that has to be transmitted to the guidance system. To transmit this information to the system, three optical landmarks are used: branching points, end-points and encoded milestones. As shown in Figure 9.6, the branching points are characterized by the presence of multiple guidelines, while the end-point is characterized by the end of a single guideline. Milestones are composed of narrow retro-reflectors assembled to create encoded patterns. The signalization function identifies the milestones that are present near the guideline.

In a network of guidepaths, the vehicle requires more than optical landmarks to reach the right destination and to perform appropriate actions at stop-point stations;

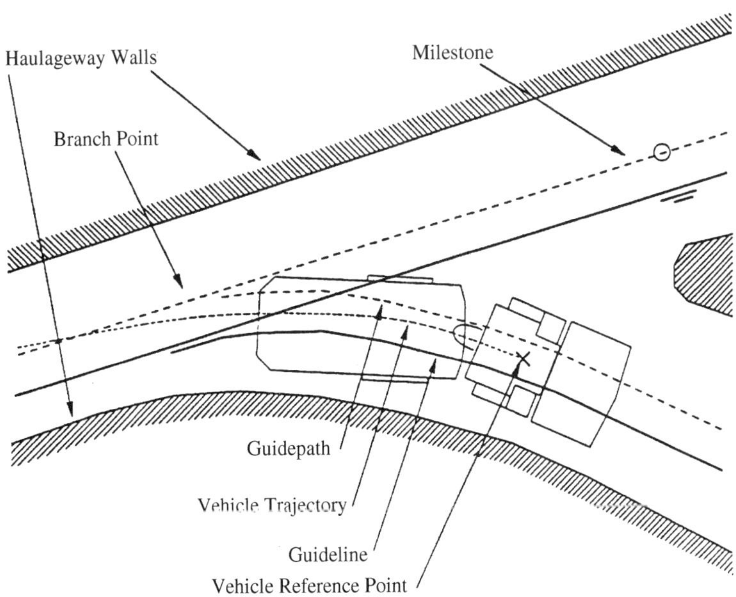

Figure 9.6. LHD vehicle guidance: concepts and definition

it requires a route description. This description is simply a sequence of decisions to be made and operations to be performed at each landmark. The navigation function is responsible for continuously situating the vehicle in the mine.

9.2.1 Demonstration Prototype

A demonstration prototype of the optical guidance system has been implemented in a mine at Matagami, Québec.[9,10] The basic materials in terms of guideline and optical landmarks were installed for this experiment. The fundamental functions of the haulageway and the vehicle modules were also integrated into an existing vehicle.

The plan of this mine including the various components of the guidance system are illustrated in Figure 9.7. The guideline is located on the left side of the drift. Two branch points, four end-points and six milestones are present; the milestones indicate a change in drive characteristics. The milestones are identified by an easily recognizable bar code pattern. Code 0 identifies a nearby dead-end, and code 1 identifies a change of curve. At this mine level, there are no crossing intersections and no loops; it is an arborescent type map of haulage drifts with a main guideline and short secondary guidelines.

The guidance, signalization, navigation and motion-control functions have been tested on a Wagner ST-5 diesel powered LHD. This vehicle was equipped with a Nautilus remote-control system, various sensors and a monitoring system. As shown in Figure 9.8, the guidance system installed on the vehicle is composed of an IBM-compatible microcomputer and a commutation unit through which the remote control receiver, the sensors and the actuators are connected. The actuators used were the throttle, the steering cylinders and the brakes. A steering angle sensor and an engine RPM sensor were also used. Some of the remote control receiver outputs were connected to the microcomputer; they allowed the operator to start remotely

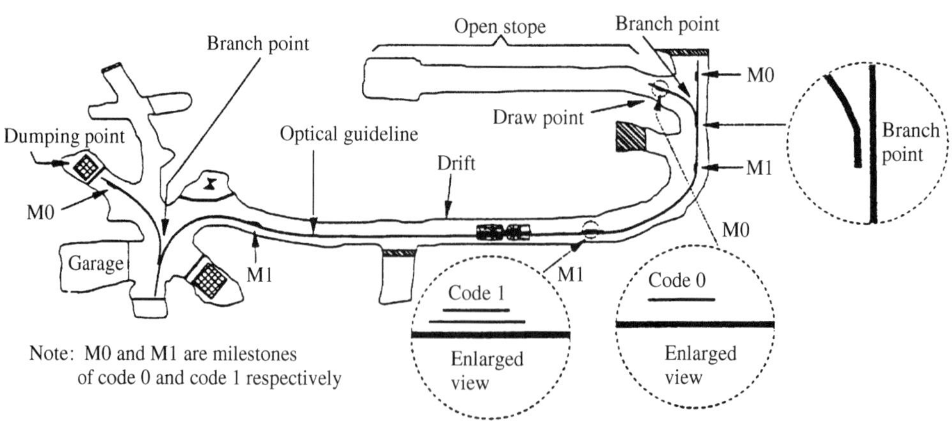

Figure 9.7. Norita Mine: the exploitation zone

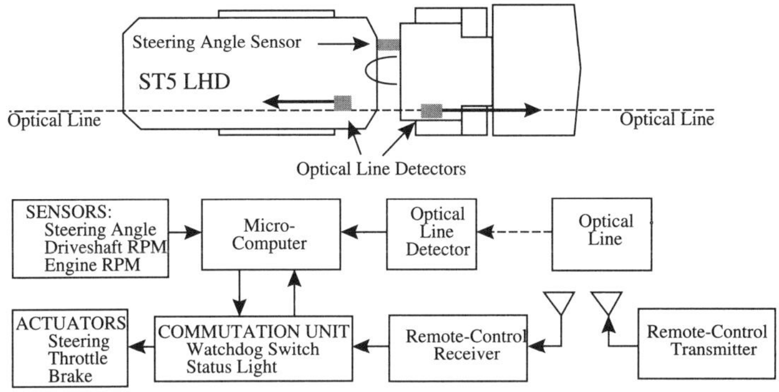

Figure 9.8. Hardware architecture of the demonstration prototype

or stop the automatic guidance program by generating a special sequence of signals. The emergency stop procedure is activated if the operator presses on the remote control, or if the watch dog circuit generates a stop command because a computer halt is detected. A status light panel, installed at the rear of the LHD, confirms to the operator the mode of operation of the vehicle (manual, remote control or automatic guidance mode). This commutation unit proved to be a useful safety element for the test sequence.

9.3 OPTICAL DETECTOR

To test the optical guidance system, an optical detector capable of operating in the hostile mine environment and of giving continuous and reliable information on the location of the vehicle has been designed and built. In terms of performance and safety, this detector is the most critical component of the guidance system. For this reason, the choice of material for the guideline and the milestone, the mechanical design of the guideline detector, and the development of the image analysis algorithm implemented in the system have been analysed in various studies. Tests in underground mines have also been conducted in preliminary phases to validate the technological choice involved.[4]

In this section, the resulting optical detector is presented. The hardware component of the guideline, the milestones and the guideline detector are given. The image analysis and the geometric model are presented. Finally, the performance of the detector for guidance is analysed and measured experimentally and by simulation.

9.3.1 Guideline and Milestones

The optical guideline can be made of any visible material such as a light source, fluorescent material or any reflective material. For mine applications in which kilometres of guidelines are required, the solution using reflective material as the guideline and a light source on board the vehicle has been selected. Different reflective materials such as painted lines, a white ribbon or a retro-reflective ribbon were evaluated in typical mine situations.[4] This study indicates that many commercial retro-reflector ribbons are at least 100 times more reflective than any white surface. To obtain this optical property, the light source has to be mounted as close as possible to the detector objective. A typical mounting of the components is shown in Figure 9.9(a). Figure 9.9(b) illustrates the importance of mounting the light source as close as possible to the camera objective, but outside its field of view. The smaller the view angle [the angle formed by the light source point (*L*), the retro-reflector surface element (*R*) and the camera objective (*C*) as shown in the figure], the higher the reflection efficiency. This angle must be kept smaller than 1°.

Figure 9.10 illustrates the relationship between reflection efficiency and observation angle for a very efficient retro-reflector material that has been selected for underground vehicle guidance purposes. The reflection efficiency is defined in relation to a white matted surface reflexion with an incidence angle of 0°. The observation angle is the angle formed by the retro-reflector normal vector (*N*), the retro-reflector surface element (*R*) and the camera objective (*C*) (Figure 9.9a). As shown by the curve, a good reflection ratio (over 20) is maintained for all observation angles between 0° and 80° along the ribbon axis.

These interesting optical characteristics of retro-reflectors and their relatively low cost made them suitable as guidelines and also for the encoded milestones in mine applications. To differentiate between them, a ribbon 5-cm wide is used for the guideline, and a ribbon 1 cm wide for milestones. On the site, the retro-reflective ribbon is glued to straight and rigid strips of aluminium which are easily suspended with brackets from the back of the drift.

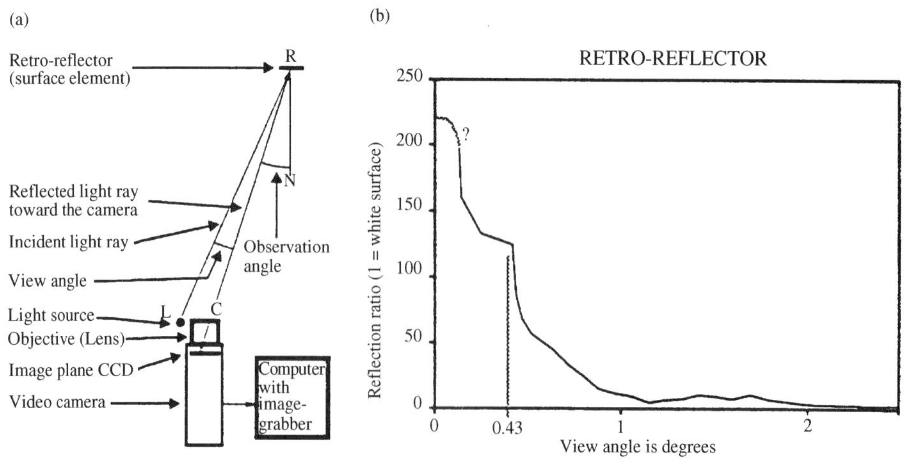

Figure 9.9. Optical guideline detector. (a) Mounting Principle. (b) Effect of the view angle on the retro-reflection ratio

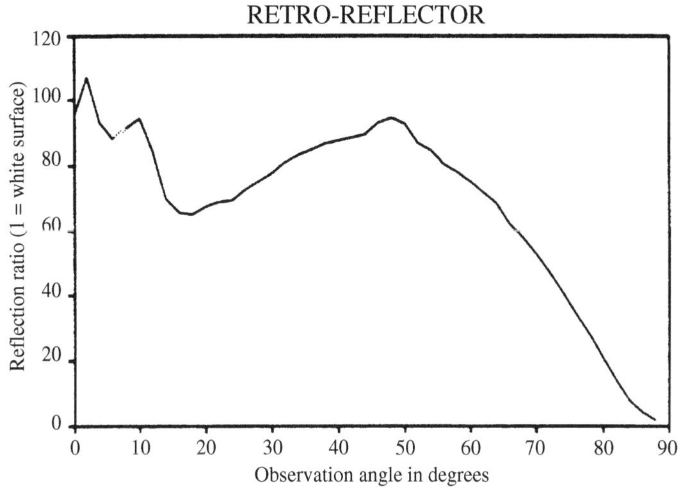

Figure 9.10. Effect of the observation angle on retro-reflection ratio

9.3.2 Guideline Detector

The guideline detector consists of a light source, a CCD video camera, a frame grabber and image-analysis software. As shown in Figure 9.4, the two hardened cases, including the light source and the camera, are installed close to the vehicle's articulation in order to be protected from falling rocks during loading.

The use of a video camera and image-processing techniques is justifed by the fact that such a detector has to be adapted to a variety of situations which can be encountered in mines. Image processing provides great flexibility in handling complex situations such as images that include multiple guidelines and milestones. Imperfect situations such as missing or hidden segments of the optical guideline can be handled easily by image processing and more detailed information can also be extracted from the image. For example, the measurement of the angular and lateral deviation of the vehicle can be obtained simultaneously and independently through proper analysis of the image.

A standard industrial-grade CCD camera was selected. It provides an analogue video signal at a frame rate of 30 per second, a 7-lux sensitivity, an area resolution of 512 by 480 pixels, and a signal-to-noise ratio of 46 dB that is just sufficient for an 8-bit resolution (256 gray levels). Each image is integrated during the 1/30 s allocated for a frame. For vehicle guidance, the frames are taken while the vehicle is moving.

9.3.3 Image Analysis[4]

The light reflecting from the gallery roof rock or from other installations is much lower than the guideline and the milestones made of retro-reflectors; these other light reflections are filtered out by a simple intensity thresholding filter. Consequently, the image obtained by the camera has low clutter, since only retro-reflectors, occasional shiny metallic objects (rare in a hard-rock underground mine) and very few light sources have sufficient intensity to pass through the threshold filter.

A typical image of the optical guideline, as taken by the video camera, is illustrated in Figure 9.11. As shown, a branching point composed of two guidelines and a milestone are present in the video image. To find the position of the optical line on the screen and identify the milestone, three video scan lines S1, S2 and S3 are selected from the digitized image; two scan lines are necessary to evaluate independently the orientation and the lateral deviations of the guideline (usually scans S1 and S3), and the other one, S2, is used for redundancy purposes.

The light intensity profile of each scanned line is interpreted as a collection of peaks of various width. A peak is any unbroken pixel sequence whose brightness value is higher than a specified threshold. The collection of peaks are then classified by their width and compared with an expected width: wide peaks are associated with the guideline while narrow peaks are related to milestones. A peak is rejected if it is too narrow or too wide. The position and alignment pattern of the peaks of every scanned line in the picture are checked against possible image patterns expected from the optical guideline installation. Ambiguous or non-credible patterns are rejected. The position of the points associated with the optical guideline are measured by the (x, y) location of the centre of their peaks in the image. These points are then used in the geometrical model of the detector to situate the optical guideline in relation to the vehicle.

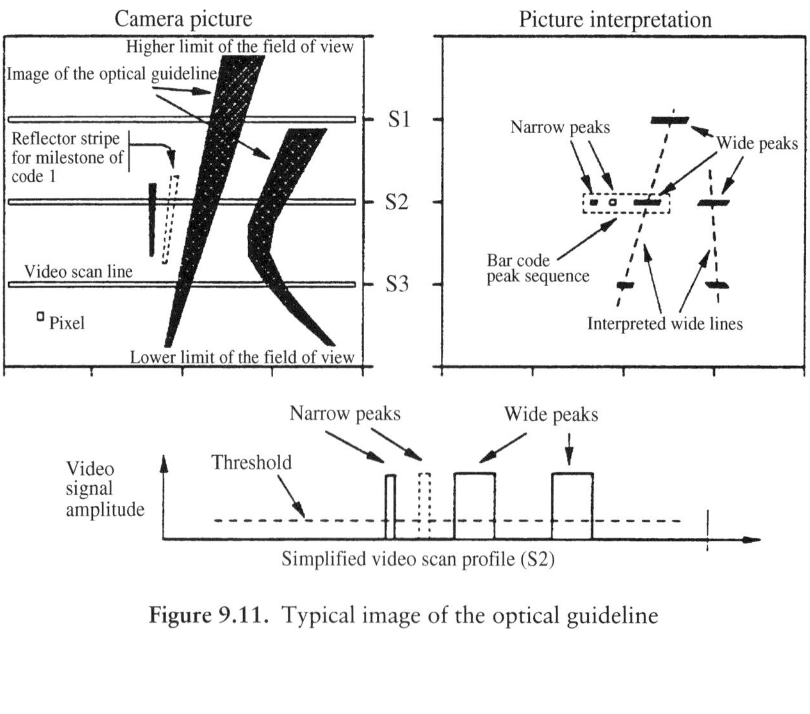

Figure 9.11. Typical image of the optical guideline

During the field tests in an underground mine, this simple image-analysis procedure was sufficiently powerful to detect the guideline in the presence of dust and water condensation. Some losses of reflection were observed occasionally in one scan only, and rarely in two scans.

9.3.4 Geometrical Model of the Optical Detector

Image interpretation

Once the location of the guideline has been identified and measured in the image plane of the camera, the next step is to calculate the relative location of the vehicle reference point with respect to the guideline. However, to ensure great flexibility, the reference point on each body of the vehicle is programmable (usually, the middle of the leading wheel axle is the reference point), and the detectors can be located, for each body of the vehicle, anywhere under the guideline axis. In consequence, the geometric model of the detector has to be known precisely to permit the computation of the navigation errors associated with a given reference point, for any location of the detector.

The development of this model is based on the following assumptions:

(1) the guideline is parallel to a horizontal rolling surface;

(2) the guideline height is known and constant; and

(3) the vehicle roll angle is zero.

The method developed takes the locations of specific points of the guideline in the image and, by applying inverse perspective transformation, evaluates the coordinates of these points in the reference frame associated with the vehicle. Then, the relative location of the vehicle with respect to the guideline is computed.

Figure 9.12 shows the geometrical description of the optical detector. Figure 9.12(a) describes the camera (represented only by its image plane) and the scene. Frames F_2 and F_1 are attached to the image plane and share the same origin at the centre of the image. The three-dimensional space surrounding the camera is expressed with respect to the F_1 frame, while objects on the image are expressed with respect to F_2. Figure 9.12(b) shows the camera support. Frame F_0 is the vehicle frame. Its origin is fixed at the reference point of the vehicle. The symbols used in this figure are: (A point ${}^j\bar{p}_i = ({}^jx_i{}^jy_i{}^jz_i)$ is said to be the ith point of frame j, noted F_j.)

$({}^2x{}^2y)$ = coordinates of point ${}^2\bar{p}$ in frame F_2

$({}^1x{}^1y{}^1z)$ = coordinates of point ${}^1\bar{p}$ in frame F_1

$({}^0x{}^0y{}^0z)$ = coordinates of point ${}^0\bar{p}$ of the optical line in F_0 (vehicle reference frame)

$(d_xd_yd_z)$ = position of the image plane with respect to F_0

$\alpha =$ elevation angle of the camera

$\lambda =$ focal length

The geometrical model described in this chapter is similar to the one in.[11] Any point ${}^2\bar{p}$ in the image plane can be expressed in the vehicle frame, provided that the vertical height, d, between the camera and the optical line is constant. Referring to Figure 9.12(a), the inverse perspective transform[12] is:

$$ {}^1x = -\frac{{}^2x}{\lambda}({}^1z - \lambda) \tag{9.1} $$

$$ {}^1y = -\frac{{}^2y}{\lambda}({}^1z - \lambda) \tag{9.2} $$

$$ \begin{aligned} {}^1z &= \frac{d}{\sin(\alpha)} + \left(\frac{d}{\sin(\alpha)} - \lambda\right)\frac{\sin(\beta)\cos(\alpha)}{\sin(\alpha - \beta)} \\ &= \frac{d\cos\beta - \lambda\sin\beta\cos\alpha}{\sin(\alpha - \beta)} \end{aligned} \tag{9.3} $$

$$ \text{where}\beta = \arctan\left(\frac{{}^2y}{\lambda}\right) \tag{9.4} $$

${}^1z$ represents the distance between the image plane of the camera and the point ${}^1\bar{p}$ on the optical guideline.

Using Figure 12(b), the expression of ${}^0\bar{p}$ is then

$$ {}^0x = -{}^1y\sin\alpha + {}^1z\cos\alpha + d_x \tag{9.5} $$

$$ {}^0y = {}^1x + d_y \tag{9.6} $$

$$ {}^0z = {}^1y\cos\alpha + {}^1z\sin\alpha + d_z \tag{9.7} $$

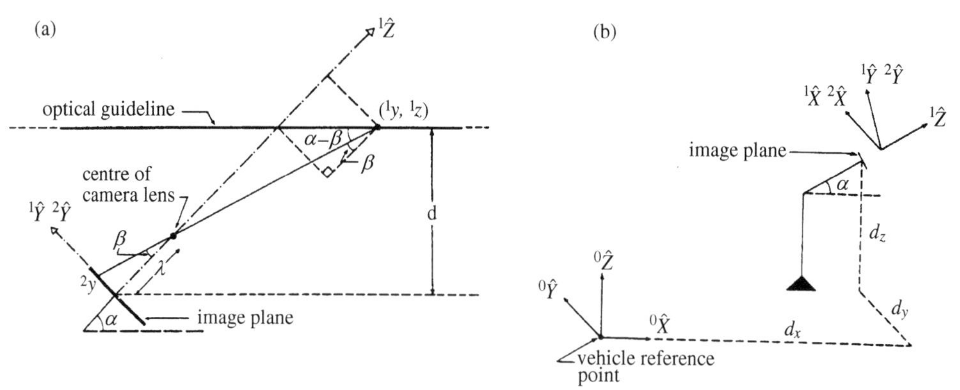

Figure 9.12. Geometric description of the optical detector. (a) Image plane and scere. (b) Camera support

The coordinates of point ${}^{0}\bar{p}$ represents the location of a specific point of the guideline detected in the video image, and expressed in the reference frame of the vehicle.

9.3.5 Navigation Errors

Navigation errors consist of two quantities: lateral offset, ε_{y}, and orientation offset, ε_{Ω}.[8] Referring to Figure 9.13, the lateral offset is the signed distance between O, the reference point of the vehicle, and S, the point of the guidepath closest to the vehicle. The lateral offset is expressed with respect to the F_{S} frame. The orientation offset is the angle between the ${}^{0}\hat{X}$ and ${}^{\mathrm{S}}\hat{X}$ axis.

With these definitions, the expressions of the navigation errors can be developed. From Figure 9.11, the intersection of the optical guideline and the three video scan lines S1, S2 and S3, gives three points noted ${}^{2}_{\mathrm{in}}\bar{p}_1$, ${}^{2}_{\mathrm{in}}\bar{p}_2$ and ${}^{2}_{\mathrm{in}}\bar{p}_3$. Using the geometrical model of the optical detector, these points can be expressed with respect to the vehicle reference frame F_0 to form ${}^{0}_{\mathrm{in}}\bar{p}_1$, ${}^{0}_{\mathrm{in}}\bar{p}_2$ and ${}^{0}_{\mathrm{in}}\bar{p}_3$. To compute the navigation errors, only two of these three points are needed. Usually, the farthest two points are chosen, i.e. ${}^{0}_{\mathrm{in}}\bar{p}_1$ and ${}^{0}_{\mathrm{in}}\bar{p}_3$. If only one of these points is missing, then ${}^{0}_{\mathrm{in}}\bar{p}_2$ replaces it. If two points (or more) are missing, there cannot be any computation. This redundancy is used to give robustness to the computation algorithm. Consequently, the expression of the navigation errors are:

$$\varepsilon_{\Omega} = \arctan\left(\frac{{}^{0}\tilde{y}_2 - {}^{0}\tilde{y}_1}{{}^{0}\tilde{x}_2 - {}^{0}\tilde{x}_1}\right) \tag{9.8}$$

$$\varepsilon_{\mathrm{y}} = {}^{0}\tilde{y}_1 \cos\varepsilon_{\Omega} - {}^{0}\tilde{x}_1 \sin\varepsilon_{\Omega} \tag{9.9}$$

$$\text{where } ({}^{0}\tilde{x}_1, {}^{0}\tilde{y}_1) = \min\{({}^{0}x_1, {}^{0}y_1), ({}^{0}x_2, {}^{0}y_2)\} \tag{9.10}$$

$$({}^{0}\tilde{x}_2, {}^{0}\tilde{y}_2) = \max\{({}^{0}x_2, {}^{0}y_2), ({}^{0}x_3, {}^{0}y_3)\} \tag{9.11}$$

$$\text{and } ({}^{0}\tilde{x}_1, {}^{0}\tilde{y}_1) \neq ({}^{0}\tilde{x}_2, {}^{0}\tilde{y}_2) \tag{9.12}$$

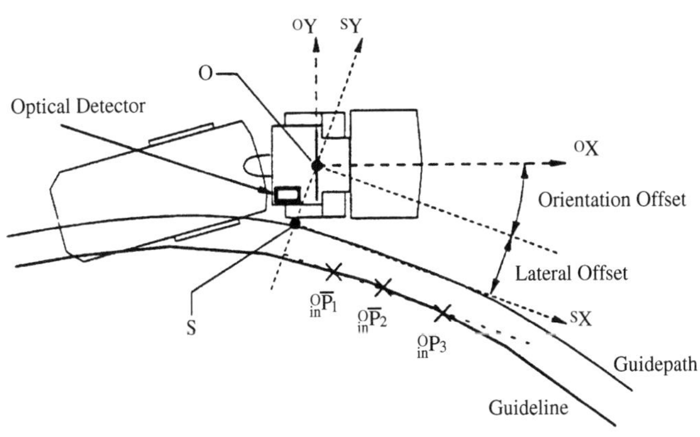

Figure 9.13. Definition of the navigation errors

9.3.6 Performance

Under static conditions, the optical detector is precise to 0.1° for ε_Ω, and to 2 mm for ε_y. Under dynamic conditions, when the vehicle is moving under the optical guideline, the precision obtained was 1° for ε_Ω and 20 mm for ε_y. In both cases, this precision is obtained after a simple calibration procedure which consists of shifting the image of the optical guideline to the centre of the screen.

The sampling period of the detector is 70 ms. This figure was obtained with an IBM PC/AT running at 12 MHz and equipped with a MATROX PIP 512B frame grabber. Image processing is done in parallel with the grabbing. Consequently, the sampling period is limited by the time required to grab an image.

When the optical detector was developed, the assumption of a constant guideline height was made. Practically, in a mine environment, this could be difficult to achieve very precisely. However, it can be shown, using a sensitivity analysis, that the guideline height variations have only limited effects on the navigation errors. To do this, assume that the vehicle is well aligned under the guideline (i.e. $\varepsilon_y = 0$ and $\varepsilon_\Omega = 0$) and that the camera is pointing right over the vehicle (i.e. $\alpha = 90°$). The starting equations of the sensitivity analysis are written:

$$\Delta\varepsilon_\Omega = \frac{d}{\varepsilon_\Omega}\frac{\partial(\varepsilon_\Omega)}{\partial d}\Delta d \tag{9.13}$$

$$\Delta\varepsilon_y = \frac{d}{\varepsilon_y}\frac{\partial(\varepsilon_y)}{\partial d}\Delta d \tag{9.14}$$

With the help of the geometrical model (equations (9.1)–(9.7)), equations (9.13) and (9.14) reduce to:

$$\begin{aligned}\Delta\varepsilon_\Omega &= 0(\%)\\ \Delta\varepsilon_y &= \Delta d(\%)\end{aligned} \tag{9.15}$$

Consequently, a height variation Δd, expressed in %, has no effect on the orientation offset, and the lateral offset varies proportionally with d.

This result is interesting because, as will be seen in the next section, the stability of the closed-loop system is more dependent on the quality of the orientation offset than on that of the lateral offset. Thus, guideline height variations will have limited effects on vehicle motion control.

The effect of vehicle roll has also been studied. The main effect of the roll angle is to shift the image of the guideline laterally on the screen. The overall effect on navigation error can be shown to be directly proportional to the shifting values of the image. The same kind of assumptions and analysis that were used to develop equations (9.13) and (9.14) have been made to demonstrate this result.[13]

Experimental data, shown in Table 9.1, were obtained for a vehicle well-aligned under the guideline, for roll angles γ varying between −15° and 15°, and for three

Table 9.1. Effect of vehicle roll on navigation errors

Roll angle γ	Elevation angle α 85°	75°	60°
Orientation offset ε_Ω (degrees)			
−15°	−1.7	−4.0	−9.1
−10°	−1.1	−2.8	−5.7
−5°	−0.5	−1.1	−2.8
0°	0.0	0.0	0.0
5°	0.5	1.7	2.8
10°	1.1	4.0	6.3
15°	1.7	5.1	9.1
Lateral offset ε_υ (metres)			
−15°	−0.36	−0.31	−0.34
−10°	−0.23	−0.22	−0.22
−5°	−0.11	−0.02	−0.11
0°	0.00	0.00	0.00
5°	0.10	0.12	0.10
10°	0.22	0.25	0.23
15°	0.35	0.33	0.35

camera elevation angles α. The table shows that when the camera is nearly vertical, the roll angle has little effect on the orientation offset. It worsens as the elevation angle decreases. For the lateral offset, the error is proportional to the roll angle, but remains constant for all elevation angles.

Under automatic guidance conditions, the vehicle roll angle remains well under 10°. For this reason, the roll has only little effect on the navigation errors.

9.3.7 Simulation of the Optical Detector

As explained previously, the optical detector selected three points on the image of the guideline, which is composed of a set of straight segments. This process is implemented at a certain sampling rate while the vehicle is in motion. Consequently, the guideline perceived by the detector is different from the guideline installed on the roof of the drift. To analyse this situation, a simulation model of the optical detector, including a representation of the guideline in the plane and the implementation of the image-processing algorithm, has been developed. With this model, the dynamic behaviour of the optical detector can be explained.

Figure 9.14 illustrates a typical situation where the optical detector perceives only the section of the guideline that intersects two of the three scan lines. The resulting perceived guideline is an instantaneous approximation of the guideline.

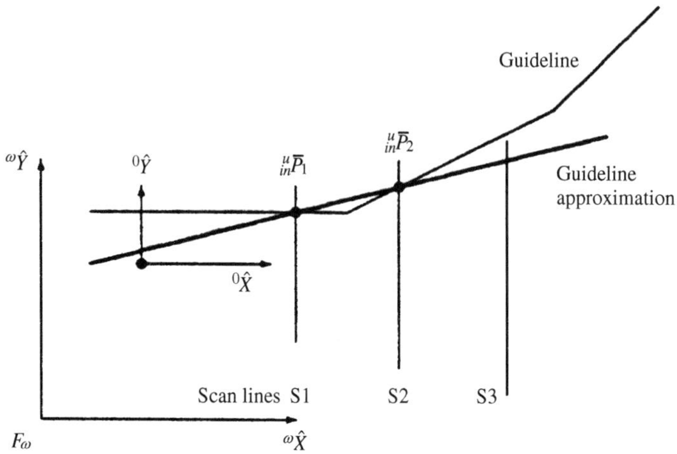

Figure 9.14. Detection of the guideline by the optical detector

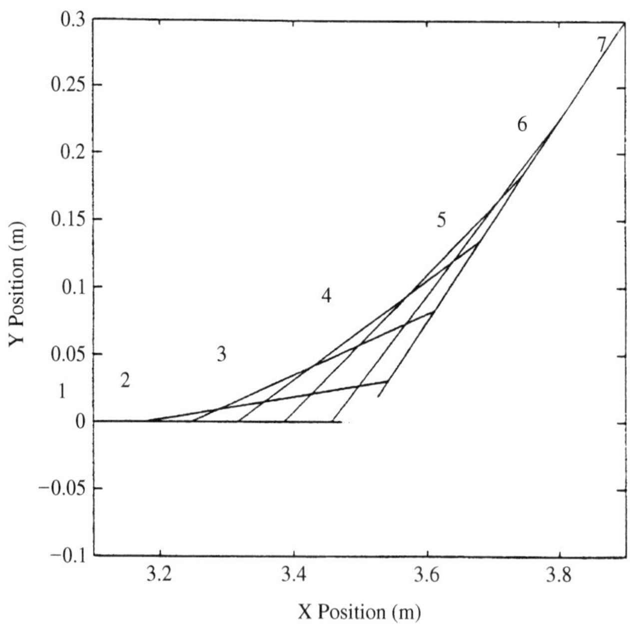

Figure 9.15. Dynamic behaviour of the optical detector

Figure 9.15 shows a guideline composed of two segments with different orientation. The original guideline is the dotted line. The other line segments represent the guideline as it is perceived by the optical detector while the vehicle is moving. Each segment numbered 1–7 corresponds to a single reading of the detector and constitutes an approximation of the guideline. The optical detector computes the navigation errors with respect to these guideline approximations. The change in orientation of the guideline is then perceived by increment. The optical detector smoothes the curve and the vehicle gradually follows the different segments.

9.4 MOTION CONTROL OF THE VEHICLE

Owing to the mine environment, the vehicle motion controller has to operate in the presence of major disturbances and must maintain the vehicle in the middle of the drift with good precision and small tolerance. Consequently, a systematic procedure has been applied to the design of this motion controller. This procedure is based on theoretical analysis, simulation and experimentation.[8] A dynamic model of the guidance system, including the vehicle and the optical detector, was developed and implemented in a simulator. The validation of this simulator was carried out using a test bench composed of a small-scale articulated vehicle. This vehicle was designed and assembled in our laboratory. It has the same geometry as a real LHD, i.e. four-wheel drive with an actuator controlling the articulation angle between the two bodies of the vehicle. Finally, the resulting motion controller was tested on an LHD in an underground mine. In this section, the design procedure is explained and the characteristics and the performance of the guidance system are illustrated in specific situations.

9.4.1 Model of the Vehicle

In this particular application, the speed of the vehicle is relatively low, and for this reason, simplifying assumptions can be made in developing the model of the vehicle. The roll and pitch dynamics of the vehicle, the slippage of the wheels, and tyre deformation are neglected. Consequently, the orientation of the vehicle is a direct result of the steering action. Following these simplifications, the model of the vehicle is written by considering the kinematics and by adding simple dynamic models for the translation motion and the steering action.

Kinematics of the vehicle

Figure 9.16 illustrates a schematic model of the vehicle. The mid-point between the wheels of each body of the vehicle are denoted, respectively, by points P_a and P_b. The articulated point between the two bodies is denoted as P. The coordinates of these three points P_a, P_b and P are therefore (X_a, Y_a), (X_b, Y_b) and (X, Y). The orientation of each body with respect to the X axis is denoted by Ω_a and Ω_b.

The kinematic model is developed according to the methodology described in Barraquand and Latombe[14] and Laumond.[15] The vehicle is supposed to be non-symmetric in the sense that $L_a \neq L_b$. The haulage trucks have this characteristic whereas LHD vehicles are symmetric.

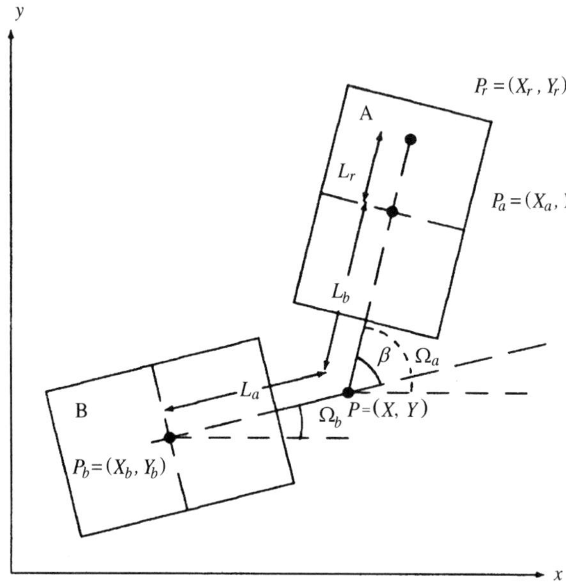

Figure 9.16. Schematic representation of the articulated vehicle

The configuration of the vehicle is represented by $(X_a, Y_a, \Omega_a, X_b, Y_b, \Omega_b)$. The velocity parameters are $\dot{X}_a$, $\dot{Y}_a$, $\dot{\Omega}_a$, $\dot{X}_b$, $\dot{Y}_b$, $\dot{\Omega}_b$. The control parameters of the articulated vehicle are the speed V_a of the front wheels, and the steering angle β, which is the angle between body A and body B of the vehicle.

For each body, assuming that there is no wheel slippage, the speed of points P_a and P_b is always parallel to the main axis of the body. The following equations can be written:

$$\dot{X}_a = V_a \cos(\Omega_a) \qquad \dot{Y}_a = V_a \sin(\Omega_a)$$
$$\dot{X}_b = V_b \cos(\Omega_b) \qquad \dot{Y}_b = V_b \sin(\Omega_b) \tag{9.16}$$

The following constraints are obtained by eliminating speeds V_a and V_b.

$$-\dot{X}_a \sin(\Omega_a) + \dot{Y}_a \cos(\Omega_a) = 0 \tag{9.17}$$
$$-\dot{X}_b \sin(\Omega_b) + \dot{Y}_b \cos(\Omega_b) = 0 \tag{9.18}$$

Now the geometric constraints between body A and body B are written:

$$X_b = X_a - L_a \cos(\Omega_a) - L_b \cos(\Omega_b) \tag{9.19}$$
$$Y_b = Y_a - L_a \sin(\Omega_a) - L_b \sin(\Omega_b) \tag{9.20}$$

The derivatives of these equations are:

$$\dot{X}_b = \dot{X}_a + \dot{\Omega}_a L_a \sin(\Omega_a) + \dot{\Omega}_b L_b \sin(\Omega_b) \tag{9.21}$$

$$\dot{Y}_b = \dot{Y}_a - \dot{\Omega}_a L_a \cos(\Omega_a) - \dot{\Omega}_b L_b \cos(\Omega_b) \tag{9.22}$$

Constraint equation (9.18) can be written using equations (9.21), (9.22) and (9.16):

$$V_a \sin(\beta) - \dot{\Omega}_a (L_b + L_a \cos(\beta)) + \dot{\beta} L_b = 0 \tag{9.23}$$

with $\beta = \Omega_a - \Omega_b$.

Thus, the configuration of the vehicle can be entirely described by four parameters (X_a, Y_a, Ω_a, β), and the kinematic equations for the articulated vehicle are:

$$\dot{X}_a = V_a \cos(\Omega_a) \tag{9.24}$$

$$\dot{Y}_a = V_a \sin(\Omega_a) \tag{9.25}$$

$$\dot{\Omega}_a = \frac{V_a \sin(\beta) + \dot{\beta} L_b}{L_b + L_a \cos(\beta)} \tag{9.26}$$

Dynamics of the vehicle

The two control variables for this set of equations are the speed V_a and the steering angle β. The speed is related to a propulsion system which is a diesel engine and transmission for the LHD (amplifier and electric motors for the small-scale vehicle). The steering angle is controlled by a hydraulic cylinder (a linear electric actuator for the electric vehicle). For the purposes of this study, the dynamics of these two systems are represented by a static gain and a time constant. An identification procedure has confirmed this representation experimentally.

$$\frac{V_a}{U_m} = \frac{K_v}{S\tau_v + 1} \tag{9.27}$$

$$\frac{\dot{\beta}}{U_v} = \frac{K_\beta}{S\tau_\beta + 1} \tag{9.28}$$

where U_m and U_v are the input of the propulsion and of the steering system, respectively.

The resulting model, including the dynamics of the actuator and the kinematics of the system, is represented in Figure 9.17.

Finally, if for the controller, the vehicle reference point is located in X_r and Y_r as shown in Figure 9.16, then kinematic equations (9.24) and (9.25) become:

$$\dot{X}_r = V_a \cos(\Omega_a) - L_r \dot{\Omega}_a \sin(\Omega_a) \tag{9.29}$$

$$\dot{Y}_r = V_a \sin(\Omega_a) + L_r \dot{\Omega}_a \cos(\Omega_a) \tag{9.30}$$

9.4.2 Design Model and Motion Controller

The objective of the motion controller is to select the appropriate actions in the propulsion and the steering actuator so as to maintain the vehicle close to the

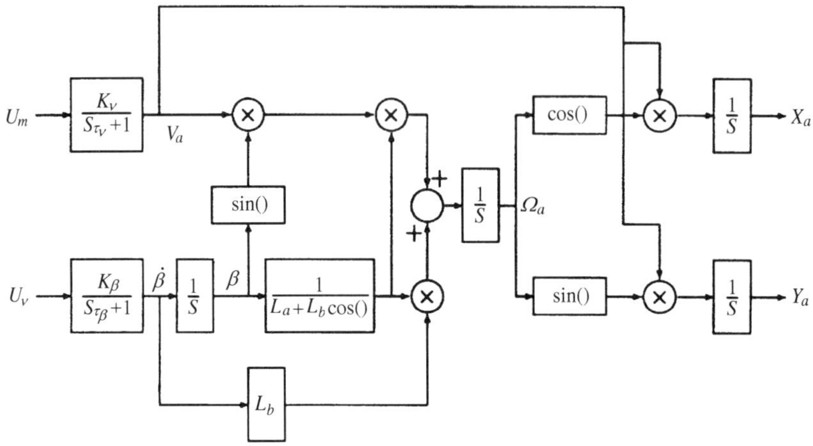

Figure 9.17. Dynamic model of the articulated vehicle

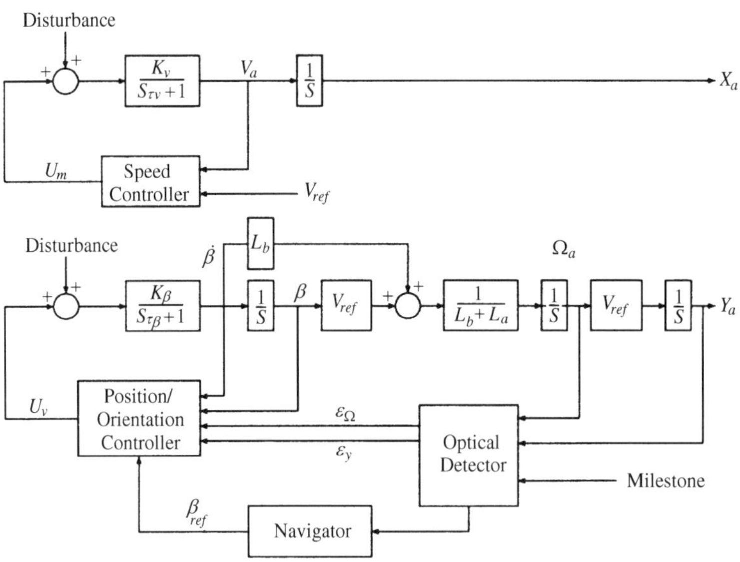

Figure 9.18. Design model of the articulated vehicle

guide-path, as defined by the guideline. Considering the concept of the optical guidance system, two remarks can be made on the specific characteristics of the motion control problem.

1. The optical detector measures only the relative location of the vehicle with respect to the guideline, as defined previously by the lateral and orientation offsets.

2. The characteristics of the different sections of the drift are available to the controller via the signalization and navigation modules. These characteristics are the desired speed and the radius of curvature. The desired speed of the vehicle is related directly to the zone of operation of the LHD in the drift network: if the vehicle is in a straight section of the drift, the speed can be set at its maximum; in a tight curve, the speed is lowered. For a given vehicle, the radius of curvature is transformed into an equivalent desired articulation angle, which is obviously a function of the operating zone.

The remarks, in line with the structure of the model, suggest that two independent controllers can be applied to guide the vehicle: a speed controller and a position/orientation controller. These controllers can be designed using a linearized version of the model based on the following assumptions: the speed of the vehicle is constant, and the orientation angle Ω_a and the steering angle β are kept small.

The linearized model (evaluated at the desired speed V_{ref}) and the structure of the two controllers are illustrated in Figure 9.18.

9.4.3 Speed Controller

The U_m control of the speed controller has two components:

1. U_{m1}: a feedforward action which corresponds to the control action required to obtain the desired speed under nominal conditions and in the absence of disturbance. The signal is obtained by inverting the model of the system.

$$U_{m1}(t) = \frac{\tau_v}{K_v}\dot{V}_{ref}(t) + \frac{1}{K_v}V_{ref}(t) \tag{9.31}$$

2. U_{m2}: a linear PI (Proportional Integral) feedback component which reduces the influence of parameter variations, external perturbations and model uncertainties.

$$U_{m2}(t) = k_{tra,1}\varepsilon_v(t) + k_{tra,2}\int_0^t \varepsilon_v(\tau)d\tau \tag{9.32}$$

with $\varepsilon_v = V_{ref} - V_a$.

The values of the gains $k_{tra,1}$ and $k_{tra,2}$ are determined by the pole placement method, permitting a pre-assigned dynamics to be imposed on the closed-loop system.

9.4.4 Position Orientation Controller

The objective of the position/orientation controller is to ensure that the vehicle follows the trajectory defined by the guidepath. This is achieved by reducing to zero the lateral and the orientation offsets measured by the optical detector. A suitable

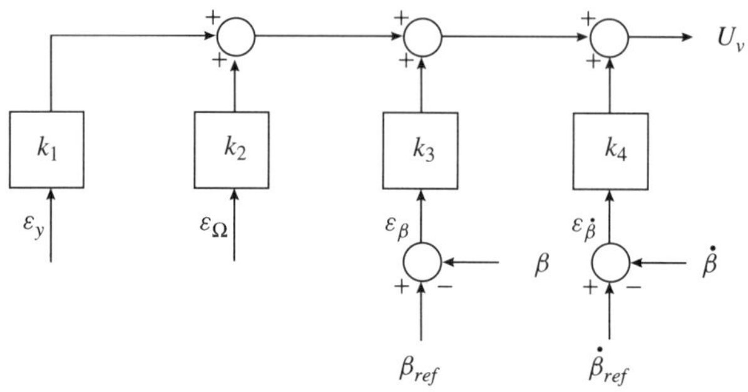

Figure 9.19. State feedback controller

corrective action has to be applied to the steering system. A state feedback controller can be used to achieve this objective; it can be easily implemented because all the state variables can be directly measured. Figure 9.19 illustrates the structure of the controller.

The control law is written:

$$U_{\mathrm{v}}(t) = k_4(\dot\beta_{\mathrm{ref}} - \dot\beta) + k_3(\beta_{\mathrm{ref}} - \beta) + k_2\varepsilon_\Omega + k_1\varepsilon_{\mathrm{v}} \tag{9.33}$$

with β_{ref} being a signal coming from the navigator and a function of the radius of curvature of the zone where the vehicle operated, and $\dot\beta_{\mathrm{ref}}$ which is equal to zero.

The gains k_1, k_2, k_3 and k_4 are determined using the pole placement method, by imposing the dynamics of the closed-loop system. The transfer function of the closed-loop system is written:

$$\frac{Y_{\mathrm{a}}(S)}{Y_{\mathrm{a,ref}}(S)} = \frac{\dfrac{k_1 K_\beta}{\tau_\beta(L_{\mathrm{a}}+L_{\mathrm{b}})}(L_{\mathrm{b}}S + V_{\mathrm{ref}})}{S^4 + \dfrac{k_4K_\beta+1}{\tau_\beta}S^3 + \dfrac{K_\beta}{\tau_\beta}\left(k_3 + \dfrac{L_{\mathrm{b}}k_2}{L_{\mathrm{a}}+L_{\mathrm{b}}}\right)S^2 + \dfrac{K_\beta V_{\mathrm{ref}}(k_2+k_1L_{\mathrm{b}})}{\tau_\beta(L_{\mathrm{a}}+L_{\mathrm{b}})}S + \dfrac{k_1K_\beta V_{\mathrm{ref}}^2}{\tau_\beta(L_{\mathrm{a}}+L_{\mathrm{b}})}} \tag{9.34}$$

If the polynomial $\Delta_{\mathrm{d}}(S)$ corresponds to the desired dynamics of the closed-loop system:

$$\Delta_{\mathrm{d}}(S) = (S-p_1)(S-p_2)(S-p_3)(S-p_4) \tag{9.35}$$

$$= S^4 + a_1S^3 + a_2S^2 + a_3S + a_4 \tag{9.36}$$

Then the gains of the controller are given by:

$$k_1 = \frac{a_4\tau_\beta(L_{\mathrm{a}}+L_{\mathrm{b}})}{K_\beta V_{\mathrm{ref}}^2} \tag{9.37}$$

$$k_2 = \frac{a_3\tau_\beta(L_{\mathrm{a}}+L_{\mathrm{b}})}{K_\beta V_{\mathrm{ref}}} - k_1L_{\mathrm{b}} \tag{9.38}$$

$$k_3 = \frac{a_2\tau_\beta}{K_\beta} - \frac{L_b k_2}{L_a + L_b} \tag{9.39}$$

$$k_4 = \frac{a_1\tau_\beta - 1}{K_\beta} \tag{9.40}$$

9.5 SIMULATION AND EXPERIMENTAL RESULTS

The main characteristics and the dynamic behaviour of the optical guidance system will be illustrated by discussing some selected results from a certain number of simulations and experiments on the electric vehicle. The motion controller was implemented on an IBM-PC/AT computer equipped with a standard input/output (I/O) interface. All the simulations and the tests were performed with a sampling period of 70 ms. The values of the dynamic parameters used for the experimentations are listed in Table 9.2.

9.5.1 Gain Adaptation with the Speed

As discussed previously, the model of the vehicle is a function of the speed. Consequently, the tuning of the position/orientation controller and the dynamic behaviour of the closed-loop system in relation to the lateral offset and the orientation offset have to be analysed for a given range of speed. To illustrate the design procedure, the following experiment is considered: the vehicle is required to follow a guideline that has a step change in the lateral position of 0.5 m. The test is performed at a speed varying from 0.5 to 1.5 m/s.

The objective is that the dynamics of the lateral and orientation offsets be independent of the speed of the vehicle. In that case, the time response for a lateral change of 0.5 m is required to be the same for the range of speed considered. This strategy can be achieved through a proper selection of the poles of the closed-loop system.

Considering the expression of the transfer function of the closed-loop linearized system given by equation (9.34), the following procedure is applied to select the closed-loop poles: first cancel the effect of the zero by superimposing a pole at $-V_{\text{ref}}/L_b$, and second, select the remaining three poles to obtain a desired dynamics in terms of response time.

Multiple simulations have indicated that the set of poles given in Table 9.3 meet the objective of having a fixed time response for the closed-loop system for the

Table 9.2. Dynamic parameter values for the experimentations and simulations

parameters	K_v	τ_v	K_β	τ_β	L_a	L_b	L_r
values	0.36	0.44	−0.03	0.1	0.585	0.585	0

Table 9.3. Poles and gains for various speeds

speed (m/sec)	poles			gains			
	$p_1 = z_1$ $-\frac{V_{\text{ref}}}{L_a}$	p_2	p_3, p_4	k_1, ε_y	k_2 ε_Ω	k_3 ε_β	k_4 $\varepsilon_{\dot\beta}$
0.5	−0.9	−1.5	$-1.2\pm 1.2j$	−57.6	−43.2	−11.1	17.5
1.0	−1.7	−1.5	$-1.2\pm 1.2j$	−28.8	−43.2	−22.2	14.6
1.5	−2.6	−1.5	$-1.2\pm 1.2j$	−19.2	−43.2	−33.3	11.8

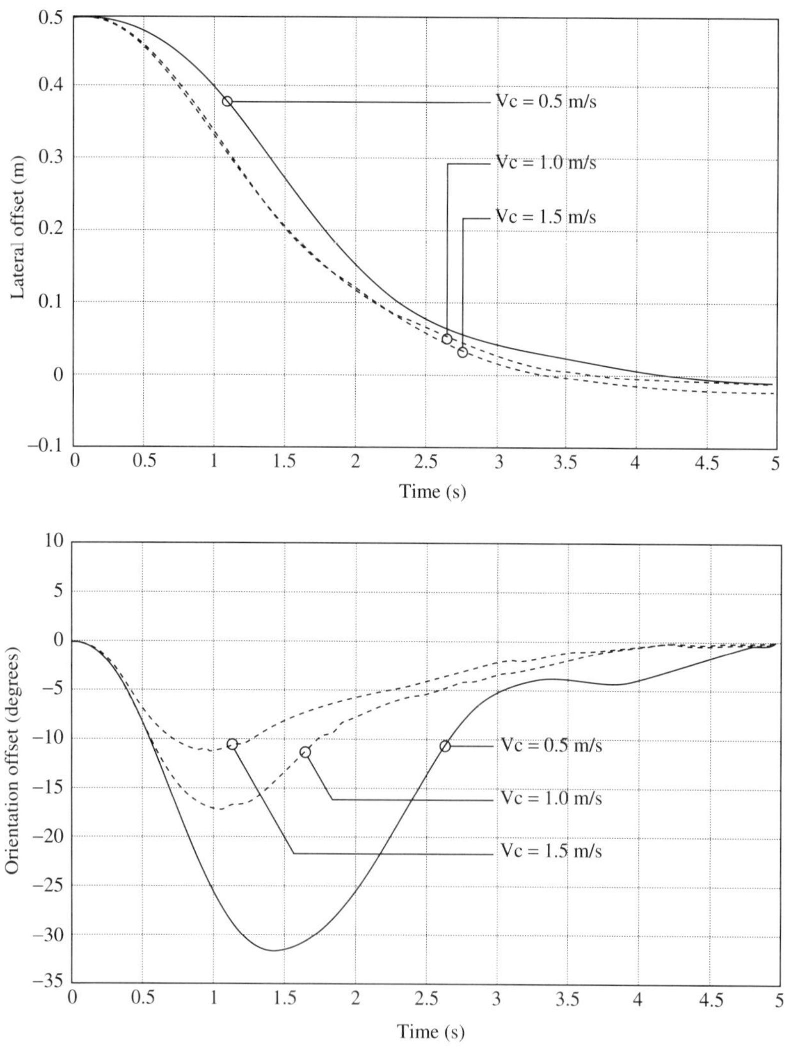

Figure 9.20. Trajectories of the navigation errors for various speeds

different speeds. It is interesting to note that the value of the gain k_1 associated with the lateral offset decreases at a rate that is directly proportional to the speed, while the value of the gain k_2 associated with the orientation offset is constant. The value of the gain k_3 increases with the speed. Hence, as the speed increases, more weight is given to the orientation offset as compared with the lateral offset.

Figure 9.20 shows the lateral and orientation offset trajectories. In line with the pole placement strategy, the dynamics of the lateral offset and the orientation offset are independent of the speed of the vehicle. This strategy can be modified so as to have, for example, the dynamics of the orientation offset independent of the speed, while the lateral offset is not, or is a case in between. In both cases, the gains of the controller have to be adapted to the speed of the vehicle.

9.5.2 Tracking Properties

The tracking properties of the optical guidance system have been extensively tested in simulation and on the electric vehicle. A typical guidepath composed of three straight segments with a step change in orientation of 27° and 60°, respectively, has been used to test the system. This guidepath is illustrated in Figure 9.21.

Figure 9.22 shows the trajectories for the lateral and orientation offsets obtained by simulation and by experimentation. A close correspondence is observed between the trajectories. In fact, the system has a better acceleration than the simulation model, and a time delay between the trajectories is introduced. For example, segment 2 is perceived by the vehicle at 7.5 s and by the simulator at 8 s. The discontinuity observed at 15 s is due to the fact that in the presence of a large change in orientation, the optical guide is partially out of the image, introducing an inconsistent measurement which results in an erratic motion. The robustness of the

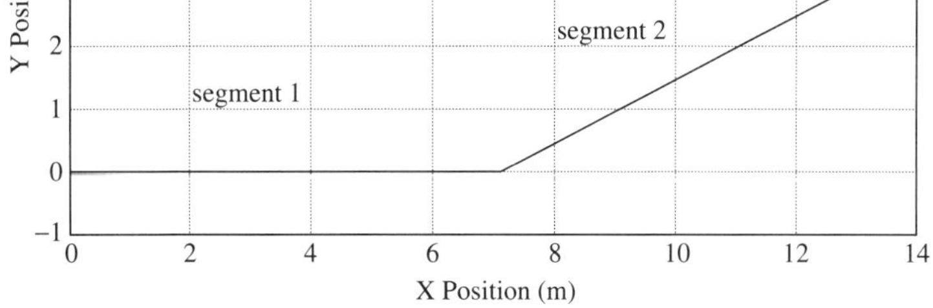

Figure 9.21. The guidepath

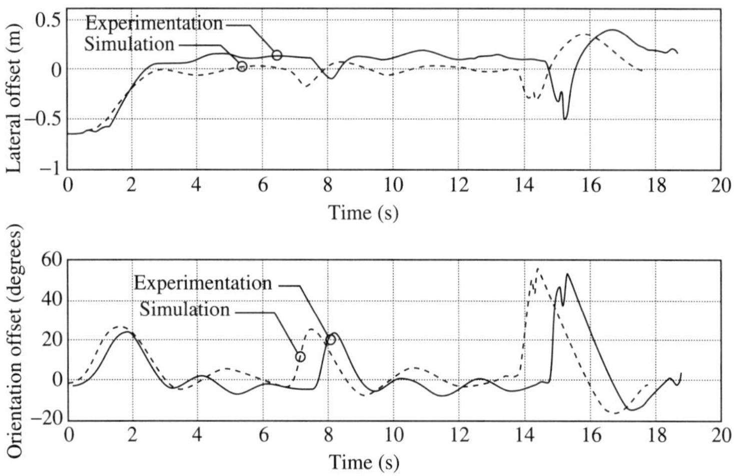

Figure 9.22. Tracking properties of the controller: simulation and experimentation ($\alpha = 60°$)

image processing algorithm and of the motion controller allows the vehicle to stay under the guideline.

Effect of the elevation angle of the camera

One of the interesting features of the optical guidance system is the versatility offered by the vision system in detecting the guideline. Practically, the elevation angle α of the camera allows the viewing angle of the detector in front of the vehicle to be selected. If α is set equal to 90°, the detector looks at the guideline directly over the vehicle. If α is equal to 50°, the detector looks at the guideline in front of the vehicle, which is the oncoming path to follow.

Two tests illustrate the beneficial effect of properly selecting the elevation angle of the camera over the tracking properties of the system. Figure 9.23 illustrates the improvement in the tracking of the guideline by decreasing the elevation angle of the camera. With an angle of 85°, the system has to stop at 15.5 s because the guideline is out of the field of view of the camera. This situation occurs when the vehicle arrives at the intersection between segments 2 and 3. With this angle, the vehicle starts to turn when it is located at the intersection. The limits imposed by the dynamics and by the constrained radius of curvature of the vehicle result in navigation errors too large to be corrected. The guideline gets out of the detector field of view and the vehicle has to stop. When the angle is equal to 60°, the detector compares the actual location of the vehicle with the oncoming guidepath, anticipating future errors. In this case, the vehicle starts to turn before the intersection and is able to follow the guide correctly.

This property is important for mine applications where the location of the detector is limited by various practical factors. For example, on an LHD, the detectors

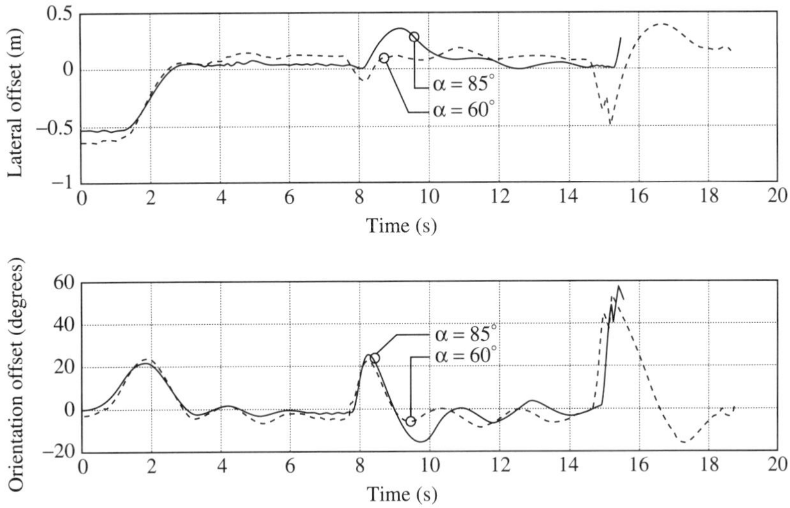

Figure 9.23. Effect of the elevation angle α of the camera on tracking performance

are installed as low as possible and close to the vehicle's articulation in order to be protected from stones falling from the bucket. A proper choice of the camera elevation angle allows a good tracking behaviour to be obtained by measuring the navigation error at the front of the vehicle.

The detector's elevation angle can also be adjusted to compensate for time delays present in the system. Measurement of the navigation error by the image-analysis algorithm introduces a time delay equal to one sampling period (70 ms). In the LHD, another delay of approximately 70 ms was present in the actuator control. When the vehicle enters a curved section composed of straight segments, the time delay in detecting the navigation errors can result in situations where the detector will become optically disconnected from the guideline. The vehicle then has to stop. Again, a proper choice of the camera elevation angle allows the navigation error in front of the vehicle to be measured and the curve to be anticipated. The phenomenon is illustrated in Figure 9.24. With an elevation angle of 75° and a time delay of 140 ms, the vehicle leaves the guideline at the intersection of segments 1 and 2. If the elevation angle is set at 50°, the vehicle is able to negotiate the two intersections and successfully follow the guideline.

Steering angle set-point

Finally, the tracking performance of the optical guidance system can be improved by using an a priori knowledge of the radius of curvature of the incoming path and of the maximum possible speed. This information can be transformed into a set-point for the steering angle system. As mentioned in Section 9.3, the use of milestones and of

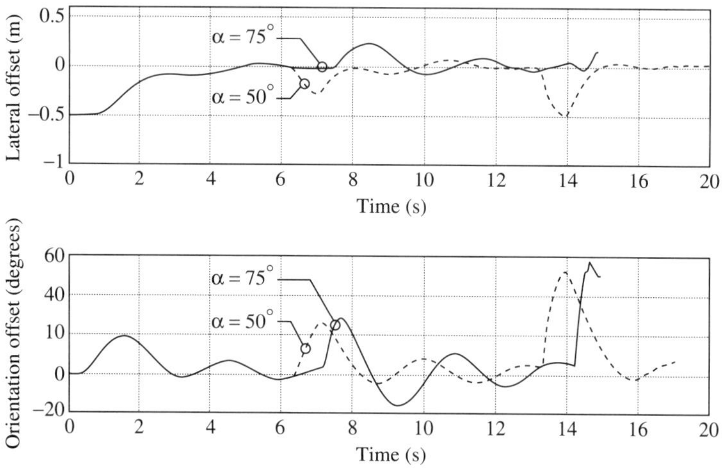

Figure 9.24. Effect on tracking of the camera elevation angle α in the presence of a time delay 140 ms

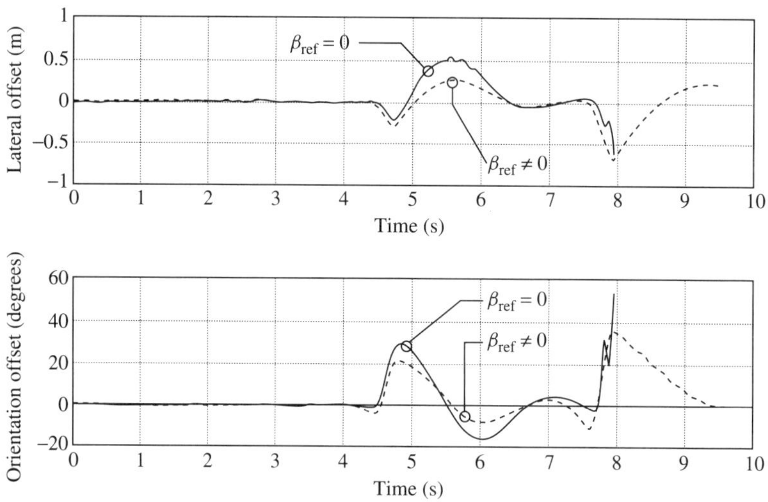

Figure 9.25. Effect of β_{ref} on tracking performance

a navigator allow this information which is stored in a database to be extracted. Set-points for the translation speed and for the articulation angle are then transmitted to the motion controller. The performance of the motion controller with and without these two set-points is shown by the trajectories in Figure 9.25. As indicated, when the set-point β_{ref} is maintained equal to zero, the vehicle stops in the intersection between segments 1 and 2 at 8 s. If a proper β_{ref} and V_{ref} are used by the controller, the vehicle follows the guideline correctly.

Table 9.4. Set-points for the motion controller

distance (m)	V_{ref} (m/sec)	β_{ref} degrees	k_1	k_2	k_3	k_4
0	2.0	0.0	−14.4	−43.2	−44.4	8.9
5.5	2.0	7.0	−14.4	−43.2	−44.4	8.9
7.0	2.0	0.0	−14.4	−43.2	−44.4	8.9
12.5	1.5	40.0	−19.2	−43.2	−33.3	17.8
14.0	2.0	0.0	−14.4	−43.2	−44.4	8.9
17.0	0.0	0.0	0.0	0.0	0.0	0.0

The set-points of the speed and the steering angle can be tuned to obtain a better performance for the guidance system for each section of the guideline. The objective is to maintain the appropriate maximum speed for each section. Table 9.4 summarizes these set-points and the controller gains used to follow the guidepath considered for the experimental test. These set-points are listed as a function of the distance in metres from the starting point.

9.6 TEST ON A LOAD HAUL DUMP VEHICLE

Most of the LHDs actually operating in mines are equipped with solenoid valves to control the steering actuator. The control action is limited to only three levels: left or right turn and neutral. In this case, the state controller previously studied, which gives a proportional control action, cannot be applied to existing LHDs without a major modification of the hydraulic circuits. In this context, an output feedback controller with a bang–bang control law has been developed.

As shown in Figure 9.26, this controller is composed of two loops, a minor loop which sets the steering angle at a desired value, and a second loop which computes this value from the navigation errors. A non-linear bang–bang-type controller is implemented for the minor loop. The feedforward signal β_{ref} coming from the navigator is injected as a supplementary signal for the minor loop. This loop is tuned so that the transfer function between the steering angle and the angle setpoint can be approximated by a first-order transfer function:

$$\frac{\beta}{\beta_{ref}} = \frac{1}{S\tau_{int} + 1} \tag{9.41}$$

with τ_{int} being the time constant of the minor loop.

The two parameters of the loop, the dead zone Δ and the gain k_{int} of the steering speed, can be selected by a standard analysis of the behaviour of a non linear controller (phase plane).

The main loop is then analysed by considering that a proportional reference can be applied to the minor loop. A simplified linear model is then used by substituting the transfer function (9.41) into the model. The resulting model is of order three

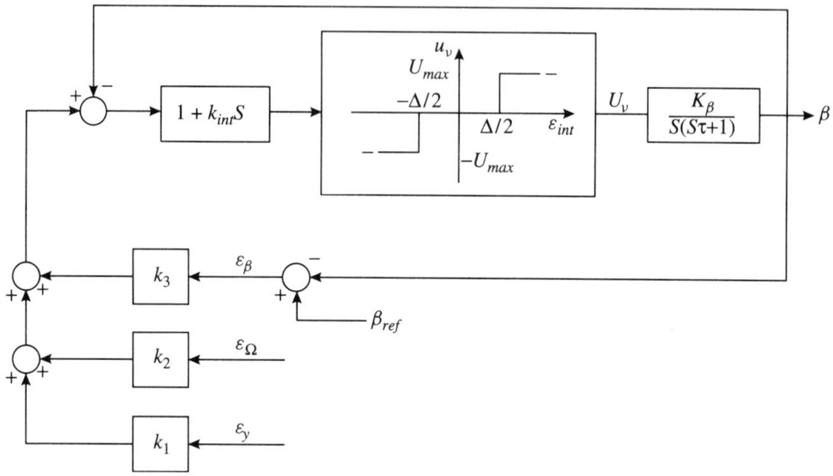

Figure 9.26. Output feedback controller with a minor loop

and the transfer function is given by:

$$\frac{Y_a(S)}{Y_{a,ref}(S)} = \frac{\dfrac{k_1}{\tau_{int}(L_a + L_b)}(L_b S + V_{ref})}{S^3 + \dfrac{1}{\tau_{int}}\left(1 + k_3 + \dfrac{L_b k_2}{L_a + L_b}\right)S^2 + \dfrac{V_{ref}(k_2 + k_1 L_b)}{\tau_{int}(L_a + L_b)}S + \dfrac{k_1 V_{ref}^2}{\tau_{int}(L_a + L_b)}} \tag{9.42}$$

The gains k_1, k_2 and k_3 are computed using the pole placement method. The gains are given by:

$$k_1 = \frac{a_3 \tau_{int}(L_a + L_b)}{V_{ref}^2} \tag{9.43}$$

$$k_2 = \frac{a_2 \tau_{int}(L_a + L_b)}{V_{ref}} - k_1 L_b \tag{9.44}$$

$$k_3 = a_1 \tau_{int} - 1 - \frac{L_b k_2}{L_a + L_b} \tag{9.45}$$

with a_1, a_2 and a_3 being the coefficients of the characteristic polynomial $\Delta_d(S)$ which corresponds to the desired dynamics of the closed loop system.

$$\Delta_d(S) = (S - p_1)(S - p_2)(S - p_3) \tag{9.46}$$

$$= S^3 + a_1 S^2 + a_2 S + a_3 \tag{9.47}$$

Simulation shows that good performances are obtained by using this control strategy. However, a certain degradation in performance is observed as compared with the state feedback controller. Using this motion control strategy, the optical guidance system was under test on a 3.8 m^3 (5 cubic yards) Wagner ST5 LHD at the Norita Mine.[10,16] Tests were done in both directions at various speeds ranging from 0.3 to approximately 1.0 ms and for different radii of curvature. Overall, 2 km were covered during the test period. Field tests have shown that the optical guidance system has the capability of guiding underground mine vehicles in a drift in both directions. Docking manoeuvres and navigation in a network of routes can be easily performed.

9.7 CONCLUSIONS

Ore transportation by man-controlled vehicles such as the LHD can be automated at least partially by using the optical guidance system presented in this chapter. For many years, the guidance of underground mine vehicles has been investigated by many researchers,[1,5,7] but no such system is in operation on a continuous basis. The optical guidance prototype described in this chapter has a number of technical characteristics which are well adapted to the hostile environment of underground hard rock mines. The system has a high level of intrinsic security because a physical guideline is used and the optical detector maintains continuous visual contact with the guideline. The detector has no mobile parts, so it can support shock and vibrations.

Use of a video camera and an image-processing algorithm for the detector make it possible to detect guidelines and milestones on a single image. The navigation error in terms of lateral offset and orientation offset are measured independently, and a navigation module can situate the vehicle in the network of drifts and refer pertinent information on the characteristics of the oncoming drift to the motion controller. Using this information, a motion controller can be easily tuned so as to maintain the vehicle close to the guideline at the cruising speed required by mine operation.

Simulation and experimental tests have confirmed these interesting features of the optical guidance system. To have the same performances at various speeds, the controller gains have to be adapted. A good choice in the elevation angle of the video detector makes it possible to anticipate the navigation errors that occur on curves. This feature allows better performances in curved sections and makes it possible to compensate for delays in the control loop. Overall performance of the guidance system can be tuned systematically for each section of the haulageway.

The implementation on an existing LHD vehicle at the Norita Mine, Matagami, has confirmed the potential of the system for mine applications. However, automation of such vehicles involves the implementation of a set of complementary functions, such as monitoring, tele-operation and obstacle detection. In the context of the hostile environment of undergound mines, each of these functions has to be developed and tested extensively. A production prototype of an automated haulage truck is under development.

REFERENCES

1 Amdahl K, Lundstrom M, Automatic truck saves money underground. *World Mining*, 1972; November: 40–44.
2 Giralt G, Chatila R, Vaisset M, An integrated navigation and motion control system for autonomous multisensory mobile robots. In *Autonomous Robot Vehicles* (Cox IJ, Wilfong GT, eds). 1990 Springer-Verlag, 1990 New York 420–443.
3 Hennessy SJ, King RH, Future mining technology spinoffs from the ALV Program. *IEEE Transactions on Industry Applications*, 1989; **25**(2): 377–384.
4 Saint-Amant M, Piché A, Hurteau R, Chevrette G, Sauriol G, A proposed optical guideline for mine vehicle automation. *Mineral Resource Engineering*, 1990 3(4): 281–294.
5 Kitchener LC, Vollenwyder K, Development of a 70-ton capacity automated underground trolley truck. *CIM Bulletin* 1988; **81**(909): 51–55.
6 Saint-Amant M, Piché A, Hurteau R, Evaluation of automatic guidance systems and their applicability to load haul and dump (LHD) Vehicles. *3rd Canadian Symposium on Mining Automation*, Sept. 1988, Montréal, 177–186.
7 Vagenas N, Wikstrom S, Granholm S, A Painted-line guidance system for automatic load-haul-dump (LHD) vehicles in underground hard rock mines. *Mining Research Engineering* 1988; **1**(4): 363–370.
8 De Santis RM, Hurteau R, Piché A, On the motion control of electric AGVS, *Second Workshop on Military Robotic Applications*, Kingston, August 1989, 306–315.
9 Saint-Amant M, Hurteau R, Laperrière Y, Chevrette G, A simple robust vision system for underground vehicle guidance. *International Symposium on Mine Mechanization and Automation*, June 10–13, 1991, Colorado, 6.1–6.10.
10 Hurteau R, St-Amant M, Laperrière Y, Chevrette G, Piché A. Optically guided LHD: a demonstration prototype. *International Symposium on Mine Mechanization and Automation*, June 10–13, 1991, Colorado, 6.11–6.20.
11 Drake KC, McVey ES, Iñigo RM, Experimental position and ranging result for a mobile robot. *IEEE Journal of Robotics and Automation*, 1987; **RA-3**(1): 31–42.
12 Duda RO, Hart PE. *Pattern Classification and Scene Analysis*. New York: Wiley-Interscience, 1973, 386–394.
13 Laperrière Y. Etude d'un système de guidage optique pour un véhicule minier articulé, MScA thesis, *Ecole Polytechnique de Montréal*. April 1991.
14 Barraquand J, Latombe J.-C. On nonholonomic mobile robots and optimal maneuvering. *Proceedings of the IEEE International Symposium on Intelligent Control*, September 1989, Albany, 340–347.
15 Laumond J-P. Controllability of a multibody mobile robot. (*I.C.A.R. 91*, June 1991, Pisa, Italy.
16 Hurteau R, St-Amant M, Chevrette G, Desrochers Y, Piché A, Optical Guidance system for underground mine vehicles. *Robotics and Automation: 1991 Conference Video Proceedings*, May 1991, Sacramento, 3 minutes.

10

Application of Neural Networks to Guidance of Autonomous Vehicles

10.1 INTRODUCTION

The technologies that are associated with autonomous vehicle (AV) development have wide ranging applications from automated manufacturing and underground mining to unmanned exploration of the oceans and space. These technologies have been researched in detail for many years with limited success. With recent developments in neural networks (NNs) and NN/fuzzy logic controller (NN/FLC) hybrid systems, has come the motivation for building neural controllers to control these AVs. NNs have been applied to a fairly wide variety of information-processing problems, but application to pattern classification/recognition has perhaps the best potential for practical systems. Recent advances in analogue VLSI chips have made it possible to implement these algorithms in real-time hardware. Their ability to adapt, learn and integrate information from several input sources, could result in many practical applications in the field of AVs.

Fuzzy logic controllers (FLC) have been used in many applications over the past decade, allowing the application of heuristic information to be used in the control of complex systems. The heuristic information in a FLC consists of a set of linguistic IF–THEN rules, comprising the fuzzy logic rule base. Inputs to the FLC are in the form of linguistic variables which belong, in varying degrees, to various fuzzy membership sets. These variables are matched to conditions defined by the individual members of the fuzzy rule base and the response of each rule is weighted according to the degree of membership of its inputs. Finally, the weighted average of the responses of all the rules in the rule base is calculated to generate

Advanced Vehicles and Infrastructure Systems, Edited by C.O. Nwagboso

an appropriate output signal. Currently, there is no systematic way of designing FLCs. A typical approach is to define membership functions and the fuzzy logic rule base by studying and extracting the required information from human-operated controllers, existing controllers or experts in a particular field. The performance of the FLC is then measured, and the rules and membership functions are adjusted until a desired performance is achieved. It is clear that the learning and generalizing characteristics of NNs may have a substantial impact on the application of these controllers. An approach for applying NN techniques to the development of fuzzy system controllers has been presented by Kosco.[1] These can be referred to as NN/fuzzy logic adaptive controllers. These hybrid controllers show great promise in the area of very complex control problems, where conventional methods are difficult to apply. Typically, NNs are used to adapt the membership functions or adjust the weighted average of the rule base output responses, or both. Traditional approaches use task descriptions in the form of state variables, production rules and conditional branching, which place an overwhelming burden on the programmer. Even for a moderately complex task, the programmer must account for every possible interaction between a large number of possible events. NN and NN/FLCs provide a radically different approach as their computational behaviour depends on a collective property resulting from many simple processing elements interacting with one another in a richly interconnected system. Two distinct features of this approach are: (i) complex non-linear transformations like coordinate transforms, trajectory control, inverse kinematics and inverse dynamics can be performed implicitly, as the NNs transform the inputs into desired outputs non-algorithmically, using the stored pattern of weights; (ii) NNs can learn complex non-linear control functions without explicit programming by updating the weights in response to an internal or external teacher. This offers an intriguing possibility of training a NN for task-level control by observing a human expert perform the task. In the case of NN/FLCs, expert information, or a priori information can be applied directly to a control problem.

At the heart of an AV control system is the guidance function. This subsystem is the basic feedback control which keeps the AV on the desired path and helps to meet the primary goal of reaching a path end-point from a path starting point. Path development can take various forms from general mission planning statements to highly defined paths. The sensors/pathways which have been employed include sonar, TV cameras, buried wires, optical roads, lines or paths, infra-red, radar, ultrasound, magnetic paths and combinations of these. These approaches run the spectrum of being completely autonomous (in the case of general mission planning statements with sonar, radar, TV camera, etc.) to being completely dependent on the environment (highly defined paths using for example buried cables). As the degree of autonomy increases the amount of sensory data which must be processed also increases rapidly. An example is that of visual data from a TV camera. Vision systems typically process a pixel array of 512×512 pixels which can have a mixture of three colours and various colour levels or 256 grey levels for each pixel. These pixel arrays can be presented to a guidance control system many times a second. It becomes obvious that sensor fusion should be an important part of an AV control system. NNs have been shown to handle large amounts and various types of sensory data while maintaining some fault tolerant attributes due to the parallel nature of the NN internal processing.

10.2 AUTONOMOUS LAND VEHICLE APPLICATION

Autonomous land vehicles (ALV) consist of highly non-linear systems and ALV control systems are subjected to non-linear disturbances (hills, wind, etc.). Autonomous navigation systems based on traditional vision and robotic techniques perform reasonably well under certain conditions, where processing remains fixed under various driving situations. However, noise and variability associated with real world scenes has been a difficult problem for such traditional approaches. NNs offer the ability to design vehicle control systems without requiring complex system models. In this section we shall describe several applications where neurocontrollers adaptively learn state–space path trajectories and encode control strategies for vehicle motion.

The problem of generating appropriate state vector responses to provide for desired AV trajectories and manoeuvres is posed here as one of input/output pair associations. Multilayer perceptrons are the simplest types of feedforward NN models, which can form any decision region and can approximate any desired non-linear input/output mapping. In this case the input features consist of state vectors with corresponding state corrections as output feature vectors. Supervised training using a learning paradigm such as the back-propagation algorithm enables the multilayer perceptrons to synthesize effectively a set of rules from a body of training exemplars. After sufficient training the neurocontroller outputs a set of changes corresponding to the present input state vector, which are used to update the state vector, thus enabling the vehicle to navigate accurately to the desired position.

Pomerleau describes an ongoing project at Carnegie Mellon University.[2] The project, ALVINN (autonomous land vehicle in a neural network) uses an NN with images from a video camera and a laser range finder as input and produces as output the direction the vehicle should travel in order to follow a road. ALVINN's NN architecture consists of a 30×32 unit "retina" on to which a video camera image is projected. Each of the 960 units in the input retina is fully connected to a hidden layer of five nodes, and each of these are in turn fully connected to the output layer of 30 nodes.

Training was originally performed using simulated road images. The network was presented with 1200 road images and corresponding correct output vectors. The network required 30–40 presentations of the images to train the NN. Generating these synthetic road images was very time consuming, and actual road scenes were not always well represented by the synthetic images. As a result training is now done by having a human driver steer it over a portion of the road to be followed. The system inputs the camera image and the steering angle at each time step. The image is enhanced to enhance road contrast. The enhanced image and the steering angle are fed to a back-propagation algorithm that adjusts weights in a NN. The training continues until the network converges, typically requiring less than 100 input images, and takes less than 5 min. Instead of training the network to activate a single output unit, ALVINN is trained to produce a Gaussian distribution of activation centred around the steering direction that will keep the vehicle centred on the road. This Gaussian desired output can be thought of as representing the probability density function for the correct steering direction, in which a unit's probability of being correct decreases with distance from the Gaussian's centre. This training scheme improves the training characteristics of the NN and provides a means for interpolating between the output units.

Two additional problems are presented using this training scheme:[2] (i) as the human driver steers the vehicle down the centre of the road during training, the network will not be presented with situations where it must recover from misalignment errors, and (ii) naively training the network with only the current video image and steering direction runs the risk of over-learning repetitive inputs, such as straight road segments. Both of these problems are a result of the fact that back-propagation requires training data that are representative of the full task to be learned. The first of these problems is solved by taking each original image and shifting it laterally with the proper perspective to create 14 new images. The correct steering for each of these images is generated to account for the extra vehicle lateral displacement. The second problem is addressed by replacing 15 of a 200 image set with the 14 new shifted images, and the original image, and using this new set for training the network. Ten of the images to be replaced are chosen as the images which have the least training error. The other five images to be replaced are chosen randomly. The training requires approximately 50 iterations through the set.

The performance of the system using simulated images, indicates that the system correctly indicates the correct turn radius within 2 units approximately 90% of the time. Also the network can accurately drive the NAVLAB at a speed of 1.8 m/s along a 400 m path through a wooded area of the CMU campus under sunny fall conditions. Using the alternate training algorithm, speeds of 20 miles per hour have been reported which is the top speed of the original NAVLAB. Recent results, using the NAVLAB II indicate that highway speeds have been achieved.

10.3 DESIGN RATIONALE

When designing guidance control systems for AVs using video image information as the sensory input for the controller, several important issues can be identified. In this section we will briefly discuss the determination of road from non-road portions of the image data and then a general framework for AV guidance will be presented.

10.3.1 Determination of Road from Non-road Portions of Image Data

A major problem associated with road following is in determining the road from the non-road portions of a video scene as described by Thorpe *et al.*,[3] in reference to the Carnegie Mellon NAVLAB Project. The NAVLAB project used adaptive colour threshold techniques for the overall image to make this distinction possible. Waxman *et al.*[4] and Dickmanns and Mysliwetz,[5] and Kuan *et al.*,[6] do have done detailed analysis of small areas of interest of the video image, then track these areas over time.

Most research in this field,[3–7] involves the use of a model of the environment (road models for instance), then uses these models to predict the actual location and features of the road. Schneiderman and Nashman,[8] describe a road edge detection system for AV control which uses three stages of computation: (i) extracting edges; (ii) matching extracted edge points with a geometric model of the road; and (iii) updating the geometric road model. The algorithm requires that lane markings be present, and attempts to track the lane markings on both lane boundaries. This system is an integral part of the AV effort at the National Institute of Standards and Technology, which involves the automation of a US Army high-mobility multipurpose wheeled vehicle (HMMWV), as mentioned previously. This effort differs from most other systems in that the road model is based on the information provided by the two-dimensional image and not a reconstructed model of the three-dimensional world.

10.3.2 Determination of Guidance Controller

A variety of techniques have been employed for the guidance control function of AV. Waxman *et al.*,[4] use a technique which decomposes a curved path into a set of short straight line segments, which in turn are sent to the AV, thus simplifying the trajectory planning. Kuan *et al.*,[6] describe a method where a set of candidate subgoals are positioned on the visibility limit (horizon). Each subgoal is determined to be reachable or not while staying on the road surface. The reachable goal which maximizes an objective function is selected for execution. Szabo *et al.*,[9] report using a method called the pure pursuit method. In this approach, a goal-point is selected, and a turn radius is calculated, which will result in the vehicle turning in a circular are which will intersect the goal.

Pomerleau,[10] as described above used an NN to integrate the functions of feature extraction and vehicle guidance. In addition to NNs, FLCs have also been used as intelligent controllers for the guidance of AVs, see for example Kamada, Brown, and Banta.[11–13]

More recently, NN/FLCs have been used for controlling AVs.[14] These NN/FLCs have certain advantages over either NN or FLC controllers by themselves. For example, expert and a priori information may be used with these hybrid controllers and the system is easily adapted using traditional NN training laws. This hybrid type of control model is used for the basis for the development below.

10.3.3 A General Framework for an Autonomous Land Vehicle Control System

The framework for the ALV control system presented below was developed as part of a project at the National Institute of Standards and Technology for a US Army HMMWV. This development is documented in Luebbers.[15] The general outline here is described in greater detail in Section 10.4 below. Although developed for a HMMWV, this design can be adapted to any AV guidance function which uses video images as the sensory input to the controller.

1. *Edge extraction/edge modelling*. The edge extraction and edge modelling functions are used to determine the lane boundaries and to determine if information coming from the edge detection algorithms has a high confidence level based on road models.
2. *Feature extraction*. Features must be extracted from the edge data to be used for control. For this effort we use current vehicle lateral error, current vehicle orientation angle and a future road curvature measure. A brief discussion and advantages for extracting the features using NNs is described below in Section 10.4, and a detailed treatment can be found in Luebbers.[16]
3. *Feature filtering*. This is used for filtering any high frequency noise that is irrelevant to the control algorithm.
4. *Feature modelling*. A model is used to determine how the various features vary with time as a result of various operating conditions of the AV and at different operating points (velocities). From this model a set of predicted features can be generated.
5. *Input fuzzification*. The features extracted from the edge data must be fuzzified for incorporation into the fuzzy inference engine.
6. *Fuzzy rule base and inference engine*. The rule base and inference engine produce the actual control law and is the object which is to be adapted. This stage also includes a de-fuzzification module.
7. *Controller adaptation module*. A NN error back propagation module for adjusting the control law to minimize an error criterion function for adjusting the fuzzy inference engine.
8. *Vehicle model*. A multilayer feedforward NN module used to model the AV and to predict the values of the features that are extracted from the video image.

9. *AV model adaptation module.* A NN error back propagation module for adjusting the AV model is used to minimize an error criterion function.

10. *Output modelling module.* This model is used to determine if the data being sent to the vehicle are valid and includes hard limits and velocity limits on the outgoing control signals.

10.4 DESCRIPTION OF NEURAL NETWORK/FUZZY LOGIC CONTROLLER FOR HMMWV

The NN/FLC, consists of a FLC in a form which can be adapted over time (can learn a control law while minimizing an error criterion) to control the steering and velocity of an AV, a US Army HMMWV. This adaptation is accomplished by using

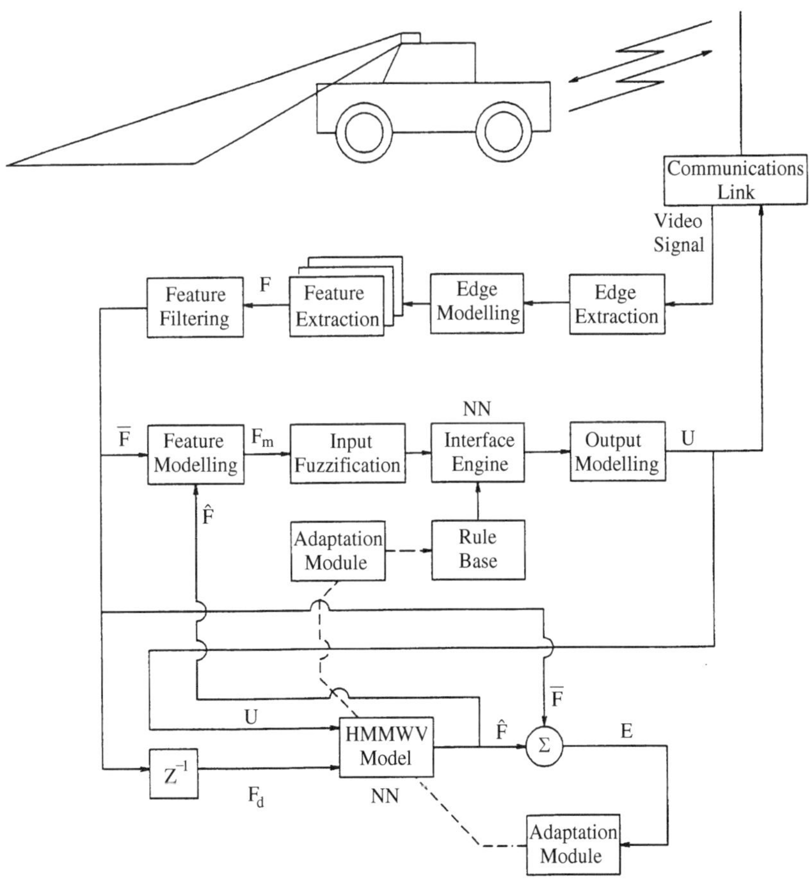

Figure 10.1. Overall neural network (NN)/fuzzy logic controller control system for HMMWV

a NN to learn the control law. In addition NNs are used for feature extraction and for modelling the system to be controlled (thereby providing a method of predicting the features extracted from the video image and for use in back-propagating an error signal through the NN controller). The controller consists of several parts (see Figure 10.1).

10.4.1 Edge Extraction and Edge Modelling

The edge extraction module was discussed briefly above and is described in detail in Schneiderman and Nashman.[8] The output of this module consists of two quadratic polynomials for the left and right road lane edges. These quadratic polynomials are a best fit to the two-dimensional road lane edges in the video image. A problem associated with this algorithm as well as edge extraction algorithms in general, is that often one or both of the road edges are partially or completely obscured. This shortcoming will be addressed in the next section.

10.4.2 Feature Extraction

In general, at any instant in time, the state of an AV system with respect to its environment consists of a lateral error, an orientation angle (we define orientation angle as the angle between the vehicle axis and the tangent to the lane centre at the location of the camera), and a future curvature of the path to be traversed (see Figure 10.2). In addition, the vehicle has an instantaneous velocity, which may be

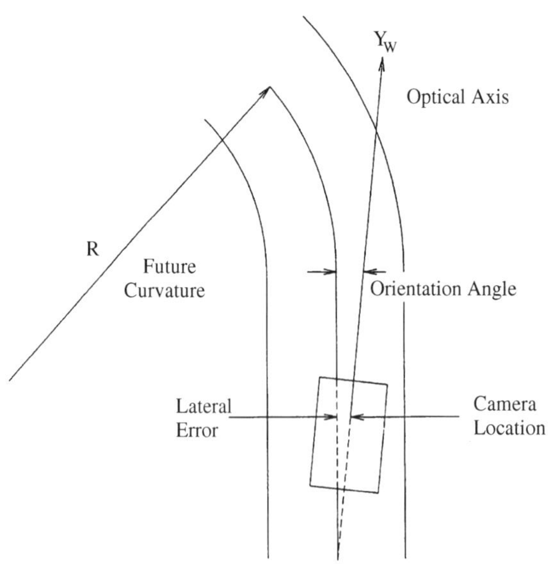

Figure 10.2. Definition of extracted features for neural network (NN)/fuzzy logic controller for HMMWV

separated into longitudinal, lateral and vertical velocities. The first three of these states (features) can be extracted from a static video image. For a detailed treatment of the feature extraction see Luebbers.[17]

A major problem associated with edge extraction is that at times one or both of the lane edges is extracted with low confidence. This is due to missing centre lines at intersections, missing or obstructed shoulder lines, objects in the field of view that

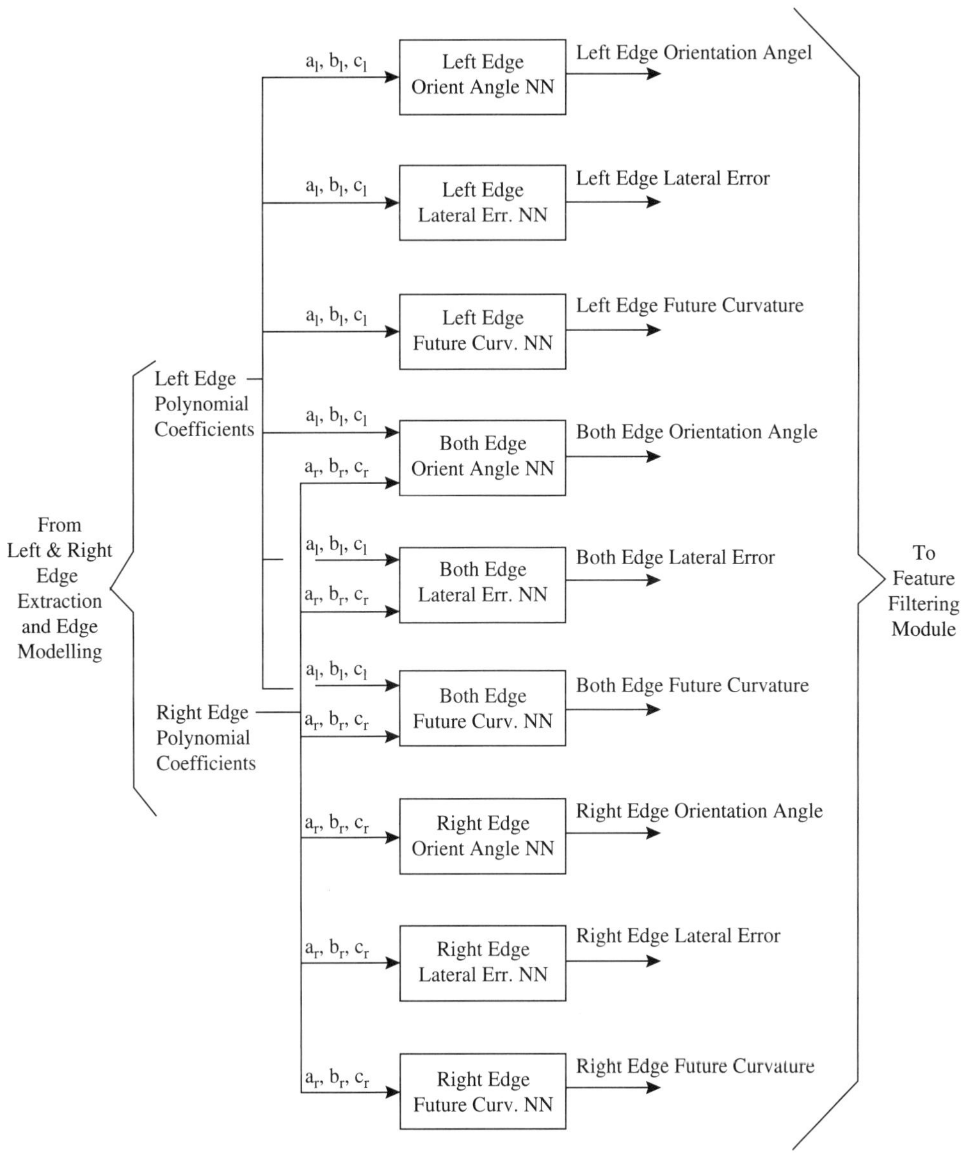

Figure 10.3. Neural network modelling module for estimating image features

confuse the edge extraction algorithms, etc. As a result, it was decided to train a set of NNs to do the feature extraction based on the left road edge, the right edge and both of the road edges (see Figure 10.3). The networks are feedforward NNs and were trained using the familiar error back propagation training algorithm. The input to the NNs are the quadratic equation coefficients of either the left or right road edges, or from both. The training sets were developed by doing feature extraction using data from various test runs at the NIST test site. The determination of which set of features to use is determined in the feature modelling module of the controller.

10.4.3 Feature Extraction Results

Extensive ground truth data were taken to verify the lateral error and orientation angle features. Figure 10.4(a) and (b) show a comparison of the extracted orientation angle and the extracted lateral error versus the actual calculated ground truth data. The ground truth data were taken at various locations along the NIST test path. A typical data set consisted of a static collection of video data, that is the vehicle was stopped during the collection phase. Measurements were taken of the relative location of the vehicle to the lane edges. These measurements were converted to true lateral error and orientation angle. Figure 10.4(a) and (b) show that the feature extraction data is fairly accurate for all sections of the NIST test track.

10.4.4 Neural Network Feature Approximation Results

NN feature approximation programs were written for training and approximation of all three features and by all three methods, that is by using the left edge, right edge and both edge information. The performance of these networks was very good. Figure 10.5(a) is a comparison of the extracted lateral error and the lateral error approximated by a NN, which was trained using both sets of edge polynomial coefficients. Figure 10.5(b) is a comparison of the extracted lateral error and the lateral error approximated by a NN which was trained using only the left edge polynomial coefficients. Both plots show good approximations to the extracted data.

10.4.5 Neural Network/Fuzzy Logic Controller

The NN/FLC consists of six modules. These are the input fuzzification module, inference engine, rule base, controller adaptation module, HMMWV model and the model adaptation module. Figure 10.6 presents a detailed diagram of the NN/FLC.

1. *Input fuzzification module*. The features from the feature modelling module are fuzzified using a set of membership functions depicted in Figure 10.6 as A_i, B_i, and C_i, where $i = 1, 2, \ldots n$. The subscript i represents a rule input grouping. The membership functions represent linguistic variables for each of the features, such as: far left, moderately left, centred, moderately right and far right; for the

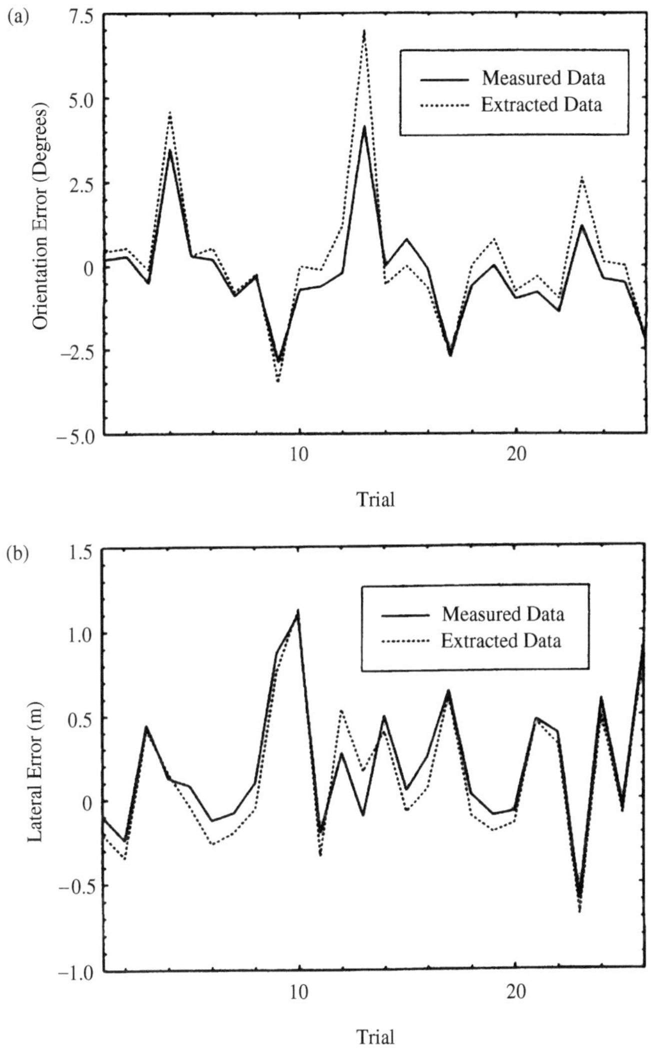

Figure 10.4. (a) Comparison of extracted orientation angle and ground truth data. (b) Comparison of extracted lateral error and ground truth data

lateral error feature. This fuzzification procedure lets us incorporate knowledge of the input space directly into the design of the controller. For instance we may be willing to tolerate a small deviation from the middle of the lane for the lateral error. In addition, we may be less likely to tolerate a moderate deviation to the right of the middle of the lane, and have zero tolerance for moderate deviations to the left or large deviations in either direction. This information can be incorporated in the design of the controller by judiciously choosing the membership functions for the lateral error feature.

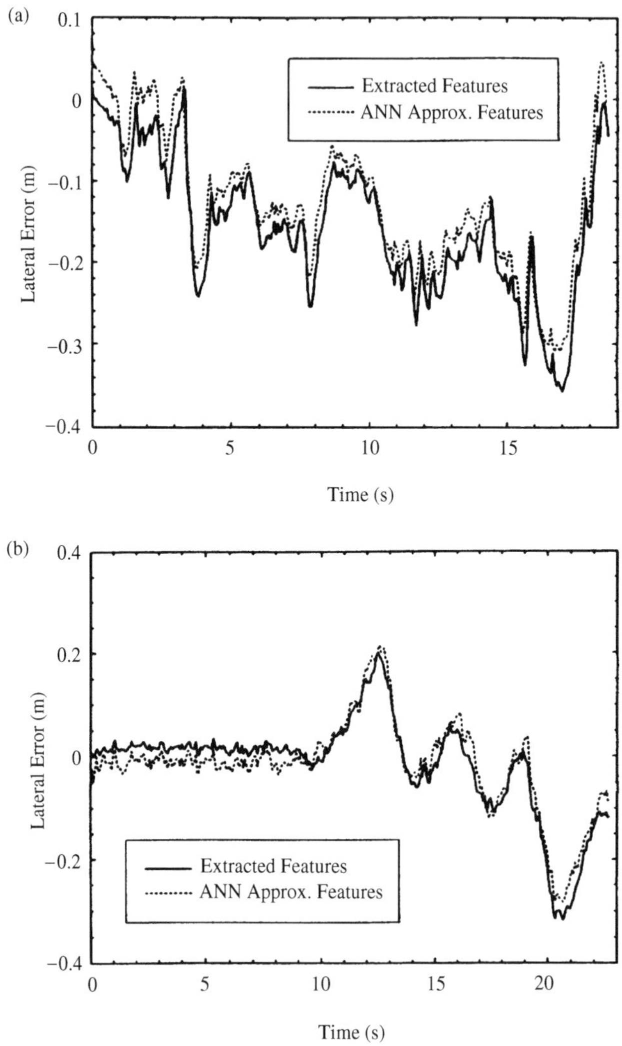

Figure 10.5. (a) Comparison of the extracted lateral error and the lateral error estimated by a neural network using both sets of edge polynomial coefficients. (b) Comparison of the extracted lateral error and the lateral error estimated by a neural network using only left edge polynomial coefficients

2. *Inference engine and rule base*. The inference engine and rule base of a traditional FLC is replaced in this implementation by a feedforward NN. This substitution allows for the system to be adapted over time by back propagation of some error measurement and the minimization of an error function.

3. *Controller adaptation module*. The NN inference engine and rule base are adapted by the controller adaptation module. The lateral error is used in an error

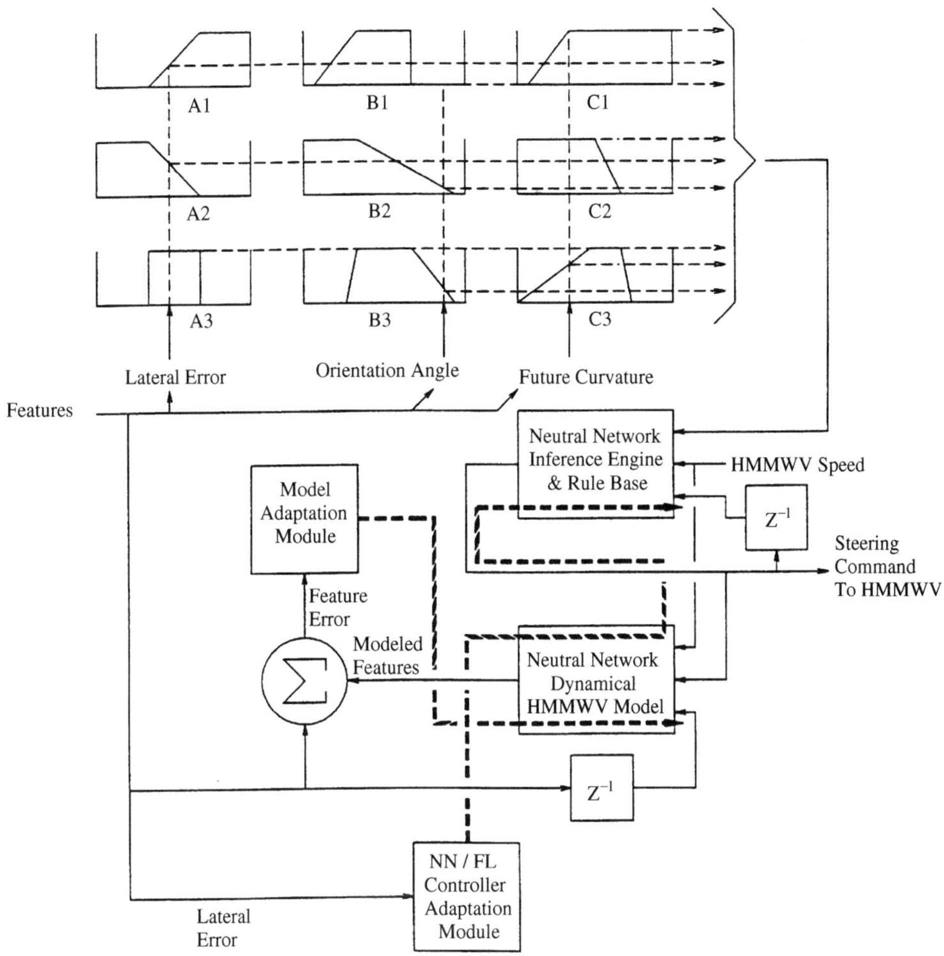

Figure 10.6. Block diagram of neural network (NN)/fuzzy logic (FL) controller for HMMWV

function that is to be minimized. The integral of the squared error (ISE) can be taken as the error function, or a reference signal can be generated which can be compared with the lateral error. This reference signal can be used to teach the controller a particular trajectory that we want the lateral error to follow. As our objective is strictly to track the lane middle, the ISE may be the preferred error function. This error signal can be back propagated through the NN dynamical HMMWV model to determine a desired HMMWV steering command. This desired steering command can be compared with the steering command generated by the NN inference engine. The deviation of the desired and actual steering command can further be back propagated through the NN inference engine. In this way the weights of the controller can be updated by the typical steepest descent learning algorithm as an example.

4. *Neural network dynamical model and adaptation module*. The NN dynamical HMMWV model was described previously and depicted in Figure 10.6. The model consists of a feedforward NN (similar to the inference engine described above). For training the model, we compare the actual extracted features with the modelled features. This error is used to generate an error function just as we did for the controller adaptation module. This error signal is back propagated through the NN model and the weights can be updated by the typical steepest decent learning algorithm.

There are several issues to be addressed, such as at what rate do we update the controller network and at what rate do we update the model network. It is assumed that the dynamical HMMWV system changes gradually, unless some severe catastrophe occurs such as having a flat tyre, sudden rain storm, etc. In this light it would appear that the model would be updated at a slower rate than the controller. This issue, as well as others, requires further research.

10.5 CONCLUSION

The NN developed for the ALVM is currently being adapted for the autonomous control of a HMMWV, in cooperation with the National Institute of Standards and Technology. An overall architecture has been presented for implementation of a NN/FLC for HMMWV.

An accurate method has been developed for the extraction of control features which can be used for steering control as well as velocity control of AVs. A method has been devised, utilizing NNs, for extracting these features using all or partial data from the edge extraction modules. This provides a means for increased confidence level in the extracted features, and results in a more robust controller design. A simulation tool has been developed for visualizing extracted edge data and its relationship to extracted features. This tool can be expanded to include the HMMWV modelling function described and can be used for designing accurate, robust feature extraction algorithms with minimum noise.

Powerful learning algorithms are one of the main strengths of NN approach. Use of this learning ability could result in two main advances for robot navigation, automatic knowledge acquisition and continuous system refinement. NNs have a natural ability to acquire the necessary information from the environment which could aid system construction and save the user from spending the enormous amount of time necessary to derive a rule-based database for objects and environments. System requirements could then be incorporated to make any necessary changes to improve the performance of the controller.

Although they have been successfully implemented as computer simulations, NN hardware is also feasible. Robots with NN controllers have the ability to learn complex control functions and to generalize in unexpected situations, which could provide tremendous benefits over traditional approaches.

REFERENCES

1 Kosko B. *Neural Networks and Fuzzy Systems*. Englewood Cliffs, NJ: Prentice Hall, 1992.

2 Pomerleau DA. Efficient training of artificial neural networks for autonomous navigation. *Neural Computation* 1991; **3**(1): 88–97.

3 Thorpe C, Hebert MH, Kanade T, Shafer SA. Vision and navigation for the Carnegie-Mellon Navlab. *IEEE Trans. on Pattern Analysis and Machine Intelligence* 1988; **10**(3): •••.

4 Waxman AM, LeMoigne JJ, Davis LS, Srinivasan B, Kushner TR, Liang E, Siddalingaiah T. A visual navigation system for autonomous land vehicles. *IEEE Journal of Robotics and Automation*, 1987; **RA-3**(2): •••.

5 Dickmanns ED, Mysliwetz BD. Recursive 3-D road and relative ego-state recognition. *IEEE Transactions on Pattern Analysis and Machine Intelligence*, 1992; **14**(2): •••.

6 Kuan D, Phipps G, Hsueh A. Autonomous robotic vehicle road following. *IEEE Transactions on Pattern Analysis and Machine Intelligence* 1988; **10**(5): •••.

7 Turk MA, Morgenthaler DG, Gremban KD, Marra M. VITS — a vision system for autonomous land vehicle navigation. *IEEE Transactions on Pattern Analysis and Machine Intelligence* 1988; **10**(3): •••.

8 Schneiderman H, Nashman M. Visual processing for autonomous driving, 1992 (Preprint).

9 Szabo S, Scott HA, Murphy KN, Legowik SA, Bostelman RV. High-level mobility controller for a remotely operated unmanned land vehicle. *Journal of Intelligent and Robotic Systems* 1992; **5**, 63–77.

10 Pomerleau DA. Neural network based autonomous navigation. In *Vision and Navigation, The Carnegie Mellon Navlab* Thorpe CE, (ed.). Norwell, MA: Kluwer, 1990.

11 Kamada H, Naoi S, Gotoh T. 'A compact navigation system using image processing and fuzzy control. *Proceedings of IEEE Southeastcon — 1990*.

12 Banta LE, Nutter RS, Xia Y. Mode-based navigation for autonomous mine vehicles. *IEEE Transactions on Industry Applications* 1992; **28**(1): 181–185.

13 Brown M, Fraser R, Harris CJ, Moore CG. Intelligent self-organizing controllers for autonomous guided vehicles: comparitive aspects of fuzzy logic and neural nets. *IEE International Conference on Control '91*, Edinburgh, UK, 25–28 March 1991, **1**, 134–139.

14 Lin CT, Lee CS. Neural-network-based fuzzy logic control and decision system. *IEEE Transactions on Computers* 1991; **40**(12): 1320–1336.

15 Luebbers PG, Raviv D. *A Neural Network/Fuzzy Logic Controller for HMMWV*. NIST Report, 1992.

16 Luebbers PG, Raviv D, Pandya AS. 'Feature Extraction using Neural Networks for Vision-Based Autonomous Driving. (Preprint).

17 Pomerleau DA. *ALVINN: An autonomous land vehicle in a neural network*. Technical Report, CMU CS-89-107, Carnegie Mellon University, 1989.

PART 4

VEHICLE NAVIGATION AND GUIDANCE

11

Camera Telemeter Multisensory System for Obstacle Avoidance

11.1 INTRODUCTION

During the last decade, numerous computer vision methods were developed for perception in autonomous road vehicles. This expansion is due to the main projects which are ALV (Autonomous Land Vehicle) in the United States of America and then PROMETHEUS (PROgram for a European Traffic with Highest Efficiency and Unprecedented Safety) in Europe. The aim of the ALV project is cross-country navigation, but several results have been obtained on structured road with the autonomous vehicles Navlab (see[13] and[14]) and Alvin (see[15]). In Germany, Vamors[10] and VITA[9] are two efficient autonomous vehicles especially with regard to road location. In Japan, Harunubu-4[11] is an outdoor vehicle too.

The vision problems for autonomous road vehicles are road location, obstacle detection and the recognition of road signs. The road location problem is the one most often dealt with in the published material. Obstacle detection is certainly more difficult, and the recognition of road-signs is particularly dealt with in the PROMETHEUS project.

This chapter deals with road obstacle detection and tracking.* Why detection and tracking? It is obvious that detection is a fundamental aspect, but smart driving needs the knowledge of the kinematic state of the obstacle and this knowledge can only be obtained by focusing the attention of the system on this obstacle. This approach is based on the behaviour of a human driver who scans the whole road scene until

* This work was supported by the European Eureka-Prometheus project

Advanced Vehicles and Infrastructure Systems, Edited by C.O. Nwagboso

he detects obstacles and then tracks the more dangerous one. After he has observed the obstacle for long enough, he can deduce the right driving strategy to deal with the road obstacle.

We use a multisensorial approach based on a camera/3D-sensor mixed sensor which is mounted in front of our vehicle. In the first section, we present the multi-sensor. We report first its description and its major characteristics. The calibration procedure shows how it is possible to match heterogeneous data. Finally we present two specific capabilities of the sensor which make it a smart sensor. The second section deals with the multisensorial data fusion algorithm which performs the road obstacle detection and tracking.

11.2 THE MULTISENSOR

11.2.1 Description and Performances of the Multisensor

The original idea of our approach is to add a depth sensor or 3D-sensor, to a camera. So, for a same point, we will be able to determine its intensity and its 3D location. Another advantage of such a multisensor is that it combines the accuracy of the location given by the 3D-sensor with the reduced acquisition time of the video signal by the camera. For an application of on board vision, where the different parts of the multisensor are in oscillation, their motion must be identical to ensure the right match between the two kinds of data. So the camera and the 3D-sensor are locked together. Moreover, the closeness between the two sensors reduces the shadow areas. We will now briefly describe the camera and the 3D-sensor. The first presentation of the multisensor was in.[1]

The video camera

We use a CCD camera of 1/2 inch equipped with a lense equal to 12 mm. This configuration leads to a field of vision close to 30°. The standard of the camera is CCIR 625 lines and its resolution is 570 lines. The aperture is formed by an electronic shutter controlled by a CCD iris. The numerization of the video signal provides an intensity image of 512×512 pixels of 8 bits.

The 3D-sensor

The 3D-sensor is a laser telemeter based range finder. The laser beam is controlled by two flat mirrors.

The laser telemeter Its principle is the measurement of flight time. The relation between the distance and this time is straightforward:

$$v \cdot \tau = 2 \cdot r$$

where v is the speed of propagation of the signal, r the distance from the obstacle, and τ the time of flight between the emitted pulse and the received pulse. The high sensivity of the telemeter makes it possible to detect diffuse light. We can thus measure the distance of non cooperative natural targets. The main characteristics of this range finder, are its range equal to 100 m and a low divergence (the diameter of the beam is equal to 20 cm at a distance of 50 m).

The scanning system The scanning system uses two mirrors. A mechanical mounting supports two galvanometric deflectors which control the two flat mirrors. Two electronic modules permit to control the horizontal (X) and vertical (Y) deflections in a scale of $\pm 20°$ with a sensivity equal to 4 °/V. It is important to note that this system is protected by a window of silica glass. Indeed, the mirrors have to be protected against possible projections such as atmospheric precipitation or any other projectile. Its coefficient of reflection is, of course, minimum for the wavelength of the laser (905 nm).

Figure 11.1(a) shows the multisensor. The range finder is put across in a such way that the laser beam goes onto the deflection system (on the left side). We can see the window of protection. The camera is put onto the deflection system. The multisensor can be included in a volume of 26 cm × 27 cm × 64 cm.

Performances of the 3D-sensor

In this section, we present some characteristics of the sensor after the coupling between the range finder and the deflection system. The range, the precision, the resolution and a study on the mirrors are presented.

Range of the 3D-sensor This characteristic is certainly one of the most important ones. We have evaluated it at 85 m in real traffic situations. Thus the successive

(a) (b)

Figure 11.1. The multisensor (a) is mounted in front of our experimental vehicle (b) reflections of the laser beam on the mirrors, the window of protection, and the optical reduction of the distance between the emission and the reception decreased the range of 15 m in comparison with the use of the telemeter alone. This range depends, of course, on the nature and on the positioning of the object. In this way, we have tested the response of the 3D-sensor to the rear view of a car. Figure 11.2 shows the area of a red car seen by the 3D-sensor for two different distances. Nearly all the shots have received an echo. It is important to note that these measurements were carried out under normal conditions of cleanliness. Indeed, if the two windows of the car are especially cleaned for the measurement, the laser beam can pass across these two windows and so that no echo is available in this direction. The area without echo is nevertheless reduced because the laser beam often meets with components inside the car. So, for this characteristic, this 3D-sensor is particularly well adapted to our application.

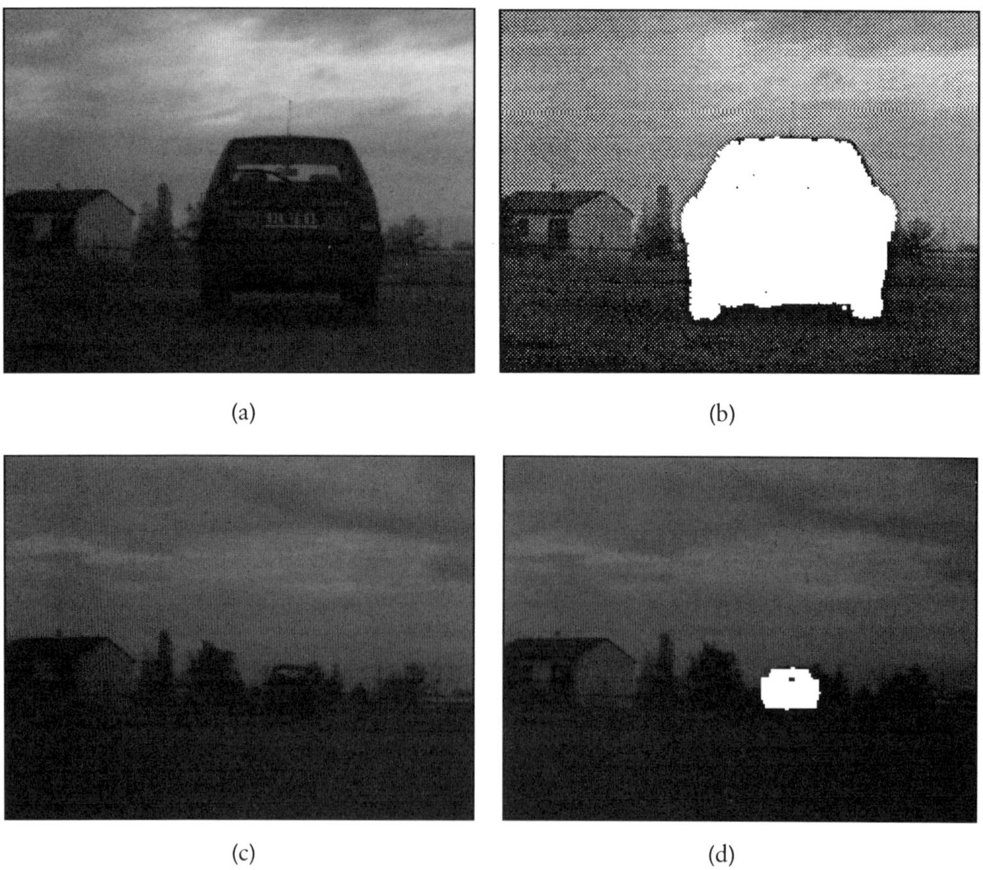

(a) (b) (c) (d)

Figure 11.2. Area seen by the 3D-sensor for a red car in normal situation of cleanliness (a) car at 8.9 m (b) response of the 3D-sensor. (c) car at 30.1 m (d) response of the 3D-sensor

Accuracy of the 3D-sensor Accuracy is an important parameter for the 3D-sensor. In Figure 11.3 we present how the distance measurement on a fixed point in the space can be reproduced. For the three distances the standard deviation is about 10 cm for 1000 measurements. That is a good level of accuracy for our application.

Spatial resolution of the 3D-sensor Spatial resolution is an important parameter to find out how it will be possible to separate two closed objects. We used a Foucault test card with a half step p of 20 cm. For several distances we measured the ratio between the width of a space without echo and the width of the panel calculated by the 3D-sensor. We found that $r_s = 0$ for $d = d_m = 45$ m. So the divergence α of the beam defined by the formula $\alpha = \arctan(p/d)$, is equal to 4.4 mrad. In other words, at 45 m the 3D-sensor is able to distinguish two different road obstacles in a plane perpendicular to the laser beam if they are more than 20 cm apart. Thus measurements without overlap are possible by acquiring no more shots than 4 per degree.

Response of the mirrors Characterization of the mirrors is useful for several reasons. Firstly, when we need to change the area of acquisition in space, the time of response to a step is a helpful parameter. Secondly their dynamic response will

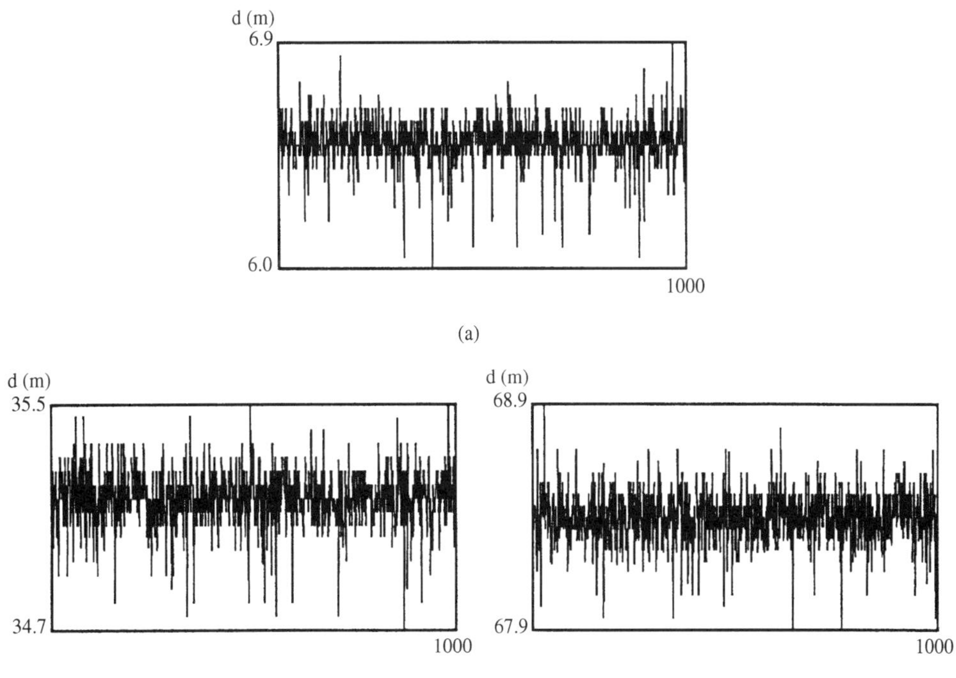

Figure 11.3. Laser shots on a fixed point (a) average distance: 6.51 m; standard deviation: 0.08 m (b) average distance: 35.18 m; standard deviation: 0.10 m (c) average distance: 68.42 m; standard deviation: 0.11 m

have an influence on the scanning of the scene. Finally, we present some results on their sensivity to the disturbances which are numerous in a car.

The responses of the mirrors to a step have shown that they behave like a second order system corrected by a perfectly PID corrector. If we define the time of response of a mirror as the time taken by the mirror to reach 90% of the input value, we obtained 8.5 ms for the *X* mirror and 11 ms for the *Y* mirror. The time is greater for the *Y* mirror because of its greater size. So the time necessary between the acquisition of two 3D images located at different positions in the space must be more or less equal to this time.

Acquisition of a 3D image is obtained by a triangular scanning in the two directions. The acquisition rate of a 3D image is limited by the *X* mirror because of the principle of acquisition which is carried out line by line. The response frequency showed that it is reasonable to limit the frequency of scanning to 25 Hz. It is important to note that we can read the output of the deflection system, so, we can find out the true location taken by the mirrors.

The last point studied here is the mirror sensitivity to disturbances. First, the intrinsic electronic noise is much reduced. Then, we measured the disturbances caused by the vibrations of the engine and due to disturbances caused by driving over ramps (Figure 11.4) which leads to high impulsional responses. Given the mirror sensivity (4 °/V) we can see that in exceptional conditions the deviation of

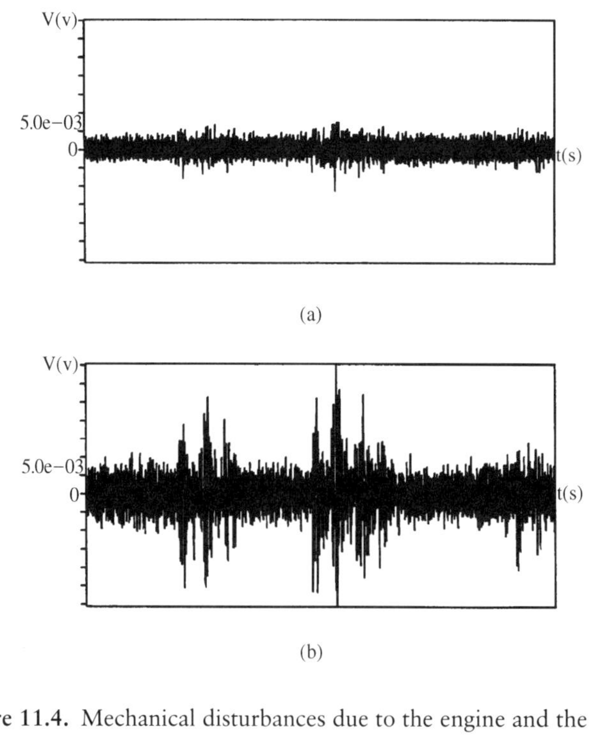

Figure 11.4. Mechanical disturbances due to the engine and the ramps

the laser beam is about ±15 cm at a distance equal to 50 m. In usual driving situations, it is reduced to ±5 cm at the same distance. These errors are acceptable for our application.

The 3D image parameters Given the characteristics described above, we use two kinds of 3D images: one is optimized in acquisition time, the other in resolution. The first is used in real traffic situation. It is restricted by the X mirror scanning rate and the rate of the acquisition of the distance (4 kHz). We have chosen a 3D image of 10 lines. So a 3D image of 10 lines of 80 shots is available every 200 ms. The second is a dense image. The only constraint is to acquire no more than 4 shots per degree.

11.2.2 The Multisensor Calibration

The aim of the multisensor calibration is to match a 3D image and an intensity image. Thus apart from the resolution, we will find out the intensity and the 3D position of each point of the space in the multisensor's field of vision.

In this section, we first studied the calibration of the 3D-sensor,[16] i.e., what are the Cartesian coordinates of a point in space from the data given by the different parts of the 3D-sensor? Moreover, these coordinates must be expressed in an absolute reference frame because the sensor reference is not *a priori* suitable. This section then briefly deals with the camera calibration. It makes it possible to link the location of a point in space and the location of its projection in the intensity image.

We denote the different frames as follows:

$\mathcal{R}_c(O_c, x_c, y_c, z_c)$ is the 3D coordinate system of the 3D sensor.

$\mathcal{R}_a(O_a, x_a, y_a, z_a)$ is the 3D absolute coordinate system.

$\mathcal{R}(O, u, v)$ is the 2D coordinate system of the video camera.

Figure 11.5 shows the different matrices we will use. M makes it possible to go from $\mathcal{R}_c$ to $\mathcal{R}_a$, T from $\mathcal{R}_a$ to $\mathcal{R}$ and H combines these two frame changes.

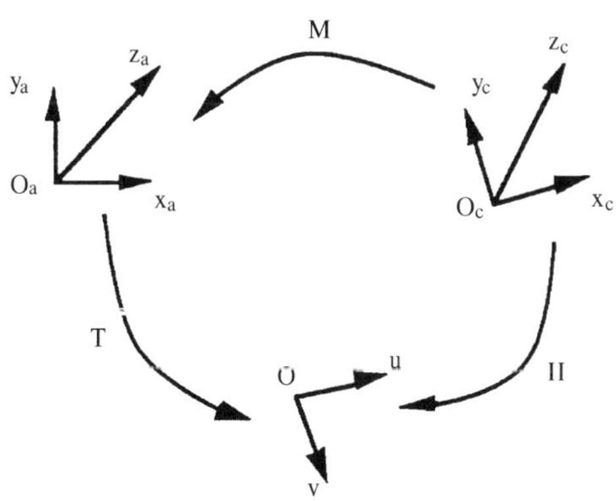

Figure 11.5. The matrix for frame changes

The 3D-sensor calibration

We propose a general method for the calibration of a 3D-sensor based on a laser range finder. The calibration of such a sensor consists in computing the Cartesian coordinates (x_a, y_a, z_a) of a point in space in $\mathcal{R}_a$ from the data $Y = (y_1, \ldots, y_d)$ given by the 3D-sensor. We write first, the Cartesian coordinates (x_c, y_c, z_c) of this point in $\mathcal{R}_c$ from the linear $B = (b_1, \ldots, b_m)$ and the non linear $A = (a_1, \ldots, a_n)$ parameters of the sensor and from the data Y:

$$\begin{pmatrix} x_c = \sum_{i=1}^{r} b_i \cdot f_i(A, Y) \\ y_c = \sum_{i=r+1}^{r+s} b_i \cdot f_i(A, Y) \\ z_c = \sum_{i=r+s+1}^{m} b_i \cdot f_i(A, Y) \end{pmatrix} \tag{11.1}$$

where the $f_i(A, Y)$ are non linear functions.

The system (1) can be expressed in a matrix form:

$$X_c = D \cdot F(A, Y) \tag{11.2}$$

with

$$X_c = (x_c \quad y_c \quad z_c)^t,$$

$$D = \begin{pmatrix} b_1 \ldots b_r & 0 & 0 \\ 0 & b_{r+1} \ldots b_{r+s} & 0 \\ 0 & 0 & b_{r+s+1} \ldots b_m \end{pmatrix},$$

and

$$F(A, Y) = (f_1(A, Y) \ldots f_m(A, Y))^t.$$

The relation (11.2) expresses the Cartesian coordinates of a point in $\mathcal{R}_c$ from the data given by the 3D-sensor and the intrinsic parameters of the 3D-sensor.

We now need to change the coordinate system between $\mathcal{R}_c$ and $\mathcal{R}_a$. Such a transformation is traditionally achieved by three rotations $R_{\alpha\beta\gamma}$ and three translations T_{xyz}:

$$X_a = R_{\alpha\beta\gamma} \cdot X_c + T_{xyz}$$

or

$$X_a = M \cdot \bar{X}_c \tag{11.3}$$

with

$$\bar{X}_c = (X_c^t 1)^t$$

and

$$M = (R_{\alpha\beta\gamma} T_{xyz})$$

By homogenizing the relation (11.2) we can write:

$$\bar{X}_c = \bar{D} \cdot \bar{F}(A, Y) \tag{11.4}$$

with

$$\bar{D} = \begin{pmatrix} D & 0 \\ 0 & 1 \end{pmatrix}$$

and

$$\bar{F}(A, Y) = \left(F^t(A, Y)1\right)^t$$

Finally, from (11.3) and (11.4), we obtain:

$$X_{\mathrm{a}} = M \cdot \bar{D} \cdot \bar{F}(A, Y) \tag{11.5}$$

This relation expresses the position of a point in the space from the parameters and the measurements of the 3D-sensor. The last step of the calibration is the computation of the coefficients of the matrix product $M \cdot \bar{D}$ and the calculation of the non linear parameters. We use a least mean squares method for the former and an iterative method for the latter.

For the particular case of our 3D-sensor, the data are three tensions $Y = (V_\rho, V_{\theta_{\mathrm{x}}}, V_{\theta_{\mathrm{y}}})$ proportional to the distance ρ measured by the range finder and to the deflections of the mirrors $(\theta_{\mathrm{x}}, \theta_{\mathrm{y}})$. The relations (11.6) link the electric variables to the geometric ones.

$$\begin{pmatrix} \rho = K_\rho \cdot V_\rho \\ \theta_{\mathrm{x}} = K_{\theta_{\mathrm{x}}} \cdot V_{\theta_{\mathrm{x}}} \\ \theta_{\mathrm{y}} = K_{\theta_{\mathrm{y}}} \cdot V_{\theta_{\mathrm{y}}} \end{pmatrix} \tag{11.6}$$

The K_{i} are the intrinsic electric parameters of the 3D-sensor.

Figure 6 schematizes the geometry of our 3D-sensors, except for the origin of the reference frame which is taken from the middle of the Y mirror. The geometric

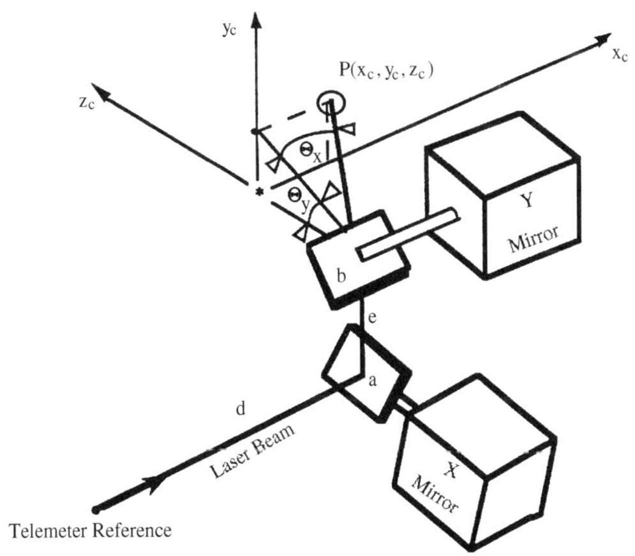

Figure 11.6. The 3D-sensor geometry

intrinsic parameters of the 3D-sensor are the distance e between the two mirrors and the distance d between the range finder reference and the location of the laser beam on the X mirror. The laser beam, sent out at the distance d from the X mirror, hits this mirror at point a. It is then deflected, with an angle θ_x, towards the Y mirror which is hit at b. The Y mirror then deflects the laser beam, with an angle θ_y, towards the point of the space $X_c(x_c, y_c, z_c)$. The equations (11.7) relate the pseudo-spherical coordinates $(\rho, \theta_x, \theta_y)$ to the Cartesian coordinates (x_c, y_c, z_c) in the frame $\mathcal{R}_c$:

$$\begin{pmatrix} x_c = (\rho - \mathrm{d}) \cdot \sin\theta_x \\ y_c = (\rho - \mathrm{d}) \cdot \cos\theta_x \cdot \sin\theta_y - e \cdot \sin\theta_y \\ z_c = (\rho - \mathrm{d}) \cdot \cos\theta_x \cdot \cos\theta_y - e \cdot \cos\theta_y \end{pmatrix} \tag{11.7}$$

From the relations 11.6 and 11.7 we can identify the model:

$$F(A, Y) = \begin{pmatrix} V_\rho \cdot \sin(K_{\theta_x} \cdot V_{\theta_x}) \\ \sin(K_{\theta_x} \cdot V_{\theta_x}) \\ V_\rho \cdot \cos(K_{\theta_x} \cdot V_{\theta_x}) \cdot \sin(K_{\theta_y} \cdot V_{\theta_y}) \\ \cos(K_{\theta_x} \cdot V_{\theta_x}) \cdot \sin(K_{\theta_y} \cdot V_{\theta_y}) \\ \sin(K_{\theta_y} \cdot V_{\theta_y}) \\ V_\rho \cdot \cos(K_{\theta_x} \cdot V_{\theta_x}) \cdot \cos(K_{\theta_y} \cdot V_{\theta_y}) \\ \cos(K_{\theta_x} \cdot V_{\theta_x}) \cdot \cos(K_{\theta_y} \cdot V_{\theta_y}) \\ \cos(K_{\theta_y} \cdot V_{\theta_y}) \end{pmatrix}$$

with

$B = (K_\rho, -\mathrm{d}, K_\rho, -\mathrm{d}, -e, K_\rho, -\mathrm{d}, -e)$

$A = (K_{\theta_x}, K_{\theta_y})$

$r = 2; s = 3; m = 8$

11.2.2.2 The camera calibration

The camera calibration is a well-known vision problem. The problem is, given a point $X_a(x_a, y_a, z_a)$ in the absolute coordinate system $\mathcal{R}_a$: What is the transformation that allows the computation the numerical location (u, v) of its projection in the image coordinate system $\mathcal{R}$? The camera model is the pin hole model and the method is described in[1] and[8]. The transformation is split up into three stages: the frame change between the absolute frame $\mathcal{R}_a$ and the 3D camera frame, the perspective transformation and the computer acquisition. The whole transformation can be written as

$$\begin{pmatrix} w \cdot u \\ w \cdot v \\ w \end{pmatrix} = T \cdot \begin{pmatrix} x_a \\ y_a \\ z_a \\ 1 \end{pmatrix} \tag{11.8}$$

where T is a 3×4 matrix and w a scale factor. The twelve matrix coefficients combine five intrinsic parameters of the camera and six extrinsic parameters for the frame change. These coefficients are computed by a least mean square method.

The equations (11.3) and (11.8) allow the same transformation from the sensor coordinate system $\mathcal{R}_c$ by writing:

$$\begin{pmatrix} w \cdot u \\ w \cdot v \\ w \end{pmatrix} = H \cdot \begin{pmatrix} x_c \\ y_c \\ z_c \\ 1 \end{pmatrix} \tag{11.9}$$

with

$$H = T \cdot \bar{M} \text{ where } \bar{M} = \begin{pmatrix} M & 0 \\ 0 & 1 \end{pmatrix}$$

Results

The measurements for the multisensor calibration are taken by putting sheets of wood, 50 cm to 2 m high, on a flat area 30 m wide and 80 m long (see Figure 11.7).

When the laser beam hits the target, we write successively the 3D-sensor data $Y = (V_\rho, V_{\theta_x}, V_{\theta_y})$ the location of the laser beam (u, v) in the image frame $\mathcal{R}$, and its location (x_a, y_a, z_a) in $\mathcal{R}_a$. We choose the absolute coordinate system with the z-axis parallel to the car longitudinal axis and the x-axis parallel to the ground.

As far as the 3D-sensor calibration is concerned, the mean relative errors are $\Delta x = 0.4\%$, $\Delta y = 2.3\%$, $\Delta z = 0.03\%$. These results are better than those we obtain by using a linear model: $\Delta x = 0.9\%$, $\Delta y = 5.2\%$, $\Delta z = 0.17\%$. The linear model considers that the parameter values given by the supplier are true, so an iterative process is not necessary for this model. We note that the results are not so good for the y coordinate because of its smaller absolute value.

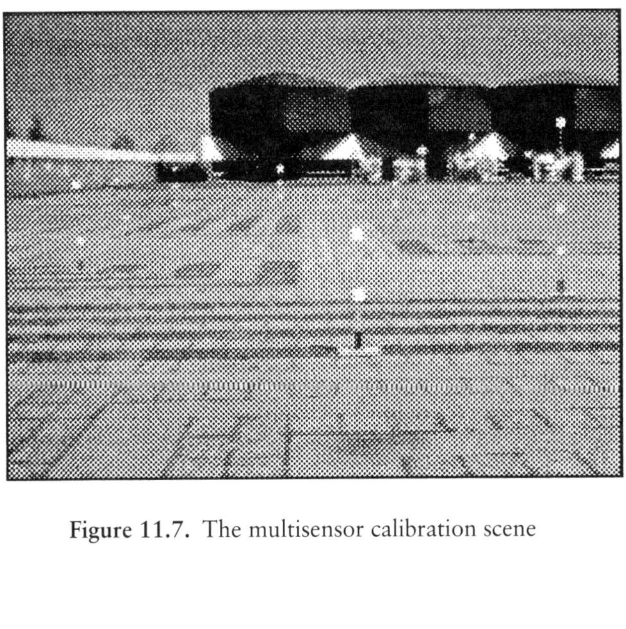

Figure 11.7. The multisensor calibration scene

Figure 11.8. Matching example: (a) the intensity image, (b) the 3D image (the closer the shot is, the blacker it is), (c) the matching of the 2 images

In the sequel, we will use a simplified 3D-sensor model:

$$\begin{pmatrix} x_c = \rho \cdot \sin\theta_x \\ y_c = \rho \cdot \cos\theta_x \cdot \sin\theta_y \\ z_c = \rho \cdot \cos\theta_x \cdot \cos\theta_y \end{pmatrix} \tag{11.10}$$

This simplified model considers that d and e are negligible in comparison with ρ. The error is always smaller than 1% for a normal road situation.

With regard to the camera calibration, the mean error is approximately one pixel. As the target location accuracy in the space is equal to 10 cm, this mean error in the camera frame is a satisfactory result.

So, we are now able to match an intensity image and a 3D image. Figure 11.8 presents a matching example using a dense 3D image with an optimized spatial resolution.

11.2.3 A Smart Sensor

Generally, a sensor is smart if it is able to change its perception strategy according to the scene. In this perspective, we have endowed our multisensor with two faculties to improve its intelligence. They allow close collaboration between the intensity world and the 3D world. 2D image processing will control the 3D sensor in two ways. The first faculty is controlled perception, the second is visual servoing.

Controlled perception

Controlled perception makes the acquisition of a 3D image possible in a window described in the intensity image. Thus the problem is to determine the mirror deflections (θ_x, θ_y) from the 2D location (u, v) of a point in the image frame. It is the opposite problem, and can be solved as follows:

According to equation (11.9), we can write:

$$\begin{cases} u = \dfrac{h_{11} \cdot x_c + h_{12} \cdot y_c + h_{13} \cdot z_c + h_{14}}{h_{31} \cdot x_c + h_{32} \cdot y_c + h_{33} \cdot z_c + 1} \\ v = \dfrac{h_{21} \cdot x_c + h_{22} \cdot y_c + h_{23} \cdot z_c + h_{24}}{h_{31} \cdot x_c + h_{32} \cdot y_c + h_{33} \cdot z_c + 1} \end{cases} \tag{11.11}$$

Let $t_1 = x_c/z_c$ and $t_2 = y_c/z_c$ be two new variables. As z_c is never equal to zero in our experimental configuration, the relationship (11.11) can be rewritten as follows:

$$\begin{cases} u = \dfrac{h_{11} \cdot t_1 + h_{12} \cdot t_2 + h_{13} + h_{14}/z_c}{h_{31} \cdot t_1 + h_{32} \cdot t_2 + h_{33} + 1/z_c} \\ v = \dfrac{h_{21} \cdot t_1 + h_{22} \cdot t_2 + h_{23} + h_{24}/z_c}{h_{31} \cdot t_1 + h_{32} \cdot t_2 + h_{33} + 1/z_c} \end{cases} \tag{11.12}$$

By extending the above equation, we can obtain the following system:

$$\begin{cases} (h_{31} \cdot u - h_{11}) \cdot t_1 + (h_{32} \cdot u - h_{12}) \cdot t_2 = (h_{13} - h_{33} \cdot u) + \dfrac{h_{14} - u}{z_c} \\ (h_{31} \cdot v - h_{21}) \cdot t_1 + (h_{32} \cdot v - h_{22}) \cdot t_2 = (h_{23} - h_{33} \cdot v) + \dfrac{h_{24} - v}{z_c} \end{cases} \tag{11.13}$$

If we define the following matrix:

$$t = (t_1\ t_2)^t$$

$$L = \begin{pmatrix} h_{31} \cdot u - h_{11} & h_{32} \cdot u - h_{12} \\ h_{31} \cdot v - h_{21} & h_{32} \cdot v - h_{22} \end{pmatrix}$$

$$S = \begin{pmatrix} h_{13} - h_{33} \cdot u \\ h_{23} - h_{33} \cdot v \end{pmatrix}$$

$$N = \begin{pmatrix} \dfrac{h_{14-u}}{\cos\theta_x \cdot \cos\theta_y} \\ \dfrac{h_{24-v}}{\cos\theta_x \cdot \cos\theta_y} \end{pmatrix}$$

and by using $z_c = \rho \cdot \cos\theta_x \cdot \cos\theta_y$, the system 11.13 becomes:

$$L \cdot t = S + \rho^{-1} \cdot N$$

and finally

$$t = L^{-1} \cdot S + \rho^{-1} \cdot L^{-1} \cdot N \tag{11.14}$$

This equation has two terms: the first term is independent of the distance ρ, and the second term is proportional to ρ^{-1}. In the first approximation, we can discard this second term and so determine (θ_x, θ_y) by solving the following system:

$$\begin{cases} \theta_x = \arctan(t_1 \cdot \cos\theta_y) \\ \theta_y = \arctan(t_2) \\ t = L^{-1} \cdot S \end{cases}$$

Moreover, the ρ value can be used, by an iterative process, equation (11.14) to improve the value of θ_x and θ_y.

We are now able to acquire a 3D image in a window defined in the intensity image.

Visual servoing

The second faculty of the multisensor is a visual servoing of the laser beam on a moving point in the 2D image. In this case, the 3D sensor is not used as an image sensor but it is focused on a single point. The problem is to express the mirror deflection variations $(\dot{\theta}_x, \dot{\theta}_y)$ from 2D point location variations $(\dot{u}, \dot{v})$ assuming the distance ρ is constant between two iterations. The key point of the approach is the computation of the 3D sensor jacobian J which allows the expression of $(\dot{u}, \dot{v})$ from $(\dot{\theta}_x, \dot{\theta}_y)$:

$$\begin{pmatrix} \dot{u} \\ \dot{v} \end{pmatrix} = J \cdot \begin{pmatrix} \dot{\theta}_x \\ \dot{\theta}_y \end{pmatrix} \tag{11.15}$$

The jacobian J can be obtained from the relations 11.9 and 10.0. Indeed from 11.9 we can write:

$$\begin{cases} \dot{u} = \dfrac{\partial u}{\partial x_c} \cdot \dot{x}_c + \dfrac{\partial u}{\partial y_c} \cdot \dot{y}_c + \dfrac{\partial u}{\partial z_c} \cdot \dot{z}_c \\ \dot{v} = \dfrac{\partial v}{\partial x_c} \cdot \dot{x}_c + \dfrac{\partial v}{\partial y_c} \cdot \dot{y}_c + \dfrac{\partial v}{\partial z_c} \cdot \dot{z}_c \end{cases}$$

with

$$\frac{\partial u}{\partial x_c} = \frac{1}{w^2} \cdot \left(\frac{\partial (w \cdot u)}{\partial x_c} \cdot w - (w \cdot u) \cdot \frac{w}{\partial x_c} \right) = \frac{h_{11}}{w} - \frac{h_{31} \cdot (w \cdot u)}{w^2}$$

$$\frac{\partial u}{\partial y_c} = \frac{1}{w^2} \cdot \left(\frac{\partial (w \cdot u)}{\partial y_c} \cdot w - (w \cdot u) \cdot \frac{w}{\partial y_c} \right) = \frac{h_{12}}{w} - \frac{h_{32} \cdot (w \cdot u)}{w^2}$$

$$\frac{\partial u}{\partial z_c} = \frac{1}{w^2} \cdot \left(\frac{\partial (w \cdot u)}{\partial z_c} \cdot w - (w \cdot u) \cdot \frac{w}{\partial z_c} \right) = \frac{h_{13}}{w} - \frac{h_{33} \cdot (w \cdot u)}{w^2}$$

$$\frac{\partial v}{\partial x_c} = \frac{1}{w^2} \cdot \left(\frac{\partial (w \cdot v)}{\partial x_c} \cdot w - (w \cdot v) \cdot \frac{w}{\partial x_c} \right) = \frac{h_{21}}{w} - \frac{h_{31} \cdot (w \cdot v)}{w^2}$$

$$\frac{\partial v}{\partial y_c} = \frac{1}{w^2} \cdot \left(\frac{\partial (w \cdot v)}{\partial y_c} \cdot w - (w \cdot v) \cdot \frac{w}{\partial y_c} \right) = \frac{h_{22}}{w} - \frac{h_{32} \cdot (w \cdot v)}{w^2}$$

$$\frac{\partial v}{\partial z_c} = \frac{1}{w^2} \cdot \left(\frac{\partial (w \cdot v)}{\partial z_c} \cdot w - (w \cdot v) \cdot \frac{w}{\partial z_c} \right) = \frac{h_{23}}{w} - \frac{h_{33} \cdot (w \cdot c)}{w^2}$$

Then from equation (11.10), if we assume that ρ is constant between two frames, it follows that:

$$\begin{cases} \dot{x}_c = \dfrac{\partial x_c}{\partial \theta_x} \cdot \dot{\theta}_x + \dfrac{\partial x_c}{\partial \theta_y} \cdot \dot{\theta}_y \\ \dot{y}_c = \dfrac{\partial y_c}{\partial \theta_x} \cdot \dot{\theta}_x + \dfrac{\partial y_c}{\partial \theta_y} \cdot \dot{\theta}_y \\ \dot{z}_c = \dfrac{\partial z_c}{\partial \theta_x} \cdot \dot{\theta}_x + \dfrac{\partial z_c}{\partial \theta_y} \cdot \dot{\theta}_y \end{cases}$$

with

$$\frac{\partial x_c}{\partial \theta_x} = x_c \cdot \cot \theta_x \qquad \frac{\partial x_c}{\partial \theta_y} = 0$$

$$\frac{\partial y_c}{\partial \theta_x} = -y_c \cdot \tan \theta_x \qquad \frac{\partial y_c}{\partial \theta_y} = z_c$$

$$\frac{\partial z_c}{\partial \theta_x} = -z_c \cdot \tan \theta_x \qquad \frac{\partial z_c}{\partial \theta_y} = -y_c$$

Finally, we have:

$$J = \begin{pmatrix} \dfrac{h_{11}}{w} - \dfrac{h_{31} \cdot (w \cdot u)}{w^2} & \dfrac{h_{12}}{w} - \dfrac{h_{32} \cdot (w \cdot u)}{w^2} & \dfrac{h_{13}}{w} - \dfrac{h_{33} \cdot (w \cdot u)}{w^2} \\ \dfrac{h_{21}}{w} - \dfrac{h_{31} \cdot (w \cdot v)}{w^2} & \dfrac{h_{22}}{w} - \dfrac{h_{32} \cdot (w \cdot v)}{w^2} & \dfrac{h_{23}}{w} - \dfrac{h_{33} \cdot (w \cdot v)}{w^2} \end{pmatrix} \cdot \begin{pmatrix} x_c \cdot \cot \theta_x & 0 \\ -y_c \cdot \tan \theta_x & z_c \\ -z_c \cdot \tan \theta_x & -y_c \end{pmatrix}$$

By inverting equation (11.15), the problem is solved and we can express $(\dot{\theta}_x, \dot{\theta}_y)$ from $(\dot{u}, \dot{v})$:

$$\begin{pmatrix} \dot{\theta}_x \\ \dot{\theta}_y \end{pmatrix} = J^{-1} \cdot \begin{pmatrix} \dot{u} \\ \dot{v} \end{pmatrix} \tag{11.16}$$

This relation allows the realization of a visual servoing of the laser beam. The reference of this closed loop control is the point location in the image. Figure 11.9 describes this loop. The 3D sensor gives the data $(\rho, \theta_x, \theta_y)$ of a space point, the 3D sensor calibration followed by the camera calibration gives the location of the

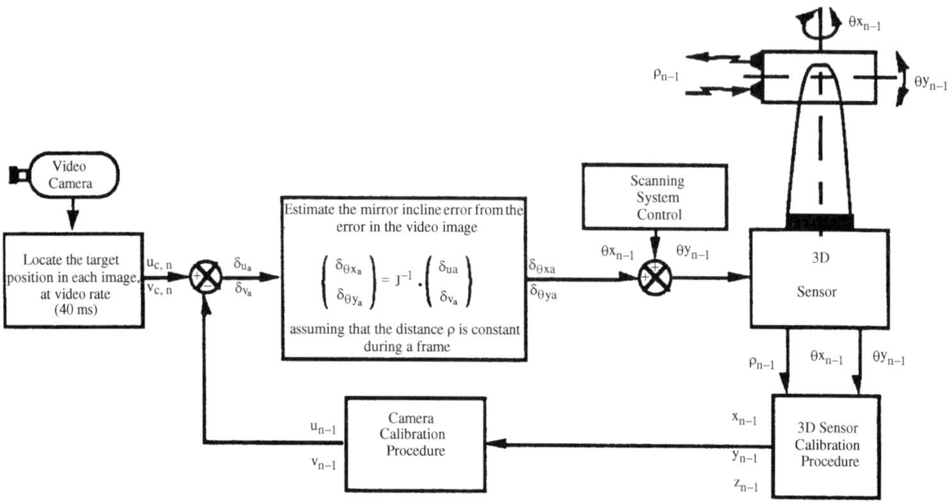

Figure 11.9. Visual servoing description

laser beam in the intensity image. The error of the closed loop is the difference between this location and the reference given in the camera coordinate system. The relationship 11.16 makes it possible to compute the new mirror deflections in order to track the moving target.

Results

Figure 11.10 gives an example of controlled perception. Four obstacles were *a priori* possible but the controlled perception by acquiring a 3D image in each window allows for the validation of only two of them.

(a) (b)

Figure 11.10. Controlled perception: (a) *a priori* interest windows, (b) validation of the objects in some of them

The visual servoing results will be presented during the global results of the application.

11.3 OBSTACLE DETECTION AND TRACKING BY A MULTISENSORIAL APPROACH

We now present a method for the detection and the tracking of a road obstacle. Indeed road obstacle detection is important, but avoidance requires knowledge of the kinematic state of the obstacle. This state can be computed after the obstacle has been tracked for long enough. This approach is based on the behaviour of the human driver: in a pre-attentive stage, the driver scans the whole scene looking for obstacles, then, after he has detected one, he focuses his attention on it to gain more accurate parameters on its kinematic state.

11.3.1 Obstacle Detection

Obstacle detection corresponds to the pre-attentive stage. The whole road scene is scanned by the laser beam to obtain a 3D image. Scanning is performed while there is no obstacle in the field of vision. The detection algorithm is split into two steps. The first step is the segmentation of the 3D image into regions, the second step consists in the recognition of the obstacles among these regions.

3D image segmentation into regions

The aim of this segmentation is to obtain areas containing spatially-closed laser shots. In other words, two shots belong to the same region if, first, they are at the same distance z_a from the experimental car with a tolerance σ_z; second, if they are connected. The segmentation criterion is the distance z_a, because we assume the road is horizontal and flat, so the obstacles appear on a plane nearly perpendicular to the road; that is all shots on a road obstacle have the same distance z_a. This segmentation method is possible because the road is not included in the 3D image. In the opposite case, computation of the surface normals, which is more time-consuming, will be necessary (see[6,3,15] for other implementations).

A region growing algorithm performs the segmentation. Regions are built from a single shot and grow by including shots located at nearly the same distance z_a. We describe the algorithm principle as shown on the following page.

The result of the segmentation is a set of regions R_i including N_i shots. Among these regions, some of them can be road obstacles, others not. We need a recognition step.

Obstacle recognition

The problem is: 'is the region a road obstacle or does it belong to the road environment?'

every shot is still not classified
⇒ **For** each shot
  → **If** distance $\neq$ zero **and** shot not classified
    • init a new region
    • the shot is classified
      ⇒ **For** each neighbour
        → **If** distance z_a equal with a tolerance σ_z **and** shot still not classified
          • put the neighbour into the region
          • put the neighbour into the search buffer
          • the neighbour is classified
        → **EndIf**
        ↦ **While** the search buffer is not empty
          ⇒ **For** each neighbour of the shot in the buffer
            → **If** distance z_a equal with a tolerance σ_z **and** shot still not classified
              • put the neighbour into the region
              • put the neighbour into the search buffer
              • the neighbour is classified
            → **EndIf**
          • Put the shot out of the search buffer
        → **EndFor**
      ↦ **EndWhile**
    ⇒ **EndFor**
  → **EndIf**
⇒ **EndFor**

Each region R_i is parameterized by a vector:

$$R_i = (X_{min,i}, X_{max,i}, Y_{min,i}, Y_{max,i}, Z_i)$$

where $(X_{min,i}, X_{max,i})$ are the minimum and maximum values of x_a, $(Y_{min,i}, Y_{max,i})$ are the minimum and maximum values of y_a, and Z_i the distance of the region from our vehicle.

The five parameters are computed as follows:

$$\begin{cases} X_{min,i} = \min \{x_{a,i,j}, & \forall j \in (1, \ldots, N_i)\} \\ X_{max,i} = \max \{x_{a,i,j}, & \forall j \in (1, \ldots, N_i)\} \\ Y_{max,i} = \min \{y_{a,i,j}, & \forall j \in (1, \ldots, N_i)\} \\ Y_{max,i} = \max \{y_{a,i,j}, & \forall j \in (1, \ldots, N_i)\} \\ Z_i = \left(\sum_{j=1}^{N_i} z_{a,i,j}\right) / N_i \end{cases}$$

From these parameters, we compute region features which will be useful for recognition:

the height h_i: $h_i = Y_{max,i} - Y_{min,i}$.

the width l_i: $l_i = X_{max,i} - X_{min,i}$.

the centre location $C_i(X_{m,i}, Y_{m,i}, Z_i)$:

$$\begin{cases} X_{m,i} = (X_{min,i} + X_{max,i})/2 \\ Y_{m,i} = (Y_{min,i} + Y_{max,i})/2 \end{cases}$$

The road obstacles are detected by comparing these features with their real measurements.

Let $\mathcal{O}$ be the set of the N road obstacles. A road obstacle is defined by the vector $O_j = (L_j, H_j)$ where L_j is the obstacle width and H_j its height. For each region R_i, we can write:

$$R_i \in \mathcal{O} \text{ if } \left\{ \begin{array}{l} h_i = H_j \pm \sigma_{H,j} \\ l_i = L_j \pm \sigma_{L,j} \\ Y_{m,i} = (H_j \pm \sigma_{H,j})/2 \end{array} \right\} \forall j \in (1, \ldots, N).$$

where $\sigma_{H,j}$ and $\sigma_{l,j}$ are the tolerances on the measurements of the obstacle O_j. When we use a line keeping algorithm, we add the following constraint:

$$X_{m,i} \in [X_{rg}(Z_i), X_{rd}(Z_i)]$$

where $X_{rg}(Z_i)$ and $X_{rd}(Z_i)$ are the x-coordinates of the road left and right sides in the absolute coordinate system $\mathcal{R}_a$ at a distance Z_i from our experimental vehicle. This constraint means that the road obstacle must be on the road.

After obstacle detection, the next stage of our approach is obstacle tracking.

11.3.2 Obstacle Tracking

Method

The goal of obstacle tracking is the accurate location of obstacles during a long enough period to obtain the exact kinematic state of the obstacle, and especially its lengthwise relative speed. We have chosen a multisensorial approach to ally video signal swiftness with the accuracy of the 3D-sensor location. Indeed, it is difficult to compute the kinematic state of an obstacle by using the obstacle detection process because the 3D image acquisition time is too long (200 ms with a 10×80 image). Moreover, obstacle detection from the video image, which is a difficult problem, will not give us an accurate 3D location, so the relative speed would not be computed with accuracy. Cooperation between the 3D world and the intensity world uses the second facility of our multisensor: the visual servoing (see Section (2.3.2)). The obstacle detection described above is the initializing step of the algorithm. The tracking consists in a search for the obstacle 2D model in an interest window with knowledge of the obstacle distance. Conservely, the new obstacle location in the intensity image makes it possible to focuse the laser beam on the obstacle. The first location of the interest window in the image is computed from its location $C_i(X_{m,i}, Y_{m,i}, Z_i)$ in $\mathcal{R}_a$ with the relation 11.8.

Let W_i be the window for the tracking of the obstacle R_i; it is defined as follows:

$$W_i = (u_{w,min,i}, u_{w,max,i}, v_{w,min,i}, v_{w,max,i})$$

where $(u_{w,min,i}, u_{w,max,i})$ are the minimum and maximum values in the u direction and $(v_{w,min,i}, v_{w,max,i})$ are the minimum and maximum values of the interest window in the v direction. If we denote proj, the transformation obtained by the relation (8), the location of the interest window W_i is calculated by:

$$\begin{cases} (u_{w,min,i} \quad v_{w,min,i}) = \text{proj} \ (X_{m,i} - g_{wl} \cdot l_i/2 \quad Y_{m,i} - g_{wh} \cdot h_i/2 \quad Z_i) \\ (u_{w,max,i} \quad v_{w,max,i}) = \text{proj} \ (X_{m,i} + g_{wl} \cdot l_i/2 \quad Y_{m,i} + g_{wh} \cdot h_i/2 \quad Z_i) \end{cases}$$

where g_{wl} and g_{wh} are the horizontal and vertical scale factors for the size of the interest window W_i.

The obstacle tracking in the interest window is a full image processing. The last step of the process is the search for a 2D model. The 2D model we have chosen comes from a simple observation: the rear view of a car has several horizontal line segments (see Figure 11.11). So, the 2D model is a rectangle including such line segments. Given this model, the processing chain is split up into edge points detection, contour chaining, polygonal approximation and search for the 2D model of the obstacle.

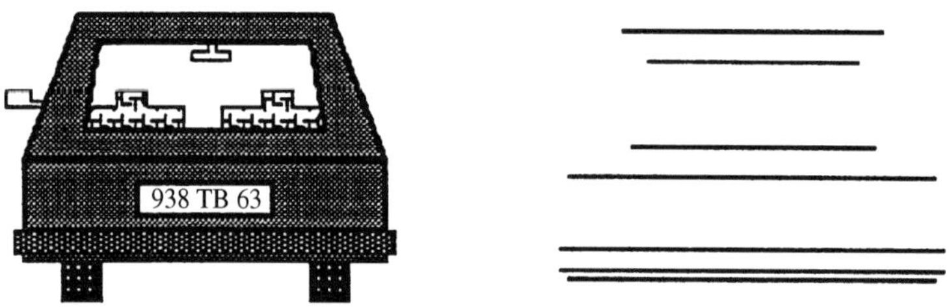

Figure 11.11. 2D model of the rear view of a car

Edge points detection

Many computer vision systems are based on edge detection, and many techniques have been developed. Edges correspond generally to abrupt change of the physical or geometrical properties of the perceived image, and are very interesting features for image analysis. In most of these techniques, we use first order (Roberts, Prewitt, Sobel, ...) or second order (Laplacien) local operators of derivation. For several years, new techniques[2,4,12] optimize a criterion which takes into account a predefined model of the edge. For our application, we have chosen a template matching edge detector. We obtain good results for the processing time which is not too long. This aspect must not be neglected for an on board system application. For each image point, we compute the gradients with the four masks described in Figure 11.12. We keep only the maximum value of the gradient linked with its direction. The edge points are the local maximum of the thresholded image, i.e., a point is an edge point if the maximum gradient is greater than a threshold ε_{ch} and if it is greater than that of its two neighbours in the associated direction.

It is possible to use an hysteresis thresholding to close the edge lines. This technique, presented by Canny, can be described as follows:

−1		1
−1		1
−1		1

−1	−1	
−1		1
	1	1

−1	−1	−1
1	1	1

	−1	−1
1		−1
1	1	

Figure 11.12. Convolution masks used for the template matching

- threshold the local maximum with $\varepsilon_{\mathrm{ch}}$ (as before).
- validate all the neighbours of the edge point if their gradient is greater than a low threshold $\varepsilon_{\mathrm{cb}}$.

Edge points chaining and polygonal approximation

Edge points chaining makes it possible to describe of the image in chained point lists. All the connected contour pixel are linked. We have developed our own algorithm.

From the edge points link, polygonal approximation allows for a description of the image with line segments.[5] Each link is divided into two sublinks until the maximum distance $\mathrm{d_m}$ of the furthest point from the line segment is less than a threshold ε_{s}. We need to know the distance d from a point $p_0(u_0, v_0)$ to a straight line which includes $p_1(u_1, v_1)$ and $p_2(u_2, v_2)$:

$$\mathrm{d} = \begin{cases} \dfrac{u_0 - a \cdot v_0 - (u_1 - a \cdot v_1)}{\sqrt{1+a^2}} \text{ with } a = \dfrac{u_2 - u_1}{v_2 - v_1} & \text{if } v_1 \neq v_2 \\ |v_2 - v_1| & \text{otherwise} \end{cases}$$

The image can be described by a set $\mathcal{S}$ of N_{s} line segments:

$$\mathcal{S} = \{(p_{1\mathrm{i}}, p_{2\mathrm{i}}), \forall i \in (1, \ldots, N_{\mathrm{s}})\}$$

where $p_{1\mathrm{i}}$ et $p_{2\mathrm{i}}$ are the end points of the line segment s_{i} defined by their coordinates:

$$\begin{cases} p_{1\mathrm{i}} = (u_{1\mathrm{i}}, v_{1\mathrm{i}}) \\ p_{2\mathrm{i}} = (u_{2\mathrm{i}}, v_{2\mathrm{i}}) \end{cases}$$

Road obstacle search

From the line segments, the road obstacle search can be split into three steps:

- the choosing of the line segments of interest,
- the building of the rectangles,
- the search for the 2D model.

We use two criteria to find the line segments of interest. The line segment must be nearly horizontal and its length must be consistent with the width of the road obstacle in the image. So, we define the subset $\mathcal{SI}$ of $\mathcal{S}$ which includes N_{SI} line segments of interest. For all line segments s_i:

$$s_i \in \mathcal{SI} \text{ if } \begin{cases} |a_i| & \leq \sigma_p \\ |u_{1i} - u_{2i}| & = w_i \pm \sigma_w \end{cases}$$

where σ_p is the tolerance on the line segment incline, w_i the width of the road obstacle in the image, and σ_w the tolerance on this width.

The width w_i is calculated for each obstacle location by:

$$w_i = \left\|\text{proj}\left(X_{m,i} + l_i/2 \quad Y_{m,i} \quad Z_i\right)\right\| - \left\|\text{proj}\left(X_{m,i} - l_i/2 \quad Y_{m,i} \quad Z_i\right)\right\|$$

We then build the set ε_r of the N_r rectangles including n_{si} line segments of interest. We notice that $N_r = C_{n_{si}}^{N_{si}}$

A rectangle r_i is defined by:

$$r_i = (u_{r,min,i}, u_{r,max,i}, v_{r,min,i}, v_{r,max,i})$$

where $(u_{r,min,i}, u_{r,max,i})$ are the minimum and maximum values in u direction and $(v_{r,min,i}, v_{r,max,i})$ the minimum and maximum values of the rectangle in v direction.

Given n_{si} line segments of interest s_i, the corresponding rectangle is calculated as follows:

$$\begin{cases} u_{r,min,i} & = \min & \{u_{1i} \text{ et } u_{2i}, \forall i \in (1, \ldots, n_{si})\} \\ u_{r,max,i} & = \max & \{u_{1i} \text{ et } u_{2i}, \forall i \in (1, \ldots, n_{si})\} \\ v_{r,min,i} & = \min & \{v_{1i} \text{ et } v_{2i}, \forall i \in (1, \ldots, n_{si})\} \\ v_{r,max,i} & = \max & \{v_{1i} \text{ et } v_{2i}, \forall i \in (1, \ldots, n_{si})\} \end{cases}$$

Finally we search for the rectangle whose size best fits that of the tracked road obstacle. We try to minimize a criterion K_i which takes into account the width w_i and a pattern factor $f_{f,i}$ such as:

$$f_{f,i} = \frac{h_i}{l_i}$$

This criterion K_i is defined as follows:

$$K_i = |w_i - (u_{max,i} - u_{min,i})| + |f_{f,i} - \frac{v_{max,i} - v_{min,i}}{u_{max,i} - u_{min,i}}|$$

So, the new location of the road obstacle o_i is calculated by:

$$o_i \in \varepsilon_r \text{ and such as } K_i < K_j \forall j \neq i \in (1, \ldots, N_r)$$

We can easily compute the new reference of the closed loop control (u_c, v_c):

$$\begin{cases} u_c = \dfrac{u_{r,min,i} + u_{r,max,i}}{2} \\ v_c = \dfrac{v_{r,min,i} + v_{r,max,i}}{2} \end{cases}$$

The process is so continued while no new road scanning is necessary.

11.3.3 Results

We now present some results on the detection and road obstacle tracking. Concerning detection, we first present (see Figure 11.13) the result of the segmentation algorithm on a dense image obtained with $\sigma_z = 50$ cm. This figure shows a parking scene with two pedestrians. On the 3D image (b), the redder the shots, the closer they are. On the segmented image (c) each color corresponds to a region.

Figure 11.14 presents an example of obstacle detection in real road situation. The following parameters were used:

$$N = 1$$

$$H_1 = 1.4 \text{ m} \quad L_1 = 1.5 \text{ m}$$

$$\sigma_{H1} = 20 \text{ cm} \quad \sigma_{L1} = 20 \text{ cm}$$

(a)

(b)

(c)

Figure 11.13. Segmentation of a 3D image: (a) the scene, (b) the 3D image, (c) the segmentation in regions of the 3D image

Figure 11.14. Obstacle detection in real situation: (a) the motorway scene, (b) The 3D image (the bigger the circles, the closer they are) matched with the 2D image, (c) result of the detection

Figure 11.15 shows the successive processes required to obtain the new position of the road obstacle in the interest window and Figure 11.16 presents an example of road obstacle tracking. The biggest window is the interest window, the other one is the rectangle which includes the new location of the car. The following parameter values were used: $\varepsilon_{ch} = 20$, $\varepsilon_s = 2$, $\sigma_p = 0, 1$, $\sigma_w = 20\% \ w_i$, et $n_{si} = 3$.

11.4 CONCLUSION

We have presented a new approach to road obstacle detection and tracking. We have shown that our multisensor, and particularly the 3D-sensor, is well adapted to this problem. The faculties of the multisensor lead to close collaboration between these two worlds, rather than just matching between the intensity image and the 3D

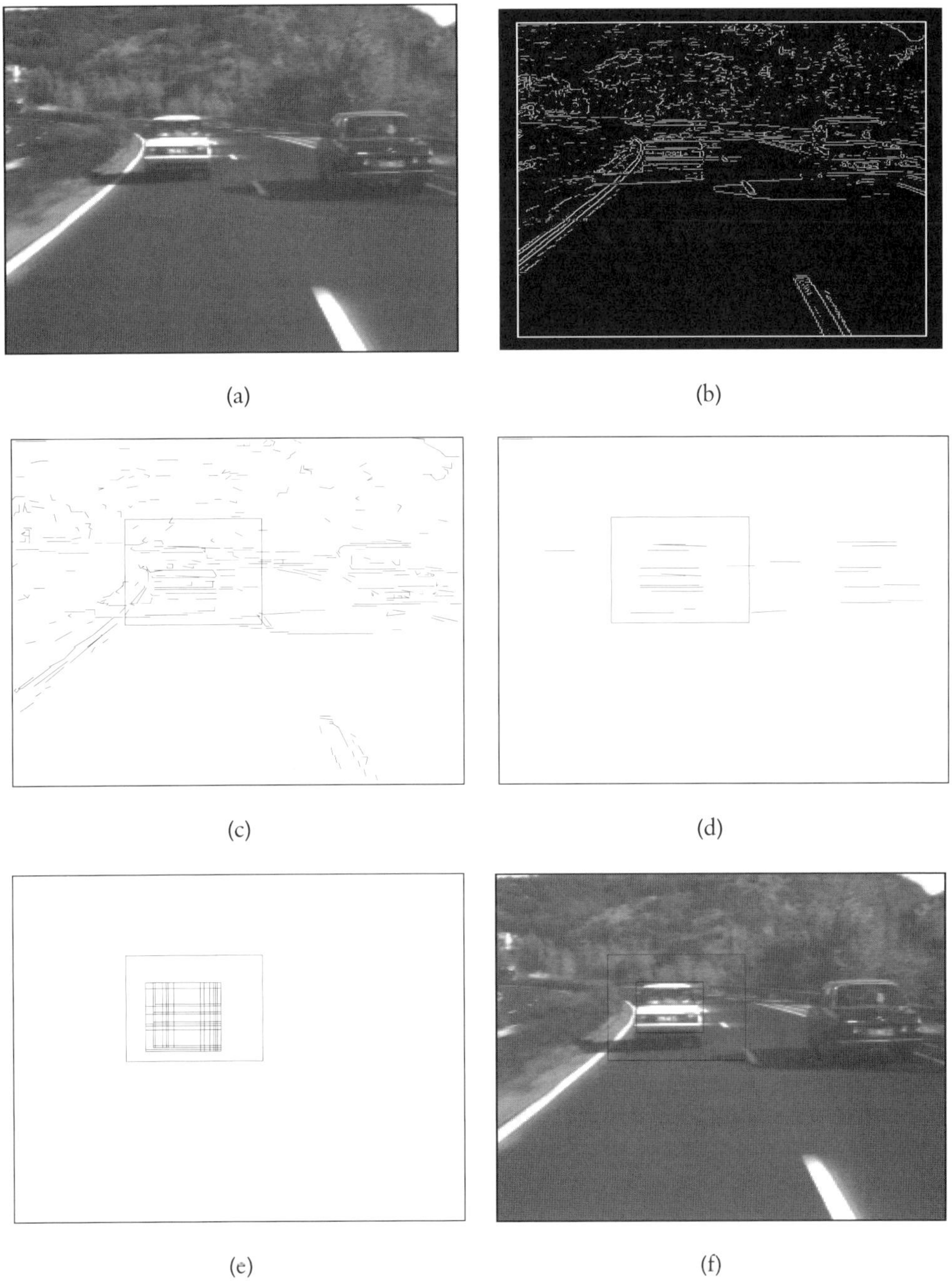

Figure 11.15. Obstacle tracking: (a) the road scene, (b) the contour image, (c) the line segments, (d) the line segments of interest, (e) the set of the rectangles, (f) the new position of the road obstacle

Figure 11.16. Road obstacle tracking example

image. The algorithm uses this collaboration to track the road obstacles. It has been validated in real situations with our experimental car.

We are currently working on the characterization of the kinematic state of the obstacle from our data. Our future work will be to improve this algorithm for more complex scenes. More generally, we will study dense 3D image, data fusion and smart sensing.

REFERENCES

1 Alizon J, Gallice J, Trassoudaine L, Treuillet S. Multisensory Data Fusion for Obstacle Detection and Tracking on Motorways, *Proc. of 3rd Prometheus Workshop*, 85-94, Torino, April 1990.

2 Canny J. A computational approach to edge detection, *IEEE Trans. on PAMI*, 8(2), 679-698, 1986.

3 Daily MJ, Harris JG, Reiser K. Detecting obstacles in range imagery, *DARPA Image understanding workshop*, 87-97, 1987.

4 Deriche R. Using Canny's criteria to derive a recursively implemented optimal edge detector, *Int. journal of computer vision*, 1(2), 167-187, 1987.

5 Duda RO, Hart PE. *Pattern classification and scene analysis*, New york, ED Willey, 1973.

6 Dunlay RT, Morgenthaler DG. Obstacle detection on roadways from range data, *SPIE Vol. 727, Mobile robot conference*, Cambridge M.A., 110-116, 1986.

7 Faugeras OD, Toscani G. The calibration problem for stereo, *IEEE conference on computer vision and pattern recognition*, Miami, 15-20, 1986.

8 Faugeras OD, Toscani G. Camera Calibration for 3D Computer Vision, *Proc. of International Workshop on Machine Vision and Machine Intelligence*, Tokyo, February 87.

9 Franke U. Real-time 3D-road modeling for autonomous vehicle guidance, *Proc of the 7th Scandinavian Conference on Image Analysis*, Aalborg, 316-323, 1991.

10 Graefe V, Kuhnert KD. A High-Speed Image Processing System Utilized in Autonomous Vehicle Guidance, *Proc. of the IAPR workshop on Computer Vision*, Tokyo, 10-13, 1988.

11 Mori H, Ishiguro H, Kotani S, Yasutomi S, Chino Y. A Mobile Robot Strategy Applied to Haunobu-4, *Proc. 9th Int'l Conf. on Pattern Recognition*, 525-530, 1988.

12 Shen J, Castan S. Edge detection by sign correspondance for zero crossings, *actes du premier colloque Image'Com*, Bordeaux, 19-21 nov., 279-284, 1990.

13 Thorpe C, Hebert MH, Kanade T, Shafer SA. Vision and navigation for the Carnegie-Mellon Navlab, *IEEE Transactions on pattern analysis and machine intelligence*, 10(3), 362-373, 1988.

14 Thorpe CE. Vision and navigation — The Carnegie-Mellon Navlab, ISBN 0-7923-9068-7, *Kluwer academic publishers*, 1990.

15 Turk MA, Morgenthaler DG, Gremban KT, M. Marra M. VITS-A vision system for autonomous land vehicle navigation, *IEEE Transactions on pattern analysis and machine intelligence*, 10(3), 342-361, 1988.

16 Trassoudaine L, Treuillet S, Gallice J, Alizon J. Tracking Front Moving Object in Outdoor Scenes by Using a Multisensory Vision/Telemetry System, *Proc. on the 7th Scandinavian Conference on Image Analysis*, Vol. 2, Aalborg, Denmark, 570-577, 1991.

12

Trends in Vehicle Navigation and Guidance Systems for Intelligent Motoring

12.1 INTRODUCTION

It has been assumed that navigating in an unfamiliar environment can be a demanding and stressful task, that can lead to an increased probability of an accident; although attempting to find firm data to support this view in the available accident statistics, is an unrewarding experience. However, the pursuit of increased road safety, and the economic stimulus of reduction of inefficiency in the traffic system make it desirable to improve the level of information available to drivers to help their decision making.

Perhaps an in-vehicle aid to navigation is desirable? There are, after all, examples of successful devices in other environments. Aeroplanes and ships have systems to ensure arrival at the expected destination, after a journey which often entails travelling across areas devoid of natural landmarks, signposts or local inhabitants to point the way. Road vehicles, however, typically operate in circumstances where these aids to navigation are in abundance. If the task environment is different, perhaps the style and type of information offered should also be different. Just because it is technically feasible to transfer a certain style of navigation device into a vehicle, is not in itself sufficient reason to justify that development.

Various autonomous navigation and guidance systems are already in the market-place, and infrastructure supported real-time systems are being developed as demonstration systems in several countries.[1] The common assumption is that perceived utility to the individual will outweigh the financial cost of purchasing the system.

Advanced Vehicles and Infrastructure Systems, Edited by C.O. Nwagboso

Putting exact figures to potential savings to society of a perfect route guidance system for vehicles, seems at present to be more of a creative art than a science. However, it is possible to be too critical of the accuracy of current cost–benefit equations. Perhaps the research and development community should be content with the notion that any potential saving is worth pursuing? Whether the research community welcomes them or not, route navigation systems will become available in increasing numbers, in varying degrees of sophistication.

Should we be complacent in the assumption that route navigation devices will lessen rather than exaggerate the problems that exist in the current traffic system? Cost–benefit analyses at present assume that the systems will work efficiently, *caeteris paribus*: i.e. if better information is provided in one part of the system, then overall system efficiency will improve. Is this necessarily true; could not a change in one component influence other components negatively, leading to reduced system efficiency? With fatal accidents currently being estimated to cost society more than half a million pounds each, in the UK,[2] it is clear that a high proportion of the cost savings accrued through reductions in fuel and pollution costs, could be quickly offset if accident rate, i.e. system failure rate, increased.

The realization that we are dealing with a total traffic system is central to a human factors perspective, and should dictate the research and development approach to the problem. It can be useful to use DRIVE as an acronym to highlight the different elements of the traffic system:

D DRIVER
R ROADWAY
I INFRASTRUCTURE
V VEHICLE
E ENVIRONMENT

It is not possible to influence only one element of the system in isolation. For example, if changes are made to certain characteristics of the vehicle, driver behaviour may adapt to the change.[3,4] It is not reasonable to suppose that the introduction of a new information device in a vehicle will not result in a change of driver behaviour of some kind. Not all behavioural change can be expected to be of benefit to the system.

In the context of the DRIVE system, the systems engineering term "catastrophic failure" can take on a very humane meaning. If an element of the system is so designed as to lead to a breakdown in the competent control of the vehicle exercised by the driver, either because it promotes inattention or information overload, it can result in an increased probability of an accident, which may have genuinely catastrophic consequences.

The literature abounds with descriptions of systems that are termed either route planning, route guidance or route navigation systems. The difference between these systems is often blurred, but a logical distinction between the functions does exist.

12.2 ROUTE PLANNING

Route planning, as distinct from other driving related tasks, is usually performed outside the vehicle, prior to a journey. Drivers may wish to perform some sort of strategic planning for several reasons:

- they are travelling to an unfamiliar destination;
- they have many destinations to reach in the same journey;
- there are problems with a particular route, e.g. traffic congestion or road works.

The effective route planner requires information not only about the most logical route but also about such things as congestion, weather conditions, traffic accidents and road network changes. The driver also needs strategies to optimize the use of this information to derive fast, safe and economic routes.

12.3 ROUTE NAVIGATION

Route navigation systems often do not offer a specific route for the driver to follow, but rather they facilitate the drivers navigation task by making information available, usually in map form, while driving. This information usually includes points of departure and destination, and updates regarding journey progress, i.e. distance and surrounding geography in relation to the drivers position. Thus a route navigation system consists of a map database, a display, a computer and a location determination system. The logic behind this approach is that the systems are much more flexible than direct guidance systems and drivers have a greater sense of control as they determine the route themselves. Indeed some decisions are inevitably left to the driver, such as which lane to take on multiple carriage ways.

12.4 ROUTE GUIDANCE

RTI (Road Transport Informatics) systems which offer route guidance, do so *en route* and in real time. The driver is given immediate instructions on what to do next with little other information. The advantage of route guidance systems are that most of the navigation tasks are performed for the driver. Thus all the "redundant" information given by maps is filtered out, so only small amounts of information are presented at one time, and this is the information which is most relevant to staying on the route. In this way the driver does not need to know exactly where they are, and may well be driving with very little appreciation of their precise location.

The starting point for deciding on the functional specification of a route information device should be an investigation of the information needs of the proposed user market. Different driver needs, capabilities and expectations should dictate different information provision, which in turn should dictate different mediums and styles of information transfer.

The information a driver may require for journey planning is varied. Obviously, information is needed about road numbers, street and town names, but also information of much more specific types might be required. For instance, a driver may wish to know where the most suitable place to stop for refreshment is or where places of interest or natural beauty are along the route. Other information may be needed on hotels, railway and bus stations, restrictions on certain roads and petrol stations. The driver should be able to access this information on demand rather than have it displayed continually, so that only the information required at a particular moment is presented.

Drivers also need information on what the system is doing at any given time. The driver might receive feedback on what function has been selected, and an indicator if the system is busy, e.g. when it is searching for a route. Too often systems leave the user mystified about which stage of interaction has been reached.

The third area of information requirement, is the information the driver needs to navigate from the starting point to the destination. Current systems differ in the "level" of information they present. In addition to location information, guidance information might concern time, distance and direction to destination, as well as fuel consumption.

How effective a new route information system will be, and how acceptable to the individual, will depend to a large degree on the method previously used for such purposes. At present there is a great variety of strategies for route planning and guidance by the general public. Routes may be chosen for various reasons; speed, distance, traffic volume, complexity, beauty of surroundings, familiarity, superstition, economy or *en-route* facilities.

There is similar variety in how individuals choose to attempt to follow the planned route once underway. Some people will use maps, though preference will be given to a variety of scales and styles of representation. Some people prefer to make detailed lists incorporating road names or road numbers. Yet others may place more importance on landmarks, compass headings or the presence of local experts. It is important to recognize that drivers vary widely in their ability to use navigation aids, particularly maps.

The results of Parkes and Martell[5] from a survey of drivers who used a PC-based route planning product, provide a clear indication of factors to be taken into account in the design of future route planning and guidance systems. Currently, most people find their way on novel journeys with the use of notes taken from maps, and confirm their progress by reference to local landmarks. Few people in younger or older age groups (<15% under 25, <19% over 55) usually follow a road map while driving to a novel destination. About 70% of users, across the age groups had either a slight or definite preference for an automated text-based route guidance system, over their previous method of route following. When making recommendations for improvements to a demonstration system, over 50% suggested the addition of landmark information and over 13% wanted the inclusion of real-time traffic information.

It is important to note that drivers information needs are not homogeneous. Different drivers have varying requirements from a system, e.g. a courier requires the quickest route, whereas a holiday maker may want to follow the most scenic roads. An expert user of a system may require much less detail than a novice user. The needs of an individual driver may also vary over time. For example, while travelling on a motorway the driver requires little information, but when entering a city centre, or nearing the final destination, much more detail is required. Longitudinal studies which have looked at the acceptability of navigation or guidance systems over long periods[6] for normal commuting and social journeys have shown that people familiar with routes require justification from the system if they are directed in what seems a counterintuitive fashion. The problem seems compounded by the fact that many drivers seem suboptimal in their normal choice of routes. Even for commuting to work, drivers may have preferences for routes that may be objectively demonstrated to be longer than routes that could be calculated by the system. Field trials in Berlin[6] have shown that drivers tend to underestimate their own route travel times, but to overestimate those produced by a traffic management system. This seems to demonstrate that obtaining market acceptance, and hence high market penetration for many of these proposed systems will depend on more than just the technical reliability if the system and the accuracy of the information provided.

In human–computer interaction (HCI), four types of user are generally distinguished: naive, novice, skilled and expert. When introducing in-vehicle systems, questions arise about the amount of knowledge that can be assumed within the user population and what training is required for efficient and safe use of the system. Training may take the form of a manual or set of tutorials or even after-sales tutoring. Whatever training is required should be based on an in-depth analysis of the knowledge of potential users and pre-set safety standards for user performance.

Once the system designer has decided on the information content of a new device, the next step should be a decision on how to present that information to the driver. Various media can potentially be used by in-vehicle systems to give information, each with their relative merits. These include text, symbols, maps and speech. Streeter[7] concluded that speech output is superior to maps in terms of errors and distraction, and systems should be voice-oriented. However, no comparison was made to alphanumeric or symbolic displays. Her conclusions were based on the premise that voice displays are less visually distracting, but this does not necessarily mean they are less demanding on attention or perception. Indeed, processing of verbal instructions such as "left" or "straight on" may be more demanding and time consuming than symbols such as arrows or text instructions. An obvious problem with speech is the possibility that instructions will not be heard in the noisy vehicle environment. If the message is heard correctly, there is still the problem that the transient nature of voice messages might mean that the instruction could be forgotten by the time the desired action is required. There is no possibility for the driver to recheck information presented in this way. Also restriction will be placed on the driver and passenger conversation and on the use of music and radio systems.

Dingus *et al.*[8] compared a number of navigation methods (memorized route, paper map, electronic navigator) and found that the electronic navigator could be used "effectively" by the driver. However, visual scan patterns and visual sampling processes were quite different from those used for either paper maps or memorized routes. For example, some tasks such as determining the name of a road made heavy demands on the driver's visual capacities. However, Dingus *et al.* found little evidence for the intrusion of the navigation method on indices of driver performance such as lane exceedences and brake actuations. These authors suggest that any additional time-sharing demands of the navigator drew largely upon spare driver resources.

In a similar fashion Labiale[9] investigated the effects of different electronic map displays as part of in-vehicle navigation systems. He varied the level of guidance information (map alone, map plus auditory message and map associated with written message) and the complexity of the map display, using memory recall techniques as a primary source of data. It was concluded that drivers preferred simple to complex map displays, that drivers reduced speed when consulting the system, and that the complexity of the display had significant effects on memory recall.

In a laboratory-based study, Parkes and Coleman[10] compared the effectiveness of directional symbols, printed text and voice simulation as methods of presenting drivers with the directional information required for route guidance. Task completion times, error scores, eye movements, blink rates and subjective measures were the criteria used for evaluation. Voice simulation was clearly considered the easiest method of route guidance to follow, and task completion times were significantly faster in this condition. Directional symbols consistently ranked higher than printed text. Up to 25% of visual scanning time was spent looking at visually presented information, instead of on the primary task (a figure supported in other research at HUSAT[11–13]). Although fewer errors overall were made when information was presented by voice, it is worth noting that the number of "uncorrected" errors increased, i.e. those where the subject made an assumption about the instruction that was incorrect, due to misinterpretation, and went on to make the wrong manoeuvre.

Given the variety of ways information can be presented and the variety of driver information needs and capabilities that exist, it seems clear that systems should be flexible and adaptable to the characteristics of the individual rather than generic. The driver has only a limited information processing capacity and is presented with an enormous amount of information at one time while driving. It is clear that selection of information is necessary for successful negotiation through the driving environment. Every piece of information competes with every other for the attention of the driver; as Rumar[14] highlighted, some will win and some will lose.

The information selected by the driver, is determined largely unconsciously, but is based on experience, expectation and motivation. Rumar[14] suggests that the selected information, is that which the driver judges to reduce uncertainty in a situation the most. Thus the argument suggests the most important information is that which is of relevance to the road. The next level of information relates to the interaction with other road users. The lowest importance would be given to navigation information. In the situation of overload, information is missed in reversed order.

The role of memory in the driving situation is a key consideration. Memory is involved in such fundamental processes as recognition and identification of stimuli. Moreover, its role is of greater significance in particular circumstances: when a stimulus is only available for a limited time; when there is a delay between recognition and recall; or when information load is high and attention is divided with another task. It is important to note;

(1) other inputs impair recall making irrelevant material harmful;

(2) very similar inputs impair recall;

(3) immediate recall of details of complex images is poor;

(4) recall is better for text plus symbol than either individually;

(5) people remember in the short term (<30 s) by scanning back along input.

Sutcliffe[15] points out that: "the fact that we are basically sequential machines should be apparent from our poor ability to do two or more mental tasks concurrently". Research into short-term memory (STM) has indicated that in situations where the individual has to store information from many sources, capacity is soon reached. Miller[16] gave the number of items as 7 ± 2, and highlighted the importance of chunking information into logical units for successful recall. More recently, Hitch[17] has suggested that there are two subsystems in STM, one language based, the other visual–spatial. Experimentation into the whole, termed working memory, has shown that distraction causes the forgetting of recently learned material. Even a small number of simple chunks are lost within 20 s, if there is distraction during input. These limitations have major implications for the design of route information systems.

Complex displays which require the driver to locate and interpret relational information, i.e. maps, will necessitate continual rechecking of the display to update progress. This is a distraction from the primary task of observing the road ahead, and likely to be inefficient and confusing. Map-based information is not amenable to verbal support, and thus an important medium of transfer is ruled out. In short, it is clear that map-based navigation systems are difficult (at least for a large subset of the population) to use while driving. Anyone who has tried to use an A–Z-type road map while driving in a strange town, can attest to the innate problems of simultaneously trying to orientate oneself on the map, find roadside place names, and fully attend to the movements of other traffic in the system.

Attention is selective, therefore there exists a problem of users switching primary focus between vehicle control and route guidance/navigation device. How does the driver cope with such overload? One means of coping is to predict, further and further ahead, what will occur in the environment. A driver may turn to look at a navigation system if he/she glances at the road ahead and perceives no risk. This assumption that while the driver has looked away from the road ahead, no event will occur to disrupt the safe progression of the vehicle, may of course lead to accidents. A vehicle travelling at 50 k.p.h. is covering the ground at a rate of nearly 14 m/s. Systems which require long glance durations for drivers to assimilate the desired information from a display, are dangerous. These systems obviously run the very real risk of distracting the driver from the primary task of monitoring the road ahead, for such long periods that the likelihood of some event occurring that would require some compensatory action, and being missed, increase. It is conceivable that drivers might misinterpret the potential benefits of an information system. Take for example, a driver who has often experienced a high level of stress while attempting to navigate in an unfamiliar environment. This driver might feel that his/her normal performance in such a situation is so poor, that there is a high likelihood of being involved in an accident. Providing such drivers with an information device that takes away a large part of the uncertainty associated with the task, may markedly reduce stress levels, making drivers feel they are coping more effectively and safely with the traffic system. This perceived increase in safety may be purely illusory, if what in fact happens is that the driver spends long periods looking away from the road ahead, and has an increased potential for being involved in a collision.

Indeed Zimmer[18] argues that: "In order to achieve safer driving behaviour and thereby less accidents, the information provided should counteract the illusory security in traffic situations, and instead should raise the perceived risk, which most drivers tend to underestimate".

Thus we see the potential fundamental drawbacks of implementing navigation systems into the vehicle. The development of in-vehicle information systems should not simply be a case of creating the technology, and installing it directly into the car. Designers must be aware that they are fundamentally altering the driving task, which is a highly complex behaviour. Systems must be tailored to meet the needs of the driver, without overloading psychological and physiological capabilities. The design must be "driver centred" maintaining efficiency, comfort and above all safety.

According to Sabey and Stoughton,[19] human factors are a component in 95% of all accidents. This figure is perhaps more interesting if turned around, i.e. about 5% due to factors outside the driver's control. It is likely that with continual technical improvements, particularly in tyre construction that this figure will have reduced. Treat *et al.*[20] highlighted improper lookout, excessive speed and inattention, as the leading causes of accidents. Smiley and Brookhuis[21] argued that attention and perception errors predominate over simple response errors in accident causation mechanisms. These findings, although not trivial, are not unexpected. They do, however, indicate the potential danger of introducing an element into the DRIVE system that demands that the visual attention of the driver is taken away from the roadway ahead, for even the briefest of periods.

For the purposes of discussion, we can accept the hierarchical classification[22] of driver behaviour being:

(1) strategic — planning;

(2) tactical — manoeuvring;

(3) operational — control.

It is clear that behaviour such as "improper lookout" or "inattention" are most highly associated with the tactical level of behaviour, where manoeuvres are enacted. Route guidance devices are designed to help a driver make decisions at the tactical–manoeuvring level of behaviour. They do this by providing the driver with information that will directly influence the control movements of the vehicle. To receive this information a certain amount of attention must be directed away from what we can regard as the primary task, which is immediate safe control and progression of the vehicle. Indeed, there are a number of potentially negative effects of new driving aids, from well understood mechanisms such as distraction or overload, to promoting over-reliance on the system by the driver, or encouraging misuse, or engendering a change in behaviour based on a real or perceived shift in the locus of control within the vehicle.

The way in which information is presented to the driver can directly influence ability of the driver to operate as an efficient element of the total DRIVE system. A prescription for information systems in vehicles would include these elements:

(1) the information should be timely;

(2) the information should be relevant to the current task;

(3) the information should be easily discerned without error;

(4) the information should be easily understood;

(5) the information should match the expectations and experience of the recipient;

(6) the information should not distract the receiver, nor compete directly for limited attention resources.

At this stage no technology-dependent decisions should be made. If systems are to be effective they must meet certain criteria. No attempt is made here to give specific answers or guidelines to problems of interface design. The following discussion is intended as an outline of what designers of in-vehicle information systems should attempt to achieve and the issues they should consider.

Systems should provide the optimum display for identification, perception and interpretation of information so that the driver is distracted no more than is necessary from the driving task. Any in-vehicle system should require a minimum amount of prior knowledge and its usage should be easily learned. Consider, for example, the complex in-car entertainment systems which possess numerous small buttons and controls and provide a multitude of complex functions. A driver should not have to struggle to operate systems when their attention is obviously needed for the driving task itself.

The psychological capabilities, such as attention and information processing, of the driver can be overloaded by poorly designed displays and inappropriate media. Displays should be easily read and understood, so issues such as glare, flicker, contrast and illumination are important physical considerations. Other issues concern the use of colour, design of graphics, symbols and text, layout of information and the amount of information on the screen at one time.

Physiological capabilities are also important to consider in design. Perhaps the most obvious is the use of visual and auditory senses to take in information. The driver can only take in a certain amount of information at one time. Thus the amount of detail presented on a screen or the length of instructions must be tailored to meet human capabilities. The most important information should be the most prevalent. Also special consideration should be given to the elderly as regards visual functions, e.g. greater susceptibility to glare and a longer time for accommodation.

At the heart of the design of a good route information system would be a full consideration of the nature of information transfer itself. Before any decisions are reached concerning hardware, the designer should be able to answer the following task-specific questions:

1. Which of the possible channels should be used for a specific task?
2. What kind of information should be chosen (natural or symbolic)?
3. How much information is necessary to guarantee simplicity, robustness and meaningfulness of the communication?
4. Is the available information directly related to concrete actions?
5. Is the amount of support to be given by the device sufficient for the information needs of the driver?

The preceding discussion has outlined the problems associated with introducing a new information system into an already complex multitask environment. The most important consideration must be the safety effects of any development. Systems that do not take this into account will sooner or later disappear from the market-place, even if they reach other goals, such as efficiency or environmental protection.[23]

How should designers and legislators proceed? Unfortunately, it is not possible to consult a fully formulated set of design guidelines or standards that are relevant to new information devices such as navigation or guidance systems in vehicles. The research findings are simply not yet available that can validate any performance standards that might be proposed. Does this mean that all development work should be halted until the framework for performance standards exists? Although there have been strong voices in favour of this view,[24] it is more realistic to attempt to support designers in their current tasks.

The first stage must be to provide a detailed checklist that can be used in a product-independent manner, and which is useful at both the product specification and evaluation stages of system design. Various initiatives are being taken within both the PROMETHEUS (*PRO*gra*M*me for a *E*uropean *T*raffic with a *H*ighest *E*fficiency and *U*nprecedented *S*afety) and DRIVE European programmes. A useful starting point has been made by Alm,[23] who has produced a behavioural checklist that details the following design considerations:

(1) impact on drivers abilities to perceive information outside the car;
(2) impact on drivers ability to identify important information;
(3) impact on drivers ability to attend to important information;
(4) impact on drivers' ability to implement judgments into action;
(5) impact on drivers' judgments of ability to drive the car;
(6) impact on drivers' motivation to drive safely;
(7) changes in the drivers' stress level;
(8) possible system misuse.

There is a growing body of literature on the application of human factors principles to the software and hardware design process. Findings show that of the nominal stages of design — specification, prototyping, build, evaluation and implementation — it is prototyping and evaluation which typically come under pressure when cuts need to be made, for whatever reason. These are the very stages where real usability trials need to be conducted to ensure acceptable safety performance, but they tend to be expensive to conduct, and the software design community in general seem unconvinced that user-centred design necessarily leads to "right first time" solutions. However, the development of safety checklists may be the first step towards providing marketing and product managers with an easily used tool, that will help them make design decisions appropriate to the safety goals required.

Once a system reaches a certain level of market penetration, it becomes increasingly difficult to make major revisions, either functional, or aesthetic. It is more common simply to offer updates, or "new improved" versions, than to effect a fundamental revision of the interface principles employed. It is therefore of paramount importance for safety and usability criteria to influence design at the product specification stage, rather than be seen as an optional extra at stages after the critical system decisions have already been made.

What predictions can be made about future navigation and guidance systems? It seems clear that we are quickly moving past the concept of simply providing an electronic A–Z map in the vehicle. Drivers require guidance information. Whether this information should be ego centred, as in simple direction systems, or world centred as in map-based systems; or what role supporting auditory information should occupy are still the concern of research and debate. As technological advances continue the range of options available to the display designer increase. Very high resolution display screens, virtual image systems, and head-up displays have all been demonstrated successfully in vehicles. Multiformat (e.g. map and direction indicators) and multimodal (e.g. visual and auditory) combinations can be provided. This allows information which needs to be acted on immediately to be presented to the driver in a simplified form in one way, for example as a combination of an auditory message saying "turn right at T-junction" with the corresponding symbol in an head-up or mid-head display. More complex contextual information which does not require instant action, and which can be consulted by the driver as the driving task allows, such as maps of the area with landmarks, and distance to arrival information can be presented separately. Other information may be available only on demand, for example certain parking availability information, or details of timetables of integrated local transport systems or road pricing structures. Some of this information will be available for interrogation by the driver while on the move, some will be restricted to times when the vehicle is stationary.

Another factor that will influence the design of navigation and guidance systems in vehicles will be the presence of many other advanced systems, all requiring some display to the driver. Not only will there be a variety of services and value-added features which relate to the driving task, such as local traffic information, weather updates, auto-debiting, collision avoidance, parking reservation and manoeuvring; but also non-task-related systems such as telephones, fax machines and entertainment systems will vie for space in the vehicle, and compete for the driver's attention. Much current research is being devoted to determining driver's needs and preferences for the range of information that can be made available. Sophisticated information management systems are being proposed that can prioritize and regulate the information flow to the driver from this battery of applications and services.

12.5 CONCLUSIONS

A final point to highlight, is that there will be increasing opportunity for drivers to influence their in-vehicle environment. Until now a driver had the choice of which vehicle to buy, and therefore had a limited influence, depending on model and style level, of the types of displays and controls that appeared. However, we have already discussed the fact that drivers are heterogeneous in their requirements. Different drivers like and require different things. The drivers will increasing be able to decide for themselves the level and style of information that they require, and the whole area of tailorable displays will grow. Although there are certain statutory requirements for display provision, for example a speedometer and a fuel gauge must appear, a rev. counter need not. With the eventual demise of traditional electromechanical displays, and the advent of computer screen technology, the drivers themselves will be able to choose between alternative layouts and formats for certain displays, e.g. rev. counter or no rev. counter, analogue or digital speedometer and fuel gauge, imperial or metric units. Even in relation to route navigation and guidance, different drivers on different journeys at different points on the route, require different information displayed to them. Indeed it must be remembered that for very many routine "no alternative" journeys, the driver does not require the presence of a sophisticated in-vehicle system at all. The driver should at the very least have the ability to turn the system off.

REFERENCES

1 DRIVE '89 The DRIVE Programme in 1989. Doc XIII/F/DR0789, Brussels, 1989.

2 Department of Transport. *Road Accidents Great Britain 1988: The Casualty Report*. London: Transport Statistics Publications, 1989.

3 Rumar K, Berggrund U, Jernberg P, Ytterbom U. Driver reaction to a technical safety measure-studded tyres. *Human Factors* 1976; **18**(5): 443–454.

4 Wilde GJS. The theory of risk homeostasis: implications of safety and health. *Risk Analysis* 1982; **2**(4): 209–225.

5 Parkes A, Martell A. A usability analysis of a PC based route planning system. In *Proceedings of the Ergonomics Society 1990 Annual Conference* (Lovesey EJ, ed.) Taylor & Francis, London, 1990.

6 Lorenz K. *A comparison of two route information systems in an urban environment: Efficiency and acceptability with special emphasis on parking guidance*. CEC DRIVE I Programme Project V1017. Report 49A. HUSAT Research Institute, Loughborough University of Technology, UK.

7 Streeter LA, Vitellis D, Wonsiewicz SA. How to tell people where to go: Comparing navigational aids. *International Journal of Man–Machine Studies* 1985; **22**: 549–562.

8 Dingus TA, Antin JF, Hulse MC, Wierwille WW. Human factors issues associated with in-car navigation system usage. *Proceedings of the Human Factors Society, 32nd Annual Meeting, 1988*.

9 Labiale G. *Influence of in car navigation map displays on driver* performance. SAE Technical Paper, 891683. Future Transportation Technology Conference and Exposition, 1989.

10 Parkes A, Coleman N. Route guidance systems: a comparison of methods of presenting directional information to the driver. In *Proceedings of the Ergonomics Society 1990 Annual Conference* (Lovesey EJ, ed.) Taylor & Francis, London.

11 Parkes AM, Fairclough SH, Ross T. (1991) Multi-level evaluations of in-vehicle route navigation systems. In *Contemporary Ergonomics* (Lovesy T, ed.). London: Taylor & Francis, 1991, 157–162.

12 Burnett GE, Parkes AM. The benefits of 'pre-information' in route guidance system design for vehicles. In *Contemporary Ergonomics* (Lovesy T, ed.). London: Taylor & Francis, 397–402.

13 Parkes AM, Ashby MC, Fairclough SH, Lorenz K. The effects of different in-vehicle route information displays on driver behaviour. *Proceedings of VNIS '91 International Conference*. Dearborn: SAE., 1991, 61–70.

14 Rumar K. In-vehicle information systems. *International Journal of Vehicle Design* 1988; **9**(4/5), 548–556.

15 Sutcliffe AG. *Human–Computer Interface Design*. Macmillan Education, Basingstoke, 1988.

16 Miller GA. The magical number seven, plus or minus two: some limits on our capacity for processing information. *Psychological Review* 1956; **63**: 81–97.

17 Hitch GJ. Working memory. In *Applying Cognitive Psychology to User Interface Design* (Christie B, Gardiner MM, eds) New York: Wiley, 1987.

18 Zimmer A. *Driver Information Needs*. DRIVE Project V1017 (BERTIE), Report No. 25 BWM AG Munich Germany, 1990.

19 Sabey BE, Staughton GC. *The Drinking Road User in Great Britain*. TRRL Supplementary Report No. 616, 1980.

20 Treat JR, Tumbas NS, McDonald ST, Shinar D, Hume RD, Stansifier RL, Castellan NJ. *Tri-level study of the causes of traffic accidents*. Report No. DOT-HS-034-3-535-77 (TAC), Indiana University, 1977.

21 Smiley A, Brookhuis KA. (1987) Drugs and traffic safety. In *Road Users and Traffic Safety* (Rothengatter JA, de Bruin RA, eds) Assen: Van Gorcum.

22 Michon JA. Explanatory pitfalls and rule-based driver models. *Accident Analysis and Prevention* 1989; **21**: 341–353.

23 Alm H. *Behavioural Checklist*. Swedish Road and Traffic Research Institute, Linkoping, Sweden, 1990.

24 Zwahlen HT, Adams CC, Schwartz PJ. Safety aspects of cellular telephones in automobiles. *Proceedings of the 18th ISATA Conference*. Florence, Italy, 1988.

25 Harker SDP, Eason KD, Poulson DF, Parkes AM. Classifying the target for human factors output. *Proceedings of Alvey Annual Conference*. UMIST. Manchester. July 1987.

13

Simulation Modelling of Route Guidance Control Strategies

13.1 INTRODUCTION

13.1.1 Development of Dynamic Route Guidance

A variety of traffic management measures have been introduced into urban areas over the past 10–20 years in an attempt to make the best possible use of existing road space: A good example is traffic responsive urban traffic control (UTC) systems such as SCOOT,[1] (Split Cycle Offset Optimization Technique) which has been shown to provide delay savings of about 12% compared with up-to-date TRANSYT[2] (Traffic Network Study Tool) generated fixed time plans. However, these measures are designed to control *existing* traffic patterns, and many studies have illustrated that inefficiencies exist in drivers' route choice such that many trips incur unnecessary extra mileage and journey time, contributing to congestion. This has led to the development of route guidance systems, aimed at reducing this "wastage" and hence making a further advance in network traffic efficiency. This chapter describes the role of simulation modelling in the design and evaluation of route guidance systems.

The most advanced route guidance systems currently in operation, and under development, are those involving in-vehicle units and a communications infrastructure, so that drivers receive "optimum" guidance to their destination according to existing traffic conditions. Such dynamic route guidance (DRG) systems have been tested in Japan, the USA and Europe, using a variety of technologies. In Europe, the ALI-SCOUT system[3] has been trialled in Berlin (LISB)[4] and further applications of an enhanced system (EUROSCOUT) are planned in Europe as

Advanced Vehicles and Infrastructure Systems, Edited by C.O. Nwagboso

part of the DRIVE2 programme. (Dedicated Roadside Infrastructure for Vehicle Efficiency and Safety-second research programme). EUROSCOUT is based on two-way vehicle–infrastructure communications using infra-red beacons at key junctions in the network. Optimum routes are transmitted from beacons based on up-to-the-minute traffic conditions, obtained from guided vehicles and a variety of other sources.

An alternative DRG architecture, planned for London and elsewhere, is based on cellular radio communications technology.[5] This system has an in-vehicle route calculation capability, with two-way communication of dynamic traffic data. As such, it may be considered as one of the "dual mode" route guidance systems now emerging.

Despite considerable activity at present, DRG is still in its infancy in Europe as at 1992 and many questions remain concerning benefits, system design, control strategies and related issues. Simulation modelling offers a valuable technique for addressing these questions as discussed later in this chapter.

13.1.2 Control Strategies

Control strategies are the means by which the DRG system translates a variety of objectives into route recommendations for drivers. In early DRG installations, such as the LISB trial in Berlin, the main objective was to improve the efficiency of vehicle routeing by minimizing the journey time of individual guided drivers. This would lead to reduced travel costs and, potentially, a reduction in accidents due to reduced "exposure" to accident risk. However, such systems also offer the elements of a comprehensive closed-loop traffic control system, including:

(1) monitoring of the present state of a network, allowing direct measurement of journey times and delays, with additional information about origins and destinations;

(2) a strategy computer, which determines a pattern of routeing to achieve certain pre-set criteria;

(3) a system for influencing the routeing of individual vehicles.

In the early years of DRG operation, the proportion of vehicles equipped for DRG (the "penetration rate") will be low, so that the routes advised to individual drivers can be computed without regard for the effects of this routeing on the network state (i.e. interaction effects can be ignored). In these circumstances, key aspects of control will be related to issues such as dynamic journey time estimation/forecasting and the control "facilities" (e.g. density of DRG beacons).

However, as DRG penetration grows, it becomes increasingly important for control strategies to reflect interaction effects and to consider system optimization rather than user optimization. It is also necessary to appreciate that DRG must operate in the presence of other dynamic processes, such as:

(1) Other traffic management and control systems, particularly signal controlled systems in urban areas, such as SCOOT;[1]
(2) the route choice of other road users (public transport, goods vehicles, etc.);
(3) the actions of network operators.

It is clear that the derivation of optimum DRG control strategies, including inter-relation with other systems, becomes increasingly complex as DRG penetration increases, and simulation modelling can play a key role in helping formulate the most appropriate strategies.

13.1.3 The Need for Simulation Modelling

Traffic simulation modelling has become a widely used tool as an aid to the design and evaluation of many traffic management and control systems. There are obvious advantages in being able to model a traffic system in a controlled, simulated environment, to determine the key elements affecting the performance of the system as well as aiding the system design and evaluation process. Simulation models have now been developed to cover almost every requirement in traffic operations, varying from detailed modelling of individual road junctions to the modelling of traffic movements over large networks.

In the case of DRG, the need for simulation modelling is apparent for a number of reasons.

1. To illustrate the potential benefits of DRG, and the relative benefits to guided and unguided drivers. This will give authorities and/or system operators early evidence on which to judge the needs for system development.
2. To illustrate the key factors affecting system performance, as an aid to efficient design.
3. To illustrate the effects of different control strategies and, through modelling in a variety of scenarios, to produce recommendations and guidelines for these control strategies.
4. To look ahead towards future requirements, to help shape the next generation systems.

13.2 SIMULATION

13.2.1 Simulation Requirements

Simulation modelling can be undertaken at a variety of levels of detail, depending on the objectives and requirements. For the overall DRG system, the following simulation requirements can be identified:

1. The ability to model networks of sufficient size such that real routeing alternatives can be investigated, with detailed modelling of all typical network and control characteristics.
2. A dynamic traffic model incorporating time-varying demand, queueing and (as appropriate) congestion, with emphasis on a detailed modelling of junction operations.
3. A dynamic assignment process for the routeing of unguided drivers incorporating routeing criteria and optimization processes, which could also form a base for investigating routeing strategies for guided drivers.
4. A simulation structure which allows the DRG system operation to be reflected adequately (e.g. beacon operations and information transfer) particularly the time-varying guidance which drivers will receive as they travel through the network.
5. The ability to incorporate different driver behaviour models as they emerge from other studies (e.g. route choice and driver response to guidance).
6. The provision of outputs for detailed investigation (routes, link flows, delays, etc.) and summary statistics for evaluation.

In addition, there may also be requirements for simulating very detailed aspects of system operation, such as its communications/data handling processes and its detailed integration with other systems (e.g. traffic control). These aspects may require their own specific simulation models, rather than the more general traffic-based models considered here.

13.2.2 Simulation Options

A number of network modelling approaches have been developed in recent years. For very large urban networks, transportation models have been developed, which incorporate traffic assignment based on speed/flow relationships on links and equilibrium concepts. Detailed junction modelling is not undertaken. One application of such a model to study route guidance in London is described by the MVA Consultancy[6] and results are summarized by Stoneman.[7] This work was particularly useful in looking broadly at the potential benefits of route guidance, while not simulating actual systems in detail.

For modelling traffic assignment in medium-large networks, *macroscopic* models have been developed which incorporate flow-based models for predicting link speeds and junctions delays, usually within a static modelling framework. SATURN[9] (Simulation and Assignment of Traffic in Urban Road Networks) is an example of such a model. Alternatively, *microscopic* models have been developed which aim to reflect the detailed performance of individual vehicles and their interactions in a network (e.g. using distributions to reflect accelerations, decelerations, car following behaviour, etc.). Because of their more complex sub-models, these models are often restricted to the detailed modelling of smaller areas within a network and may exclude assignment processes. Examples of such models are TRAFFICQ[9] (UK), SITRAB[10] (France), SIMNET[11] (Simulation of Networks) (Germany) and NEMIS[12] (Network Microscopic Simulation) (Italy).

More recently, an intermediate level of modelling has been proposed. This includes both traffic assignment and large network coverage, as with the macroscopic models and also allows modelling of individual vehicles, or small groups of vehicles ("packets"), but in less detail than most microscopic models. An example of this family of "mesoscopic" models is CONTRAM.[13] This model has been developed significantly in the UK by the Transport Research Laboratory (TRL) and the University of Southampton for route guidance modelling. This is described in the following sections of this chapter.

13.2.3 Developments of CONTRAM*

The CONTRAM model

CONTRAM[13] is a dynamic traffic assignment model used mainly for the evaluation of road and traffic management schemes in urban areas. Its structure is illustrated in Figure 13.1. It was developed by the TRL in the late 1970s and has been continually developed since then to incorporate a range of new facilities.[14] It has found widespread use both in the UK and elsewhere in the world.

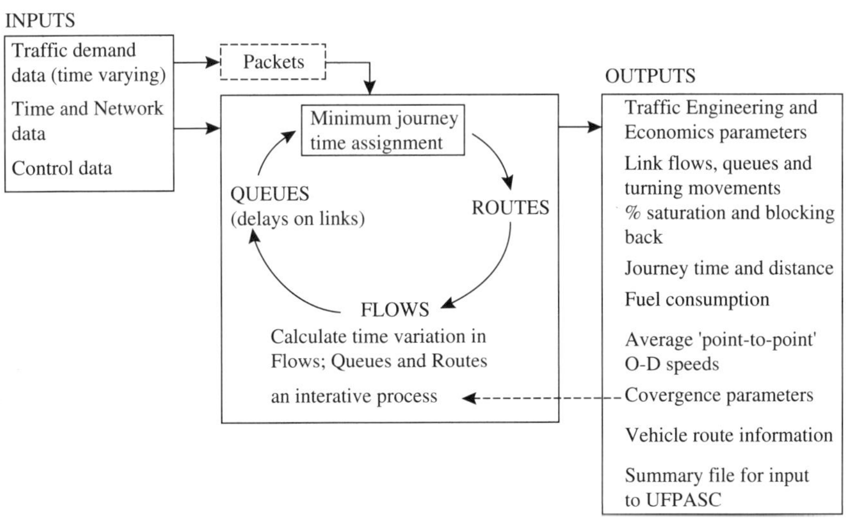

Figure 13.1. CONTRAM: structure of the model. (Source: Leonard *et al.*, Report LR178, Transport and Road Research Laboratory, Crowthorne, 1989[13])

* CONTRAM Continuous Traffic Assignment Model

CONTRAM incorporates a link/node description of a network, with each link or node being described by its key attributes which affect traffic performance. So, links are described according to their length, storage capacity (of queued vehicles), cruise speed and so on, with the option also to specify speed/flow relationships. Junctions (nodes) may be described as uncontrolled, priority controlled, signal controlled or roundabout controlled and in each case, geometric descriptors are required to enable CONTRAM to calculate traffic stream capacities and delays according to the latest TRL formulae. For traffic signals, details of the control characteristics are also required (e.g. isolated or linked, fixed time or optimized, etc.).

CONTRAM also requires a full origin–destination (O-D) matrix for traffic moving between the different zones in the network. This matrix is usually time-varying (e.g. a 2-hour peak period might be divided into 15-min intervals) and categorized into different vehicle classes (e.g. cars, buses and lorries). In addition, the facility exists in a related program ("COMEST"-CONTRAM Matrix Estimation Program) for O-D matrix estimation using link flows, which is useful where a full and up-to-date O-D matrix is not available.

In addition to the above, CONTRAM has a key attribute which makes it particularly suitable as a base for enhancement for route guidance modelling (according to the requirements of Section 13.1.1). This is its packet-based dynamic assignment with time-varying routes reflecting traffic conditions on downstream links at the time drivers would arrive on those links. This modelling structure is amenable to enhancement such that a packet/vehicle can be ascribed any desired characteristics (guided, unguided, behavioural characteristics, etc.) and given any specified route. The possibility also exists for enhancements such that drivers can change route during their journey, perhaps in response to unexpected queues or route guidance advice.

These attributes of CONTRAM have led to its development for route guidance modelling, as described in the following sections.

ROGUS

The *RO*ute *GU*idance *S*imulation model, ROGUS[15] has been developed by the TRL with assistance from Southampton University, in aspects of model specification and testing and from Roke Manor Research, in software production. ROGUS has been developed from a CONTRAM base, but has a range of route guidance functions and a different structure for assigning guided and unguided drivers. Unguided drivers are assigned as in "normal" CONTRAM, i.e. the complete route of each vehicle packet is reassigned in sequence. However, guided drivers are all moved simultaneously within prescribed "update" periods, their routes only being modified if a beacon is passed and changed advice received. Guided and unguided drivers are loaded separately, with an iterative process being used to allow for the interaction of these vehicle categories. At the end of ROGUS, the final assignment of routes is saved, and can be used as the start point for a new ROGUS run.

RGCONTRAM

The *R*oute *G*uidance CONTRAM model, RGCONTRAM[16] has been developed by Southampton University, mainly within a "rolling programme" of research for the Science and Engineering Research Council (SERC).[17] Although concerned with the evaluation of route guidance systems, as with ROGUS, somewhat different modelling techniques have been developed with RGCONTRAM, and emphasis has been placed on different aspects of the system. For example, particular facilities have been developed with RGCONTRAM to study DRG at traffic incidents and to allow different driver responses to route guidance. A "single day" (non-iterative) version of RGCONTRAM has also been produced, to enable investigation of day-by-day dynamic effects of route guidance. In some areas of evaluation RGCONTRAM and ROGUS can be applied to the same issues and this can be particularly useful for model cross-validation.

13.2.4 Model Features and Applications

Unguided routes

The first requirement of a DRG model is to reflect the route choice "inefficiency" of unguided drivers. Such inefficiency has been shown from measurements to be typically 6–8% in wasted mileage/journey time (average), a characteristic not reflected in most assignment models (including CONTRAM) where "perfect" route choice is assumed! If that occurred, there would be no need for route guidance! The mechanisms adopted in ROGUS and RGCONTRAM for reflecting route choice inefficiency are similar, being based on random distortions applied to link costs, so that unguided drivers have some misperception of true link costs. Routes are chosen based on minimization of perceived cost, with the level of "distortion" being user controlled, to produce the required level of route choice inefficiency. It is accepted that this process does not accurately reflect the complex route choice decisions made in practice, but is considered to be an acceptable mechanism pending results from further research.

Beacons

Route guidance beacons are the means by which two-way communication is achieved between guided vehicles and the control centre. At each beacon, journey time information on links traversed from the previous beacon is transmitted to the beacon by guided vehicles, while the beacon transmits predicted optimum routes to guided vehicles updated typically every 5 min. These routes are based on both historic data and dynamic data obtained from guided vehicles and from other sources, such as traffic detectors within UTC systems.

Beacon update frequency is likely to be an important issue affecting the dynamic benefits of guidance: the greater the update frequency, the greater should be the dynamic response of the system (e.g. to incidents), although there will be heavier communications/processing requirements and too small a time aggregation interval can compromise data accuracy. User-defined update periods are available in the models to allow investigation of these effects. Other facilities available in ROGUS are (i) "beacon attraction" parameters where guided vehicle routes can be biased towards links containing beacons, and (ii) the specification of transmission delays associated with receiving information from guided vehicles and sensors. Both models provide full flexibility for specifying beacon densities and locations.

An example from ROGUS of the effects on journey time savings of different beacon densities is given in Figure 13.2. Similar trends are seen for these different models/scenarios, with benefits increasing with increasing beacon densities. Results such as these will enable the balance to be investigated between the benefits and costs of providing different beacon densities.

Routes

The routeing criteria currently adopted in the models are minimum perceived cost for unguided drivers and minimum predicted journey time for guided drivers, depending on the control strategy being evaluated. It is known that, in reality, drivers choose their routes (and might wish to be guided) according to a range of criteria in addition to, or instead of, journey time/cost.[18] For example, "reliability of journey time" and "avoidance of congestion" are often cited as route choice

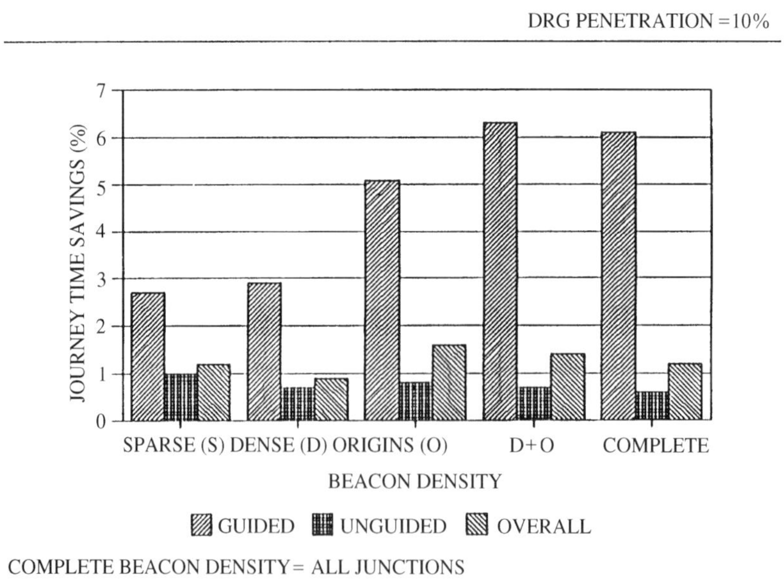

Figure 13.2. Example of effects of beacon density on journey time savings (ROGUS). DRG, dynamic route guidance

criteria. CONTRAM's perceived cost equation which is used for assignment, is well structured to allow a variety of criteria to be investigated. It has the following form:

$$C = aL + bT + cLV^2 + dS + eD + fP + gR + hM \qquad (1)$$

where C = perceived cost of travel
L = link length
T = travel time
V = speed (LV^2 is a term related to fuel consumption)
S = number of stops
D = delay
p = price
R = risk
M = marginal cost
a–h = coefficients

Equation (13.1) is also incorporated within ROGUS and RGCONTRAM and its use is being investigated[19] for guiding drivers according to community criteria (safety, environment, etc.) rather than individual criteria (e.g. journey time). For example, Figure 13.3 illustrates the effects of applying increasing penalties (costs) to the use of "environmental" links in a network. (These might be minor/residential roads.) As expected, increasing penalties produce the benefit of reducing traffic flows

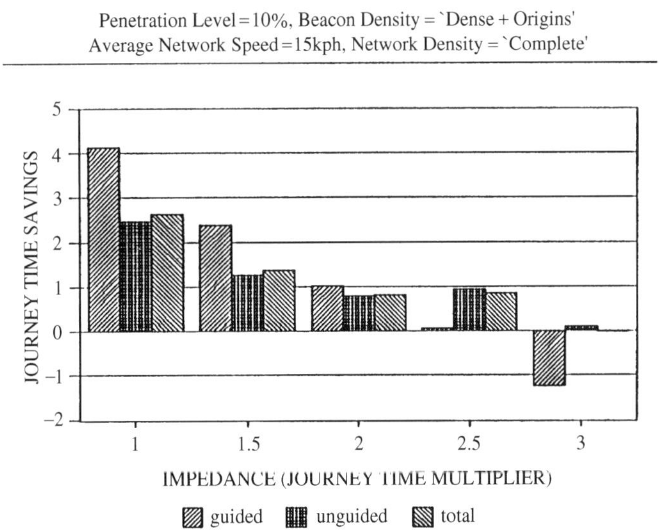

Figure 13.3. Example of effects of link impedance on minor roads for guided vehicles on journey time savings (RGCONTRAM)

on minor roads, but at the expense of increasing journey times for guided drivers as they are increasingly "confined" to the major road network. This type of analysis can also be extended to illustrate the effects of different DRG network densities, relative to the complete public road network.

Other routing features of the models are that:

1. Unguided drivers can be allowed either to fully respond to guided driver routes (using normal iterative processes) or they can be fixed on to their "normal" routes. These two options represent the opposite extremes of what is likely to occur, and emphasis in RGCONTRAM has been on incorporating other driver behaviour models to seek a more realistic intermediate "solution".
2. A modified Dykstra minimum cost algorithm is used for route calculation, which, although not necessarily the algorithm used in DRG systems, should produce compatible routes.
3. Guided routes from any beacon are fixed within the beacon update period (e.g. 5 min). This is satisfactory at low levels of DRG penetration, as such routeing will have little effect on link flows and delays. However, it becomes less satisfactory at higher levels of penetration because the assignment of a large volume of vehicles to one route causes congestion on that route. This is shown by the example from RGCONTRAM in Figure 13.4. Clearly, at higher levels of penetration some form of multi-routeing will be required. CONTRAM provides user optimum multi-routeing and this "solution" will be

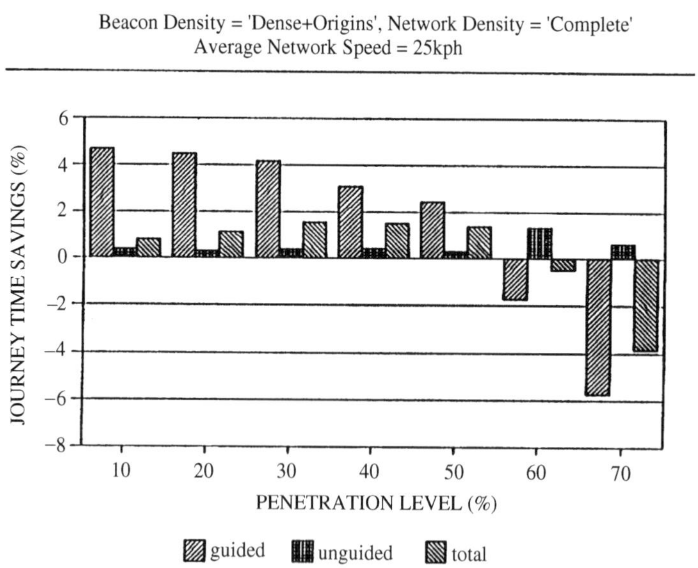

Figure 13.4. Example of effects of penetration level of journey time savings, maintaining single routing (updated) for guided vehicles (RGCONTRAM)

particularly useful in determining requirements for, and developing, practical multi-routeing methods.

4. User-optimum routes are produced. These are most appropriate for "low penetration" DRG scenarios as they should be most acceptable to DRG users and because routeing on this basis has little effect on overall network "performance". However, at higher penetrations where many more routes can be influenced, a "system optimum" solution is likely to be better in which individual routes are optimized having regard for the system (network) as a whole. This will warrant further investigation.

Journey time forecasting

The route calculation process requires estimates to be made of journey times on links for the (future) time at which vehicles will arrive on the links. This requires good measurements of the current traffic state and an effective travel time prediction algorithm. The process is illustrated in Figure 13.5. Current facilities in ROGUS and RGCONTRAM have been designed to reflect the requirements in Figure 13.5. Taking ROGUS as an example:

1. Link travel time data can be obtained from any user-defined combination of: (i) historic data; (ii) "sensor" data (e.g. from traffic detectors); and (iii) current data from guided vehicles. The weight given to each source of data can be

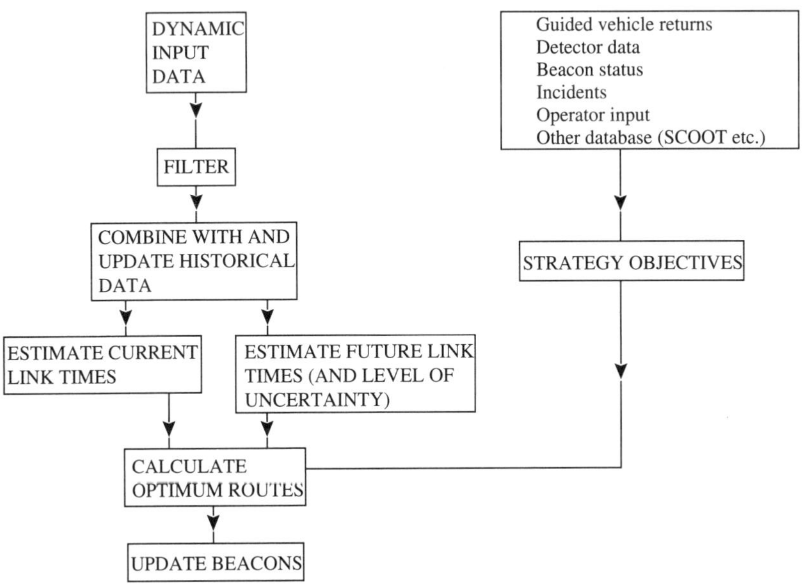

Figure 13.5. Route calculation process

varied (e.g. to reflect its accuracy/"immediacy") and procedures are available for coping with missing data (e.g. if no guided vehicles use a link).

2. Delays and errors associated with data from sensors can be represented.
3. Travel time forecasting is similar to procedures adopted within EUROSCOUT (e.g. in the LISB trial) where historic trends are followed, but updated to reflect current conditions. A "convergent forecast" is also available, where the weight given to current data decreases with increasing time horizon for the forecast.

This link-based "time-series" forecasting process is likely to be satisfactory where traffic conditions are "normal" (i.e. without incidents or high congestion) and where DRG penetration is low, so that there is no interaction between routes advised and journey times on those routes. Forecasting processes for more complex situations are currently being investigated.[19]

Traffic incidents

DRG is expected to be particularly beneficial in guiding drivers around unpredictable incidents (accidents, breakdowns, etc.) where unguided drivers are faced with unexpected queueing situations and have to make difficult routeing decisions. Such incidents can be characterized in ROGUS and RGCONTRAM by their location, duration and severity (e.g. reduction in link capacity). Modelling then requires unguided drivers routes to be fixed, although a new "incident module" has recently been developed by Southampton University for TRL and its principles are incorporated into RGCONTRAM. This module is called CONTRAMI:[20] It incorporates improved logic for drivers route choice in incident conditions, based on the use of fixed routes with diversion possibilities at any junction if a driver encounters an unexpected queue ahead and a reasonable alternative route exists.

Incident management strategies for guided drivers are also currently being researched,[19] particularly the aspects of journey time forecasting, as time-series forecasting procedures based on historical trends, which are used for "normal" traffic states, are inappropriate in incident conditions.

Integration with traffic control systems

Research within the DRIVE programme[21] has identified the benefits of integrating route guidance and traffic control systems. These range from one- or two-way cooperation (e.g. data sharing) to higher level integration at high levels of DRG penetration, where there may be a joint optimization of signal settings and routes. Figure 13.6 illustrates some potential benefits of integration, taking the SCOOT traffic responsive UTC system as an example: on-line traffic data from SCOOT improves DRG data and routeing, while guidance around congestion helps SCOOT maintain efficient network control. ROGUS and RGCONTRAM retain the range

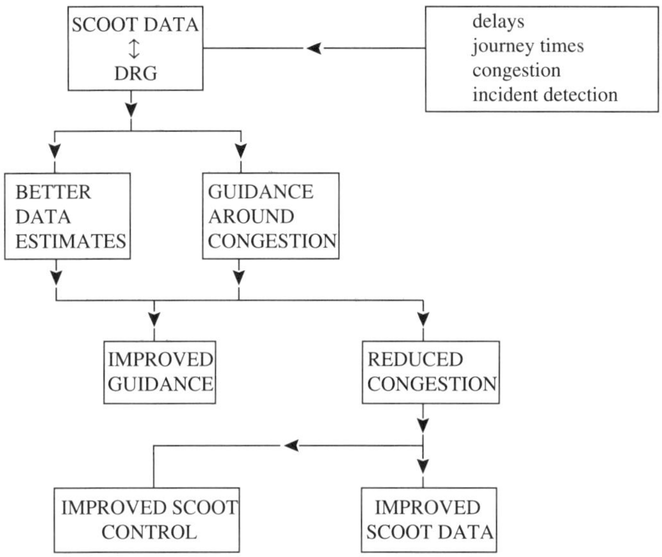

Figure 13.6. Example of integration between route guidance and SCOOT-based UTC. (Source: Reference 21)

of CONTRAM signal control facilities, including fixed time and traffic responsive signals, both isolated and co-ordinated within a UTC system; these form a good basis for modelling integration possibilities. Facilities exist in the models for studying aspects of data integration, and further developments are planned[19] to allow investigation of integrated control strategies.

Driver response to guidance

The modelling of route guidance operations requires a sub-model describing driver response to the guidance advice. In the preliminary modelling, it has been assumed that all guidance advice is followed. However, such perfect compliance is unlikely to be achieved in practice, and research at the University of Leeds[22] has led to behavioural models describing the circumstances where guided drivers may not follow advice. To some extent, system manufacturers may be able to reduce non-compliance by, for example, providing additional information to drivers when an unusual turn is recommended (e.g. if caused by an incident). However, it will still be necessary for simulation models to reflect adequately driver response, if realistic modelling is to be undertaken.

Key behavioural issues which have been found[22] to affect drivers response to guidance include:

1. The familiarity of the driver with the network. Familiar drivers are more likely to "reject" certain items of route guidance advice than unfamiliar drivers.

Examples here would be: (i) where guidance is restricted to the main road network when a better minor road route exists, and (ii) where guidance advice is to join a link which is seen to be unusually congested.

2. The perceived quality of both the current and previous guidance advice. If previous advice has been poor, current advice may be rejected, even if it is good. Alternatively, if previous advice has been good, then drivers are likely to accept some poorer current advice.

The above issues are being explored, particularly in RGCONTRAM, for their effects on system benefits and control strategy formulation.

13.3 CONCLUSIONS

DRG systems are likely to emerge in a variety of forms over the coming years and offer the potential for improved efficiency both to individual guided drivers and to the network as a whole. Simulation modelling is considered to be an important tool to help systems realize this potential, by helping with aspects of system design, the identification of key factors affecting performance and with the development of optimum control strategies. High DRG penetration scenarios can also be investigated long before they occur on-street, aiding efficient system development.

With few DRG systems currently operational, the opportunities for calibrating elements of the simulation models have been limited. However, it is essential that a strong link is established between simulation and field trial activities so that such calibration can be achieved and simulation results can be validated.

Evaluation of DRG systems in isolation will be important, to quantify their benefits relative to other systems. However, a range of other road transport informatic (RTI) systems are also being introduced into urban networks, particularly traffic information systems. These will require special attention to be paid in simulation to driver response sub-models, which are likely to be that much more complex. A further important challenge in simulation modelling will be the evaluation of individual systems in an increasingly complex urban environment incorporating many RTI systems and the modelling of integrated systems.

The evolving form of DRG systems and of the tools used for their evaluation are advancing rapidly[23], so that this chapter can only necessarily reflect the situation at one point in time. Most recently, there is a move in Europe away from dedicated infrastructure-based DRG systems towards systems building on existing infrastructure (e.g. using radio-based communications) and concentrating more on the provision of traffic information rather than route guidance. The market has also grown for autonomous navigation systems. In terms of simulation modelling in the U.K. RGCONTRAM has been extended to cover a range of in-vehicle and roadside information systems, with logic to describe driver behaviour and response to information[24]. A number of other models have also been developed across Europe to design and evaluate control strategies for route guidance, particularly as a result of the European Commission's third and fourth Framework Programmes covering Advanced Transport Telematics.

REFERENCES

1 Hunt PB *et al*. *SCOOT — A Traffic Responsive Method for Coordinating Signals*. Report LR1014. Crowthorne: Transport and Road Research Laboratory, 1981.

2 Vincent RA *et al*. *User Guide to TRANSYT Version 8* Report LR888. Crowthorne: Transport and Road Research Laboratory, 1980.

3 Von Tomkewitsch R *et al* ALI SCOUT — A Universal Guidance and Information System for Road Traffic. *Second International Conference on Road Traffic Control*. London: Institution of Electrical Engineers, 1986.

4 Von Tomkewitsch R. *The LISB field trial forerunner of AUTOGUIDE*. Paper delivered at Institution of Mechanical Engineers, London (Reference C391/060), 1989.

5 Catling I *et al*. SOCRATES: system of cellular radio for traffic efficiency and safety. *Proceedings of the DRIVE Conference*, Brussels, February, 1991.

6 MVA Consultancy. *Estimates of AUTOGUIDE Traffic Effects in London*. Contractors Report 129, Crowthorne: Transport and Road Research Laboratory, 1989.

7 Stoneman B. *The Effects of Dynamic Route Guidance in London*. Research Report 348, Crowthorne: Transport Research Laboratory, 1992.

8 Logie DMW. TRAFFICQ: a comprehensive model for traffic management schemes. *Traffic Engineering and Control*, 1979; November: •••.

9 Van Vliet D *et al*. Equilibrium traffic assignment with multiple user classes. *Proceedings of PTRC Summer Annual Meeting*, University of Sussex, 1986.

10 Charbonnier C. *et al*. Models and strategies for dynamic route guidance. Optimal control approach. *Proceedings of the DRIVE Conference on Advanced Telematics in Road Transport*. Brussels: Elsevier, 1991.

11 Hoffman G. State and future development of traffic signal control in the Federal Republic of Germany. *67th TRB Annual Meeting*, Transportation Research Board, Washington DC, 1988.

12 Beccaria G. *et al*. Models and strategies for dynamic route guidance: a decentralised, fully dynamic, infrastructure supported route guidance. *Proceedings of the DRIVE Conference on Advanced Telematics in Road Transport*. Brussels: Elsevier, 1991.

13 Leonard DR *et al*. *CONTRAM: Structure of the Model*. Report RR178. Crowthorne: Transport and Road Research Laboratory, 1989.

14 Taylor NB. *CONTRAM5: An Enhanced Traffic Assignment Model*. Research Report RR 249, Crowthorne: Transport and Road Research Laboratory, 1990.

15 Harris SP *et al*. ROGUS — a simulation of dynamic route guidance systems. *Traffic Engineering and Control*, 1992; 33(5), 327–329.

16 Njoze SR *et al*. *The Development of RGCONTRAM*. University of Southampton, Working Paper 12 of Project "Fundamental Requirements of Full-Scale Dynamic Route Guidance", 1991.

17 May AD *et al*. Factors affecting the design of dynamic route guidance systems. *Sixth International Conference on Road Traffic Monitoring and Control*. London: Institution of Electrical Engineers, 1992.

18 Bonsall PW, Parry T. Drivers requirements for route guidance. *Third International Conference on Road Traffic Control*, London: Institution of Electrical Engineers, 1990.

PART 5

INFORMATICS SYSTEMS

14

Development of Safe Road Transport Informatic Systems

14.1 INTRODUCTION

The following chapter is based on the results of the project V 1051 "DRIVE SAFELY", a project of DRIVE. *DRIVE* is the short name for Dedicated Road Infrastructure for Vehicles in Europe and is a Research Programme of the Commission of the European Communities (CEC Directorate — General XIII Telecommunications, Information Industries and Innovation).
DRIVE has the following three primary objectives:

(i) improving road safety;

(ii) maximizing of, and breakthrough in road transport efficiency; and

(iii) contributing to environmental improvements

The project "DRIVE SAFELY" is a research project intended to ensure the programme of the CEC to have the highest safety standard of programmable and non-programmable electronics systems in the integrated road transport environment (IRTE) in Europe. The final project report is on a way towards a European standard: "The development of safe road transport informatic systems". It includes road transport informatic (RTI) systems build-in vehicles on roadsides or centralized stations.

There is a marked increase in the use of computers and programmable electronic systems (PES) in the road transport industry. The principal reason for this is the

Advanced Vehicles and Infrastructure Systems, Edited by C.O. Nwagboso

increased functionality now possible, which enables systems to do something that would otherwise be impractical with traditional techniques. Systems are being developed that are more complex than ever before, and perform functions that used to be impossible. Large volumes of data can now be processed in a short time resulting in systems that are easy to use, with improved error detection and safety features. In such situations the overall safety of vehicles and their occupants depends on the correct functioning of the computer or PES, and therefore comes under the category of safety-critical computing.

Current research programmes — in Europe DRIVE and PROMETHEUS (*PRO*gra*M*me for a *E*uropean *T*raffic with a *H*ighest *E*fficiency and *U*nprecedented *S*afety) — are thinking about future systems such as intelligent cruise control, anticollision systems, automatic distance control, etc. These systems are also safety-critical. At present, there is no concrete information included in the existing regulations for type approval of vehicles or roadside systems in Europe on how to ensure safety systems with computers and PES being used. The same situation is in true for usual electronics without any memory.

The work of the CEC research project "DRIVE SAFELY" is a contribution towards a European Standard: "The Development of Safe Road Transport Informatic Systems".

14.2 NOTES ON TERMINOLOGY

14.2.1 Choice of Jargon

In the writing of any technical document it is necessary to be consistent in one's use of jargon, both within the document itself, as well as with the accepted usage among other professionals. The production of the DRIVE SAFELY document has caused a number of problems in this respect for two particular reasons. The first was already known before starting, and that was that there was still no standard usage for certain jargon words between all workers in the field. The second came as a surprise during the production of the document (although in retrospect perhaps it should not have done) and was that workers in different languages did not have a one-to-one translation of all jargon words.

Two sets of words caused particular difficulty, namely: (i) fault, error, failure, and (ii) verification and validation. Therefore, two issues had to be addressed. The first was what precisely did the words mean, and the second was how should they be used in the main text of the document so that they would be intelligible to readers with a variety of first languages. It did in fact take a certain amount of argument and talking before it was finally appreciated that these two issues were related. The result is that it is now possible to understand these words and their issues much better than before.

The rest of this section will discuss the problem in general using these two sets of words as examples.

14.2.2 Fault, Error, Failure

The use of these three words in English is not consistent throughout the literature. (A discussion of the subtle differences between American English and UK English will not be undertaken here, although that is not to say that they are unimportant.) A detailed review was undertaken by Hobley.[14] He showed that not only "fault" and "error" have been used interchangeably, but also that there is more than one model of how systems fail, resulting in a different use of these words again. The following is a description of the major usage in the literature, which is also used throughout this document.

Any mechanism that maintains a pattern of behaviour between itself and its environment is a system. It consists of a set of interacting components, or subsystems together with a design. When a system deviates from its specification it is said to have failed. If a system is in an erroneous state, then any part of the state which differs from a valid state is an error. A failure therefore occurs because the system is in error; however, it should be noted that an error does not always cause a failure. A defect in a component or the design of the system is a fault. The cause of an error is therefore a fault, although once again a fault does not always cause an error. Thus, a fault produces an error which creates a failure, or causes an effect (event). (A mistake is a fault produced by human action.) This should be clarified by the following example.

Consider a system which specification is "the ability to see the road when driving a car by night or day". The two subsystems of interest here are the "headlights" and a "fuse" (assuming a single fuse for all headlights). If the fuse has blown there is then a fault in the subsystem "fuse"; however, this does not cause an error in the subsystem "headlights" until the driver turns them on. Even then this error in the headlights does not cause a failure in the system specification unless the car is being driven at night.

Thus, as far as the English language version of these jargon words are concerned, once a model requiring three different terms has been determined, three different words from the dictionary are chosen, whose normal meaning is close to that required, and these are then given a precise definition. This is quite normal in scientific activity, and the fact that colloquial usage makes synonyms of some terms is accepted (c.f. speed and velocity).

14.2.3 Verification and Validation

These two words proved more elusive to clarify, both because the concepts that they represent are still often not appreciated as being distinct, and also because their translations into both Dutch and German served to cause even more confusion (see below).

Verification is concerned with ensuring that, as the design and implementation develop, the output from each phase fulfils the requirements specified in the output of the previous phase.

Validation is concerned with demonstrating the consistency and completeness of a specification with respect to the initial ideas of what the system should do. Once the system is complete, validation is reduced to the process of determining the level of conformance of the final operational system to the original requirements.

For some time now the distinction between these two words has been summarized using the following two sentences. Verification provides the means of answering the question: "Are we building the thing right?" Validation provides the means of answering the question: "Have we built the right thing?" This succinct playing with English words can, however, be lost on one whose first language is not English. For this reason some more thought was given to the processes implied by the use of the terms to see whether a more helpful distinction could be made between them. The distinction eventually made was in the nature of the proving mechanisms that are used when performing the tasks.

It is only possible to verify one thing — that a statement is true, or not. Thus "I have verified it" means "I have checked it and it is true", or not. Verification therefore implies a comparison against some reference. A design can be good or bad, but it cannot be true or false; consequently, a design cannot be verified. What can, however, be verified is the truth of a statement: e.g. "the design meets the specification", or "the bits that went into the ROM correspond to the source code". A verification is always based on a comparison between something known to be good, true or valid, and something else. The outcome is a statement of conformity that shows that the second thing is good, true or valid.

In contrast a validation is the making of an absolute statement: "this is good, valid and acceptable". Making such a statement without some sort of (hidden) reference point is practically impossible, but stating what is being used as the reference point is not required. Often the reference point is incomplete and so it may be a matter of judgement as to whether a system should be validated or not. Thus validation requires a decision to be made, which may differ from one person to another.

Therefore, the word verification is used in those cases where a "comparison" is made, and the word validation in those cases where a "decision" is made.

14.3 THE SAFETY TASK FORCE

During the CEC DRIVE I programme members from a group of projects met regularly during Concertation Meetings and ultimately formed a "Safety Task Force" to provide, for the first time, a unified front to the problem of road traffic safety. They identified three basic areas:

1. *Systems safety*: Concerned with the safe functioning of equipment under all conditions, and where response times are usually less than 1 s.

2. *Human–machine interface (HMI) safety*: Concerned with how an individual reacts to the information displayed by the equipment and provides responses back to the equipment. Response times are usually a few seconds.

3. *Traffic safety*: Concerned with the whole road traffic infrastructure (e.g. road layout) where it can often take a very long time to demonstrate the effectiveness of a new feature.

The Safety Task Force is an expression of the fact that it is not always possible to categorize the safety aspects of any proposed system under a single heading. Indeed sometimes an application can involve all three safety areas simultaneously. One such is intelligent automatic route guidance. In this case *system safety* is concerned with ensuring that safe instructions are given so that, for example, a driver is not instructed to go the wrong way along a one-way street. *HMI safety* is concerned with ensuring that the mechanism by which the driver is instructed to get to the destination does not create too much of a distraction from the primary task of driving. Meanwhile, *traffic safety* can be involved in the following scenario: the centralized control of the guidance system discovers that there is an obstruction on a particular road which has produced major congestion; in order to keep the rest of the traffic flowing, drivers are diverted on to another set of roads. If these roads are in a residential area then a relatively safe situation for pedestrians and playing children, suddenly becomes less safe.

14.4 SYSTEMS CONSIDERATION

14.4.1 Definition of Systems

A system is defined to consist of a set of components which interact according to a design. A component of a system can be another system (called subsystem). Such components (subsystems) may be, depending on the level: (i) a controlling or controlled system; (ii) hardware, software, human interface. An alternative definition of system is: any mechanism that maintains a pattern of behaviour between itself and its environment.

14.4.2 Systems Configuration

Distinctions have to be made between the technical process, the control systems and the safety-related systems. The technical process creates the hazard. The

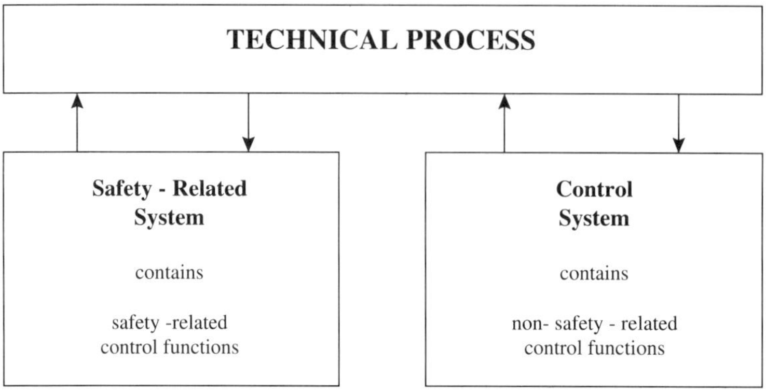

Figure 14.1. Example for a total system configuration

safety-related system is used to reduce the effects of hazardous events. Primarily safety-related systems are technology independent. That means, its construction can contain electrical, mechanical, pneumatical, hydraulical, electromechanical or electronical components as well as a combination of them. A safety-related system can also be an integral part of the normal control system (Figure 14.1).

14.4.3 Systems Safety Considerations

System safety general

Safety is a system issue — this means that when assessing safety, the entire system must be considered, not just individual components. A hazard analysis of the entire system should be performed to identify those subsystems/components whose failure will affect the safety of the system. This process is described later. If any such subsystem/component contains software then particular design measures for safety critical software development should be followed. The basic tenet to be observed is that a subsystem/component is safety-critical unless it can be shown not to be so.

It is important, when performing a hazard analysis, that the boundaries of the system are specified sufficiently to encompass all the factors that may affect the system, while not making the system so large as to render the exercise intractable. In particular, relevant human or environmental activity must be paid attention to.

System safety

Electronics is also used currently working more and more in safety-critical parts. These electronics have the possibility of an uncontrolled access via actuators to the mechanic, hydraulic and pneumatic systems. Concerning vehicles the answer to the question why electronics in the vehicle is safety-critical (Figure 14.2) is: it has by now become an integral part of safety-critical systems in vehicles.

WHY ARE VEHICLE ELECTRONICS SAFETY - CRITICAL?

Because it has by now become an integral part of safety-critical systems in vehicles. The electronic functions are not separable and therefore electronics is jointly responsible for the overall safety of vehicle subsystems, such as steering, brakes, speed control, seat adjustment, passenger safety, e.g. air bag, etc.

Figure 14.2. Safety relevance of vehicle electronics

The electronic functions are not separable and therefore electronics is jointly responsible for the overall safety of vehicle subsystems, such as steering, brakes, speed control, seat adjustment, passenger safety, e.g. air bags and the like.

Definition of hazard

A hazard is a physical situation with a potential for human injury, damage to property, damage to the environment or some combination of these. With each hazard a risk can be associated.

Definition of risk

In the project risk is defined as the product of the seriousness of the effects of a failure and the probability of that failure:

$$\text{risk} = \text{effect} \times \text{probability } (R = E \times P).$$

It must, however, be appreciated that this equation is really a qualitative description of the meaning of risk, rather than a mechanism for obtaining a value. It indicates that a given risk can be reduced either by decreasing the consequence of the hazardous event, or by decreasing the probability of the occurrence of the event. It also indicates that for a given risk there is an inverse relationship between consequence and probability.

Effects of possible electronic faults

Informatic systems consist of computers or programmable electronic systems. Both electronics include hardware and software. Therefore, such systems have the

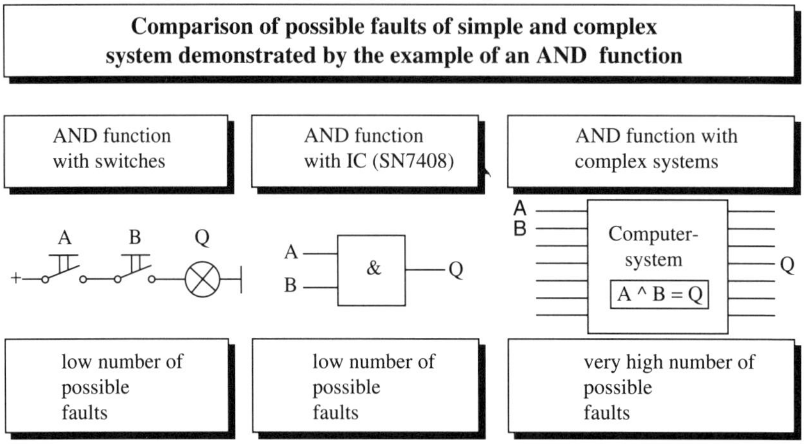

Figure 14.3. Possible faults of simple and complex systems

possibility of having faults in hardware and software. Also the occurrence of random hardware failures is not excluded. Such failures can disturb the correct running of the computer program. The complexity of an informatic system becomes very high if the number of functions to control the process is increased.

The comparison of possible faults of simple and complex systems in Figure 14.3 clearly demonstrates by the example of an AND operation that the number of possible faults is limited when using switches or integrated circuits with a small degree of integration.

If, however, a complex system as shown on the right in Figure 14.3 is used for realizing an AND operation, there will be a great number of possible faults. So it might for example happen that inputs A and B have no logic signal, but output Q is activated. Another possible fault consists in a conjunction of the inputs, but another output than Q being used for the result. Consequently, the number of possible faults is much greater when using complex electronics such as a microcomputer than in the case of the other two hardware types.

Effect of electronics on safety

The effect of electronics on the safety result is illustrated in Figure 14.4 in the form of a pair of scales.

This illustration clearly shows that when using electronics, the scales at first tip in the direction of less safety. The more components and the higher the system complexity, the heavier becomes the pan for less safety. However, to weigh the scales in favour of more safety, special safety measures are required when using electronics, the minimum requirement being that with respect to the safety result electronics in safety-relevant systems must be comparable with the current safety standard of the systems without electronics.

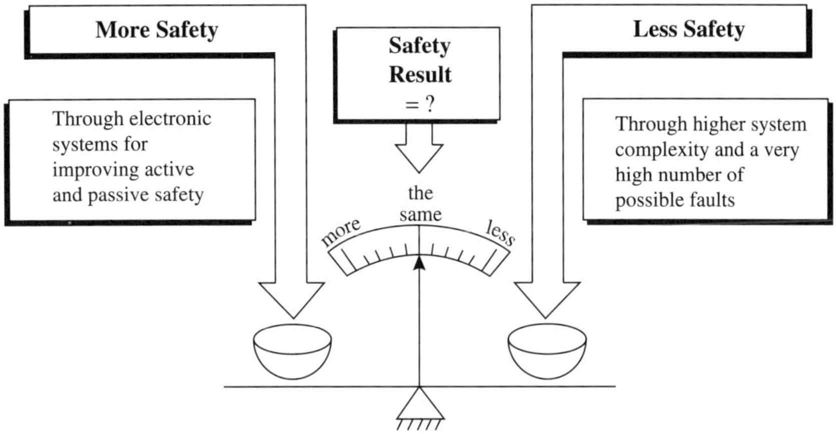

Figure 14.4. Increase in safety through electronics

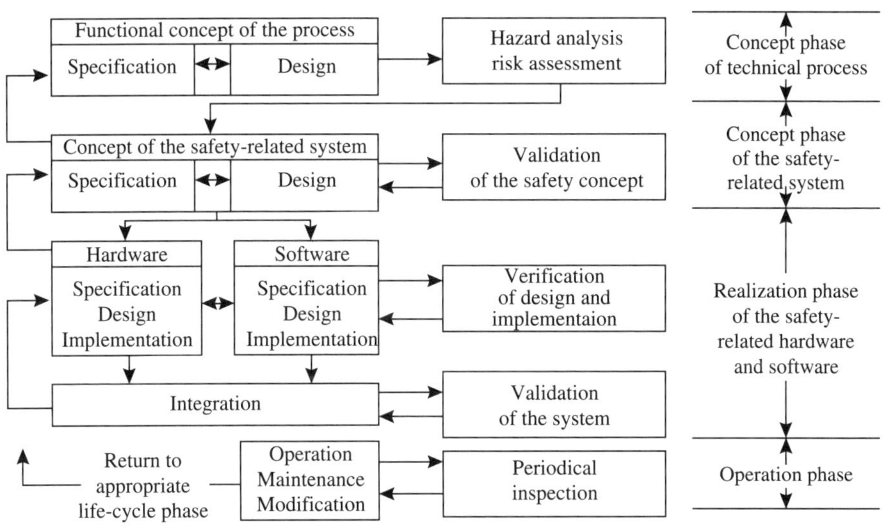

Figure 14.5. Life-cycle model

Safety life-cycle model

The life-cycle can be divided into four main phases:

(1) concept phase of the technical process;

(2) concept phase of the safety-related system;

(3) realization phase; and

(4) operation phase.

In addition to the normal system life-cycle hazard and risk analysis must be carried out and a concept for the safety-related system must be made.

On the life-cycle model it is noticeable, that hardware and software are so closely related that interaction between the two tasks is often necessary during their realization phase.

Validation and verification should be carried out parallel to the developing phase of the system by independent persons, who were not involved in the creation process. The concept must be validated before the hardware and software can be constructed. The validated concept is the basis for further phases of the life-cycle including verification and validation of the final operational system.

14.5 PROCESS OF INTEGRITY LEVEL ASSIGNMENT

The process to use the integrity levels is shown in Figure 14.6. A new system is proposed and, once the full requirements are known, a top-down hazard analysis (Section 14.5.1) is performed by a group containing as wide an expertise related to the system as possible (e.g. systems engineers, traffic engineers, behavioural scientists, etc.). The result of this analysis will be the placing of each hazard into one of the five controllability categories (Section 14.5.2). This will then dictate the integrity level required for the design of the new system or subsystems. During the design phase decisions will be taken as to which items are to be implemented in hardware and which in software. Once the system or sub-system has been built it should have a final level of assurance that corresponds to the integrity level used for its design.

14.5.1 Hazard Identification and Specification

Accidents or dangerous situations need not be the result of a failure. The other possibility is an unfortunate occurrence of circumstances that was not foreseen when the system was designed. In order to eliminate or at least greatly reduce the latter possibility a hazard analysis shall be made at a very early stage. The purpose of a hazard analysis is to find every situation that may be hazardous. At the stage the hazard analysis shall be made, only the functional requirements and the operating conditions are known and consequently the hazard analysis is concerned only with the

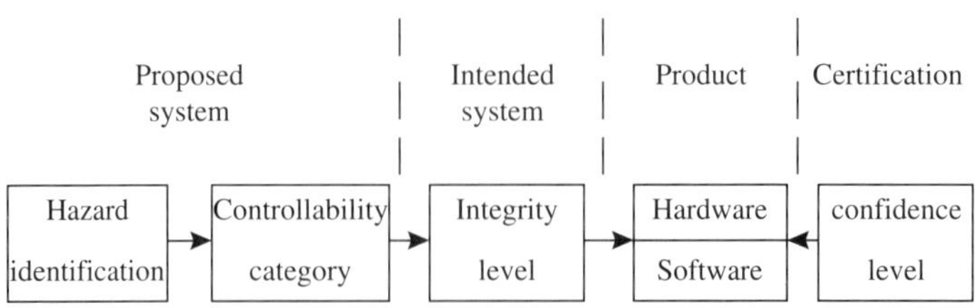

Figure 14.6. Process of integrity level assignment

functioning of the system as a whole. What is required is a list of all the consequences of any failure of the system to perform its specified functions. The list shall be as complete as possible and also include external causes of malfunctioning, such as human error and mechanical damage. A hazard analysis is not the same as a failure mode and effects analysis (FMEA). A FMEA is made during or after the design and lists all the possible failures at a low level of integration and evaluates the effects. As an example, it lists all the wires in the system and the effects of an open or short circuit. This obviously can be done only when the design is (almost) completed. The result of the FMEA is then used to improve the design. A hazard analysis, on the other hand, is made before the design starts and follows a top-down approach. Taking an air-bag system as an example, it lists possible dangerous situations, such as inflation when this is not required, without considering the underlying cause. Together with the functional requirements, the hazard analysis forms the basis for the specification. Again using the air-bag as an example, the specification shall, among other things, contain the following requirements:

(1) it shall inflate when required;

(2) it shall not inflate when not required;

(3) it shall operate for X seconds following an interruption of the power supply.

14.5.2 Assignment of Controllability Category and its Integrity Level

Following the hazard identification the evaluation of all identified hazards and their effects has to be carried out. For assignment to a controllability category the controllability needs to be defined.

Definition of controllability categories

The possible effects of a failure are divided into five categories of controllability as follows:

1. The top category is one that is *not normally controllable* by the road user and is likely to result in a serious accident (usually including fatalities) (e.g. the jamming of the steering mechanism, break of a trailer drawbar, traffic-lights simultaneously showing green at two intersecting roads).

2. A next category is one that is *very difficult to control* and hence has a relatively high probability of a serious accident (e.g. partial brake failure, unjustified inflation of an air-bag at high speed).

3. The *debilitating* category is one that gives a large reduction in safety margin but, with an experienced driver and under favourable circumstances does not necessarily result in an accident (e.g. the sudden loss of both headlights, unjustified inflation of an air-bag at a low speed).

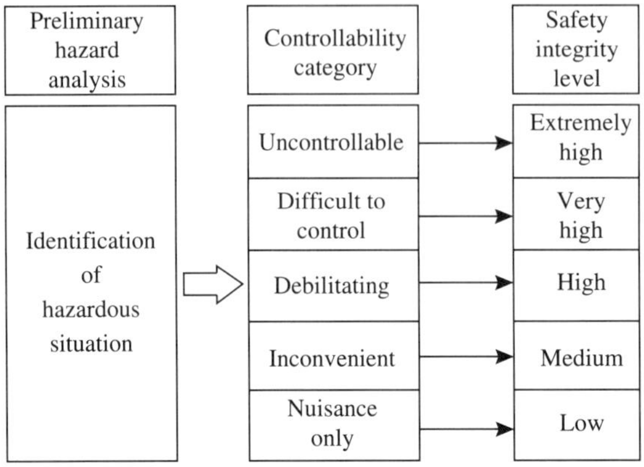

Figure 14.7. Assignment of controllability category and integrity level

4. The *inconvenient* category gives operational limitations, but normally does not result in an accident (e.g. the failure of windscreen wipers, the loss of power assistance to the brakes without a warning signal).
5. The lowest category is *only a nuisance* and is unlikely to affect safety (e.g. the impossibility to adjust the seat or the mirrors, the loss of power assistance to brakes with a warning signal).

It will be noticed that the categories have been given descriptions, not numbers. This has been done to avoid confusion. In other standards various numbering systems are used, some using '1' for 'uncontrollable', others using the same number for 'nuisance only'. It is for this reason that self-explanatory denominations are used.

It should also be noticed that only a small number of subsystems are expected to be categorized at the highest level.

Now it is possible to assign each hazard to its controllability category. That means it is automatically assigned to its integrity level. The connection is shown in Figure 14.7.

14.6 APPROACH TO ATTAIN SAFETY

In order to ensure safe operation of safety-critical electronic systems, it is necessary to recognize the various possible causes of failures and to ensure that adequate precautions are taken against each failure. The main safety objective is to avoid hazardous malfunctions of the electronics system with sufficient probability for a particular application. This goal could be attained, if the following approach is applied. In a preliminary hazard analysis the hazards must be identified and evaluated. The results place each hazard into one of five controllability categories. This will then dictate the integrity level required for the design of the new unit. According to the integrity

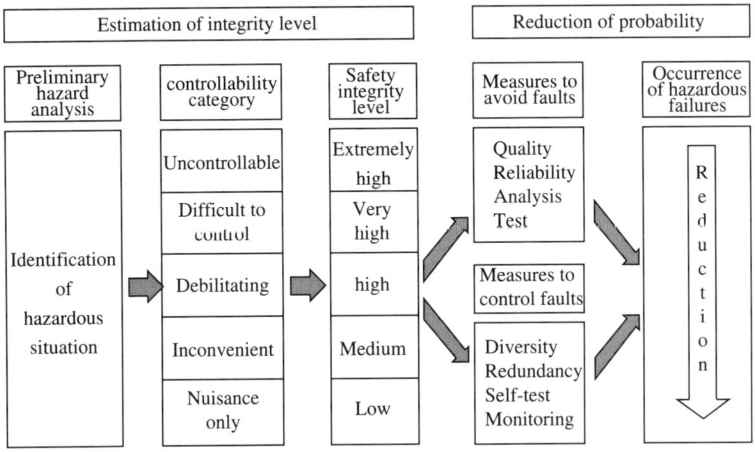

Figure 14.8. Approach to avoid hazardous malfunctions

level a combination of measures with suitable effectiveness must be taken to avoid faults and to control failures. This approach is shown in Figure 14.8.

14.7 SAFETY REQUIREMENTS

The first basic safety requirement on RTI systems for all levels of integrity is, that a single fault may never lead to a hazardous situation. To attain this goal faults must be detected and an appropriate protection measure must be carried out.

The second requirement is, if a fault is undetected, it should never lead to a hazardous situation even in combination with one or more other faults.

The connection between the integrity level and the number of non-hazardous faults which must be considered according to the fault assessment flow chart are shown in Table 14.1.

Figure 14.9 shows the fault assessment flow chart itself.

Table 14.1. Connection between integrity level and number of non-hazardous faults to be considered

Integrity level	Number of considered non-hazardous faults
Extremely high	Up to infinity
Very high	Up to 3
High	Up to 2
Medium	Up to 1
Low	Only the absence of hazardous feedback must be considered

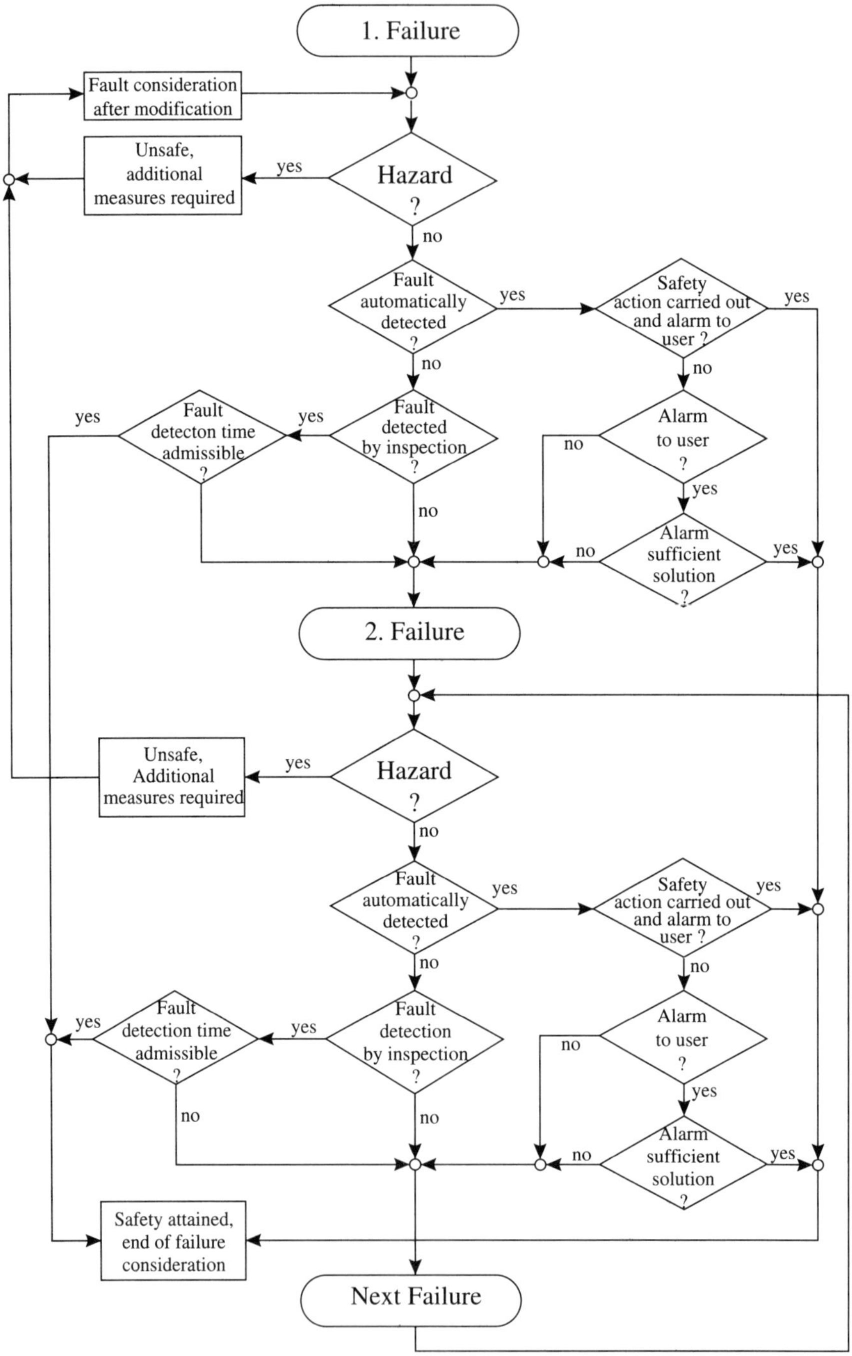

Figure 14.9. Fault assessment flow chart

14.8 CONSIDERATION OF FAILURE

We have two categories of failure:

(i) random hardware failure, and

(ii) systematic failures of software and hardware.

1. *Definition and explanation of random hardware failures*: Random hardware failures are those failures which result from a variety of normal degradation mechanisms in the hardware. As there are many such mechanisms occurring at different rates in different components and as manufacturing tolerances cause individual components to fail due to degradation after varying lengths of operation, failure of equipment comprising many components occurs at unpredictable, i.e. random, times.

2. *Definition and explanation of systematic failures*: A systematic failure is the result of a fault in some stage of the life cycle (i.e. specification, design, implementation, production, installation, operation, maintenance, or modification phase) of a system.

Owing to systematic failures a system fails every time if a particular set of conditions occurs. It is not feasible to test a system under every possible combination of operating conditions, faults may remain hidden until a particular set of circumstances arises; therefore, faults must be avoided at each stage of the life cycle through quality control measures.

Consideration of systematic failures is particularly important in situations in which redundancy is used in order to increase safety. Homogeneous redundancy systems are only effective against random hardware failures. It is possible that each homogeneous channel will suffer a systematic failure under some particular conditions. These failures are known as "common mode failures" and they can be a severe limitation to safety.

14.9 MEASURES AGAINST FAILURE

Precaution must be taken against both random hardware failure and systematic failures in hardware and software. Suitable measures must be taken to avoid faults during the development and production phase and to control failures during the operation phase of the system.

Software faults are always the result of systematic failures. They can only be avoided by quality control measures during the development process and controlled by diverse redundancy during operation.

Hardware failures can be caused by both systematic faults made during the development and production processes, as well as by random faults in components.

Systematic faults can be avoided through quality measures which must be applied throughout the life cycle of the system.

The occurrence of random hardware failures can be reduced by improving reliability, but there is often a limit as to what one can do especially with complex PES.

Therefore, additional design measures (configuration) are necessary to control this kind of failure. In contrast to the probabilistic approach (failure rate calculation) used in must of the industry standards, a deterministic approach is applied to evaluate the electronic systems in a prescribed manner. The most important requirements of this approach are:

(i) a single failure should never lead to a dangerous situation;

(ii) if a failure is undetected it should never lead to a dangerous situation even in combination with one or more other failures.

The number of simultaneous failures that are to be considered depends on the integrity level required.

The measures against software and hardware failure are summarized into three safcty elements:

(1) quality — tackles mainly systematic failures;

(2) reliability — tackles random failures;

(3) design — tackles mainly random failures.

An overview about this approach is given in Figure 14.10.

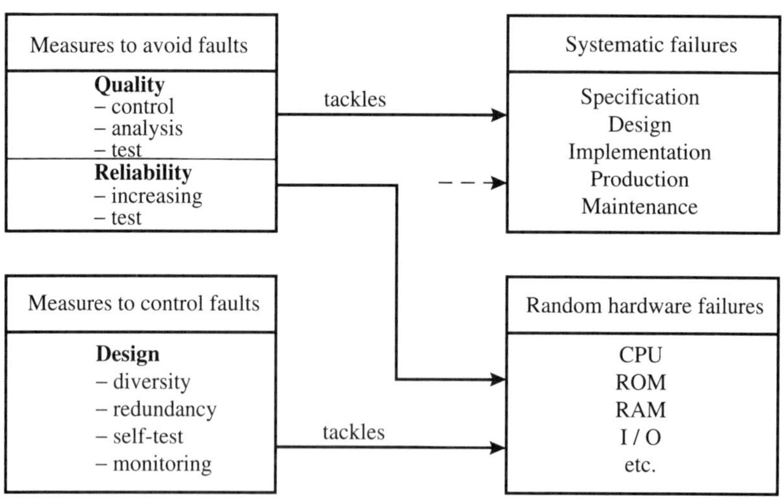

Figure 14.10. Measures against failure

14.9.1 Measures to Avoid Failure

(1) use of development tools;
(2) structured design;
(3) different development teams;
(4) increasing reliability;
(5) fault tree analysis (FTA);
(6) failure mode and effect analysis (FMEA);
(7) simulation;
(8) type-testing;
(9) good quality.

14.9.2 Measures to Control Faults

As safety cannot always sufficiently be attained by quality and reliability measures, additional measures are necessary to control failures during the operation phase.

The first basic safety requirement on RTI systems for all levels of integrity is, that a single fault may never lead to a hazardous situation. To attain this goal faults must be detected and an appropriate protection measure must be carried out. The second requirement is, if a fault is undetected, it should never lead to a hazardous situation even in combination with one or more other faults.

These safety requirements can be fulfilled by the use of three kinds of measures:

The choice of the system structure:

(i) homogeneous redundancy;
(ii) diverse redundancy;
(iii) single redundancy;
(iv) multiple redundancy;
(v) single-channel computer with underlying safety level.

The comparison of provisional and final results of redundant systems.
Use test procedures to detect faults in systems.

14.9.3 Fault Detection Measures

Besides structural measures, measures are required to detect faults in components, such as the CPU, memory sensors, actuators, etc. The effectiveness of these measures could generally be ranged from simple to high. The measures are divided into technical and non-technical measures.

Technical measures

1. *CPU*: (i) standard CPU test; (ii) CPU test with high effectiveness.

2. *Monitoring units*: (i) watchdog; (ii) comparator; (iii) voter; (iv) other units.

3. *ROM*: (i) parity bit technique; (ii) multi-bit redundancy technique; (iii) longitudinal redundancy check (LRC), simple check word; (iv) cyclic redundancy check (CRC), simple check word; (v) cyclic redundancy check (CRC), multiple check word; (vi) double ROM with hardware and software comparison.

4. *RAM*: (i) RAM test checkerboard; (ii) RAM test March; (iii) RAM test Walkpat; (iv) RAM test Galpat; (v) RAM test transparent Galpat; (vi) parity bit technique; vii) multi-bit redundancy technique; (viii) double RAM with hardware or software comparison.

5. *Input and output data*: (i) cyclic test of input and output units; (ii) code check of input and output data; (iii) redundancy check; (iv) multi-channel parallel output; (v) rewrite of outputs; (vi) comparison of inputs; (vii) fault detection through monitoring of the technical process; (viii) plausibility check.

6. *Internal data transfer*: (i) one-bit hardware redundancy; (ii) multi-bit hardware redundancy; (iii) double BUS structure; (iv) cyclic test of data paths; (v) time redundancy; (vi) information redundancy.

7 *Supply units*: (i) power supply monitoring; (ii) clock monitoring; (iii) cooling system (ventilation) monitoring.

8. *Other devices*: (i) sensors test; (ii) actuators test.

9. *Software*: (i) plausibility check; (ii) time monitoring of program sequences (watchdog). (iii) logical monitoring of program sequences; (iv) combination of time and logical monitoring of program sequences; (v) diversity of software.

Non-technical measures

1. *Periodical inspection, preventive maintenance*: (i) effect test; (ii) component test.

2. *Organizational measures*: (i) qualified personnel for maintenance; (ii) technical manual; (iii) user manual; (iv) modification; (v) plausibility test of parameter input; (vi) other.

Table 14.2.

	Integrity level				
	L	M	H	VH	E × H
Structural measures					
One-channel structure					
with operational checkout or	x				
with self-test or	x				
with self-test and monitoring or		x			
with self-test and independent protective subsystem			x		
Two-channel without safe state					
with self-test or	x				
with self-test and monitoring or	x	x			
with mutual monitoring	x	x			
Two-channel with safe state					
homogeneous or		x	x	x+	x+
diverse				x	x+
Three-channel without safe state			x	x+	x+
three-channel with safe state					
homogeneous or			x	x	x+
diverse				x	x+
Fault detection measures					
CPU					
Standard test	x	x			
High value test			x	x	x
Monitor units					
Watchdog	(x)	x	x	x	x
Comparator (only in case of 2 channel structures)					
Voter (only in case of ≥ 3 channel structures)			x	x	x
ROM test					
Little effectiveness	x				
Average effectiveness		x	x		
Great effectiveness				x	x
RAM test					
Little effectiveness	x				
Average effectiveness		x	x		
Great effectiveness				x	x
Input and output test	(x)	(x)	x	x	x
Internal data transfer test			(x)	x	x
Supply unit test	(x)	(x)	x	x	x

Measures: (x) could be required; x required; x+ required in conjunction with additional effective measures. L, low; M medium; H high; VH very high; E × H Extremely High.

14.9.4 Effectiveness of Measures

The necessary measures to avoid and to control failures have different classes of effectiveness. Three classes are used:

(1) little effectiveness;

(2) average effectiveness; and

(3) great effectiveness.

The effectiveness classes are established for each integrity level. The required effectiveness of the measures to avoid systematic and to control random hardware failures ranges from little to great. Measures to control systematic failures are only required at higher integrity levels. Lower integrity levels only require measures with little effectiveness. These measures are able to tackle failures with a relatively high probability of occurrence. In contrast, higher integrity levels require measures with great effectiveness, which are also able to tackle failures with a low probability of occurrence.

14.10 ASSIGNMENT OF MEASURES TO INTEGRITY LEVELS

A set of measures is necessary to avoid and to control failures of (programmable) electronic systems. A combination of measures with suitable effectiveness must be selected according to the assigned integrity level.

Possible measures to avoid and to control failures are listed and evaluated in the second draft of a European standard. The list should be an aid for the developer to find suitable measures for a particular application; however, further adequate measures with the same effectiveness could be applied. Different combinations of measures are possible for a safety integrity level, where the selected combination for a particular application depends strongly on the considered process. The total effectiveness of all measures should be examined carefully to ensure that the safety requirements are fulfilled.

Examples are given for structural measures and for fault detection measures.

14.11 CONCLUSION

Electronics has meanwhile become an integral component of safety-related systems located in vehicles or on roadsides. Particularly in motor vehicles, the overall safety of vehicle subsystems such as steering, breakes, speed control, seat adjustment, etc. is influenced by actively intervening in mechanical, hydraulic and pneumatic systems via actuators and other devices. Certain safety requirements must be fulfilled to ensure that electronics does not represent a safety risk. Through the execution of a preliminary hazard analysis an identification of hazardous situations is possible. To graduate these hazardous situations there are five controllability categories from uncontrollable to nuisance only which lead directly to five safety integrity levels. Depending on the safety integrity level special measures are to be used. There are two main groups of measures: those to avoid faults and those to control faults. It is necessary to use measures from both groups. If this requirement profile is taken into account during development, statutory requirements on safety should be considered as having been fulfilled.

REFERENCES

1 *Software for Computers in the Application of Industrial Safety Related Systems* (Draft). IEC/SC 65/WG. 65 A (Sec) 94-I, August 1989.
2 *Functional Safety of Programmable Electronic Systems. Generic Aspects* (Draft). IEC/SC 65 A WG 10 (Sec) 96, October 1989.
3 *Measurement and Control; Fundamental Safety Aspects to be Considered for Measurement and Control Equipment* DIN V 19 250, 1989.
4 *Principles for Computers in Safety Related Systems* DIN VDE V 0801 1989.
5 *Dependability of Critical Computer Systems*, Vols. 1–3. EWICS TC7 1988–1990.
6 *Programmable Electronic Systems in Safety Related Application*. HSE, 1987.
7 Hölscher H, Rader J. *Microcomputers in Safety Technique*, TUEV 84.
8 Hatley DJ, Pribhai IA. *Strategies for Real-Time System Specification*, 1987.
9 Bell R. Functional safety of programmable electronic systems. *ACOS Workshop on Safety-Related Control Systems*, 8–9 March 1990, London.
10 Trier H, Stall E. Proposal for a Draft Guideline: *Safety requirements for Automotive Electronics* Final Report VdTÜV Research Project No. 232, 2 November 1989.
11 Trier H. Method for planning of system safety in the use of electronics in the motor vehicle. *Automobile-Industrie* 1990; 35: 65–72.
12 *Testing of electronic systems in motor vehicles*. Final Report VdTÜV Research Project No. 302, 13 October 1988.
13 DRIVE-Safely Consortium. Towards a European Standard: *The Development of Safe Road Transport Informatic Systems*. (Draft 2) March 1992; TÜV Rheinland, Cologne (Germany): Heuser G, Trier H, Sonntag J, Asmuth W; TNO Industrial Research, Delft (Netherlands): Elink Schuurman R, Giezen J, Wackie Eijsten JL; University of Leeds, Leeds (UK): Buckley TF, Jesty PH, Kara M, West MM; Program Validation Limited, Southampton (UK): Clutterbuck DL.
14 Hobley IP. *Techniques for enhancing the reliability of non-critical system*. PhD Thesis, School of Computer Studies, University of Leeds, 1989.

15

Information Systems for Public Transport Management

15.1 INTRODUCTION

15.1.1 Scope and Contributions

This chapter presents the state of the art concerning computer-aided operation systems for public transport management. It is especially based on recent developments performed in the EC DRIVE (*D*edicated *R*oad *I*nfrastructure for *V*ehicle safety in *E*urope) Programme (mainly through the Cassiope, EuroBus, Bartoc and Phoebus Projects). These projects deal with the building of a conceptual and physical architecture for public transport information systems (including vehicle scheduling and control systems, passenger information aids, fare systems, etc.). Other systems developed worldwide are also reviewed (cf. bibliography below).

This chapter was originally written at the end of 1992. For publishing reasons, it had to be updated 3 years later, through a compilation of the main results of public transport-related projects. All these projects have to be thanked, especially the EuroBus partners (Transmodel and Popins areas).

15.1.2 Some Definitions

First, it is necessary to specify the terms. Several acroynms are used to describe the various kinds of systems, but not one really covers the whole field:

AVM	automatic vehicle monitoring
AVL	automatic vehicle location

Advanced Vehicles and Infrastructure Systems, Edited by C.O. Nwagboso

AVLC	automatic vehicle location and control
VSCS	vehicle scheduling and control system
SAE	Système d'Aide à l'Exploitation (a French term often used for computer-aided operation system)

There is obviously a need for a new general term, maybe the following, chosen for the present chapter. APTS, advanced public transport system.
The other main following acronyms are used below:

ATT	advanced transport telematics
CDM	conceptual data model
DBMS	database management system
DRS	demand responsive system
EIS	executive information system (also called 'management information system' or MIS)
GIS	geographical information system
GPS	global positioning system
IRTE	intelligent road and transport environment (DRIVE)
IS	information system
ITS	intelligent transport systems
PT	public transport
RDS	radio data system

15.2 WHAT IS AN ADVANCED PUBLIC TRANSPORT SYSTEM?

Basically, an APTS *provides aid to public transport management through an automated vehicle location* (AVL) tool. The other functions (vehicle monitoring, passenger information, demand counting, strategic planning, scheduling, etc.) derive from this essential tool. Mainly developed for urban road transport (buses), the technology is also applicable, with some amendments, to rail transport (underground) or to interurban services (coaches).

Several kinds of AVL tools exist, depending on the location system (continuous or discrete) and the existence of a link with the central computer control room (centralized monitoring or autonomous system). In the latter case (e.g. bus laboratory), there is no data transfer between the vehicle and the control room; the system is used to collect sample data on running times or to check vehicle adherence to its theoretical schedule.

15.2.1 The APTS Rationale: Where Supply Meets Demand

The development of APTSs is linked to public transport characteristics: the organization of services implies that *supply*, which is discrete and non-fractionable, meets *demand*, which is continuous and atomized (in time, space, level of requirements). The satisfaction of this constraint means:

1 An off-line planning process allowing a design of services according to demand, which implies some "reduction" in the supply response; for instance, the average fleet capacity amounts to about 50 seats, and the necessary search for higher occupancy rates implies a good adherence to patronage requirements in time, space and quality of service.

2 An on-line adaptation of the supply to traffic conditions, operations constraints, or patronage requests (frequency, interchanges). DRS applications go even further (route adaptation to demand). A major user (and operator) need is *regularity*, as most large networks experience unrespected schedules and bunching (set of two or three buses running together, then nothing during 1 hour!).

This should be done, of course, at the lower possible cost, and with the best possible quality of service for the user. For instance, in the operations management domain, APTSs look for two kind of improvements (cf. Figure 15.1): a gain of running time on the various routes (in order to save driver and vehicle time, hence money); a decrease in running time dispersion, so that the hazards linked to approximative timetable adherence are limited.

15.2.2 The APTS Response

In this context of a permanent need for adaptation of supply to demand, APTSs are automated tools or computer-aided systems, helping operators in the complex management of their production (design of the network, organization of schedules

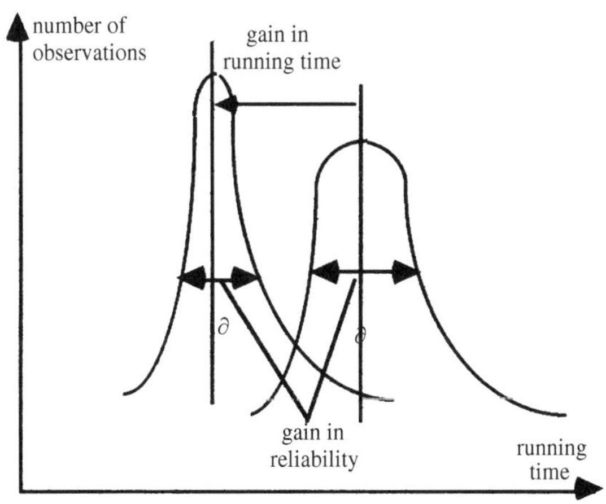

Figure 15.1. The control issue (Pierre-Yves Texier, INTRETS, France, 1992

and duties, monitoring of operations, management information, etc.). They use a complex set of equipment (odometers, computers, displays), voice and data communications (ground/vehicle and vehicle/central room), data (collected through sensors, validators, tags, etc.), processes (methods, algorithms, etc.), etc.

Physical architecture

Figure 15.2 shows the kind of physical solution proposed at the beginning of the nineties by the Bartoc/Phoebus project for medium-sized of networks.
This architecture is based on permanent or discrete links between two main actors:

(1) *the operator control centre*, which gathers all the information on the planned service (Cassiope-like company database) or vehicle locations, controls and co-ordinates current operations with other actors (e.g. traffic control centre managing traffic light priorities) and displays information at home or at bus stops, etc.;

(2) *the driver and their vehicle* (bus, tramway, etc.), which exploit this information in order to perform the best possible service.

The main functions linked to this equipment are: (i) real-time control (made by the driver himself, or by controllers); (ii) schedule adaptation; (iii) traffic-light priority;

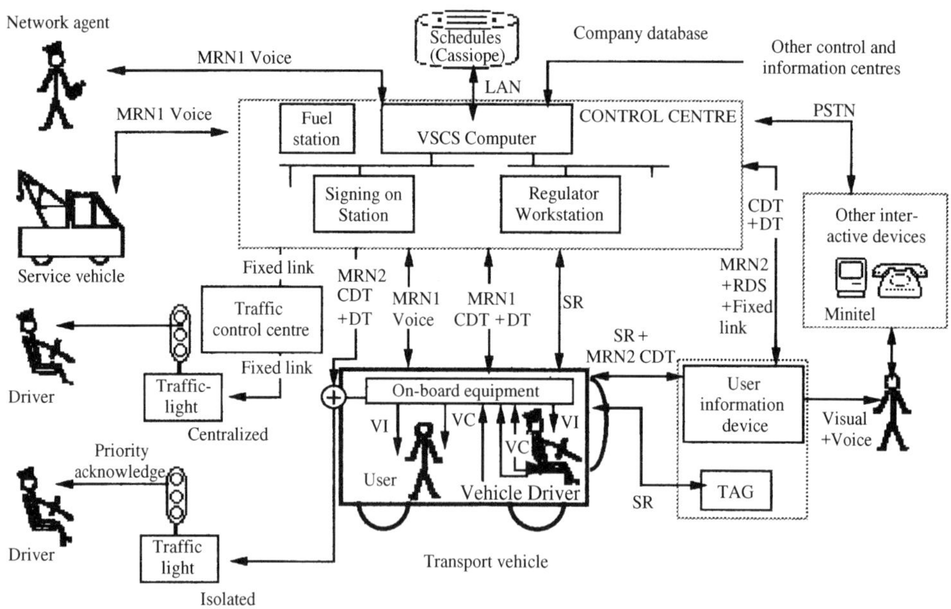

Figure 15.2. Example of APTS physical architecture (Bartoc/Phoebus Consortium Drive 1991–92)

(iv) management of interchanges; (v) passenger information; (vi) driver security; and (vii) off-line production follow-up.

In large networks, specialized staff operate the APTS, which rapidly extends to new functions linked to company management (resource planning, budgeting, results), network "tuning" (routes, frequencies, schedules, interchanges), or social relationships (performance indicators, negotiations, etc.).

The above example is only theoretical. Several architectures and possible technical solutions exist among operators. For instance, simple AVM systems are based on a discrete positioning of vehicles through beacons and tags, while more ambitious systems use a permanent (in fact cyclic) location. This implies (in the case of two-dimensional positioning) dead reckoning, radio communication (triangulation), or (more and more) satellite positioning (GPS); a frequent alternative is the odometer, plus door opening sensors (one-dimensional positioning). Similarly, the passenger load may be known through a counting device (counting steps, shock absorber sensor) or (more precisely) through fare collection validators.

Software organization

This initial view of a physical system made of sensors, onboard units and communication equipment is increasingly replaced by the consideration of a company "nervous system", where *data* is the main resource. In this view, an APTS is designed as an *information system*, where information is collected, stored, processed and displayed.

The public transport planning and operations problem may therefore be summarized by Figure 15.3, where each main "box" represents a domain (e.g. strategic planning) and only a few of the possible functions and data exchanges are shown. This is a very simplistic view. The EuroBus project has developed a whole reference data model for public transport (Transmodel), based on a much more detailed analysis of these domains (or functions), presented below.

The first basic need of operators is to design the network. Several constraints limit the possible solutions: non-continuous supply due to vehicle capacity (cf. above); limited network of streets and roads; limited number of stops; rigidity of user behaviour (they do not easily adapt to new routes); profitability (or non-excessive cost) of new routes (or route extensions); etc. Taking into account demand figures, route profiles and fleet characteristics, it is nevertheless possible to define the basic features of the new service: route, frequency, area served and population, estimated demand.

In the second step, these broad service requirements, compared with the operator means (vehicles, drivers, budget) are used in the scheduling domain to prepare detailed operational plans and timetables, to assign vehicles and drivers and to define driver duties. This is still theoretical and may change throughout the day ("daily assignment"). These planned timetables take into account the available fleet (from the maintenance domain) and are the off-line basis of operations management

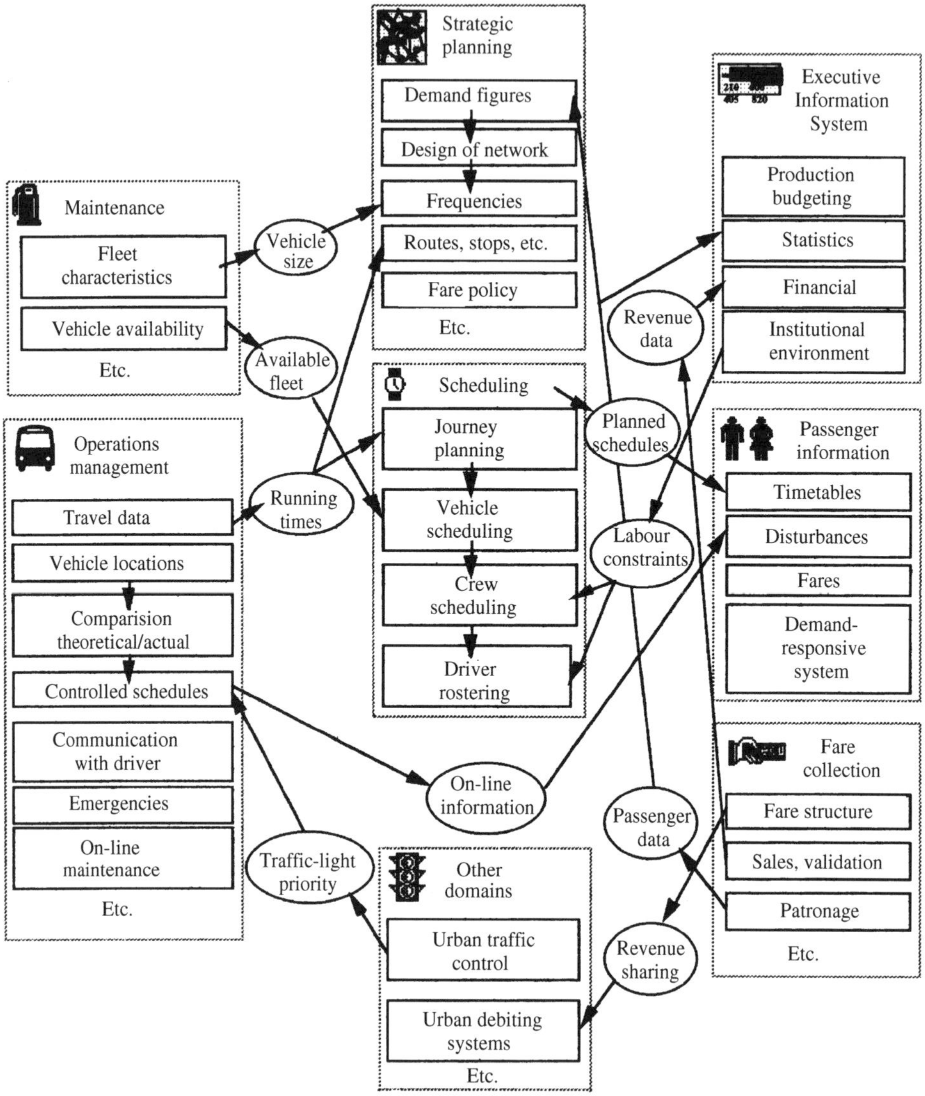

Figure 15.3. Some interrelations between the APTS domains (Cassiope/EuroBus 1992)

and passenger information, sometimes modified according to on-line adaptations (advance, delays, vehicle breakdowns and other events). These plans are also used by the passenger information or management information domains.

In a third step, the record of actual operations is used to feed back the strategic planning domain, in terms of observed running times or passenger load. It provides information for improving the next set of plans and schedules to be prepared. Similarly, fare collection data provide input to an executive information system (EIS) in terms of revenue, or to strategic planning in terms of detailed demand figures.

Some other company functions are not considered here, but they are generally not specific to public transport (e.g. payroll), although they may be linked or related to APTS functions (general accounting, payroll, electronic mailbox, exchange of commercial messages, EDIFACT, etc.).

15.2.3 Achievements of APTS so far

The development of the first APTSs were linked to a technological breakthrough in the two main domains: scheduling/strategic planning (mainly off-line aspects based on computer programming); and real-time control/AVM (mainly on-line aspects based on vehicle location and telecommunication technologies). Before the cheap solutions brought by microcomputers and GPS, some networks pioneered remarkable implementations: their starting point was often either the need for resource optimization and automated service definition (through scheduling) in large networks, or the need for congested route monitoring and driver security (through radio communications) in "difficult" cities.

The APTS extensions to passenger information (printed schedules, telephone enquiry offices, display of waiting time, trip optimization, etc.) or management information (EIS) came as a result of scheduling or AVM. Fare collection and maintenance rely on more specific systems, but tend to be also integrated in APTSs. Fare collection for instance may be linked to executive information systems (revenue data per route), strategic planning (fare policy and its consequences on company balance), scheduling (entry of the network topology in the ticketing system), or operations management (current passenger load estimated in vehicles).

After a test implementation in London (1958, simple automatic vehicle location), the first integrated systems were implemented mostly in Europe at the beginning of the seventies (Hamburg 1969, Chicago 1971, Zurich 1971, Paris 1973, Bristol 1973, Tokyo 1973, Besançon 1974, Nottingham 1974, Dublin 1974, Nagoya 1974, Toronto 1975, Toulouse 1975, etc.). During the decade, APTSs were widely disseminated over Europe (especially France, Germany), where more than one hundred systems are currently installed. North America is less advanced, with isolated implementations in a few cities. The main suppliers are CGA-Alcatel (France), Siemens (Germany), Marconi (UK) and Motorola (US). More information on the market can be found in reference (1).

15.2.4 Evaluation of APTS Impact

Limited and biased results

Though evaluation is a weak point in the APTS (short) history, some assessments made on implemented systems have shown their profitability.

In public transport companies, there is a direct relationship between journey times and operating costs (which are effected by driver and vehicle rotations). The detailed knowledge of running times allows for a significant decrease of layover time (at each terminus).

The following list (Table 15.1) proposes a set of APTS "products", with their main impacts, as known from existing studies. Impacts are rarely described in terms of patronage increase, and even if they are, these results should be considered cautiously. For instance, concerning passenger information systems, the use of the system (e.g. number of connections per day) is sometimes recorded; the origin of users (former PT users, car users) is also sometimes recorded, as well as the number of fully completed transactions; but the impact on modal shift is hardly known, or too optimistic (22% of Videobus users in Nice, a passive information display, declare they ride the bus because of the system). Some manufacturers' results are optimistic and biased. Only data that seem reliable have then been selected. More details on evaluation aspects can be found in references (2) and (3).

Table 15.1 also includes some ITS products which are not directly applied to public transport, but which affect public transport patronage (such as payment and access systems constraining car use and favouring PT use).

There are some limits to be considered when using these impact results. Most APTS and ATT measures are complementary. Modal shift percentages cannot be added. Very often, the existence of a location system (AVM) is a compulsory need before other functions (e.g. user information). When implementing a new APTS device in a bus company, the new system will not only replace former manual functions, but also create new functions or needs; it is then very difficult to compare comparable things. Most effects are qualitative, or *diffused* within the company, over a long period of time. This is the case of organizational effects. Finally, some effects *cannot easily be quantified*; for instance changes in social relationships within the company, due to an "objective" system, that can be either viewed as an impartial judge or as a "Big Brother".

Macro-economic effects of APTS on modal shift

The Tosca Project in Sweden has defined several scenarios of advanced transport telematics implementation in the Gothenburg region and attempted to assess the possible impact on car and PT use. Table 15.2 below provides a summary of the effects concerning PT modal shift.

Compared with changes in the basic PT supply itself (new metro routes, increase in frequency, etc.), APTSs have an impact of a *a second-order of magnitude on modal shift*.

15.2.5 Limits of Existing Systems

Unsolved questions

Some questions remain unsolved or insufficiently solved in the existing APTS supply:

1. Most of the systems are able to produce a large amount of data (mainly through on-line records, such as AVM or fare collection data), few of which are really

Table 15.1. APTS and ATT set of products and their impact (EuroBus)

Domain	ATT Product	Main impact on PT users	Patronage increase according to	Other impact
Vehicle scheduling and control systems (VCS)	Voice/data link bus — central computer room	No direct impact	0%	Indirect (e.g. better treatment of emergencies)
	Bus laboratory	Impact linked to changes in supply (better adherence to schedules)	Patronage increase expected with a better regularity	Profitability (network optimization); gain in driver productivity (2%) and personnel cost savings (1%, Perpignan, 1988)
	Automatic vehicle monitoring (AVM)	Better regularity; improved reliability of services	Indirect (quality of service)	Savings in driver and vehicle costs
	Scheduling system	Impact linked to changes in supply	Depends on reallocation of saved resources (e.g. drivers)	Enables info on timetables etc., savings in driver and vehicle costs (6% decrease in bus × km, Besançon)
	Full VSCS	Better overall network operations, time saved by passenger, reduced running times (3.1%, Strasbourg, 1982; 4–5%, Nancy, 1983)	1.5–3%	Gain in driver productivity (15%, Nancy, 1983), better management of staff, energy savings (2.2%, Nancy, 1983)
	Bus priority system	Better regularity; increased commercial speed (+2–4%, Caen 1989); gain in regularity (+75%, Caen, 1989; 21%, Torino, 1984)	The negative effects on private cars will contract the positive effects for PT	Incentive to car drivers! Gain of 10 seconds per bus and intersection (Paris); eight of 10 buses get green light
	Demand responsive system	Good adaptation to user needs	?	Solution for low density areas

continued overleaf

Table 15.1. (*Continued*)

Domain	ATT Product	Main impact on PT users	Patronage increase according to	Other impact
Advanced traveller information systems (ATIS)	Passive ATIS	Facilitation of trips, transfers, etc.; increased confidence in PT; waiting time less boring (60%, Nice, 1990)	0.25–1%	Time saved by passenger; increase in advertising revenue
	Interactive ATIS	Generation of new trips (3% for Digiplan, Lyon, 1989); trip optimization	0.5–2%	Time saved by passenger
	Multimodal ATIS	Possible mode combination	?	PT information to non-users (drivers)
Node management systems	Park and ride management	Incentives to car drivers (150 000 London commuters are daily users of the "park and ride" system)		Better allocation of space (PT downtown, car outside)
	Security management	Increased perceived security	Up to 20% during off-peak hours	
	Maintenance system	No direct impact	0%	Maintenance savings
PT and car debiting systems	Access control system	Shift from cars to PT	2–6% for PT (Gothenburg); 45% of the car users transfered to car-pool or PT (Singapore)	+6000 PT users (Milan); +6000 to 24 000 PT users (Turin)
	Road pricing system	Shift from cars to PT	7% (Trondheim)	
	Advanced fare system (mag, and smart card)	Fare adjustment and policy	Depends on sensitivity to tariffs, marketing	Knowledge of patronage, PT revenue increase
	Remote payment system	Ease of use	?	Possible counting when alighting
	Integrated payment system (IPS)	Multiservicing capabilities; ease of use; generation of trips	0.5–1%	Knowledge of patronage

Table 15.2. ATT set of products considered by the Tosca Project and their impact

Domain	Product	Assumptions	Patronage increase according to studies	Other impact
Vehicle scheduling and control systems (VSCS)	Automatic vehicle monitoring (AVM)	Introduced in 1995 and 100% equipped in 2000 for RTC (all buses and trams)		The total travel time for PT due to irregularities will be reduced by 1% in 2000/2005.
	Scheduling system			Reduced travel time
	Full VSCS	Public transport regularization system completed by 2000/2005	Improvement in regularity will increase the use of public transport and a transfer from car to PT. Estimated to be 1%, due to the improvement in travel time for PT	50% reduction in accident risk per trip and a positive effect in CO_x emissions
	Bus priority system	100 intersections in the city of Gothenburg in 2000/2005		Negative travel time effects from PT regularization could dominate the positive effects for PT if 250 main signalled intersections are installed. 15% decrease in travel time for PT, 3% increase for cars with 100 signals in the main streets
Advanced traveller information system (ATIS)	Multimodal ATIS	Videotex terminals at office (50% in 2005), at home (35% in 2005) and for in-vehicle use (25%)	Long-term improvement of the use of PT as the knowledge of PT facilities grows; +0.5% increase in PT passengers (1.8% in 2005) = + 0.1% increase in modal split (in 2000)	1,7% reduction on total PT trip time (2005)

continued overleaf

Table 15.2. (*Continued*)

Domain	Product	Assumptions	Patronage increase according to studies	Other impact
Node management systems	Park and ride management	Network parking guidance concerns directly the cars but it can help the "park and ride" system	A transfer to public transport at times when parking spaces are fully occupied	
PT and car debiting systems	Access control system		Expected increase in the use of PT (especially at peak hours) : +6% in Gothenburg	Reduced delay due to reduced car flows. Reduced travel time for cars by 0.5%/0.6% in 2000/2005. 42% of the residents agree for a road pricing
	Road pricing and integrated payment system	For cars: 75% equipped vehicles in 1995, 90% in 2000, 95% in 2005	2.6% change in public transport volumes at peak hour (for 2000); +1.3% (2005)	Reduced travel time, due to less 110 vehicles at peak-hour (2005). User fee collection/automatic debiting system fur fuel, parking, taxes, use of road infrastructure will effect modal shift

used; numerous systems are designed on the basis of the capacities of technology, not from operator needs; as a result, they do not really exploit their potential; for instance, more accuracy (for vehicle positioning) is expensive and not always justified; as said before, only a few existing systems have been seriously evaluated.

2. The link between on-line and off-line aspects is not exploited sufficiently; in most systems, on-line and off-line are two separate worlds; "planners" reason on theoretical data (e.g. running times, patronage, etc.) that deserve to be fed back by recorded data.
3. While the first APTSs implied a centralized architecture, there is now a discussion on what should be centralized, what could be decentralized; at the conceptual level, centralization is helpful (unique data definition, consistency between data of the various domains); at the physical level, decentralized data stores or servers (e.g. information server duplicating the central files) cannot be avoided and are often more efficient.
4. APTS are rarely integrated into the overall city traffic management scheme (link with the traffic computer); for such functions as park and ride management, traffic-light priorities, multimodal user information, a real integration is necessary.

Despite these weaknesses, public transport is probably the most advanced segment of road traffic in terms of experience in computer-aided solutions. However, the isolation of the industry (1000 operators with a fleet exceeding 100 buses in western Europe, most of them facing specific geographical, political, economic constraints) makes it difficult to generalize a given solution, or even architecture.

The need for cheaper solutions

Current technologies, operating practices and computerized equipment in the European public transport companies were analysed in 1990 by the Cassiope Project.[1]

A review of the main operator needs, through an extensive survey of the needs and methods of 119 European bus operators (cf. reference (4) and Table 15.3 below), showed:

(1) a very large variety of methods, equipment and needs throughout Europe;

(2) a strong interest for the possible benefits in introducing new computer-aided tools;

(3) a very great difficulty in implementing such tools.

The products currently offered by suppliers appeared as technically efficient, but as too expensive and too difficult to integrate within the company and in the transport environment. "Cheaper" solutions are not only a matter of money, but also of

Table 15.3. Some results of the state of the art on European APTSs Cassiope 1990

87% of the networks use radio for voice communication, especially for on-line operations.

28% of the networks use an AVM system. The main types of systems are:

1. *Continuous detection, radio data transmission.* This kind of equipment is mainly supplied in Germany (Siemens, AEG, MBB, etc.), France (CGA., Thomson, Matra, Sema-Metra, SEREL, etc.), the Netherlands (Philips) and Italy (Italtel, OTE, etc.), with an important contribution of national suppliers. Several combined techniques may be used for the detection: a dedicated odometer, door opening sensors, beacons on the street. The data exchange between the bus and the central process computer uses a radio channel.

2. *Punctual detection and other systems.* Detection is operated only by beacons, or sometimes by the driver himself. Several types of data transmission are used. The "Wayfarer" system, which provides a graduated fare ticketing system, is often linked in the UK with radio data transmission equipment, allowing continuous monitoring. Numerous networks use moveable or fixed beacons (to measure running times). "Laboratory buses" are generally able to measure both times and passengers, a small part of the fleet is then equipped and travels on all the (main) routes for temporary surveys.

29% of the networks use counting steps and/or infra-red counting cells for passenger measurements (more for collecting the number of trips than for the current load of the bus, because of insufficient accuracy); the networks generally use data collected with validation systems on-board or at stops for patronage recording. In most of the cases, these data should be corrected with models (from surveys), because all the passengers are not obliged to validate their ticket (season cards).

80% of the networks use machine validators (versus 30% manual validation by driver). Fare structures, ticketing and validation methods, and equipment vary greatly in Europe, even sometimes in the same network. Magnetic tickets have been experimented with in about 10% of the sample.

58% of the networks use odometers, of which 26% are integrated into AVM systems, and 32% are simple tachometers.

16% of the networks use onboard information displays, and 14% information displays at stops (electronic display or video systems). In France, numerous operators (32%) use telematic information ("Minitel"), which may be also linked with the AVM system.

Concerning software, more than 70% of the networks use specific scheduling software; numerous maintenance software are used; most of the management information treatments are made on small computers with software developed in-house. The trend for large networks seems to be to implement operation databases on large systems, to dispatch data in the various operational tools and finally to link them (e.g. scheduling and AVM when existing, or strategic planning and scheduling).

simplified implementation within the company, better social acceptance and a more visible return.

The need for integrated systems

The advent of cheap and powerful microcomputers has brought a rapid advance in the use of computer-aided systems. This is, in part, due to the greater willingness of operational staff than the data processing managers who control the larger computers. Most of the public transport packages require detailed knowledge of the planning and operational processes and, therefore, there can be of enormous benefit for the users of the systems. The context of deregulation of the bus industry (e.g. in UK) sometimes boosted the sales of scheduling systems, because of the need to bid speedily for services ahead of competitors.

However, this decentralization of computer procurement within companies, brought about by the availability of cheaper systems, has not been without problems. It has been one of the reasons why companies have found themselves with a collection of incompatible equipment and information systems. Different systems could only be linked, if at all, by cumbersome transfer files or manual data re-entry procedures. The functional analysis realized in the first stages of the Cassiope Project in 1990 showed the current complexity of this kind of links (Figure 15.4).

This problem becomes more critical with the emergence of new techniques allowing access to new data (vehicle location, actual fuel consumption, data on passengers) and new functions (sharing revenues between operators in multivendor systems, fare products combining several services, etc.). Most operators currently

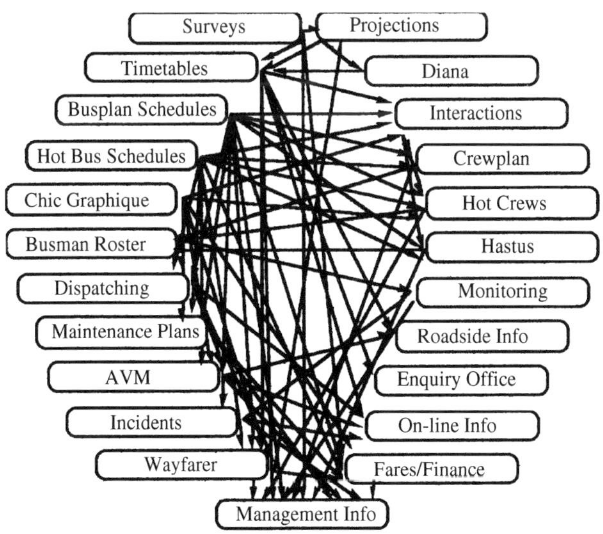

Figure 15.4. Compatibility problems between existing software (Tony Wren, University of Leeds/Cassiope)

face difficult compatibility problems when combining applications. Even with a limited number of functions implemented, combinatory rules rapidly increase the number of interfaces to be built in order to avoid redundancy or inconsistency in data collection and processing. In the current approach, most operators face critical compatibility problems when combining software programs.

The interlacing of data flows makes it interesting to define a new generation of *integrated systems*, where: (i) all data used by separate functions are part of a general database, and (ii) all applications or equipment items "plug" together, allowing the operator to build their customized modular system.

15.2.6 New Issues

Future APTS systems will take advantage of the current R&D progress in some strategic fields and will allow the treatment of new problems, such as complex trip optimization in a multimodal environment:

(1) definition of the ITS road and transport environment (e.g. RDS — radio data system — short range communications, messaging, traffic control schemes, etc.);

(2) database management systems (DBMS), currently improved in terms of techniques (object-oriented programming) or concepts (Transmodel reference data model);

(3) ergonomics (for instance, easier data retrieval in company files, well-designed on board unit, etc.);

(4) digitized cartography (GIS);

(5) new communication and positioning technologies (mobile telephone, GPS, etc.);

(6) smart cards and multivendor debiting systems (including electronic money purse).

15.3 COMPUTER-BASED ADVANCED SOLUTIONS

APTSs rely on the progress and availability of technologies. For instance the marketing of cheaper and cheaper GPS solutions makes it possible to develop inexpensive bus monitoring applications. However, the core of future systems is clearly *information* and *software*.

15.3.1 The Cassiope/EuroBus/Transmodel Framework in DRIVE

During 1989–95 an ambitious European process developed a set of standards for public transport systems within the DRIVE Programme. While the Cassiope Project designed a first version of an information system framework, the

Bartoc Project developed proposals for the physical system (hardware, sensors, communications, etc.).

Starting in 1992, the EuroBus Project concentrated on more practical achievements: test and validation of the tentative information framework (with another project, PHOEBUS and with the rest of the DRIVE/ATT community); development of applications, mainly passenger information terminals; and production of a reference data model serving as the logical core of future applications and systems.

Functional breakdown

The *functional specifications* (cf. list in Table 15.4 below) have been oriented by the following questions: What is the information necessary to operate activities of public transport companies? What are the main areas and functions? Which data are necessary to operate every function? Which commonalities, used in the different areas, should be standardized?

Data modelling

During the first part of the process, a top-down functional approach was pursued to identify what PT systems were required to do. This has been followed up by greater emphasis on data modelling which focuses on the data which systems operate on. Data rather than functions are the more stable aspect of systems, particularly systems as diverse as those which are needed to cover different operating practices across Europe. Sharing data is the key to successful interoperability of systems.

The EuroBus sub-team specialized in data modelling (Transmodel) issued in 1994–95 a remarkable set of documents ("Reference data model for public transport"), which are proposed, studied and appraised as a pre-standard by European standardization bodies (CEN). The Transmodel set of proposals is a sort of PT handbook, to be used by operators and suppliers for their information technology projects. Its adoption means: a large number of savings, an improved data reliability (no double data input for different systems); the availability of numerous PT applications (without being bound to one single supplier); and easier data exchanges with other ITS applications (for instance, multimodal information servers).

The purpose was clearly *not* to define new algorithms or to design applications but to *provide a framework* within which different application systems can operate on a single bus company common database, thereby allowing companies to select the individual systems most suited to their needs, and to enhance their operations by updating or replacing individual modules as new technology becomes available, without disrupting other aspects of their information processing.

15.3.2 Implementation Aspects

The proposed European reference data model, Transmodel, provides the foundations for compatible software developments by suppliers, and for the

Table 15.4. Proposed common functional specifications for APTSs Functional Areas proposed by EuroBus-Transmodel (deliverable 8, 1994)

Study passenger behaviour and determine the demand
All actions necessary to generate usable data for network and service improvements, using input from internal (transportation process) and external (city planning) sources
- Plan passenger surveys
- Process data from operations (trip O-D, passenger load, etc.)
- Process data from marketing and city planning
- Process data from surveys (customers)
- Determine and present demand (station load, network load, etc.)
- Determine and present data on passenger behaviour

(Re)design the network
Take all global decisions regarding the network structure, taking into account the defined quality of service that has to be offered and the guidelines from management (budgets, policy)
- Define the area that will be served
- Define the stop points
- Define the routes
- Choose the transport system to be used
- Propose road adaptations (priority strategies, bus lanes, etc.)

Plan the service to be offered
Take all global decisions regarding the quality of the service that has to be offered, within the limits of budgets and other guidelines from management (define minimum and maximum travel times, interchange points, etc.)
- Define the service quality level and quality objectives
- Define the frequency on the line(s)
- Define the type of service (demand responsive, line oriented, etc.)
- Calculate the global running times
- Calculate the global need for vehicles and drivers

Define fare policy
Define a specific fare system to be used (distance, time or area related, etc.)
- Define the fare system(s)
- Define the fare system parameters (zones, etc.)
- Plan sales (policy, type, etc.)

Plan detailed journeys
Translate the broad service requirements (derived, for instance from the functional area "plan the service to be offered") into detailed journeys that give information on the exact timing of the service(s) on each type of day and for each time band
- Define the journeys (timing, instructions)
- Optimize synchronization between lines

Schedule vehicle blocks
Allocate the type(s) of vehicles to journeys, optimizing the use of the vehicles by switching them between lines
- Define blocks of vehicle operation (chain journeys into blocks)
- Allocate type of vehicles to blocks
- Determine the precise number and type of required vehicles

Table 15.4. (*Continued*)

Schedule driver duties
All activities necessary to define duty elements (parts of working days) that match the planned vehicle schedules, taking into account the general labour agreements. Cut duty elements from blocks
- Build duties from duty elements
- Determine the required number of drivers
- Analyse duties

Prepare driver rosters
Compose sequences of duty elements into duties (weeks, months) according to rules governing the number of hours worked in a period of several days, the allocation of rest days, the rotation of evening duties, etc. and finally assign names of the drivers to the blocks
- Define roster structures
- Assign duty types to roster structures
- Assign duties to the roster matrices
- Analyse duty sequences

Manage driving personnel
Optimize the use of available drivers. "Personnel management" makes available the required number of drivers according to the rosters. "Transportation" is responsible for the optimal use of the drivers
- Plan mid-term assignment of drivers to duties
- Plan short-term assignment of drivers to duties
- Control the drivers
- Collect and process the driver data

Manage vehicles
Manage the use of vehicles according to the planned vehicle schedules. "Equipment management" makes available the required number and type of vehicles according to the planning. "Transportation" is responsible for the optimal use of the vehicles
- Assign vehicles to blocks
- Monitor the availability (status) of the vehicles
- Control depot (location in the depot, entry and exit)
- Prepare the vehicles for the next block (tank, read status, clean, etc.)

Perform and control the driving process
Drive and transport passengers according to given instructions (schedule), including all activities that support the driving process (traffic-light priority, track switching, bay selection, advance/delay advice, etc.)
- Monitor the driving process and control it in case of deviations
- Drive and wait according to the given instructions (schedule)
- Support the driving process
- Monitor position (in place and time) of vehicles and drivers
- Control the driving process and update the schedule
- Inform on actual operational situation

continued overleaf

Table 15.4. (*Continued*)

Plan and organize passenger information
Take all global decisions regarding the passenger information process. Decide who is responsible for informing the passengers (the transport authority and/or the operator) and define the information strategies, including the quality level of the information

- Define the responsibility breakdown between authority and operator
- Define the passenger information strategy
- Prepare and organize the information media (info desk organization, supply of documents, etc.)

Provide passenger information on the planned service
Give information on the planned service (i.e. information available before the hours operational cycle) to passengers, either by display/video or audio (incl. phone desks). This information can be passive or delivered on request

- Provide passive information on the planned service
- Deliver on request information on the planned service (trip preparation help)

Provide passenger information on actual service
Give passive and on-request information on the actual service on the actual (traffic) conditions and different events related to the use of public transport

- Provide passive information on the actual service
- Deliver on request information on the actual service
- Exchange and record additional information
- Give additional information

Plan maintenance work
Ensure that maintenance work on vehicles and infrastructure is spread evenly over the available capacity (personnel and tools). The capacity has to be optimized, taking into account timetables and vehicle schedules

- Define all work that can be planned in advance
- Determine working steps
- Chain working steps to blocks
- Combine blocks to production sheets
- Define mean time to repair
- Set up production schedules
- Plan capacity
- Forecast material consumption
- Estimate costs

Plan and process maintenance orders
Collect the work that cannot be planned and combine this work with the planned work into workshop orders. Plan a date for each order after verifying the availability of personnel and materials. Update the orders after carrying-out the work and make data available for statistics

- Open up and close orders
- Assemble order contents from planned work activities
- Assign required material to order and check availability
- Collate orders according to priority
- Check capacity of vehicles, staff and workshops
- Call for vehicles

Table 15.4. (*Continued*)

Call for material
Deblock order
Collect the work that cannot be planned
Combine work (planned and unplanned) into orders
Edit order forms
Acquire and evaluate order processing data
Acquire and evaluate actual vehicle data

Control maintenance work
Carry out maintenance work according to planning, considering the actual capacity of personnel and materials. Collect vehicles from "transportation" and return them after the work has finished
Balance workshop capacity
Check vehicle availability for maintenance
Assure availability of tools and material

Manage incidents
Support all actions that are necessary to maintain the right number of vehicles available to "transportation" and record and process all data relevant to accidents and damages
Collect incident alerts and data
Establish diagnosis and action to be undertaken

Analyse maintenance actions
Collect and evaluate all data generated in maintenance actions and generate usable ratios
Collect and analyse data for damage statistics
Collect and analyse data for consumption statistics
Collect and analyse data for performance statistics
Collect and analyse data for vehicle and object history
Evaluate inspection and corrective maintenance actions

Manage material
Assure the availability of spare parts and material, in a quantity and quality as required at the time and place needed.
Plan material demand
Purchase material
Control stock
Manage spare parts and material distribution

Organize sales
Define and manage the sales network including sales outlets and contracts with retailers (if any)
Determine fare products
Manage customers (data, rights, cards, etc.)
Manage sales network

Operate sales
Perform the sales transactions of fare products and manage the issuing of travel documents (retail shops, vending machines)
Sell fare products
Account sales
Collect funds

continued overleaf

Table 15.4. (*Continued*)

Validate/check and charge
Verify that the passenger is entitled to ride. Then, determine fare product prices, taking into account the given fare structure and the corresponding transport network and either collect and/or mark the travel document, or debit an amount of fare, corresponding to the planned trip and to the acquired passenger rights (e.g. user profile)
- Authorize access
- Collect service consumption data
- Manage fraud

Manage money transactions
Collect the money paid for the delivered service by passengers or organizations (clearing house, subsidizing authority) and distribute these revenues between the service suppliers.
- Manage revenue sharing
- Manage clearing house

Manage statistical results
Manage storage and administration of the data collected during the operational processes, manage the database(s) in order to provide relevant data for the other functions (e.g. "strategic planning") and issue periodic reports on the operations for various levels of management
- Manage operational database (operations, fare use, equipment and personnel statistics)
- Manage connection with external databases (traffic, urban planning, law
- Issue regular reports (daily, weekly, etc.) for management control (board, depots, etc.)

Manage personnel
Perform all personnel management tasks not directly linked to the transportation process
- Plan personnel policy (contracts, regulation/law, instruction)
- Follow up driving personnel (working time recording, instruction driving attitude)
- Manage non-driving personnel
- Manage instruction
- Manage social requirements

Define marketing policy
Take all decisions regarding the corporate image and define guidelines on how to promote this image and (new) products to the public
- Define the image the company wants to have to the public
- Define unique selling points that have to be communicated

Improve and maintain relations with the public
Undertake all activities that affect the image of the company and public transport in general
- Design advertisements (brochures, posters, etc.) according to house style
- Advertise
- Perform direct mail actions
- Launch new products (new lines, new cards, etc.)
- Launch special actions

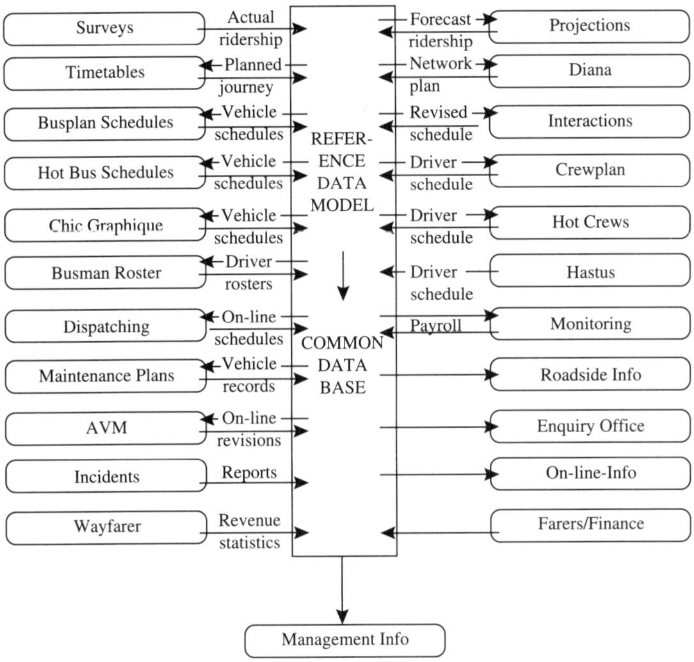

Figure 15.5. Example of links between application programs with the EuroBus Architecture Tony Wren, University of Leeds/Cassiope

implementation of integrated databases within public transport companies. Using the same (dated) examples as in Figure 15.4 above, Figure 15.5 illustrates the simplification of data organization and processing when a common core database is developed in the company.

There is a very wide variety of physical implementation solutions for the APTS system framework: hardware and operating system, DBMS and query language, distribution of intelligence, network management, etc. The choice in this matter should comply with the existing (and short-term planned) systems in the company.

Whatever this choice, *a company "central" database should include all the data managed in public transport operations, allowing a consistent updating of the various company files, whatever the domain.* The term "central" database is obviously excessive, even from a logical point of view. Most of the information systems will include intermediate databases, linked to one or two functional modules, in order to limit data flows between applications and the central database. Another recommended solution consists in applying the recent principle of distributed database systems.

Considering this partition of the company database, one of the main requirements is to secure data consistency. As a result of experience, a common database can be updated and managed by one and only one entry system. Therefore, the recommendation is that the data administration system be consistent (cf. example

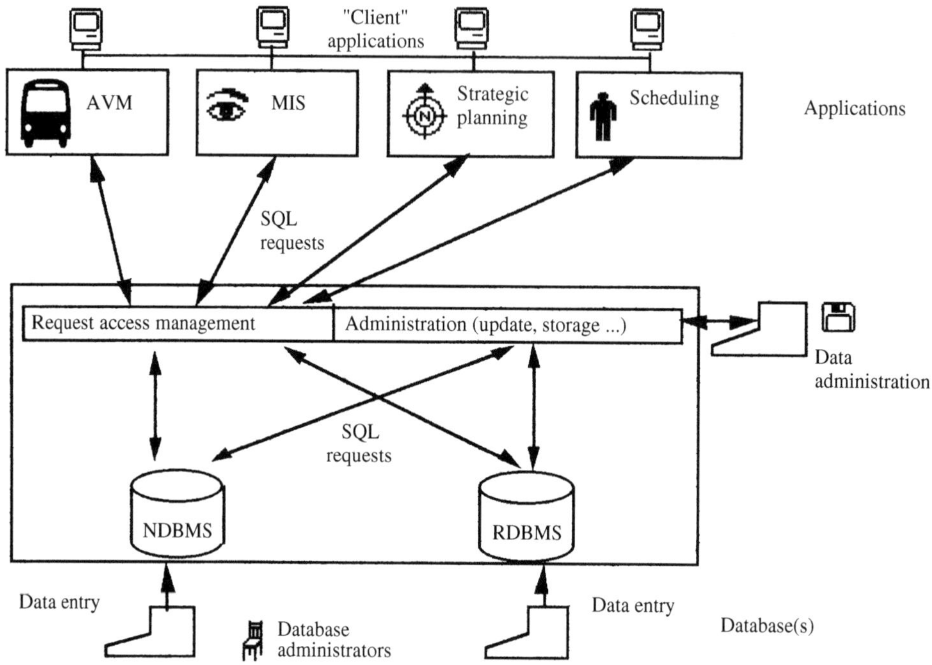

Figure 15.6. Example of database implementation

in Figure 15.6); wherever data are collected, it should obey the same rules and procedures. The data administration system should control the updating of all data, manage priorities in updating, organize access limitations (authorized users, confidentiality of data concerning persons) and verify that the rules concerning security of data are followed by all components of the system.

15.3.3 Other Needs for Standardization

The Cassiope Project proposed in 1991 a set of standards, not only in the information system field (company "nervous system", functional specifications, data models), but also for several other interfaces: communications, payment equipment, human–machine interface for user or operator terminals, etc. Figure 15.7 summarizes the proposed domains of standardization.

15.4 A VITAL ISSUE FOR PUBLIC TRANSPORT

A recent survey realized by the European Quartet Project identified the most important factors influencing the modal choice between car and public transport in Europe, i.e.:

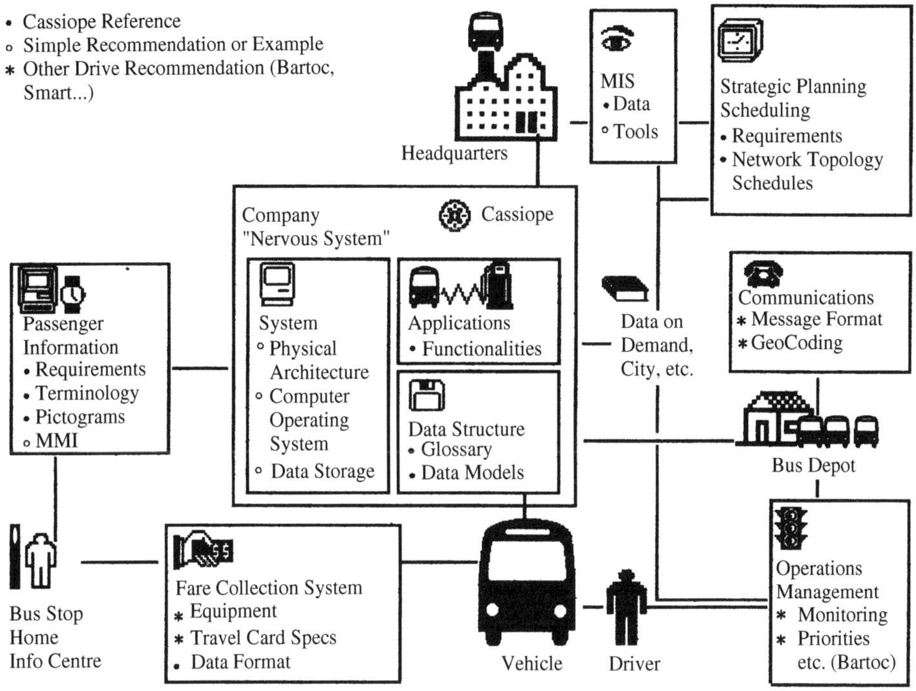

Figure 15.7. Standardization proposals in Cassiope

(1) overall travel times by car and public transport;

(2) reliability and punctuality of public transport;

(3) waiting interchange time for public transport;

(4) public transport fares compared with car fuel and parking costs;

(5) walking time for public transport;

(6) comfort and safety factors; and

(7) past experience and perception of car travel versus public transport.

This list shows the issues at stake when implementing an APTS in a city or a public transport network. Of course, background factors, which are largely outside the control of transport operators, have a particularly significant effect on public transport patronage (car ownership, income, land use, congestion in city centre, etc.). But all what concerns operations and user perception (accountability of the system, global care of the user, etc.) is strategic and can be strongly improved by the use of APTS solutions.

Public transport, which is a difficult and (generally) subsidized industry in developed countries, has to meet the critical challenge of integrating *new technologies* into its usual procedures. This is necessary for both economic reasons (automatic systems should allow a decrease in operating costs) and marketing reasons (quality of service, more stringent requirements of multimodal users).

More efficient decision-making, planning, management, control and information will contribute to improving the operating conditions and to encouraging the use of public transport.

This is a matter of financial survival of buses, trolleys or tramways in traffic; they need absolute regularity, sufficient speed, improved user information, etc. to attract customers, which in turn is necessary to contribute to solving congestion problems in large cities. only efficient computer-aided operations can provide a modern and reliable transport supply competing with cars. APTSs is part of this fight.

BIBLIOGRAPHY

Les systèmes d'aide à l'exploitation dans les réseaux de transport en commun. Papers presented at the *Ottawa International Conference on AVM Systems*. Ottawa: ACTU/CUTA, 1988.

1 *State of the Art on Computer-Aided Technology in Public Transport*. Cassiope Consortium, 1989.

2 *English-French-German Glossary on Advanced Technology in Public Transport*. Cassiope Consortium, 1989.

3 *Glossary of Terms*. EuroBus/Transmodel Consortium, 1994.

User's Guide. EuroBus-Transmodel Consortium, February 1994.

Guidelines for User Information Interface (Standardization). Eurobus/Popins Consortium, September 1993.

Cooperation with Melyssa. EuroBus/Transmodel Consortium, June 1994.

Case Study on PT Contribution to Solving Traffic Problems. Eurobus/Popins Consortium, July 1994.

Reference Data Model For Public Transport. European Pre-Standard. CEN EuroBus — Transmodel Consortium, 1995.

Rousseau JM. *Operations research in Transit: a survey*. Paper presented at the *11th Conference on Operations Research, Buenos Aires*, 1987.

Rousseau JM. (ed.). *Computer Scheduling of Public Transport — 2*. Twenty-nine papers presented at the *3rd International PT Computer Scheduling Workshop*, Montreal, May 1983. North-Holland, 1985.

TOSCA 1 project — RTI Impacts — the 2000/2005 year scenarios. Final Report. TRANSEK/ Swedish Transport Research Board, 1992.

TOSCA II — Socio-economic evaluation of advanced transport telematics. Report. Swedish Transport Research Board/TRANSEK, 1994.

A Review of the Factors Influencing Modal Split. Jupiter Project. Deliverable 1. Transport and Travel Research Ltd, May 1994.

Wren A (ed.). *Computer Scheduling of Public Transport*. Twenty-two papers presented at the *Second International Workshop on Urban Passenger Vehicle and Crew Scheduling*, University of Leeds, July 1980. Amsterdam: North-Holland, 1981.

4 *Operator Needs & Overall Requirement Report. Cassiope Consortium*, 1989.

Macro-Economic Assessment of RTI Impact on Public Transport. Cassiope Consortium, 1989.

Studies of Functionalities in the User Information Domain. Cassiope Consortium, 1990.

Studies of Functionalities in the Scheduling Domain. Cassiope Consortium, 1990.

Studies of Functionalities in the Strategic Planning, Maintenance, MIS Domains. Cassiope Consortium, 1990.
Studies of Functionalities in the Fare Collection Domain. Cassiope Consortium, 1990.
Possible Contribution of Artificial Intelligence, Esprit Tools and Ergonomics. Cassiope Consortium, 1990.
Functional Requirements, Data Modelling, Architecture. Cassiope Consortium, 1991.
Bus-Guide: an Interactive Information Terminal. Prototype Definition and Demonstration Scenario. Cassiope Consortium, 1991.
Cassiope Architecture and Standardization. Final Recommendations. Cassiope Consortium, 1991.
Results of the Passenger Information Demonstration Project (Bus-Guide). Cassiope Consortium, 1991.
Results of the Scheduling Demonstration Project. Cassiope Consortium, 1991.
Final Report. Cassiope Consortium, 1992.
Computer-aided operating systems (AVL systems) and traffic in the cities. papers presented at the Paris International Conference on AVL systems CETUR, 1992.
Daduna JR, Wren A (eds). *Computer-aided transit Scheduling.* Springer-Verlag, 1988. (This contains twenty-eight of the forty-two papers presented at the Fourth International Workshop on Computer-Aided Scheduling of Public Transport held in Hamburg in July 1987.)
Definition of the Main Functions. Eurobus/Transmodel Consortium, January 1994.
Consistency and Integrity Conditions. EuroBus/Transmodel Consortium, December 1993.
Guidelines for Adaptation of Pilot Site Database to Datamodel. EuroBus/Transmodel Consortium November 1993.

16

Analysis of Road Database Management Structure

16.1 INTRODUCTION

All countries have a need to provide an acceptable level of infrastructure to aid the movement of goods and people. Each country spends a considerable proportion of its budget in providing these facilities.

Getting the best value for money, planning a maintenance programme and forecasting future requirements, all require a detailed understanding of the transport network that the public authority is responsible for. This chapter takes stock of the questions and answers relating to the analysis of road database management structures.

Public authorities responsible for road networks have been developing road databases (RDBs) for many years to perform this function. At first these RDBs relied heavily on information stored on paper, providing information for the road network managers. There was little information available and it was difficult to exploit.

As soon as electronic data processing became available to road managers using computers, information could be stored in digital form thus allowing much more sophisticated forms of data processing.

Generally speaking about RDBs public authorities have not found that the development of the hardware and software tools has been the most difficult part of the operation — the problem area has been (and still is) the collection and updating of the enormous volume of data needed within the RDBs which service highway information systems. Such systems are only of value if the data are reliable. The cost of collecting and updating such a system can amount to some 80% of the overall cost.

Advanced Vehicles and Infrastructure Systems, Edited by C.O. Nwagboso

To date systems have generally relied on information collected and updated manually. A considerable amount of personal effort is involved as well as financial expenditure.

The thrust of current research is aimed at maximizing automation of data collection for the smooth operation of the databases. Enormous progress has been achieved in this rapidly developing field. The expanding needs of public authorities are now (potentially) added to by the emerging demands of a private sector wishing to purchase related data in order to improve their own recently developed peripheral product (e.g. vehicle navigation systems).

RDBs should not be considered as being of unique interest. They should be considered as representing one part of a vast set of database systems. Many of these systems, of course have no connection or only have a limited need to link together but others would gain significant benefit from sharing information.[1]

This chapter aims to take stock of the question.

16.2 ROAD DATABASE INFORMATION

16.2.1 Data Levels

When speaking about road information we first need to know which kind of functionalities and purposes we intend to reach. These functional categories are grouped in data levels.[2]

Each level must have different standards of quality assigned to it. Data from different functional levels may have different spatial and time dimensions but can be linked by a shared spatial database. The functional groupings help to clarify the different levels at which data are aggregated.

There are four primary functional levels:

1. *Aggregate statistics level.*
   This level includes only macro-data without any location. The road system for a complete geographic area is viewed as a whole at this level, although there may be smaller subdivisions depending on the size of area covered. This level represents the highest dimension of aggregation for the road transport system and the information procured is a primary source of data for government and the public. The data at this level will be aggregated from primary data within the database. It may be combined with data from databases containing non-road-based information.
   The nature of the data at this level would be income and expenditure by type, road lengths by classification, total vehicle kilometres travelled and indications of road quality and condition by road class.

2. *Strategic planning level.*
   At this level information is available for routes and sections of a route and is used for strategic planning purposes as well as budget allocation, planning and work programming.

To plan the development and maintenance of the road network information is required viewing the network as a series of individual sections each with details of traffic flows, physical characteristics and the levels of expenditure and resource requirements to facilitate "need based" budgeting. This network level information is used for the processes of forecasting, allocating, scheduling, spending and controlling the use of public funds on the road network.
Within this level it is possible to identify both strategic and work programming aspects that have slightly different data requirements. Strategic planning operates typically over a 5-year time span. This level helps to identify objectives and allows a capital investment programme to be developed. Summarized financial information is used to create broad budget allocations.

3. *Operational level.*
Here an exact location is identified, detailed data are available and the information is used for day to day maintenance and operations management.
The design and construction phases of road works operate with information at an operational level. The design of new construction, all maintenance activities needs a level of detail to identify specific inventory items.
This level includes facilities and operations management systems working either in real time or very short time scales, operating and maintaining traffic control devices, maintaining roads and structures in functional operation through routine and emergency maintenance operations, responding to and resolving incidents such as accidents, hazardous material incidents and road blockages.

4. *Research and development level.*
This contains very detailed data which are collected for specific projects.
The data needs for research are still more detailed and precise than for the operational function level. The data studied are usually specific and may reside only on an independent database.
However, the data will often be used to establish coefficent or parameter values that are used in the main system, such as calibration factors, and they will be used directly in sophisticated project analysis.

Only the operational level needs what may be called a typical RDB structure. The first two levels need extracts or aggregated data from the database. The R&D level may use some data coming from the RDB or may serve as an input into the RDB after special treatments of the data.

16.2.2 Data Domains of a Road Database

A typical RDB may be split up into several data domains, each of them being built up by detailed entities and attributes.

The project RIMES (Road Information and Management Euro System) belonging to the DRIVE I (Dedicated Road Infrastructure for Vehicle safety and Efficiency) programme (RTI programme supported by the Commission of the European Communities) detailed the following domains after an in-depth study through different existing RDBs in Europe.[3]

These domains are:

1. *Equipment.*
   This domain covers all items of road furniture situated on or alongside the highway (including kerbs, fences, etc.) and all roadside installations providing service for the road user.

2. *Quality.*
   This domain describes the general condition of the highway.

3. *Passage.*
   This domain covers the restriction on vehicular movement occasioned by the classification of the structures (e.g. weak bridge) and traffic restrictions (e.g. one-way street).

4. *Traffic.*
   This domain details counting methods used and includes the handling of traffic flow data in respect of (for example):

   (i) volume;
   (ii) speed;
   (iii) vehicle categories.

5. *Accident.*
   This domain covers all aspects describing individual accidents. For example:

   (i) precise location;
   (ii) vehicles involved;
   (iii) persons involved.

6. *Maintenance.*
   This domain describles in detail all the actions (preventive and remedial) undertaken and means used to maintain the road network.

7. *Pavement.*
   This domain deals with the structural composition of the road.

8. *Cost and finance.*
   This domain provides information on highway construction and maintenance costs, possibly with individual information about action undertaken. It includes details relating to sources of funding.

9. *Geometry*.
   This domain specifies the geometric characteristics of the network. It covers geographical data representing the network and also the data defining the road geometry (horizontal alignment, longitudinal section, cross-section).

10. *Optional domains*.
    Some information of partial relevance to RDB does not come under public authority jurisdiction. However, such data can contribute to a better understanding of the global environment of the highway and its surroundings, for example:

    (i) weather;
    (ii) social and economic domain;
    (iii) environment;
    (iv) hospital statistics;
    (v) location of gas, electricity water, telephone networks.

All those domains contain information that can be linked to the road network through the road referencing system.

16.2.3 Network Definition

Road referencing provides the means within a database for the precise location of a point on the road (for example "the accident happened at point X").

One of the fundamental problems associated with road referencing systems is the work involved in maintaining consistency while at the same time keeping other data (referenced to the system) updated. Some of these problems are of an administrative and pratical nature, others are related to the representation in the database.

At first sight the methods of representing locations on the road network appear to vary significantly in different countries: the way people talk about a point X on the road varies from country to country. Systems have been built according to traditional national (road administrative) procedures. In most cases the representation in the database reflects these national procedures and adds to the confusion when deciphering if systems are really different. In fact there is no fundamental difference, if correct analysis aiming at a common data description is undertaken.[3]

It is of prime importance that road referencing systems relate to the real world — in other words objects referencing in a database should be readily locatable at the roadside. To this end some systems utilize junctions as key referencing points. In other systems special markers along the road are used for this purpose. While such variations are not of fundamental importance in themselves they do give rise to different network updating procedures.

The cohabitation of different references systems and the liaison between them are unavoidable preconditions for the common exploitation of information systems.

As a result of this the network definition must be considered as one of the domains included in a RDB.

This domain is essential: it is the most important of all the different domains covered by RDBs.

It defines the reference system or systems which will be used to:

(i) collect the information;
(ii) process the data;
(iii) guarantee the accuracy of updates;
(iv) supply information to road databases users in appropriate form.

All who are involved with RDBs and especially those responsible for data collection, need to understand fully the function of the "road network identification" domain. The system must permit two-way transfer between a road-type location and a geographic-type location. In some cases administrative or general data describing the network are found included within the network referencing system. Currently, one of the major weaknesses of RDBs arises from a inadequate network identification system. The consequences of such deficiencies result in:

(i) information incorrectly located;
(ii) time-consuming and money wasting data tracking;
(iii) inability to conduct some comprehensive statistical studies (not exhaustive enough, poor quality results).

16.3 DATA COLLECTION

As already mentioned a key issue about data collection is good network identification which is the central core of the road information system and collection. Today the collection techniques are:

(i) maps;
(ii) plans;
(iii) satellite surveys;
(iv) aerials surveys;
(v) topographic surveys;
(vi) mobile vehicles;
(vii) video techniques;
(viii) visual (skilled personnel);
(ix) manual (semi-skilled personnel).

In the future automated techniques employing industry standard equipment should provide improved efficiency and quality (and contribution to lower costs). Handheld and transportable electronic equipment improve the efficiency of both the collection and downloading of data. Multipurpose vehicles can be equipped to carry out several data collection tasks during a single pass of the road network. The thirst for more and more information will be the catalyst for innovation in this specialized ficld.

It is clear that future uses of road data will require the processing of more information. As an example, traffic counts have been averaged on days or weeks, future applications will require information on traffic for each hour (or less) of the day, the information being handled in a dynamic way. This will enlarge the scope of RDBs and necessitate a further increase of the computational and storage capacities of RDBs.

Today's techniques and the new trends were also dealt with during the RIMES project.[4]

16.4 DATA QUALITY

The primary vocation of RDBs was to bring together all related to knowledge of the road network.

Now, these information systems are increasingly being used as decision-making tools. They are set to revolutionize the approach to "highway management". Because of this, the quality of the information is of prime importance.

This chapter is not intended to define a quality model for the RDBs, but merely to set out the main principles, which will ensure the quality of the information.

The subject is dealt with at two levels: (i) in the database itself, and (ii) during exchange of information.

16.4.1 Quality of Information Collection

From the start of the collection stage, it is necessary to set up a quality plan. This must ensure the following functions:

1. Check that the data entered comply with the specifications for the equipment used.
2. Check that the production of data on locations are based on and agree with the reference system used in the RDB.
3. Check data set specifically against:

    (i) data information;
    (ii) information validity date;

(iii) source used;
(iv) source design date;
(v) technique used;
(vi) instrument accuracy;
(vii) list of names used (for coding items);
(viii) scope of application.

In works on quality carried out in France by the IGN (Institut Géographique National) and the EDIGEO* group, quality of data is classified using six criteria:[5]

1. *Lineage (history).* Defines the characteristics of the information source. It specifies when, how and from where the information was obtained.

2. *Positional accuracy.* Gives elements of positional accuracy for a given item and in particular indicates the difference between the position of the item in the set of data and its real position *in situ.*

3. *Attribute accuracy.* Indicates the quality of the attributes defined in the data set.

4. *Completeness.* Defines, for given type of item, its degree of presence or absence in relation to real conditions *in situ.*

5. *Logical consistency.* Judges the quality of the model used. For example, does the method used to represent the item comply with reality?

6. *Up-to-date accuracy.* Introduces time into the quality definition. It is important to know the validity period of data.

A problem frequently met with, however, is knowing precisely how to measure these different criteria. For some of them, surveys can be carried out to compare the stored data with reality — the surveys being carried out *in situ* or in relation to a set of reference data. Based on these results, it is possible to obtain percentages or degrees of quality.

16.4.2 Quality of Database Updating

When updating the database, the quality plan will implement control processes and add information qualifying the action carried out. Some controls will not be necessary if it can be guaranteed that they are carried out at the time of acquisition. The following are considered to be the more important controls to establish:

* EDIGEO — A french standard for geographical data

(1) measures to validate the data acquisition;

(2) measures to ensure data location accords with the reference system;

(3) measures to ensure data are compatible with the information system (compatibility); and

(4) measures to ensure data stamping and nature of record (creation, update).

16.4.3 Quality During Exchange of Information

The preservation of data quality can present greater difficulties during exchange than when being handled in the database. There are several reasons for this:

(1) information will be exchanged via the use of an exchange format which has its own conceptual model;

(2) transfer from one data model to another data model may deteriorate the original data; and

(3) future users of the information may not necessarily have in-depth knowledge of the domain and may not, therefore, know exactly how to interpret the data supplied.

Problems of information source, the collection methods used, accuracy and data validity, must be looked at very carefully.

At the data exchange level, the quality model will essentially have the function of defining a product which satisfies the user's needs. The quality model can only satisfy such needs by supplying pertinent criteria to enable the user to interpret information adequately.

16.5 DISTRIBUTION OF ROAD DATA

The easy and flexible access for users to relevant, reliable and comprehensive road and traffic data is the essential purpose of road information systems.

The increasing introduction of data processing and data storage equipment at virtually any level within the road and traffic sector has made data distribution a fact of life.

The aim of this chapter is to highlight the present principles of *distribution* of data among the different tiers of road administration and of the need for the *exchange* of road data. In general, the statements in the following sections represent the *typical* situation. There exist considerable variations between different European countries and in some cases even between similar administrative units within the same country.

16.5.1 Road Databases Interested Parties

The generic structure of road administrations consists of three tiers which represent national, regional and local levels, respectively.

The tasks and responsibilities are:

1. *National level.* The national level has the overall *responsibility for the trunk road and motorway network.* This includes the construction, maintenance, planning and forecasting. However, many functions are actually handled through cooperation partners, agents or contractors. The national level *finances* trunk roads and certain activities on other parts of the road network. The national level works out and defines *standards* and *methods* (e.g. in organizations as *TRRL**, *SETRA*† or *BaST*)‡. The national level provides information to national politicians to prepare road and traffic policies and medium- and long-term planning.

2. *Regional level.* This has corresponding responsibilities for roads, but confined to the regional road network. The regional level administers the implementation of standards, methods and plans in cooperation within the national level. The regional level cooperates with the local level.

3. *Local level.* This level has corresponding road responsibilities for the local roads. The local level implements the actual construction, road repairs, etc. on the road network. The local level carries out the day-to-day survey and maintenance.

Activities with some relation to roads and traffic are in addition addressed by certain other organizations of public as well as private origin. Such institutions, which many be users as well as producers of road data bank information, include:

1. International organizations, such as:
   (i) United Nations, and
   (ii) OECD.
2. National institutions, such as:
   (i) bureaus of statistics and economics;
   (ii) geographic/geodetic institutes;
   (iii) police/gendarmerie;
   (iv) hospitals and physicians;
   (v) road user information services;

---

* TRRL Transport and Road Research Laboratory (GB)

† SETRA Service d'Etudes Techniques des Routes et Autoroutes (F)

‡ BaST Bundesanstalt für Strassenwesen (D)

(vi) automobile clubs and associations;
(vii) newspapers;
(viii) car designers and constructors;
(ix) engineering firms and contractors;
(x) cable owners (suppliers of gas, water, electricity, telephone services).

16.5.2 Road Reference System

The road reference system establishes a connection between sites on the physical road network and the information stored in the road data bank. As such it is an essential part of the correlation of road-related information from different systems. The referencing system is the skeleton on which to attach virtually all other relevant types of road-related information. Consequently, instead of assuming the referencing system to be one of the classes of distributed data, the referencing system merely *defines* which set of roads are to be covered by the particular road data bank.

1. *Geo-coordinates.* For map production purposes and in a number of other cases transformation to and from a geographical referencing system (UTM, GS84 (used in the global positioning system), etc.) is important. This transformation should be considered to be part of the road referencing system. It also should be recognized that transformation between a road referencing system and another referencing system could be handled via the geographical references.

2. *Local roads.* The characteristics of local roads and the scope of interest for these roads — particularly in built-up areas — differ in many respects substantially from those of roads which are part of a trunk or regional road network. For these reasons, the same principles and quality standards for the referencing system are generally *not* used fully throughout the hierarchical levels. As a consequence the complete referencing system will *not* be present at every level.

3. *National and regional roads.* In most countries it is the rule that the national and regional level are referenced using one common method. In France this coherence in the referencing even goes down to the local level, excluding only roads inside villages and towns. In contrast, in the UK the Department of Transport has enforced a referencing method (NIS referencing method which is defined for use in the national centralized Network Information System) for use on trunk roads. Many counties use dissimilar types of referencing, although they are urged to use the NIS method enforced for trunk roads. Some counties (using the ROMIS system) are able to handle the NIS method as well as their own method.

The road referencing system is essential for the proper handling of all other domains. Details of the acquisition of data for the other domains are detailed in the RIMES synthesis report.[4] The way in which the circumstances affecting data distribution influences the distribution of data can clearly be seen.

16.5.3 Neighbouring Data

Usually, the prevalent data exchange, flows directly up or down the road administrative structure.

Exchange with a number of external entreprises have been described, but very little has been pin-pointed about exchange between neighbouring local, regional or national managements. These sorts of exchange are virtually non-existing on a systematic basis. (Only user information is exchanged on a nation-to-nation basis.) This is altogether surprising. Discussions have shown that some exchange do in fact exist, but it is usually treated on an *ad-hoc* basis, often using informal information channels. Some possible reasons for this:

(1) frequently there exist differences in definitions and practices used;

(2) no standard exchange formats and methods have been established; and

(3) the need for and the volume of exchange have been too small till now to urge the establishment of a more systematic exchange.

However, the development of international or interregional cooperation will certainly open a new era of data distribution between the existing entities.

16.6 DATA EXCHANGE

Road administrations collect and store data to meet their operational needs and as an essential aid to the effective management and planning of the road infrastructure. The costs of collection, processing and storage of the vast volumes of highway-related data amassed is immense and represents a signifiant expenditure for the road administrations. Such expenditure represents a major investment on the part of the public organization tasked with road management. In many cases, almost all data relating to the highway network will be collected and owned by a single organization, and in these cases the granting of access to the data to other interested organizations and individuals may be seen as a way to maximize the financial return realized from the investment.

In these circumstances the road administrations may wish to give consideration to the collection of additional data or modification to the existing data collection practices to meet external demand for the data and consequently to maximize the financial returns.

In contrast to the scenario described above there are circumstances where the various functions of the road administrations within a geographic area are shared between a number of separate organizations (including possible specialized private companies) with the consequent danger of fragmented data bases allowing no individual organization an overview of the data.

In this case the exchange of data between organizations offers great advantages. Data may be passed between organizations and made available within an organization's database, or a single database may be established jointly owned which may be helped and managed on behalf of the various organizations.

Whatever the chosen solution could be the high cost of collection and maintenance of the data prohibits one single institution from keeping it to itself and the duplication of the data among completely independent organizational structures.

16.6.1 Liaison with Other Bodies

The potentiel users of road data bases may be related to group of domains as showed in Table 16.1.[4]

In Table 16.1 the groups of domains are:

(1) group 1 (assets): network–equipment–pavement;

(2) group 2 (service provision): accident–quality–passage–geometry;

Table 16.1. The use of road data domains

	Group of data domains			
	1	2	3	4
Central governement agencies	*	*	*	*
Other local authorities	*	*		
Emergency services, incl. police	*	*	*	
Public utilities	*		*	
Map making organizations	*			
Public transport bodies	*			
Motoring organizations	*	*	*	
Road freight organizations	*	*	*	
Commercial and trade organizations	*	*	*	
Building/construction industry	*			*
Heath authorities		*		
Vehicle navigation organizations	*	*	*	
Press, television, etc.		*		
Tourism organizations	*	*	*	
Climatic agencies		*		
Environmental/ecological agencies		*	*	*
Local amenity groups		*	*	*
Research bodies	*	*	*	*
Educational institutions	*	*	*	
Clients	*	*	*	*
Contractors	*	*	*	*

(3) group 3 (demand/use): traffic; and
(4) group 4 (management): maintenance–finance and cost.

16.6.2 Data Exchange

The exchange of data between suppliers and clients from a road data bank is becoming more frequent and important. An exchange standard based on a conceptual data model will help to resolve the following problems:

1. Information cannot be understood after been exchanged between client and supplier.
2. The transmission, in digital form, of complete data sets allows clients and suppliers to remain independent.
3. Rationalization of the interfaces required and that the users and producers need to develop to exchange disparate information.

The basis of an exchange is a clear representation of the data environment referred to by the conceptual data model, which must be common to all data users and suppliers.

To enable transfer of data to take place between a supplier and a client it is important that a clear understanding of what data can exchanged is available. Each item of data needs to have an exact definition common to both supplier and client.

The data dictionary needs to include:

(1) a list of entities and their codes;
(2) a concise and clear definition of each entity;
(3) a list of attributes and their codes;
(4) a concise and clear definition of each attribute;
(5) any relevant comments on data use.

There are several recognized standards to facilitate data transfer by magnetic medium. Standard ISO 8211 is the European standard for the transfer of geographical digitized information and is used by DIGEST (system for defining geographical data) in the NATO military sector and the EDIGEO standard in France. The UK NTF standard (for geographical data) is now recognized by a British Standard and has undergone change to provide ISO 8211 standardization.

Regarding road data significative work has already been produced by the projects RIMES, EDRM 1* and STRADA† included in the DRIVE programme[6–8] or in the framework of the GDF‡ product.[9]

* EDRM 1 European Digital Road Map 1 (a DRIVE 1 Project)
† STRADA Standardization of TRAffic DAta transmission and management (a DRIVE 1 Project)
‡ GDF Geographic Data Format

16.7 THE WAY FORWARD

The introduction of new RTI technologies thanks to programmes, such as EUREKA* and DRIVE in Europe or IVHS (*I*ntelligent *V*ehicle–*H*ighway *S*ystems) in the USA enlarges considerably the interest on road data and the potential use of RDB. Several new developments such as vehicle navigation or route guidance on the one hand or driver information systems as RDS-TMC† on the other hand approached the road data area, the network referencing and the location systems.

All these new techniques started thanks to private initiatives. Therefore, the concerned companies developed their own products close to the needs of their application and taking advantage from the new performances of computer tools, such as geographic information systems in the case of digital road maps.

If Traffic Message Channel (TMC) only forecast a special location system to be included in the global message coding system (through the ALERT-C‡ protocol under development), the already mentioned STRADA project[6] aimed to standardize the exchange of traffic data with real-time and dynamic actions.

The GDF exchange format is the product which may influence the most the RDB world. GDF rose from the DEMETER§ EUREKA-project. Mostly dedicated to route guidance and digital road maps it defines a road network thanks to GIS tools and to a level-based structure varying with the accuracy needed by route guidance. Although being already usable GDF is still in an improvement phase aiming to enlarge its functions in order to better embrace the RTI needs on road data exchange. Thanks to the forecast improvements GDF may be considered as a good settlement to build up a CEN¶ standard in this area.[10,11]

Nevertheless, project RIMES showed GDF does not completely fit with the public authorities needs when thinking about the development of a new RDB structure for their own purposes.[3,4] Further action still needs to be taken to open completely the exchange of road data and therefore to improve drastically the potential of RDBs.

16.7.1 Future Actions

The recent and forecast improvements of RTI and of computer science (hardware and software) may considerably revolutionize the RDB world. Nevertheless, further co-ordination work is required to resolve the conceptual differences between RDB referencing requirements and the standards developed for vehicle navigation and information systems. RIMES advocates the development of the road administration data exchange format (RADEF), which would both embrace GDF as far as possible but extend its capacities to match the demands of public administration RDBs.

* EUREKA EUropean REsearch Coordination Agency)
† RDS-TMC Radio Data System — Traffic Message Channel
‡ ALERT-C the proposed standard for RDS-TMC
§ DEMETER a EUREKA project researching digital mapping
¶ CEN Comité Européen de Normalisation (European Standard Committee)

The development of RADEF aims to resolve the difficulties mentioned by RIMES with the use of GDF for public authorities tasks and the need to address other issues (independent of GDF uses), such as the connection between RDBs with different referencing systems.

The logical concept is illustrated in Figure 16.1, and an overall physical arrangement (PARDBs, DRIVE projects and links between each other) in Figure 16.2.

The complete definition of RADEF will also be influenced by other overlapping initiatives from both the topographical and military areas. Figure 16.3 illustrates the situation (the kernel of the figure will be of fundamental importance to the final versions of RADEF and GDF). Therefore, the conceptual data model and the data dictionary already prepared in the framework of projects, such as EDRM I (GDF) or RIMES may be considered as starting points for further actions still to be undertaken.

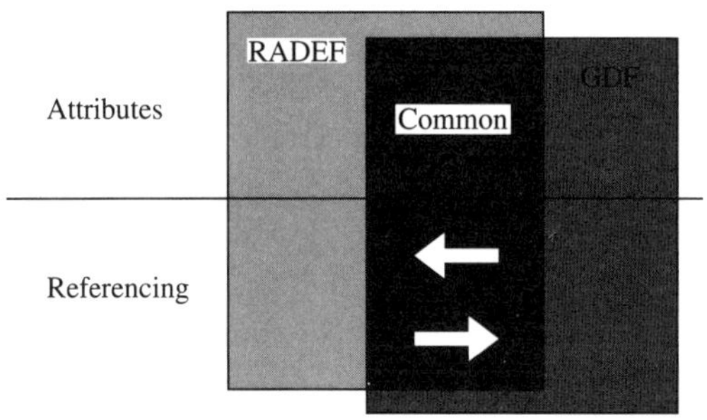

Figure 16.1. Logical liaison between RADEF and GDF

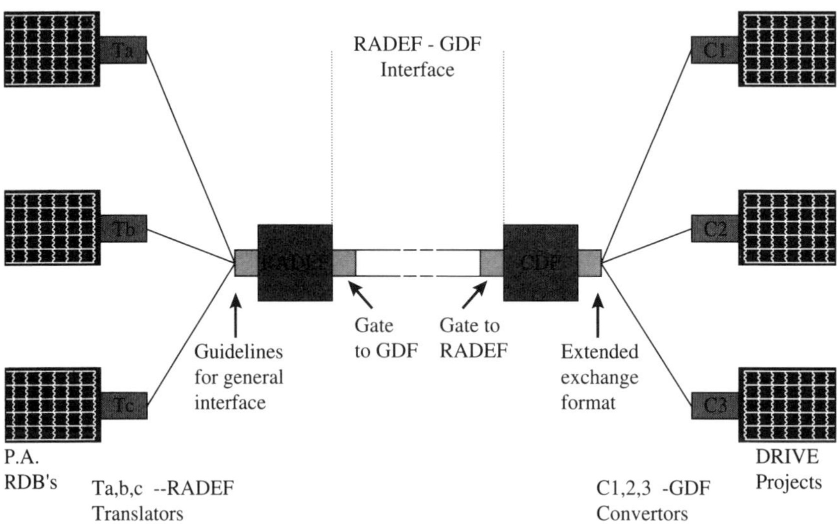

Figure 16.2. Archictecture for the interface RADEF-GDF

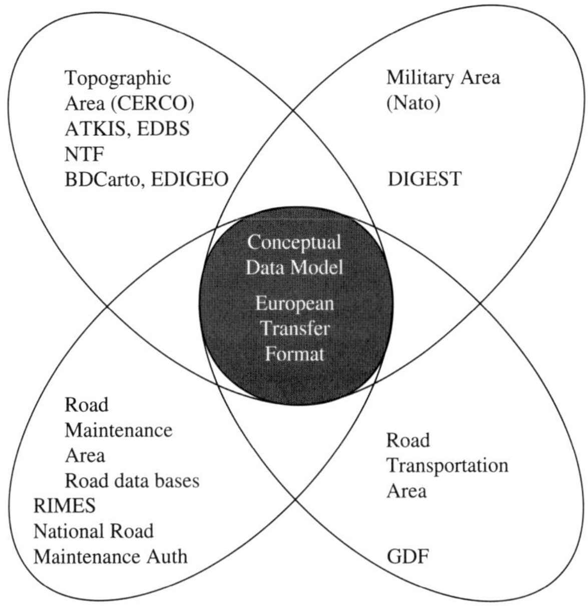

Figure 16.3. The 'flower' of geographic databases

16.8 CONCLUSIONS

Traditionally, decisions have been made on the maintenance of the road infrastructure based largely on a mix of subjective reasoning, common sense and experience. There has for many years been the need to move away from these methods for a variety of reasons.

In particular, the increase in traffic and the length of high-speed roads has meant that there has been a need to improve the cost-effectiveness of financing road maintenance. To ensure that the best use is made of scarce financial resources available to the public sector, reliable, exhaustive information must be readily available to aid the maintenance and planning of public roads. The availability of sophisticated computer hardware and relational database software has meant that the techniques are now available to carry out the necessary comparisons to help the decision-making process.

Each set of data that is included in the database is there because of operational need. It may be traffic flows, accident statistics or maintenance data. Individually these data have great value but in flexible combination such data afford a revolution in the decision-making process.

Modern advances in database technology have resulted in the development of extremely flexible database systems, which are ideal tools for the implementation of RDBs.

As such databases grow in scope so they become more useful in aiding strategic decision making taking advantage of geographic information system technology, which allows data to be displayed in map form. This is a form understandable not only to practitioners but also to politicians. This allows artificial intelligence and expert system techniques to be applied to the data. Using these methods, coarse sieving of data can highlight those areas to which the practitioners can turn their attention to greatest advantage.

The enormous potential afforded by this sophisticated mix of wide ranging data, flexible databases, graphical representation and expert systems has fuelled the thrust of the activity and investment in the RDB area over the past few years.

On the other hand the fast developing RTI techniques give a new dimension for the importance of RDBs in the future.

RDBs can and should be a source of information that can be used by many emerging RTI developments. The data can be made available by connecting the RDBs with other systems through the road referencing system. Other tools for connecting systems can be imagined but the fact that all RDBs have developed road referencing systems which can be exploited in this way is a fundamental consideration.

In order for this method to become efficient it is necessary to establish common levels of standards for the quality of different systems and of the data to be exchanged or distributed.

In any case good knowledge of the needs of the RDB structure is fundamental. It must be concretized through the referencing systems and through the data model. A powerful conceptual data model and exchange format must be developed and generalized so that the enormous potencials of RDBs — whatever they may be — really could be profitably used by anyone who has responsibilities in road or traffic management.

ACKNOWLEDGEMENTS

The writing of this chapter could only be possible thanks to the work accomplished by the colleagues who have worked for the achievement of the RIMES project. In particular constructive assistance has been given by representatives of the following companies or institutions: Ministère de l'Equipement, du Logement, du Transport & de l'Eau — Service d'Etudes Techniques des Routes & Autoroutes (SETRA) (F); Road Directorate — Road Data Laboratory (DK); S.I.A. Limited (GB); Somerset County Council and their associates (GB); Telinfo Integrated Systems (B).

REFERENCES

1 Loyaerts Y. Road data bases: state of the art, needs and uses. *International Workshop on European Digital Road Map, Programme DRIVE*, Stuttgart, March 1992.

2 Patterson WDO, Scullion T. *Information for Road Management*. Draft Guidelines on System Design and Data Issues, Washington: World Bank Infrastructure and Urban Development Department, Report INU 77, September 1990.

3 Programme DRIVE, Project RIMES. *Road Information & Management Euro-System, Deliverable*. No. 10, Brussels, October 1990.

4 Programme DRIVE, Project RIMES. *Road Information & Management Euro-System, Deliverable*. No. 11. General Synthesis Report, Brussels, April 1992.
5 AFNOR. *Projet de Norme Expérimentale EDIGEO*. Paris, Mars 1992.
6 Programme DRIVE, Project STRADA. *Standardization of Traffic Data Transmission & Management*. A Final Project Description, May 1992.
7 Mohlenbrink W. Digital maps — basic data source for RTI-systems. *Advanced Telematics in Road Transport*, Vol. 2 Amsterdam: Elsevier, 1991; 780–799.
8 Loyaerts Y. Standardization in road data bases. *Advanced Telematics in Road Transport*, Vol. 2. Amsterdam: Elsevier, 1991; 849–866.
9 Philips NV and Robert Bosch Gmbh. *Geographic Data Files*, Release 2.0. Eindhoven: Philips Consumer Electronics — Divisional Standardization Department, 1992.
10 Heres L. Standards for Driver Information Systems. *International Workshop on European Digital Road Map, Programme DRIVE*, Stuttgart, March 1992.
11 Claussen H. GDF 2.0 — Towards a pan European Standard for Geographic Information. id.

PART 6

AUTOMATED TRAFFIC CONTROL SYSTEMS

17

Video Image Understanding for Automatic Traffic Monitoring and Control

17.1 INTRODUCTION

While machine vision must still be considered as an emerging technology, there is a growing body of experience with it in the area of traffic monitoring and control. This chapter reviews the main areas where machine vision techniques have been investigated, namely loop emulation, queue detection and vehicle classification.

Loop emulation was the first application investigated. Here image processing technology is used to obtain the same form of data as that delivered by conventional vehicle detectors, namely vehicle count, speed measurement, headway and lane occupancy. In the Transport Operations Research Group (TORG) of the University of Newcastle upon Tyne a system referred to as TULIP (see Section 17.3.1) has been developed.

The second application considered by researchers was automatic queue detection and monitoring. In TORG a system known as FAST_Q has been developed to monitor queueing traffic. The algorithm makes no attempt to identify individual vehicles but does indicate the length of road occupied by vehicles and their state of motion, i.e. moving or stationary. This form of data is not collected directly by conventional detectors.

The third application is concerned with vehicle classification. Here there are a wide range of techniques that can be applied, and research is active to identify the most appropriate. This is probably the most challenging of the applications and,

Advanced Vehicles and Infrastructure Systems, Edited by C.O. Nwagboso

although combinations of other sensors can be used to classify vehicles, is one where image processing is likely to have a significant role.

17.2 PRACTICALITIES OF IMAGE PROCESSING FOR TRAFFIC MONITORING

Throughout the wide field of image processing there are certain common themes, brought about by the limited speed of the current computer technology and the limited understanding of the visual process. These are:

(i) a reduction in the amount of data obtained from the camera before processing; and

(ii) use of simple algorithms that can be efficiently implemented in real-time.

For road traffic monitoring many systems have concentrated on processing only key regions of the image, relying heavily on a priori assumptions based on particular characteristics of the application. These regions may be lines parallel or perpendicular to the roadway or small areas distributed around the scene to be monitored. They are generally referred to as windows, as shown in Figure 17.1.

For road traffic monitoring one of the attractive features of image processing is its ability to cover a wide area. The distance over which a scene may be monitored is

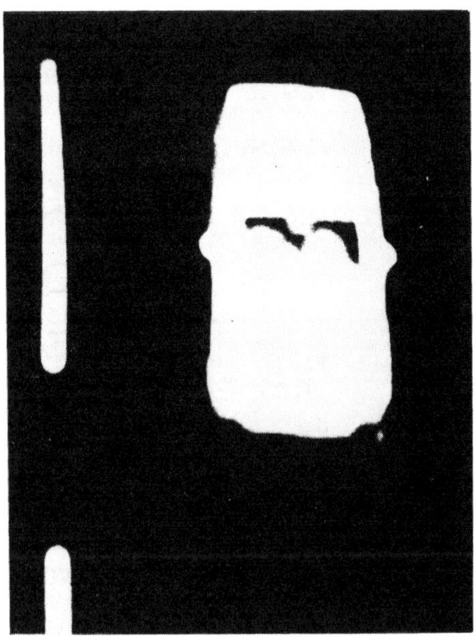

Figure 17.1. TULIP operation. Two windows (lines) projected into the scene

governed by the resolution of the camera, the height of the camera above the ground and the angle of the optical axis with respect to the vertical. From Figure 17.2 it can be seen that

$$d_n = h * \tan[(\alpha - \phi/2) + (n * \phi/r)] \quad (17.1)$$

$$D_n = h * (\tan[\arctan(d_n/h) + (\phi/2r)] - \tan[\arctan(d_n/h) - (\phi/2r)]) \quad (17.2)$$

where d_n is the distance of the projected image of line n from the camera
D_n is the length of roadway covered by image line n
h is the height of the camera above the road
α is the angle of the optical axis w.r.t the vertical
ϕ is the field of view (FOV) angle of the camera lens
r is the number of lines in the image.

In general, the image of a vehicle must cover approximately 5 × 5 pixels to be detected reliably. From equations (17.1) and (17.2) it is a straightforward matter to calculate the length of roadway, which a given camera arrangement can usefully monitor.

Errors in counting vehicles can be caused by one vehicle partially obscuring or occluding another (Figure 17.3). A human operator viewing the scene may be able to detect this situation, perhaps by observing another part of the image in later frames. However, most current image processing systems concentrate on small sections of the image, so they are unable to detect occlusions in this way. The severity of this problem can be reduced by correct positioning of the camera. A high camera location,

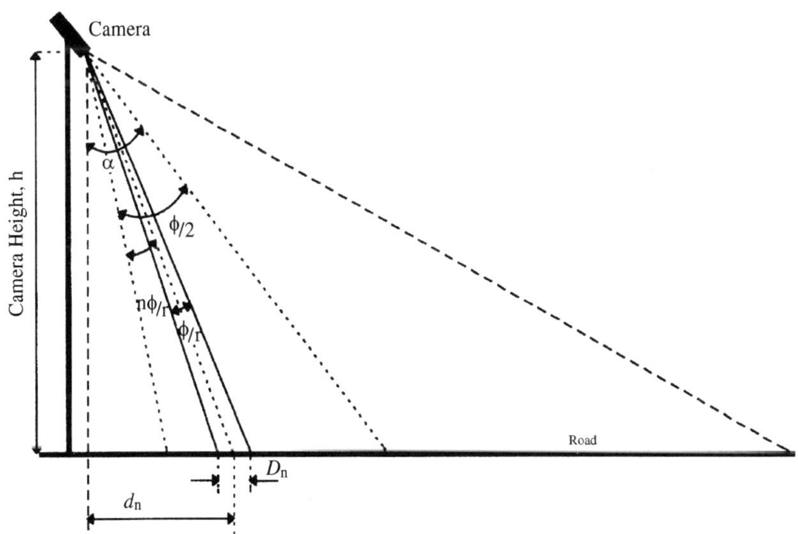

Figure 17.2. Camera arrangement for road traffic management. α, angle of optical axis w.r.t. vertical; ϕ, field of view angle; n, video image line n; r, number of lines in video image

Figure 17.3. Occlusion of two vehicles due to their close proximity

placed centrally above the roadway with a steep angle pointing down on to the road scene is generally recommended.

Speed measurements can be made by detecting the presence of a vehicle at two separate windows set a known distance apart along the carriageway. The video frame rate provides a convenient time base for timing the passage of vehicles between the two windows. It is important to note both the accuracy of this method and possible causes of error. One source of error is the point at which the vehicle is detected. From Figure 17.4 it can be seen that the distance between the projections of the two windows is dependent upon the height of the plane of detection. For accurate speed measurement the distance between the windows must be known, therefore the detection point should be set at the ground plane.

The other major source of error is due to the spatial and temporal quantization of the image. The discrete nature of the image, in both space and time, limits the resolution of the system. Thus, any speed measurements will have an associated uncertainty. For the straightforward camera geometry shown in Figure 17.2, the

Figure 17.4. Distance between windows varies with the height of the plane in which it is measured. d_t, true distance between windows; d_m, distance moved by vehicle

spatial resolution along the carriageway can be calculated from equation (17.2). This gives the accuracy with which the position of a vehicle from the camera can be measured. If two fixed windows are used for speed measurements, the temporally discrete nature of the image will also affect the measurement. The accuracy with which the distance between the camera and a vehicle can be measured will be limited by the distance the vehicle can travel in the time it takes to acquire one full frame (i.e. 40 ms). In practice, the major source of error in speed measurements will be due to the temporal quantization, except at low speeds where the spatial quantization will become more significant. To reduce the effects of the discrete images, the distance between the two windows should be made as large as possible, but not so large as to allow vehicles to become confused with one another, due to lane-changing and overtaking manoeuvres.

The lighting of the scene is also very important for correct operation of the system. This is dependent upon the weather conditions and the time of day. The system should be designed to cope with rapidly varying light levels, the existence of shadows (from vehicles or buildings outside the area of interest) and dawn and dusk periods.

17.3 LOOP EMULATION

In operation, the basic vehicle counting systems emulate inductive-loop detector installations. These systems are attractive for basically two reasons. First, the data they provide (vehicle count, speed, headway and lane occupancy) are identical to those provided by inductive-loop detector systems. Therefore they can be incorporated into existing road traffic monitoring systems very easily. Second, the loop-emulator approach allows a vast reduction in the amount of processing to be done by the computer, fitting naturally into a windowing data reduction scheme. Low-cost general purpose microprocessors can thus be employed for the processing, and yet real-time performance, by which is meant that every video frame is processed, can be achieved.

In addition to providing the usual traffic data, these systems will also provide basic vehicle classification, based on vehicle length and/or width, and they will detect periods of congestion. However, as a consequence of emulating inductive-loop configurations, they suffer from many of the same fundamental limitations. If the intervehicle gap becomes too short, for example, vehicle images will merge together and counting errors will occur. Similarly, if more than one vehicle occupies the detection area the additional vehicles may go unrecorded, while lane straddling vehicles may be recorded twice. This can be overcome by software, as in a digital image it is possible to determine the lateral position of a vehicle on the carriageway. The occurrence of simultaneous events in adjacent lanes can also be employed as an indicator of lane straddling, as it is in conventional vehicle detection systems.

17.3.1 The TULIP System

The TULIP (traffic analysis using low-cost image processing) system was developed by TORG at the University of Newcastle upon Tyne in the late 1980s.[1] It was conceived as a replacement for inductive-loop systems and as such employs a windowing technique to detect vehicles in a binary image. It uses a minimum of specialized hardware; all processing is performed by an IBM PC-AT computer in software. The binary image is produced by grey-scale histogram segmentation, which is continuously updated to compensate for ambient light changes.

To produce the binary image, all pixels with grey levels in the ranges identified as being characteristic of the roadway and unwanted shadows are set to zero and all other pixels are set to 1. This is done by reference to the grey-level histogram, a typical example of which is shown in Figure 17.5. The roadway and shadows are effectively filtered out of the image by this operation. A binary image is produced in which vehicles appear as large clusters of 1-valued pixels. Noise and other image artefacts, such as remnants of shadows thrown from vehicles in adjacent areas of the image, appear as small clusters of 1-valued pixels. Therefore a spatial threshold is applied to differentiate between vehicle presence and these noise pixels. A portion of the very darkest shadows, which occur underneath all vehicles in daylight, can also be passed into the binary image. This strengthens the vehicle's binary image

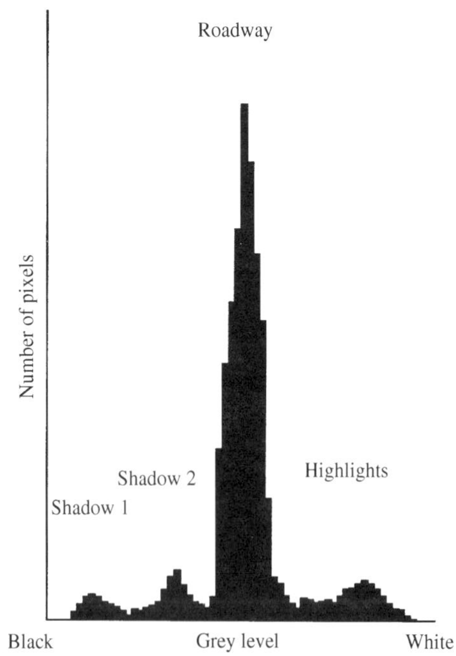

Figure 17.5. Typical grey scale histogram. Shadow 1: produced by shadows underneath vehicles. Shadow 2: produced by vehicles in adjacent lanes

and, as these shadows are known to be at the ground plane, they ensure that the observation windows are located at a known plane for calculations of vehicle speed (Figure 17.1).

If the lighting changes drastically during the operation of the system problems may be encountered with the segmentation of images, because the threshold levels chosen at one point in time will become inappropriate. The ambient lighting must be monitored continually and the segmentation values updated to maintain the best image segmentation. This requires some decision-making before the values can reprogrammed with confidence, to eliminate the possibility of the presence of stationary vehicles producing inappropriate segmentation values. This is done by reference to the initial image segmentation values obtained from the image under operator supervision. During a short set-up procedure characteristic data are obtained. The frames monitored should contain a number of vehicles passing both through the observation windows and adjacent to them so that their shadows fall in the windows. The data collected are used to determine the grey scale value of the roadway and suitable threshold values to remove noise and shadows.

For vehicles to be counted automatically two observation windows per lane must be located in the scene to be analysed. This is done by the operator simply moving a cursor across the video image to define the position of these windows. The first window in each lane defined by the operator must be upstream from the second. A vehicle must be counted in both observation windows before it will be included in the final database.

17.3.2 Results of TULIP Processing

The results reported here used a segmentation procedure with a 3-s updating period to account for ambient light changes. The tests were conducted on a library video of a motorway scene shot from a footbridge approximately 7.5 m above the carriageway. An inductive-loop system was also monitoring the traffic, recording data in the form of 1 min averaged speed statistics for all three lanes. Table 17.1 shows that TULIP can provide vehicle counting to an accuracy of 95%.

The speed statistics obtained with the TULIP system are compared with those produced by an inductive-loop system in Figure 17.6. This shows a straight line graph with a slope close to 1 and a correlation coefficient 0.94, indicating the

Table 17.1. Comparison of manual and automatic counting

Lane	Manual count	Automatic count	Error (%)
1	1529	1475	3.5
2	2261	2155	4.7
3	2693	2561	4.9

Video duration 89 min.
Lighting and weather conditions: stable, overcast

$$\text{Error} = \frac{\text{manual count} - \text{automatic count}}{\text{manual count}}^{*} 100\%$$

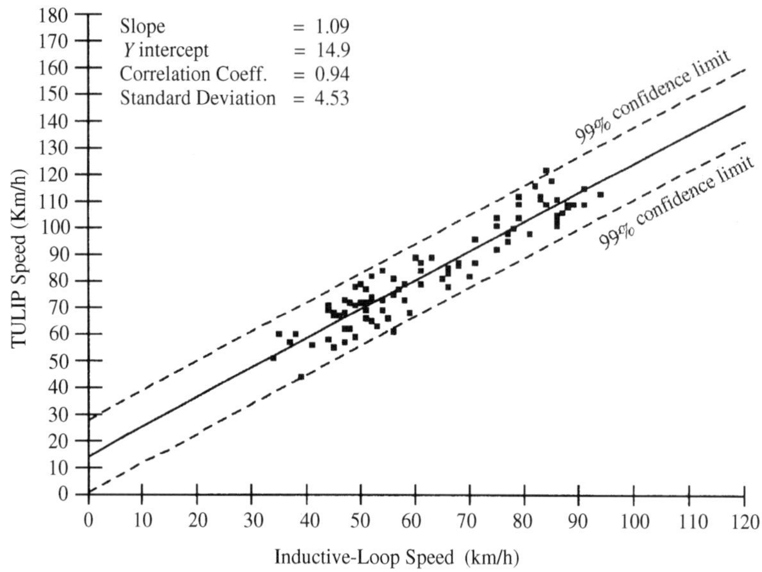

Figure 17.6. One minute averaged speed comparison between TULIP and inductive-loop systems

data obtained from TULIP are in good agreement with those obtained from the inductive-loop system. The intercept on the *y*-axis indicates that a systematic error has been introduced into this analysis. This is due to an error in estimating the distance between the windows used for speed measurement. It was estimated to be 4.5 m. A better estimate would perhaps have been 3.8 m. However, the r.m.s. error of 4.5 km/h implies that the TULIP system is accurate to a level of between 5 and 10% over this range of averaged speeds.

17.4 QUEUE AND CONGESTION MONITORING USING IMAGE PROCESSING

The inductive-loop systems and their image processing equivalents can be used to detect queueing and congested traffic situations from the speed and lane-occupancy data they acquire. However, it may take a period of minutes to collect enough data to determine the state of the traffic reliably. A human observer would be able to discern that queuing was taking place by noting the presence of vehicles over a length of roadway and noting the fact that they were not moving very quickly. The person would also be able to observe the length of the queue and thus the position of the rear of the queue. All of this information would be of great benefit to drivers approaching the scene. With point-based sensors, these data could only be obtained from a very high density of sensors placed along the carriageway. The video camera, however, provides precisely this level of coverage within its field of view.

Using image processing techniques it is possible to determine the existence and extent of queuing traffic using very similar reasoning to that used by a person viewing the scene. Vehicle presence along a roadway can be detected and the fact that they are moving or not can also be determined. Thus, without counting the actual number of vehicles or measuring their speeds, queue detection can be performed automatically. This approach has been taken in the FAST_Q (Fourier analysis of slow traffic queues) system developed at TORG.

17.4.1 The FAST_Q System

The process of queue detection can tolerate a degree of both spatial and temporal data reduction, due to the nature of traffic movements along designated lanes and the (relatively) long period of time required for a queue to form. In this system data reduction in the spatial domain is achieved by defining a window, consisting of a single line of pixels, along the traffic lane, parallel with the direction of vehicular motion (Figure 17.7a). Data reduction in the temporal domain is obtained by processing the window only once every few seconds. The system does not attempt to count the number of vehicles in the field of view. It is required to determine only the presence of vehicles, the length of roadway they occupy and the amount of movement they are undergoing. Problems of occlusion and lane straddling are therefore unimportant.

Three distinct stages of processing are employed to achieve queue detection. First, the pixel data in the window are regarded as the samples of a digitized one-dimensional waveform. Each pixel represents a sample of the wave and its grey value represents the wave's amplitude at that point. These data can, therefore, be processed using the well known fast Fourier transform (FFT) to reveal its frequency spectrum.

Figure 17.7 shows an example of an empty window. The frequency spectrum shows a strong DC component, as one would expect from such a sample. There are also some low-power high-frequency components, caused by electronic and image noise. Figure 17.8 shows the effect of introducing vehicles into the window. The frequency spectrum changes significantly. This provides the basis for vehicle presence detection. Similar changes in the spectrum can also be caused by the appearance of shadows, which must be differentiated from vehicles. This is achieved by reference to the nature of shadows, which have uniform bodies and two distinct edges, while vehicles are characterized by "high-lights" and "low-lights" which tend to be unstable temporally.

The analysis of the spectrum is restricted to a narrow band of AC components towards the low-frequency end of the spectrum. This acts as an anti-aliasing filter and it makes the analysis insensitive to the ambient light-level, which merely determines the DC level of the waveform. The system is, therefore, unaffected by the normal short-term fluctuations in the scene lighting, due to changes in cloud cover, etc. Large changes in light-level, in dawn and dusk periods, can have an effect due to the operation of the camera. In very low-light conditions (night-time) the contrast of the

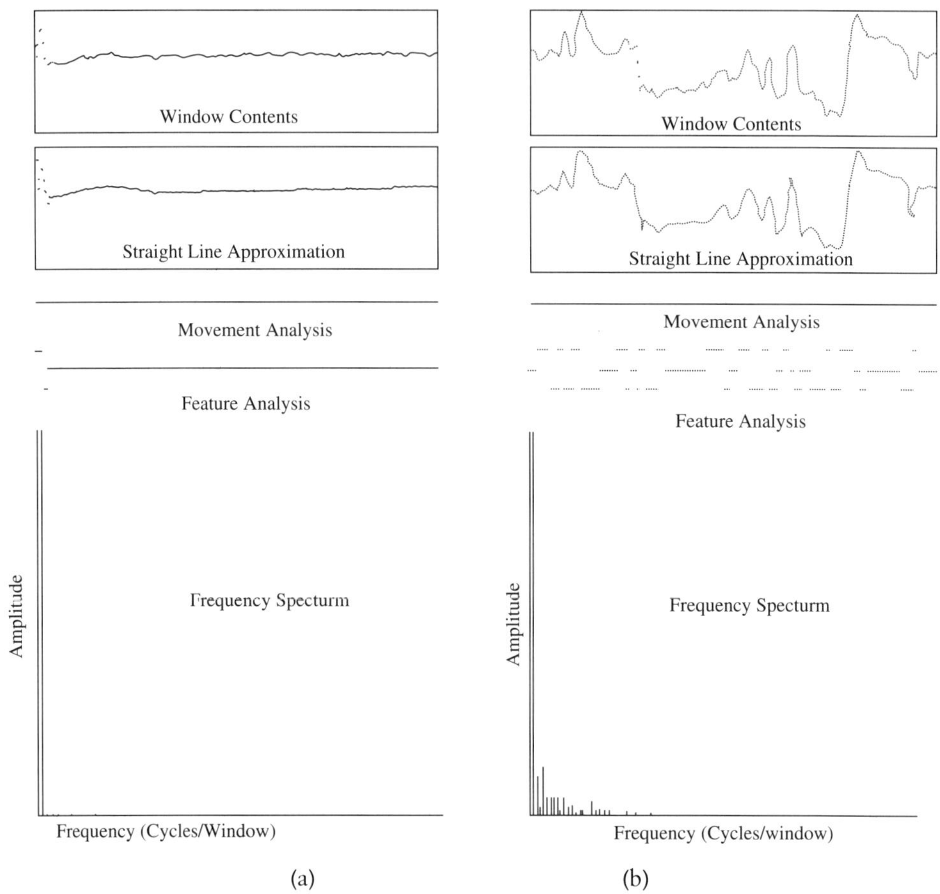

Figure 17.7. (a) Empty window. (b) Analysis of empty window

image is reduced and the signal-to-noise level deteriorates. Noise is characterized by low-power high-frequency signals. A large proportion of the noise is thus filtered out by ignoring high-frequency components of the spectrum. To reduce the effects of noise in the frequency range of interest, a threshold is applied to the amplitude of the frequency components.

Once the presence of vehicles in the window has been determined satisfactorily, the next stage of processing determines the positions of vehicles in the image by reference to features within the window which have been defined as being characteristic of vehicles. The features are characterized by rapidly changing data, which can be identified by differentiation of the waveform. However, the process of differentiation is sensitive to noise and to the size of the features under investigation. Therefore, the waveform is approximated by a series of straight-line segments, with gradient thresholds applied to highlight areas of rapidly changing data. The

(a)

(b)

Figure 17.8. (a) Occupied window. (b) Analysis of window occupied by stationary traffic

application of gradient thresholds provides a scale-invariant means of detecting significant features.

The first approximation to the waveform is a straight line, drawn from the first point of the waveform (p_0) to the last point (p_1). A test is made for the point of maximum deviation of this line from the waveform. If the maximum deviation is greater than a noise threshold, the point of maximum deviation (p_m) is substituted for the end-point (p_1) and a new test is conducted on this new line segment. The line is successively subdivided in this way until either the maximum deviation falls below the noise threshold, when the straight-line segment is considered to be a good approximation to the waveform segment, or the line segment reaches the minimum allowed length. At this point the line is considered to be an acceptable approximation to the waveform segment.

The waveform is, therefore, subdivided into its perceptually significant straight-line segments (Figure 17.7b and 17.8b). The gradient of each line segment is then calculated and compared against a threshold to determine the position of significant features in the window. From this information, queue length and the location of the back of the queue can be determined. If the camera geometry is known, these measurements can be used to provide estimates of the actual extent of the queue on the road.

The amount of movement the vehicles are undergoing is determined by a temporal-differencing operation performed on a pixel by pixel basis. The absolute difference in the grey levels of the pixels in the window are determined at two instances a short time interval apart. If there has been no movement a "zero" result will be obtained, neglecting random noise. Distinct "movement features" will be produced if there are any moving vehicles present. These movement features can be matched to the vehicle features and a flow profile over the window obtained. By setting suitable movement and occupancy thresholds for the window, a queue or congestion alarm can be set. An inter-frame period of approximately 0.2 s is sufficient time for movement to be detected, but short enough to be unaffected by fluctuations in the ambient light level.

The whole algorithm performs reliably within a 3–4-s cycle period on an 8 MHz IBM PC-AT with a maths coprocessor, comfortably within the generally accepted useful limit of 3–5 s identified by Hoose.[2]

17.4.2 Experimental Results

FAST_Q provides a quasi-qualitative description of the state of the road network under surveillance. It is, therefore, somewhat difficult, and perhaps misleading, to assess its performance in purely quantitative terms. None the less, it is useful to compare the system's performance with manual observation. The figures in Table 17.2 show the percentage of time FAST_Q and an observer were in agreement. The results in this table encompass the mean-time-to-detect and the false alarm rate, two important parameters in assessing an alarm system. The mean-time-to-detect has been observed to be approximately one to two cycles (approximately 3.5–7 s) for "incidents" which result in stationary traffic.

The frequency of false alarms must be low for any acceptable system. Arguably the false negative rate (the number of times the system failed to raise an alarm when one was required) should be lower than the false positive rate (the number of alarms raised when none was required). If the system fails to detect an incident it could potentially be fatal, whereas a false alarm call will result only in operator inconvenience. The average values in Table 17.3 suggest that this will be the case.

These results are based on the assumption that the observer always makes the correct assessment of the scene. Closer inspection of the periods where FAST_Q

Table 17.2. Comparison of manual and automatic queue analyses

Trial site	Agreement* (% of survey)	Traffic flow (vehicles/h)	Distance covered (m)
Traffic-lights	99	320	20
Junction	98	400	200
Traffic-lights	95	320	30
Motorway	100	1300	200
Motorway ramp	99	800	30
Average	98	—	—

$$\text{*Agreement} = \frac{T_s - (N * P)}{T_s}$$

where

T_s = total survey period
N = number of periods of disagreement
P = processing cycle time (3.5 s)

Table 17.3. False negatives and false positives

Trial site	False negatives (percentage of survey period)	False positives (percentage of survey period)
Traffic-lights	0.0	0.0
Junction	0.0	1.55
Traffic-lights	1.55	1.55
Motorway	0.0	0.0
Motorway ramp	0.77	0.77
Average	0.6	0.97

differed from the observer revealed two causes for the discrepancies. In the case of the false positives, the system was observed to be triggered by single slow moving vehicles at the distant end of the window from the camera position. This was largely due to perspective distortion effects which can be reduced by the use of multiple movement thresholds for different regions of the window.

The second cause of disagreement between the system and the manual analysis occurred as a result of stop/start queuing. The current implementation of the algorithm requires a cycle time of approximately 3.5 s. Vehicles stationary for less than this time can be missed. In general, this will result in an increase in the mean-time to detect and not a reduction in the overall detection rate. Serious incidents will generally result in traffic becoming stationary for very much greater periods of time than this.

17.5 VEHICLE CLASSIFICATION

The preceding systems do not attempt to recognize vehicles in any way. They merely detect features which the system's designer has defined as being characteristic of vehicle presence. They rely heavily on the a priori assumption that they are viewing a road traffic scene, which is typically of all systems currently available.

The loop emulator system can classify vehicles into one of three groups, small, medium and large. Cars, motorcycles and small commercial vans will be given a small label, large commercial vans and rigid lorries will fall into the medium category and HGVs and buses into the large category. Vehicle width is determined directly from the video picture by counting the number of pixels the vehicle's image occupies across the lane. Length measurements are calculated from lane occupancy and speed measurements. The distortion of vehicle images, caused by the process of projecting a three-dimensional objects on to two-dimensional image plane, results in quite poor classification performance for all but a "square on" view of vehicles.

A system capable of performing automatic vehicle classification based on three-dimensional information is therefore under investigation in TORG. The system has the following basic structure:

(1) a three-dimensional model representation of the classes of vehicles to be recognized, incorporating characteristic features which can be observed in a video image;

(2) a method of image segmentation which will highlight features characteristic of the vehicles to be recognized; and

(3) a model matching strategy for object recognition which is robust in the face of noisy or partial data.

Here the first two of these aspects will be described with examples of the necessary computer processing. Strategies to be employed in model matching will be discussed towards to the end of this chapter.

17.5.1 Model Creation

Difficulties in creating reliable models for object recognition in real world scenes have led a number of researchers to investigate the adaptation of pre-existing CAD models for vision applications. Models, created in three-dimensional CAD databases for manufacturing purposes, are adapted, or augmented, for vision tasks. The term CAD-based computer vision has been coined to describe this approach. In some cases CAD models have been used to synthesize images of objects which are then used in training by example exercises for building object recognition models. A number of other researchers have diverged from this path, considering the CAD models to be sufficient for recognition directly. As a source of "features" the models are used to interrogate images, looking for matching features in the images based on size, shape and the relationship between features as defined by the CAD model instantiated into the scene from the camera viewpoint.

The problem with this approach is that CAD standards, such as the IGES (initial graphics exchange specification) standard, are not compatible with vision requirements. The output must, therefore, be modified to reveal the object features of interest. As a result, attention has turned to the production of a "unified system" in which models can be shared easily between vision and CAD systems via a common database. Here a three-dimensional CAD system has been developed specifically to provide a machine vision database.

17.5.2 Three-dimensional Modelling

Three-dimensional CAD modelling requires the transformation of data from a three-dimensional world (the model world) to a two-dimensional world (the viewscreen). This is done by introducing perspective projections that transform three-dimensional objects on to a two-dimensional projection plane.

Consider Figure 17.9, with the viewpoint at some arbitrary point in the space defined by the world coordinate system. The display screen is perpendicular to the line joining the viewpoint O_e, to the world origin O, and is a distance d from O_e. The eye coordinate system is oriented so that the axis Z_e is directed towards the origin of the world coordinate system. To identify the screen coordinates (X_s, Y_s) associated with the point $P(X, Y, Z)$ the following axes transformations are required

$$T_1 = \begin{bmatrix} 1 & 0 & 0 & 0 \\ 0 & 1 & 0 & 0 \\ 0 & 0 & 1 & 0 \\ -X_v & -Y_v & -Z_v & 1 \end{bmatrix}$$

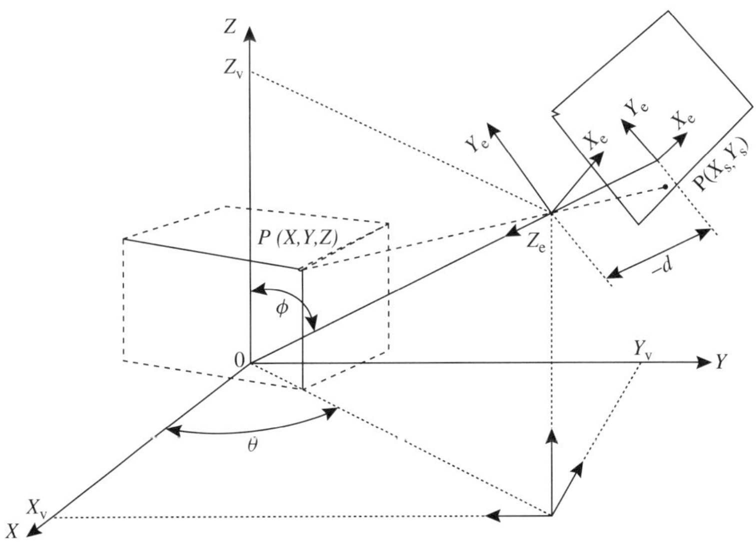

Figure 17.9. Perspective projection through an arbitrary point

$$T_2 = \begin{bmatrix} \cos\alpha & \sin\alpha & 0 & 0 \\ -\sin\alpha & \cos\alpha & 0 & 0 \\ 0 & 0 & 1 & 0 \\ 0 & 0 & 0 & 1 \end{bmatrix}$$

$$T_3 = \begin{bmatrix} 1 & 0 & 0 & 0 \\ 0 & \cos\delta & -\sin\delta & 0 \\ 0 & \sin\delta & \cos\delta & 0 \\ 0 & 0 & 0 & 1 \end{bmatrix}$$

$$T_4 = \begin{bmatrix} -1 & 0 & 0 & 0 \\ 0 & 1 & 0 & 0 \\ 0 & 0 & 1 & 0 \\ 0 & 0 & 0 & 1 \end{bmatrix}$$

where

(X_v, Y_v, Z_v) are the viewpoint coordinates in the world coordinate system

$$\alpha = (90 - \theta)$$

$$\delta = (180 - \phi)$$

To identify the eye coordinates (X_e, Y_e, Z_e) of the point $P(X, Y, Z)$

$$[X_e, Y_e, Z_e, 1] = [X, Y, Z, 1]T$$

where $T = T_1 \times T_2 \times T_3 \times T_4$

Screen coordinates are determined by projecting the eye coordinates on to the viewscreen.

From Figure 17.10 it can be seen by similar triangles that

$$X_s = -d\frac{X_e}{Z_e}$$

$$Y_s = -d\frac{Y_e}{Z_e}$$

17.5.3 Camera Calibration

To match models to objects in a scene, the relationship between the image and the scene must be known. This can be achieved with a camera calibration, matching reference points in a scene to their pixel locations on the screen.

In the calibration, a number of parameters must be determined by the system, or supplied by the operator. These parameters fall into two categories: extrinsic parameters and intrinsic parameters. The extrinsic parameters consist of the location and attitude of the camera in the world coordinate system. These parameters can be supplied by the operator and the matrix transformations used for perspective projection as defined in the previous section.

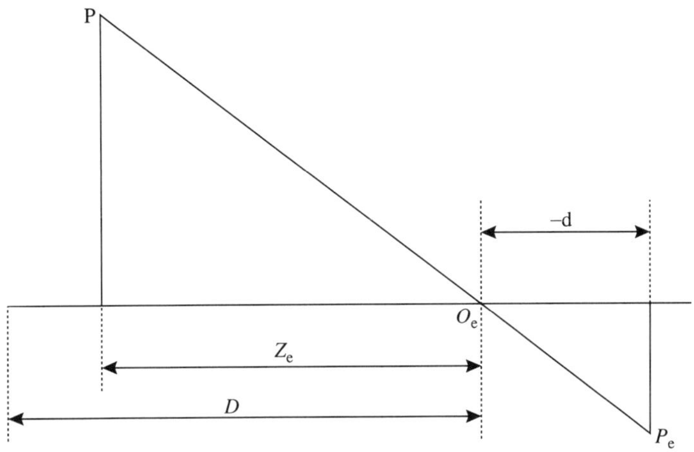

Figure 17.10. Eye coordinate to screen transformation

To determine the image pixel locations of scene reference points, the intrinsic camera parameters must be determined. These parameters account for the differences between the ideal image, as produced using a perfect pin-hole camera and the perspective transformations described above, and the actual image obtained from a practical camera and displayed on a video screen.

The differences between the ideal and practical images occur because of lens distortion, the CCD imaging device matrix dimensions, the number of sensor elements on the imaging device along the x-direction and the number of frame buffer samples in the x-direction. For highly accurate work an additional factor must be taken into account to compensate for timing signal mismatches between the camera and the frame grabber.

Tsai has shown that for off-the-shelf video cameras the distorted image coordinates are given by:

$$X_f = S_x \frac{N_f}{d_x N_c} X_d + C_x$$

$$Y_f = \frac{Y_d}{d_y} + C_y$$

where

(X_d, Y_d) = lens distorted image coordinates

d_x = centre to centre dimension of CCD sensor elements along the x-direction

d_y = centre to centre dimension of CCD sensor elements along the y-direction

N_c = number of sensor elements along the x-direction

N_f = number of frame buffer pixels along the x-direction

S_x = correction factor for line scan timing mismatch between frame buffer and camera. (Similar inaccuracies in field scan timing are much less important because of the much greater time scale involved)

(C_x, C_y) = coordinate of frame buffer centre, required to translate screen origin to centre of frame buffer.

17.5.4 Vehicle Modelling

A variety of methods of creating three-dimensional CAD models are in use today. These include wire frames, constructive solid geometry, surface models, volumetric models and sweep representations. Wire frame models are perhaps the simplest to construct. They consist of the vertices of polyhedra defined in three-dimensional space. Connections between vertices are used to represent the physical edges of an object, while the vertices themselves typically represent corner points. The representation is, however, incomplete as surface information (such as opaqueness or transparency) is not defined.

Constructive solid geometric (CSG) modelling, as the name suggests, includes surface information. Models of objects are constructed using a finite set of predefined primitive shapes. Such a set typically includes cubes, spheres and right cylinders. These are combined using regular Boolean conditions to produce the desired solid object model.

The collection of primitive shapes comprising the model can be manipulated as a single entity, allowing it to be viewed in any orientation. The method is, however, limited in the modelling which it can perform. Difficulties can arise when modelling complex curved surfaces and complex component unions, but the majority of road vehicles do not exhibit these features. They can be represented by regular rectangular structures. There are, however, a number of exceptions to this rule, most notably cars such as the VW beetle (still a common sight on many roads), the Morris Minor and the Citroen 2CV. Most buses are rectangular in shape, as too are HGVs with either large rectangular or cylindrical trailers. The internal features of vehicles also tend to be rectangular in appearance. Windows, windscreens, headlamps (although many are also circular) and bumpers, can all be seen to be basically rectangular.

For these reasons the CSG modelling method can be adopted for the construction of vehicle models. A small number of primitive shapes consisting of various polyhedra, such as rhomboids and rectangular planes, have been used to construct vehicle models. Right cylinders could be added to this set to model wheels. However, wheels are not reliable features for vehicle recognition, due largely to the fact that they are frequently obscured in video pictures when taken from the typical overhead camera angles used for traffic monitoring.

17.5.5 Model Construction

Vehicle models are constructed from a collection of the primitive shapes by specifying the primitive type, giving it some dimensional values and positioning it with respect to the other components of the object. Realistic representation of road vehicles can be produced and instantiated into a roadway scene when the three elements, perspective transformation, camera calibration and model creation, are brought together.

A model of the roadway scene may also be constructed, which can have two important uses. During system initialization it can be used for camera calibration, using the single plane method described by Tsai.[3] During system operation, it can provide a reference frame from which to determine vehicle location and orientation. Some a priori assumptions must be made for this to work, for the determination of the locations of objects in three-dimensional space from a single two-dimensional image is an ill-posed problem, i.e. it does not have a unique solution. The essential assumption to be made therefore is that the object under investigation is situated on the ground plane in the scene.

17.5.6 Image Segmentation

In addition to providing a means of vehicle detection, a segmentation procedure is required which will highlight characteristic features of the various vehicle classes. The features that the chosen procedure best detects in the video image may appear in ranked order in the CAD model database.

Two methods of image segmentation are being evaluated for this purpose, with the initial aim of identifying suitable features for modelling. In operation, the segmentation procedure is required to provide information on the presence, location and orientation of a vehicle in the scene for model matching purposes. The chosen procedure must be robust in the face of scene lighting variations. In an outdoor environment ambient light levels can vary considerably across a scene, and can also change drastically within a time scale of a few seconds.

Edge enhancement processes, such as the Sobel operator, can cope with these situations. They are unaffected by slow variations in lighting, both spatially across the image and temporally from frame to frame. They are, in general, differential operators which highlight regions of an image in which the grey scale changes abruptly. This occurs at the physical edges of objects, but can also occur as a result of reflections off interior body parts or cosmetic changes in the colour of an object. Edges highlighted by the application of an edge filter, therefore, do not relate exclusively to the structure of the object. These edges can provide clues from which the location and orientation of vehicles in the scene can be determined with respect to some reference, such as the road direction. To achieve this, the reasonable assumption that the vehicle is situated on the road surface, and is not floating somewhere in mid-air, must be made.

The edge enhancement techniques have the disadvantage of removing a great deal of the interior of an object, and generally the edges themselves do not form closed areas. To preserve the internal detail of the objects and to produce closed areas encompassing them, a multivalued thresholding operation has been developed. The procedure is applied in two stages. In the first pass, the image is segmented using a small number of thresholds to isolate vehicles and shadows from the roadway. An edge following routine is then applied to identify the locations of large image artefacts. In the current implementation, the largest artefact alone is selected for processing in the second pass, on the assumption that this is caused by the presence of a vehicle.

In the second pass, between four and 12 threshold values are allowed. The vehicle's image is then "synthesized" using average grey levels for the separate threshold regions (Figures 17.11 and 17.12). This method enables internal features of a vehicle which have approximately the same grey scale as the roadway to be retained in the final processed image. Region merging for feature reconstruction and identification is possible under the supervision of the feature model with this resultant image. Shadows, included in the segmentation, can be identified with the aid of the model.

17.5.7 Model-image Matching for Vehicle Classification

To classify a vehicle in a scene, the image of the vehicle must be matched to a model in the CAD database. Given the location and orientation of the vehicle in the scene, an hypothesis and test strategy, as adopted by Brisdon *et al.*[4] and others, can be used, in which features from the candidate model are instantiated into the scene at the vehicle's position. Given a fully calibrated scene to image transformation, the model features will exhibit all the distortion that the viewing process and the attitude of the vehicle with respect to the camera have caused to the vehicle features. The

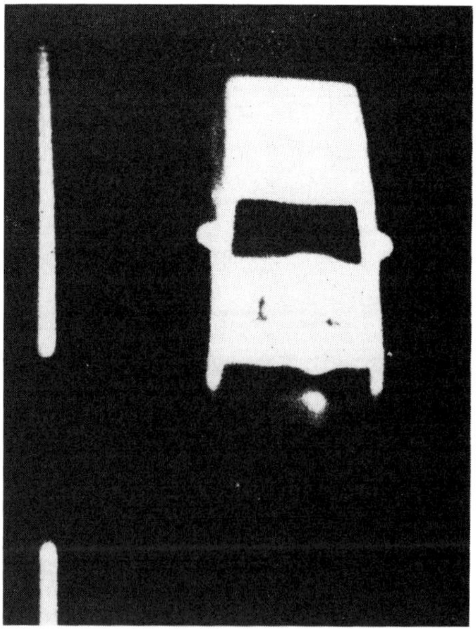

Figure 17.11. Object vehicle

Figure 17.12. Vehicle image processed with multivalued threshold procedure

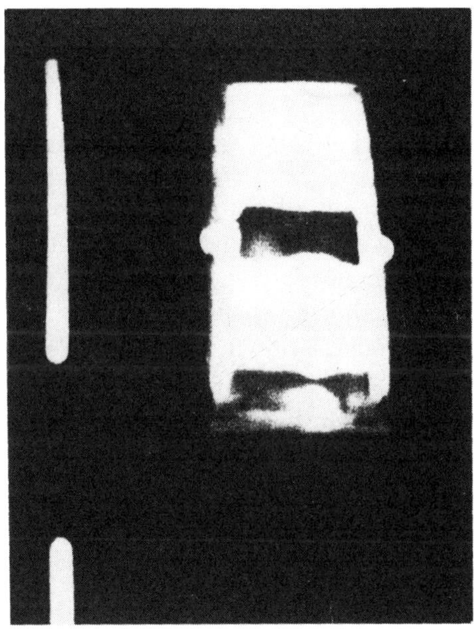

Figure 17.13. Estate car model instantiation

model features may thus be considered as templates for the features of the vehicle for recognition purposes.

Having hypothesized that the vehicle detected in the image is of a particular class, features of that class can be instantiated into the scene and their appearance appraised, in terms of shape — perhaps using a shape descriptor such as the Hough

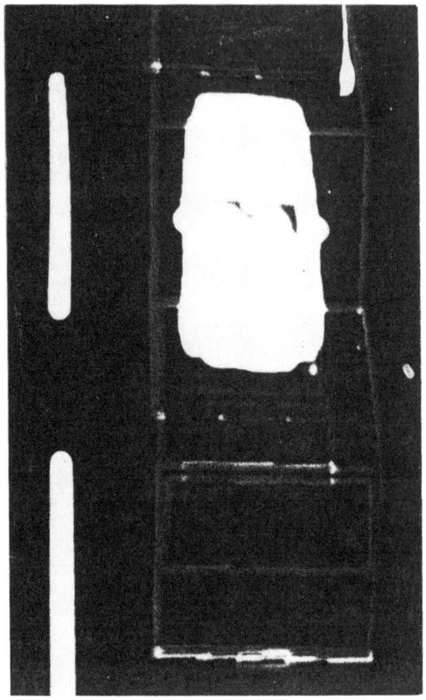

Figure 17.14. HGV model instantiation

transform[5] — and size. The testing of the image for a similarly looking feature in the region of the image predicted by the instantiation enables the hypothesis to be confirmed or contradicted. Figure 17.13 shows features from the model of an estate car projected into a scene including a vehicle of that class. A collection of consistent feature matches, with the correct relational geometry as defined by the model instantiation, will confirm the classification of the vehicle. Models of vehicles from different classes, such as the HGV shown in Figure 17.14, would produce poor matches, if any at all, and thereby contradict the hypothesis that the vehicle is an HGV.

17.6 CONCLUSIONS

Research so far undertaken into transport applications of image processing can now be assessed and lessons drawn about the appropriateness of the technology. Rather than being a substitute for existing traffic data collection devices we now perceive it as complementary, suitable for applications where conventional detectors are not appropriate. Temporary traffic surveys, for example, may employ loop emulator systems when it is difficult or costly to install devices on or in carriageways. A video taken from a convenient vantage point, such as a bridge or nearby building, can be analysed automatically in the laboratory using a system such as TULIP to obtain the information required by traffic engineers.

Although the data provided by TULIP is of comparable standard with that provided by inductive loops, operational problems are still encountered due to changes in ambient light levels. The operation of these system for long periods of time and under all weather and lighting conditions may not be sufficient to meet the needs of traffic monitoring. More robust methods of image segmentation are required, but these techniques are more computationally expensive and so require more powerful microprocessors to enable real-time performance to be attained. Even if the problems due to ambient lighting are solved satisfactorily the problem of vehicle occlusion, resulting in counting errors will still remain unless a vertical camera angle is used. This will restrict the use of the camera to one application, which is not favoured by the industry.

The FAST_Q system is less sensitive to absolute light levels, and makes no attempt to count individual vehicles. It is, therefore, a much more suitable technique for road traffic monitoring than the method used in the TULIP system. The FAST_Q method of queue detection and congestion monitoring also has distinct advantages over more conventional methods. Data are collected from a region of the network rather than a point and the processing is performed within a few seconds. Incident detection within a time scale comparable with the period with which congestion will occur on a busy road after an incident has occurred can, therefore, be performed. From Figure 17.8(a) it can also be seen that a measure of the extent of the congestion can be obtained, along with the location of the rear of the queue.

The area over which monitoring can be performed can be varied from a few metres to a few hundred metres by correct focusing of the camera and/or by defining an appropriate window size. It is therefore possible to tailor the system to suit a particular application, whether that be to monitor a small area, such as a railway level crossing, or large areas such as sections of motorway.

The power of a digital signal processing (DSP) engine can be harnessed to enable this algorithm to execute within a few milliseconds. At this rate, a single system should be capable of handling data input from up to 50 cameras, each covering four lanes of traffic. At a maximum range of 200 m per camera, a single system could monitor up to 10 km of a four-lane motorway.

Many of the difficulties which arise in acquiring conventional traffic data using image processing systems are due to image distortion caused by the imaging process. To overcome problems such as vehicle occlusion, account must be taken of the three-dimensional nature of the scenes being viewed. The problem then arises of how to describe scenes and objects in three dimensions. The approach adopted at TORG has been based on a three-dimensional CAD system for the construction of vehicle models using primitive shapes and constructive solid geometry. Three-dimensional models representing different vehicle classes can be constructed in a few minutes, containing characteristic features of the vehicles to be identified in a processed video image.

Using an hypothesis and test strategy, instantiation of models into a scene will allow automatic identification of visible and partially visible features. Using a user defined ranking system an image can be tested for the most reliable features. The region of the scene in which to test for the features will be defined by the model instantiation.

Perspective distortion and camera to viewscreen transformation will allow the features to be instantiated in a realistic fashion. The instantiated features can therefore be regarded as templates for matching purposes.

REFERENCES

1 Rourke A, Bell MGH. *Traffic analysis using low cost image processing — a real-time system for temporary traffic censuses*. TORG Research Report No. 77, University of Newcastle upon Tyne, 1989.
2 Hoose N. *Computer Image-Processing In Traffic Engineering*. Taunton, Somerset: Research Studies Press, 1990.
3 Tsai RY. A versatile camera calibration technique for high accuracy 3D machine vision metrology using off-the-shelf TV cameras and lenses. *IEEE Journal of Robotics and Automation*, 1987; **RA-3**(4): •••.
4 Brisdon K, Sullivan GD, Baker KD. Feature aggregation in iconic model evaluation. *Proceedings of 4th Alvey Vision Conference, University of Manchester*, August–September, 1988.
5 Gonzalez RC, Wintz P. *Digital Image Processing*. Wokingham: Addison-Wesley, 1987.

18

Automatic Vehicle Model Classification

18.1 INTRODUCTION

An edge-based classification (EBC) technique which is based on the principle of dynamic programming is used in this application to classify the type of car within a cued region. The cued regions are generated by a two-dimensional circle Hough transform algorithm which is used to detect line features such as car wheels and wheel arches in cluttered open-world scenes.[1] Detection of the wheels and wheel arches is achieved by searching for arcs and/or circles and is therefore intended to detect primarily the side profile of cars. This technique is used to focus a more detailed and robust object classification algorithm onto a set of the complete image.

The EBC[2] is then applied to identify various components from different cars, for example, the car roof, windows and bonnet, etc. of various car models (e.g. Citroen, Lada, Mini, etc.) within the cued region. The EBC technique is based on the one pass dynamic programming algorithm of Bridle and Chamberlain.[3] However, here it is reference data driven in that the search is carried out by looking for observed data segments to match the whole of the reference.[2] The system is designed to find a sequence of observed segments that could, when associated together, correspond to a complete reference segment.

18.2 REGION CUEING ALGORITHM

The region cueing algorithm is composed of four integrated processes: edge extraction, line type fractal discriminant, circle/arc finding Hough transformation and wheel-base detection.

Advanced Vehicles and Infrastructure Systems, Edited by C.O. Nwagboso

18.2.1 Edge Extraction

A line-edge description of the scene is first extracted using an operator based upon the Sobel edge convolution.[4] The horizontal (H) and vertical (V) Sobel convolutions give both magnitude E and direction or orientation Θ of edge gradient.

$$E = |H| + |V| \quad \text{and} \quad \Theta = \arctan\left(\frac{H}{V}\right) \tag{18.1}$$

The gradient orientation measure is required for subsequent processing both within the line extraction algorithm and for the circular Hough transformation algorithm. The Sobel was chosen because among the many simple convolution mask edge operators it has been judged to be most accurate with regard to circularity.[5]

The gradient information enables a non-maximum gradient suppression operation to thin the edge image to a single pixel wide edge map, as follows:

$$\text{if } E(x) < E(x_1) \text{ or } E(x_2) \text{ then } E(x) = 0 \text{ else } E(x) = E(x) \tag{18.2}$$

where $E(x)$ signifies the xth edge pixel, and pixels x_1 and x_2 are the two neighbours of x along the gradient orientation of pixel x. This is then followed by a hysteresis threshold similar to that employed by Canny[6] to produce a skeletal edge map. In principle this involves thresholding the edge strength at an upper threshold level and relaxing through all connected components (via an eight-connected neighbourhood) to a lower threshold. This relaxation process works as follows. Upper and lower thresholds are defined, if a pixel is above the upper threshold, i.e. a strong edge is detected, all connected pixels above the lower threshold are also accepted as edges. If a single lower threshold was to be applied many disconnected weak edge lines would be accepted, conversely if a single higher threshold was to be applied only a few short edge lines would be accepted (see Figure 18.1); however, the hysteresis approach rejects the weak lines while allowing stronger edge lines to grow and become more continuous.

18.2.2 Line Type Fractal Discriminant

In semi-urban scenes it is often found that the edge map contains an excessive line detail in regions of busy texture, such as produced by foliage and other vegetation. An operator has been developed to discriminate between man-made and natural edge lines in an image. This discrimination is based on the assumption that natural features produce wriggly edges, while man-made features are likely to be smoother and straighter. A discrimination between man-made and natural edge features can therefore be made using a simple measure of the fractal dimension of line segments.[7] The algorithm simply determines the total change in line orientation within each eight-connected pixel of a line segment. This is normalized to give a line "wrigglyness" factor, which is used as a measure of the fractal dimension of the line segment and is thresholded to remove all wriggly lines.[1]

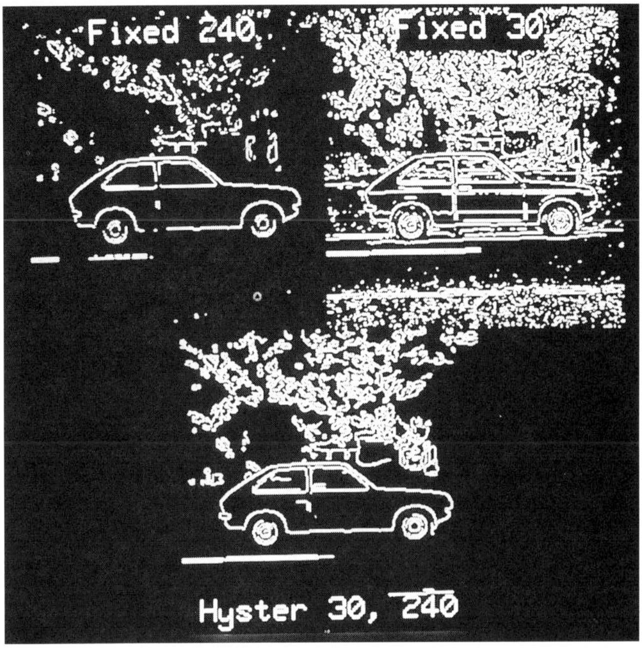

Figure 18.1. Hysteresis thresholding: (a) single fixed threshold too high (240), (b) single fixed threshold too low (30), (c) hysteresis thresholds between 30 and 240

18.2.3 Circle/Arc Finding Hough Transform

Wheel-like features are found from the retained line edge image using a version of the circle Hough transform. The normal circle Hough transform maps to a three-dimensional parameter space,[8–10] but with knowledge of edge direction a two-dimensional transform can be formed by projecting normals to edge directions.[1] The resulting two-dimensional Hough transform image will have peaks at the centres of all arcs and circles. Once the transform space has been accumulated it is necessary to find peaks which may be considered as potential wheel centres.

In this technique,[1] all concentric arcs and circles will contribute to the same peak, allowing in this case wheel arches and wheel arc segments to contribute to a peak at the nominal wheel centre. Partial evidence for the existence of a feature is therefore accumulated into global evidence for that feature.

18.2.4 Wheel-base Detection

The wheel-base detector searches for wheels and wheel arches, which are then combined to find possible wheel bases. The tangent angle to the horizontal of all candidate wheel centre lines is determined; a wheel base is assumed to be found if the tangent is less than 0.2, i.e. the cars are assumed to be roughly horizontal. A rectangular region is cued around the wheels' centre line to represent a parametrically defined car's dimensions. The car model can then be classified by applying the more computationally intensive dynamic programming algorithm within this cued region.

18.2.5 Pre-processing

Before proceeding to the model classification process, some pre-processings might be necessary in order to improve the efficiency and accuracy of the classification. The tangent angle to the horizontal of all candidate wheel centre lines was determined in the last process; if it is not horizontal but less than the predefined threshold value then the tangent angle is used to transform the putative car within the cued region to the horizontal level. Furthermore, the distance between the two wheel centres is used to measure the vehicle dimension so that appropriate magnification or reduction can be performed. Hermart, Affine and Projective transformations are used to achieve the desired rotation, zooming and correction of the object (i.e. car) in question.[11]

18.3 MODEL CLASSIFICATION

The EBC algorithm is an object identification approach to feature extraction that uses dynamic programming.[12] It is used here to classify a car model, e.g. Citroen, Mini, van etc., embedded in the large number of edges which generally result from edge extraction applied to a complex natural scene.

18.3.1 Data Representation

In this system reference templates are created consisting of a set of different car models. Each model is made up of a sequence of segments corresponding to the roof top, front and rear side windows, etc. of the car.

The observation data are selected as follows. Each cued region from the wheelbase detector provides an "edge area" in which to carry out the EBC. The edge segments in this area are then decomposed into a list format. Each segment is made up of a list of connected pixel data, each element of which has associated with it information about its horizontal and vertical positions and its orientation. Each segment is also recorded in reverse order so as to allow matching to be performed in either direction.

18.3.2 Dynamic Programming

Dynamic programming[12,17] is a graph searching technique, it has been used extensively in operational research, communications and speech recognition. Its power derives from its high stability, guaranteed convergence to a solution and its optimality based on both local and global aspects of some criterion.

Dynamic programming can be applied in curve matching. Gorman *et al.*[13] used dynamic programming to match a subset of an image contour (which is assumed to be part of a single object) to a subset of an object contour, where noise was modelled as the addition or deletion of contour segments.

Ansari and Delp[14] developed a different version of dynamic programming called hopping dynamic programming. It is used to find matching sub-curves of an input image and object's contours.

Maître and Wu[15,16,18] used dynamic programming to locate coastlines in satellite images where the coastline could be partially occluded by clouds. Their algorithm matches occluded objects to empty features, but imposes the constraint that the features immediately before and after a sequence of empty features must be subjected to the same rigid transformation; thus enforcing a consistency across the empty features. In other words the feature preceding the empty features is put into abeyance and reactivated when the object comes into view again. The complexity of their approach increases proportionally to the number of features which can be put into abeyance.

18.3.3 Edge Based Classification

In this application the EBC is based on the one-pass dynamic programming technique,[3] which is used here to find any sequence of observed segments among a large number of edges extracted from the whole image that could, when associated together, correspond to a complete reference segment. The technique is reference data driven in that the search is carried out by looking for observed segments to match the whole of the reference. Long unbroken segments are used for the reference templates, whereas the segments extracted from the natural images can be short and discontinuous.

Figure 18.2 shows the basic principle of the EBC for one reference segment matched against one input line segment.

Each element of a reference shape is matched against each element of the observed segments and a local matching cost $c(i, j)$ is evaluated.

$$c(i, j) = (\Theta_i - \Theta_j)^2 \tag{18.3}$$

where $c(i, j)$ is the local cost metric between the ith element on the reference segment and the jth element within an observed segment; Θ_i, Θ_j are the orientations of the corresponding edge pixels. This local cost is then added to the accumulated cost $C(i - 1, j)$ of the best previous sequence of matches up to the point i. In order to find the cheapest accumulated cost up to the current point the accumulated cost matrix is evaluated for each observed segment:

$$C(i, j) = \min_{pn}[P_n^{cost} + C(i - 1, j - P_n)] + c(i, j) \tag{18.4}$$

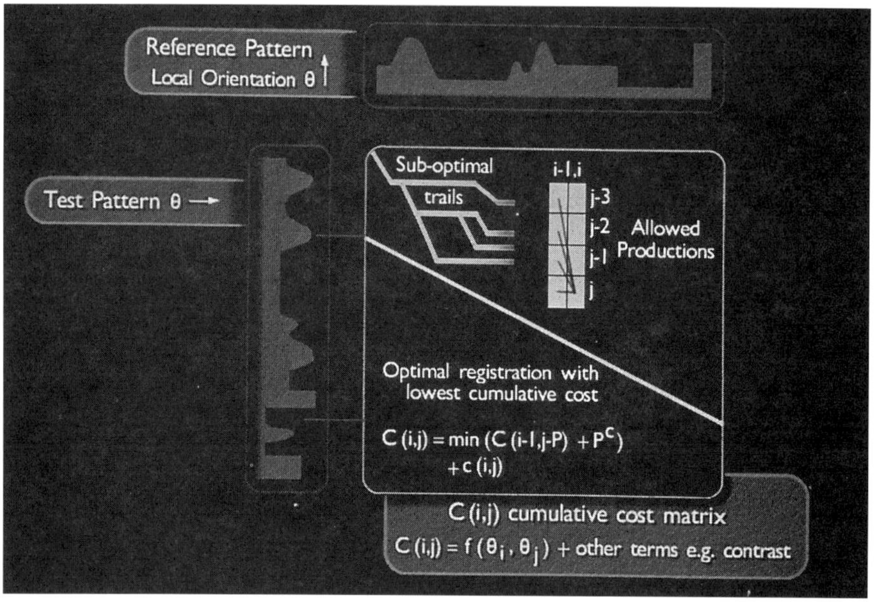

Figure 18.2. Edge-based classification algorithm

P_n is a set of transition rules and costs of the permitted transitions within the observed segment. Figure 18.2 illustrates the allowed transitions P in the j direction, P_n^{cost} is the cost associated with the nth transition. This set of transition or production rules and the corresponding production costs are defined for each edge pixel on the observed segment. The cost matrix is set up in a way that encourages diagonal transitions by assigning lower costs to transition in the diagonal directions. Thus, the cost, P_0^{cost} of a horizontal transition, $P_0 = 0$, is set to 1 and the cost of the horizontal and diagonal transitions: $P_1 = -1, P_2 = -2, P_3 = -3$ are $P_0^{\text{cost}} = 0, P_2^{\text{cost}} = 1, P_3^{\text{cost}} = 2$, respectively. These are used to determine the permitted previous edge pixels as well as to ensure a steady progress along the observed segment.

At the start of each observed segment a "linkage" matrix of allowed connectivities is searched to find the cheapest allowed connecting previous segment (corresponding to the previous reference edge pixel, i.e. $i - 1$). This matrix is based on Euclidean distance between observed segment ends, only segments within an empirically set threshold distance can be connected together. The matrix is used to reduce the search space in a way analogous to the use of a grammar. The accumulated cost at a transition between segments is thus:

$$C(i, 1) = \min_{l}[D_{m,l} + C_l(i - 1, \text{ end})] + c(i, j) \tag{18.5}$$

where $D_{m,l}$ is the Euclidean distance between the start of segment "m" and the end of segment "l".

An additional feature of this technique is that it is possible to identify explicitly edge pixels corresponding to the identified vehicle (i.e. extract the car edges). This can be achieved if during the forward calculation, equations (18.4) and (18.5), the chosen transitions are recorded in a decision matrix. When the end of the reference shape is reached the lowest accumulated cost is selected and used as the starting point for a "traceback" through the decision matrix, giving the optimal sequence of pixels that best matches the reference; these will be the edge pixels corresponding to the vehicle.

18.3.4 Modification of the Edge-based Classification

Although the vehicle model classification technique outlined above can been used, it requires a large amount of memory and CPU time. In order to reduce or minimize both without sacrificing performance several modifications have been made.

In order to reduce the storage requirements a list of permitted connecting segment identifications is generated for each segment to replace the full linkage matrix. This is based on the thresholded Euclidean distance and is used to constrain the dynamic programming search.

The dimension of the full cost matrix of $C(i, j)$ can be reduced because, when evaluating the cumulative cost $C(y, x)$ up to pixel index "y" in the reference shape and pixel "x" in the observed segment, it is only necessary to access $C(y-1, z)$, i.e. the "$y-1$" column, Figure 18.3. Therefore, only the previous column need be stored, thus the full matrix can be replaced by two columns containing the previous and current accumulated costs. The role of the two columns is swapped each time the y index is incremented.

The size of the decision matrix (the back pointer array) can be reduced by replacing the full array (which has for every reference edge pixel an entry for every observed edge pixel) with a pair of segment connection arrays which implement a linked list. The first of these arrays provides information about the identification of the previous connecting segment (the "*previous-segment*" array); while the second supplies information about which pixel index on the reference shape the start of the segment was matched against ("*entry-point*" array). If the total number of segments is S and the length of the reference template is R then only two S by R arrays are required rather than the full R by the total number of edge pixel matrix.

In detail:

$$\textit{entry-point}\ [x, r] = z \tag{18.6}$$

"*entry-point*" stores the start position z in the reference of observed segment x which finished at reference position r.

$$\textit{previous-segment}\ [x, z] = m \tag{18.7}$$

while "*previous-segment*" records the best segment from which segment x was entered at the position z in the reference. Therefore, whenever the start of a segment is encountered information as to the identity of its cheapest connecting segment, i.e.

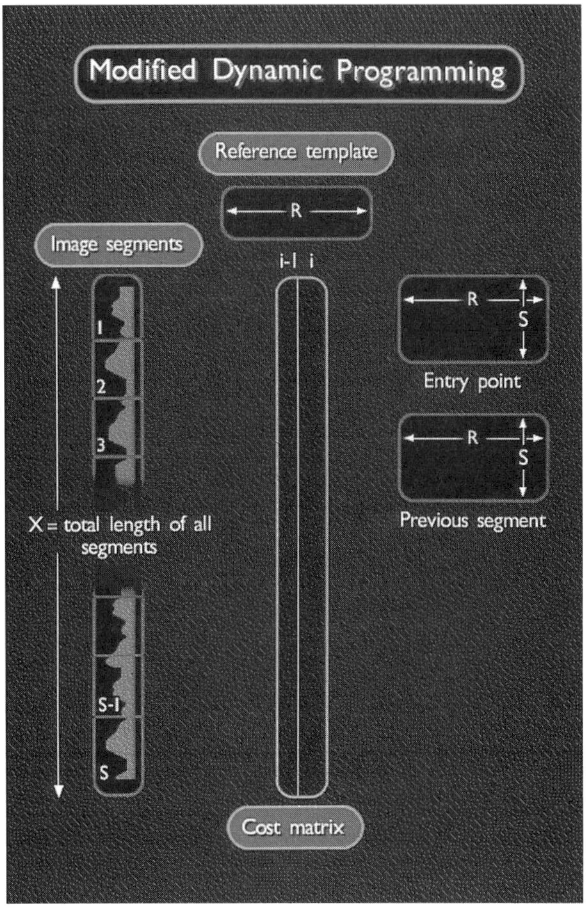

Figure 18.3. Modification of the edge-based classification algorithm

the *previous-segment*, and its corresponding position, *entry-point*, in the reference shape is recorded. Hence, the full back linkage array can be replaced by a pair of far smaller segment linkage arrays. At the end of the forward pass the traceback scans the terminating accumulated cost to find the segment with the lowest cost. Once identified the process will simply trace through the *entry-point* array, and *previous-segment* array alternatively, each providing an index for the other, until the starting location of the reference shape is reached or no further connecting segments can be found. This will provide information on the starting and ending indexes of all recognized segments. Detail within the segments is lost, thus the observed segments are identified on an all or nothing basis. The detail of the shape within each segment on the best path is not necessary, only the segment identification is of importance.

The EBC approach will always find a matching sequence of segments irrespective of whether the searched for car model is in the image. In order to ensure that only reasonable matches occur, an empirical threshold value is also set, the search is abandoned if the accumulated cost exceeds this threshold. Also, in order to speed up the program a pruning cost threshold can be set on the accumulated cost so as to kill bad paths at the earliest possible instance.

In order to eliminate ambiguities, the segments in an image are made unique by ensuring no cross-junction. This means that all the points on a segment are connected sequentially, cf. a cross-shaped segment.

The EBC algorithm is used to search for the reference model in the cued region. Each reference model, i.e. each type of car, is sought independently. The car is then classified as being of the type of the reference model with the lowest accumulated cost.

The question with this technique is whether the accumulated cost can be used as the basis for classification of car type. Experience with the technique applied to speech recognition,[3] suggests that the cost can be used for classification. Some initial highly constrained experiments have been run to test this approach. These are described below.

18.4 EXPERIMENTAL SET-UP

Initial experiments have been run with a fairly small data set; separate training and test data were used. Training was carried out using three examples of each car. These examples were images of the car from slightly varying conditions, i.e. different settings, angles, and distances. Composite templates were generated for each segment (i.e. roof top, windows, etc.) in each car using an averaging scheme. Incomplete segments were touched up by hand, i.e. any gaps in the real edge data were joined using an image editor.

The test data were similar in nature to the training data, i.e. the same car models but in different settings and at slightly different angles and distances; there were up to six versions of each of four cars.

18.5 CONCLUSIONS

This car model classification method was applied to approximately 20 versions of the cars with success. The classification was correct in all except a couple of cases in which there were close scores between very similar sets of extracted edges.

Some experimental results of the cue-generation technique approach are shown in Figures 18.4–18.8. Figure 18.4 shows the various stages of the edge extractor for a well resolved image of a car, 18.4(b) shows the edges extracted from 18.4(a), 18.4(c) shows the result of applying the fractal line discriminant to Figure 18.4(b). The background structures produce typically wriggly and contorted natural edges. The fractal line discriminant can been seen to be successful in removing natural

Figure 18.4. Line edge extraction: (a) original image, (b) line edge output, (c) fractal line discriminant output ($0 \leq R \leq 0.5$)

(a) (b)

Figure 18.5. The circle Hough transform and extracted wheels: (a) circle Hough transform (≤ 20), (b) highlighted wheel centres and cued car region

textured edges from the scene, retaining only the majority of the edges associated with the car.

Figure 18.5(a) shows the circle Hough transform derived from the lines of Figure 18.4(c), with white representing high accumulator space. It exhibits two very strong compact peaks which are extracted as the first two peaks, both of which meet the threshold sharpness criterion for a peak and subtend an almost horizontal line.

Figure 18.6. Region cueing in a more cluttered image: (a) top left — original image with edges superimposed, (b) top right — circle Hough transform, (c) bottom right — extracted peaks in Hough transform highlighted, (d) bottom left — marked wheel centres and cued car region

They are therefore extracted and identified as a putative wheel base; the resulting cued car region is outlined in Figure 18.5(b).

Figure 18.6 shows performance on a more difficult image, 18.6(a) shows the line edge input for the Hough transform superimposed on the original image, 18.6(b) shows the circle Hough transform space, 18.6(c) shows the extracted peak regions from the Hough space, 18.6(d) shows the highlighted wheel centres and cued car window.

Figure 18.7 shows performance on another cluttered and difficult image; 18.7(a) shows the line edge input for the Hough transform superimposed on the original image, 18.6(b) shows the circle Hough transform space, 18.7(c) shows the extracted seven peak regions from the Hough space. The first, third, fourth and fifth extracted regions arise from edges associated with windows in the buildings. These being rectangular rather circular features produced blurred peaks in the transform space and were excluded on the basis of not being sharp enough. Peak region 6 is derived from the edges of the bush directly above the car windscreen. These edges, although derived from foliage, were too small to be rejected by the fractal discriminant and by coincidence produced a sharp peak in the Hough transform space. However, peaks 6 and 2 subtend too steep a tangent angle to the horizontal to be selected as the wheel-base centre line. The seventh extracted peak is accepted both as sharp and

Figure 18.7. Region cueing in a more cluttered image: (a) top left — original image with edges superimposed, (b) top right — circle Hough transform, (c) bottom right — extracted labelled peaks in Hough transform highlighted, (d) bottom left — marked wheel centres and cued car region

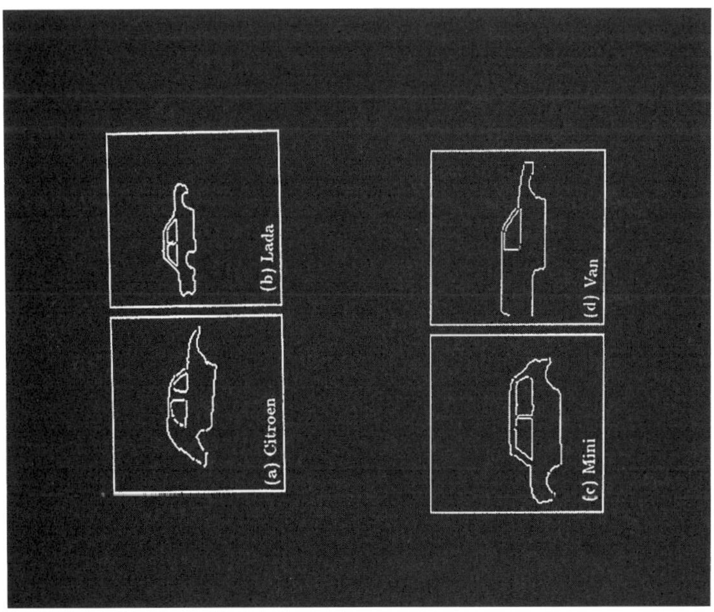

Figure 18.8. Reference models

pairs almost horizontally with the second peak to form the car wheel-base centre line as highlighted in Figure 18.7(d).

Figure 18.8 shows a set of four reference car models; each model is made up of line segments which are searched for independently within each cued region.

Figure 18.9 shows a test image, 18.9(a) shows a cued car region located appropriately, 18.8(b) shows the extracted edges for matching. Figure 18.9(c) shows the reference model with the lowest accumulated cost, the correct model for the input image (i.e. a van). Table 18.1 shows the accumulated cost of each model's components.

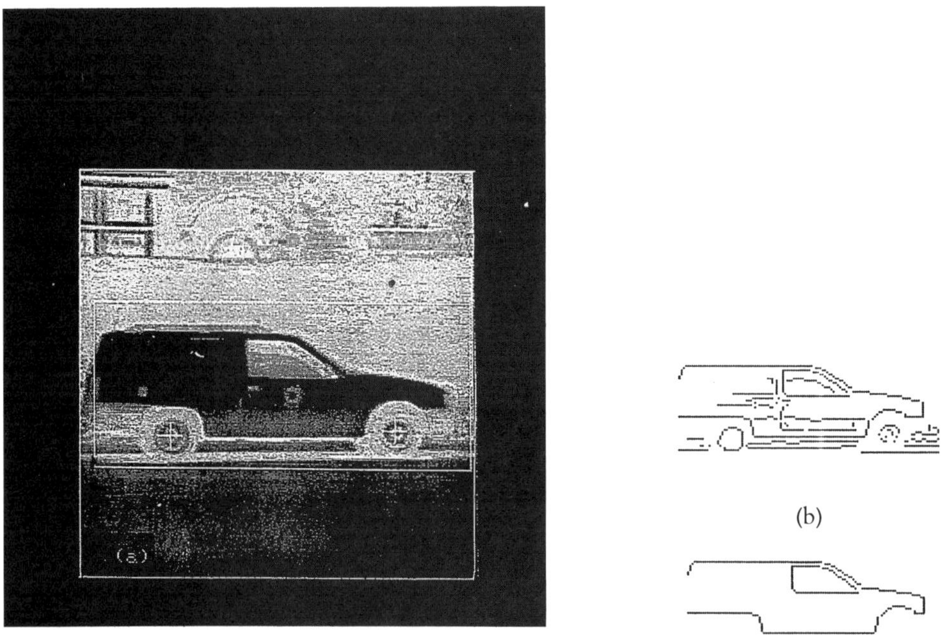

(a) (b) (c)

Figure 18.9. Model identification: (a) cued input image, (b) extracted edges within cued region, (c) identified model

Table 18.1.

REF.	RW	LW	RT	WA	Total	Average
Citroen	1408	6179	1596	2397	11580	2895
Lada	3174	4019	1353	1361	9907	2476
Mini	5285	1978	2520	1344	11127	2781
Van	170	—	80	265	515	128

RW = Right Window; LW = Left Window; RT = Roof Top and Bonnet; WA = Wheel Arch

REFERENCES

1 Radford CJ. Vehicle detection in open-world scenes using a Hough transform technique. *IEE Proceedings of the 3rd International Conference on Image Processing and Applications*, Warwick, UK, 1989; 78–82.

2 Varga MJ, Sleigh AC. The application of dynamic programming to object identification. *Proceedings of the 5th International Conference on Image and Analysis and Processing*, Italy, 1989; 365–372.

3 Bridle JS, Brown MD, Chamberlain M. An algorithm for connected word recognition. *IEEE International Conference on Acoustic, Speech and Signal Processing* 1982; 899–902.

4 Ballard H, Brown CM. *Computer Vision*. Prentice-Hall, Inc. Englewood Cliffs, NJ.

5 Davies ER. Circularity — a new principle underlying the design of accurate edge orientation operators. *Image and Vision Computing* 1984; **2**: 134–142.

6 Canny J. A Computational approach to edge detections. *IEEE Transactions Pattern Analysis and Machine Intelligence* 1986; **8** 679–696.

7 Pentland AP. *From Pixels to Predicates*. NewJersey: Ablex Publishing Corporation.

8 Hough PVC. *Method and Means for Recognizing Complex Patterns*. US Patent 3 069 654.

9 Kimme C, Ballard D, Sklansky J. Finding circles by an array of accumulators. *Communication of Association Computer Machinery*. 1975; **March 18**: 120.

10 Illingworth J, Kittler J. The adaptive Hough transform. *IEEE Transactions on Pattern Analysis and Machine Intelligence* 1987; **9**: 690–698.

11 Tang YY, Suen CY. Non-linear shape restoration by transformation models. *Proceedings of the 10th International Conference on Pattern Recognition*, USA, 15–19 June, 1990.

12 Bellman R. *Dynamic Programming*. Princeton, NJ: Princeton University, 1957.

13 Gorman JW, Mitchell OR, Kuhl FP. Partial shape recognition using dynamic programming. *IEEE Transactions of Pattern Analysis and Machine Intelligence* 1988; **10**(2): 257–266.

14 Ansari N, Delp EJ. partial shape recognition: a landmark-based approach. *IEEE Transactions of Pattern Analysis and Machine Intelligence* 1990, **12**(5): 470–483.

15 Maître H, Wu YH. On improving Dp to solve image registration, *Pattern Recognition* 1987; **20**(4) 443–462.

16 Maître H, Wu YF. The use of DP for image/map registration. *IAPR TC-7 Workshop on Analytical Methods in Remote Sensing for Geographical Information Systems, 5–58, Paris,* 1986.

17 Ney H. DP as a technique for pattern recognition. *Proceedings of the IEEE 6th International Conference on Pattern Recognition*, 1119–1125, Munchen, 1982.

18 Wu YH, Maître H. Registration of rotated pictures with hidden parts using DP. *Proceedings of the IEEE 8th International Conference on Pattern Recognition*, Paris 1986; 792–794.

19

Model Based Collaboration Architectures for Traffic Control

19.1 INTRODUCTION

The interest in applying methods from the field of artificial intelligence (AI) to various transportation problems has become an important trend over the last few years. Systems have been developed which address various problems including, for instance, infrastructure design, pavement management and rehabilitation, public transport planning and operation, and traffic signal design,[1,2] (see, references 3 and 4 for information on a multitude of AI approaches to transportation problems). Road traffic control and management, especially in urban areas, is no exception to this interest.[5–7] Recurrent congestion, increasing pollution and reduced safety, are experienced every day in most European cities. These problems require sophisticated information technology (IT) in order to extend the performance and the range of actions of conventional traffic control technologies. Contributions from the fields of advanced IT are expected along two different but interacting directions: (i) extending the capabilities of conventional, presently used software techniques in basic traffic control tasks, and (ii) supporting traffic control operators (i.e. the staff in traffic control centres) in managing the growing complexity of automated and integrated control facilities. Given the number of demanding requirements implied by the problem (i.e. large number of input/output interactions, real-time constraints, uncertainty and time-dependency of events, etc.), the traffic control domain is certainly one of the most challenging for new technologies among the various traffic engineering fields. However, it is also the sector where the benefits and the pay-offs of applying new technologies are potentially among the highest.

Advanced Vehicles and Infrastructure Systems, Edited by C.O. Nwagboso

This chapter sets out to report on generic results from several approaches followed in the DRIVE (*D*edicated *R*oad *I*nfrastructure for *V*ehicle safety in *E*urope) projects V1055 (1989–1991), and v2039 (1992–94), aimed at developing AI techniques to support traffic control operations.[8,9] First, some methodological preliminaries are described. This will lay the setting for subsequent discussions and inform the reader about what view of reality the authors have had when writing. Thereafter, in the section "The problem: a brief analysis", the traffic control problem is given a brief analysis, and a decomposition into subproblems is performed. In "Solution elements: the models", elements in a solution of the traffic control problem are described, giving problem-solving counterparts to the subproblems described in the previous section. Then, in the section "Model-based collaboration: architectures", alternative and partly disjunct approaches for architectural frameworks of the solution elements are analysed. As the main focus in this presentation is on the architectural requirements of a traffic control system, all references to instantiations of the solutions systems are suppressed until the section "Solution elements instantiated: IUTCS and KITS" in which the relationships of the analysed architectures to the implemented prototypes are described. A conclusion is drawn about the lessons to be learned from this in the "Conclusions".

19.2 METHODOLOGICAL PRELIMINARIES

When trying to solve problems that have their foundation in real life — such as traffic control — one is constantly confronted with design choices. A design choice can — rather philosophically — be viewed as the choice of looking at reality with several different glasses. Usually you cannot wear them all, and none of them is exactly right for your deficiencies. The reason for using this rather forced parallel is that we do not believe that there are "correct" solutions to open-ended problems such as real-time traffic control. The solutions put forward are inherently coloured by the glasses you are wearing — there is never one correct solution making all other solutions infeasible. As far as we are able to judge, the solutions could be better or worse, depending on what your objectives are. In this presentation we have chosen to wear computer science and AI glasses. This gives us a view of reality where the most important features are the nature of the problem being solved (what constitutes it, and how can we enhance conventional approaches by using advanced and layered models containing what we call "knowledge"), flexibility and tractability.

19.3 THE PROBLEM: A BRIEF ANALYSIS

The problem can be defined as controlling the traffic situation in an area, by means of traffic-lights. This description is not accurate enough to find a solution. It is too wide, too vague and does not give us any criteria for choosing among the rapidly growing number of techniques available from the field of IT in general, and AI in specific.

We have chosen to divide the problem of traffic control into a set of subproblems that will enable us to make better design decisions.

First, the traffic situation needs to be classified. This makes it necessary to collect traffic data from traffic sensors and to make inferences about the traffic in areas with incomplete data acquisition capabilities. Secondly, it is important to interpret these traffic data to build up a good picture of the actual traffic so that important traffic conditions can be recognized. Thirdly, good estimation about the traffic in the near future may be of good help in making control decisions. Finally, control decisions can be performed taking the three points above into consideration.

The two basic ways of controlling traffic by traffic-lights are signal plan modification, and signal plan selection, where the former changes the signal plans dynamically according to the traffic in one junction (or a small area), and the latter chooses the best out of a predefined set of signal plans either for the traffic environment as a whole, or in one singular junction. This may be looked upon as using an untyped or a typed "universe of signal plans", and the two approaches inherit the same problems as their formal language counterparts, basically ease of expression and flexibility versus complexity and security.

In this approach we have chosen to work in a typed domain, i.e. with signal plan selection. Signal plan selection can be global or local. Global signal plan selection focuses on global traffic control; green times for all traffic-lights in a traffic environment have been calculated and collected in global signal plans, which can be chosen either by a temporal or traffic-actuated mechanism. Local signal plan selection is primarily used in a traffic-actuated manner. Here, the traffic environment is divided into signal groups, each group consisting of a set of traffic lights controlling the traffic in a small area. Usually, such an area is a junction. Whenever a change in traffic situation is detected in the junction controlled by a signal group, a signal plan is selected using a criteria local to the signal group.

The problem with precalculated classes of states, is of course that the states either are too general (and the system hence is poor at discriminating and not able to react in a novel way to stimuli), or that we may have situations where no classes are applicable. Unforeseen traffic situations may arise, and the system may not be able to handle it in an adequate way. The problem with local selection of signal plans is that the result may be suboptimal. Choosing signal plans on a local level without considering global goals may seem to solve the traffic problems on a local level, but may instead worsen the global traffic situation.

In this approach we are considering all the possible signal plans that can be chosen in any one junction at a global level. In essence all possible combinations of signal plans for junctions are available for consideration in the selection process (the tractability of this task is ensured by limiting the search space by tasks conforming to the problem decomposition mentioned above).

Our approach combines the global and the local approach in the way that global requirements are given priority, and that the power of the local approach for choosing the exact right signal plan is preserved.

19.4 SOLUTION ELEMENTS: THE MODELS

19.4.1 Abstraction and Granularity

When automating problem solving in any domain, an important consideration to take into account is the granularity of the problem-solving process. Above, we discussed how the problem (signal plan selection) could be viewed globally, by directly having a set of precalculated signal plans to choose from. This is a coarse-grained solution that will not solve all problem states to the required degree of correctness. Reducing the granularity of the solution we may — if solutions elements are selected according to the requirements — solve the problem increasingly better. The price that must be paid is an increasing complexity of the solution. There are also further limits to this subcomposition not related to the computational power required to process the solutions or the intellectual power required to build and maintain it. As the solution elements become smaller, the greater is the chance of redundancy, or for some elements not to apply. The problem of finding the right level of granularity is a problem of cost-efficiency.

When the right granularity has been found, there is also the question of abstraction. Abstraction is not the same as granularity. Abstraction can be viewed as the ability of a model to give an overall view of a problem. To some extent all models of the real world are abstractions as some features are neglected and some are overfocused. Whenever modelling a complex domain it is important to have access to the model at several different levels of abstraction. One reason for this is simply that we would not like to do more work than required. If your task is to determine the colour of the sky, you would not need binoculars. However, if your task is also to determine the colour of a plane in the sky, you might need it. In traffic control, some subtasks can be compared with looking at the sky (e.g. "is there much or little traffic in the network"), and some tasks can be compared with determining the colour of a plane ("exactly what kind of signal plan would solve this traffic problem").

Small grains (a high granularity) ensures the correctness of your decisions and abstraction ensures that we attack the problem at the right level.

19.4.2 Imperative and Declarative Elements

The subproblems described in the problem analysis section must have counterparts in the program trying to solve the problem. These counterparts are elements that either attack the subproblems directly or indirectly by means of supporting some other entity that does it. The entities helping to solve the problem can be divided into two main groups: (i) elements of imperative style, and (ii) elements of declarative styles. Imperative elements perform some action, and declarative elements describe some state of problem or solution.

Of course, imperative elements can themselves be divided into declarative and imperative subelements. The imperative–declarative classification we describe only relate to the outward behaviour and purpose of an element.

The duality being formed by the declarative and the imperative elements is the same as we have between a model (declarative) and reasoning over this model (imperative). The system may therefore be understood as consisting of mainly two knowledge levels: a traffic process level and a reasoning level. It should be noted, however, that to increase the complexity of the analysis, we may view the interrelationship between the imperative elements in the same manner: as the imperative elements are dependent on each other and the results they can offer, to some degree we will have the duality mentioned (between a model and reasoning over this model) reflected here also. For the sake of simplicity, we will not consider this point in the following section.

19.4.3 Imperative Elements

Reason, raison, F, ratio, L. from Latin, ratione, The root is reor (ratus) to think, to judge.

(Robert Sullivan, *Dictionary of Derivations*, Dublin 1851)

The imperative elements are the problem solving elements — the elements that perform the low-level reasoning in a traffic control system. Imperative element constitutes a model of how to solve problems in the domain of traffic control. They can therefore be considered as active. For some purposes it can be easier to think of the imperative elements as subprograms or procedures. For some purposes we may think of them as inference engines (abstract entities processing causal information (knowledge) to obtain a solution).

The imperative elements can be further categorized in the following functional groups:

(1) data completion and consistency checking;
(2) problem identification (shallow analysis);
(3) deeper analysis;
(4) signal plan selection.

Data completion

Normally, due to economical and technical factors (cost and number of sensors, transmission faults, sensor malfunction, etc.), only an incomplete view of the composition of urban traffic is available for analysis and decision. Because of this, the adequacy of signal plans with regard to the actual traffic situation is not always guaranteed.

The general aim, at this level, is to enhance the traffic database with regard to its reliability; i.e. to improve the overall quality of traffic data acquired by sensors. This essentially entails making inferences about incomplete traffic data based on a spatial–temporal distribution of the collected series of data. Two major subproblems can be considered:

1. *Reconstruction of missing data*. In order to derive traffic data for non-equipped road sections from the actual data being registered at specific points. Basically, this is done by propagating the measured traffic volumes through the network structure, taking into account the spatial dependencies caused by confluences and their inverse, and temporal spreads. Heuristics to guide this propagation may be employed.

2. *Consistency check of detectors*. As traffic is propagated through the network from all accessible detectors, traffic volumes will be propagated from all other detectors to a link in the network that already has a detector measuring traffic. If there is a large discrepancy between the measured detector and the propagated volumes for all the other detectors, the measured detector is marked as faulty.

Both reconstruction of missing data and consistency checks are important for successfully processing towards traffic control. Everything will be dependent on the traffic volumes, and thus no measure seems too high to ensure their correctness.

Problem identification

Problem identification is a kind of traffic monitoring which is undertaken within a system to identify areas for further processing. This process can be compared with a "skimming" of the traffic environment situation. Some key nodes are polled. An overall, but inaccurately, traffic situation is determined, and focus is left on a smaller area than the full area which is under consideration. One can argue that this is something that may not always be needed. Sometimes the area under consideration is so small that a focus on some lesser aspect of it is unnecessary. We believe this is right, for specific cases in small areas. In general one cannot presuppose a ready-made decomposition into handy units. And when one cannot do this, brute force calculation on them will probably not be feasible considering the need for short reaction time in a traffic control system.

As problem identification is a "shallow" analysis of the real world, "shallow" methods seems feasible. The rule-based normalization of knowledge we find in expert systems, is well-suited for problem areas that can be attacked using empirical knowledge. Systems based on empirical cause–effect relations, do not take into account underlying reasons, and as a result of this, gives efficient short-cuts in the solution process — and hence effective systems.

Deep analysis functionality

A thing is not deep if one can see its bottom, not high if one can see its top.
(Ayn Rand)

Analysing congestion and recognizing its causes Understanding the development of congestion in order to, e.g. postpone or anticipate a given control action, is a task for which human supervision is normally required in traffic control operations. Levels of congestion can be defined according to empirical criteria, and congested sections in the network can be identified and related to each other in order to come to a global judgement about the congestion situation on the whole area. Recognizing congestion cause is often done by "following the heavy load traffic" upward through the network (against the vehicles driving direction) until the "heavy load" disappears. The congestion cause is then set to the nearest junction.

Prediction of traffic situations To arrive at a better suited selection of signal plans, each traffic situation must be seen in its chronological continuity. A desirable feature of a traffic control system would be to generate predictions about possible future traffic situations (at least in a qualitative way) based on the knowledge of the traffic situation and on recognized tendencies in the actual flow. Moreover, prediction includes the evolution of the traffic situation and the possibility of congestions. Two distinct ways of performing traffic prediction are: (i) network propagation, and (ii) traffic pattern analysis. In network propagation a set of key nodes are identified (nodes that deserve attention because of their importance regarding their traffic load, etc.). A predetermined set of nodes upwards through the network structure are similarly identified, and the relation of the load here to the key nodes are determined. Whenever a predicted traffic volume is necessary on any of the key nodes, measurements in the corresponding up-link nodes are taken, and a relational factor is multiplied with this. This kind of prediction is only suitable for short temporal horizons. In traffic pattern analysis traffic patterns for selected nodes are calculated for typical temporal sets (i.e. Monday mornings, rush-hour-inbound, etc.). A corresponding traffic pattern is accumulated for the current run of the system, and this is compared with the "typical" pattern which exist. When fitting the current traffic pattern to the "typical traffic pattern", temporal offsets (i.e. rush-hour today arrived 15 min early) and future local peaks (if today behaves as it normally does, there will be less traffic in 39 min) can be determined.

Origin destination calculations Origin destination calculations are useful to find out what streams (arteries) through a network are currently receiving most pressure, and whether this differs from "normal" situations. This kind of problem solving may give evidence to whether a situation is indeed normal, or whether something unforeseen has come up.

Traffic control

"...increased abundantly, and multiplied, and grew exceedingly..."

(Exodus 1:7, *the Bible*)

The management of control actions has to draw conclusions on the results provided by the above elements. The aim of traffic control is to improve the quality of decision making in traffic control actions management. The basic steps are: to evaluate the overall situations with respect to defined control objectives (i.e. on-line evaluation of the performance of the current signal plan), to decide about the kind and urgency of an intervention, and to select the appropriate control strategy to be implemented in the network.

Key aspects of an improved control include:

(1) the ability of managing congestion and responding to incidents and unexpected events;
(2) the capability of making decision on the basis of a temporal analysis of traffic behaviour and developments on the network (i.e. not simply reacting to actual demand).

There are of course other ways of performing knowledge-based signal plan selection, such as using requirements rules, and constraint satisfaction by a measured level of requirement in an abstract network of nodes. These are not considered here.

19.4.4 Declarative Elements

To live effectively is to live with adequate information.

(Norbert Wiener, The Human Use of Human Beings, 1954)

These elements constitute our descriptional knowledge of the world. Acting in the world is described by the behaviour of the imperative elements. For some purposes it can be easier to view the declarative entities as a data base consisting of different kinds of information relating to and describing different parts of the world.

The different declarative elements can be viewed as a level-based model representation: the model is represented in layers, where the higher levels express relationships or information in terms of the lower levels. The layers are:

(1) the network layer: structure, i.e. hard flow constraints;
(2) the control layer: flow keys, i.e. soft flow constraints;
(3) the traffic behaviour layer: flow; and
(4) the empirical layer: flow knowledge.

All the layers have determinators expressed in flow. This denote that there is a relationship between the layers which enable us to view higher level layers as expressed in terms of (and therefore dependent on) lower level layers. The layers can be viewed as an onion where the layers change with different speed.

The network layer (hard flow constraints)

> *Bypasses are devices that allow some people to dash from point A to point B very fast while other people dash from point B to point A very fast. People living at point C, being a point directly in between, are often given to wonder what's so great about point A that so many people from point B are so keen to get there and what's so great about point B that so many people from point A are so keen to get there. They often wish that people would just once and for all work out where the hell they wanted to be.*
>
> (Douglas Adams; The Hitchhiker's Guide to the Galaxy)

This layer represents the topology of the physical traffic network, i.e. links, junctions and flow control hardware such as detectors. The topologically interesting objects can be grouped into three groups:

(1) area objects;

(2) primary objects; and

(3) secondary objects

Area objects describe what we associate with an area in the real world. This can be a collection of spatially connected links and junctions, or a collection of smaller areas.

Both area objects and primary objects are what we call identifiable objects. These objects can be accessed and reasoned about by the imperative entities described above. Examples of primary objects are junctions, links, detectors and traffic-lights. When saying "identifiable" we mean that rules can be expressed that take into account relationships about these objects, e.g.: *IF the traffic flow in all dlinks leading into junction#* 11 > *X THEN choose signal plan #32*. In this example junction#11 is called by name.

Secondary objects are objects connecting primary objects or otherwise helping to ensure consistency or connectedness in the traffic model. Secondary objects are not identifiable by imperative entities. An example of a secondary objects is the path that a driving direction describes through a junction. This can be described by three identifiable objects: "from link#2 to link#3 through junction#4".

The reason for describing the elements of the network structure to such depth is that there is a lesson to learn. In the three levels, we have three levels of abstraction. All levels are referable (though not identifiable). Thus we can choose at what abstraction level we want to look at the network structure. The distinction between the area level and the primary object level is important if we handle a traffic environment that is so big that we cannot consider the totality of it in one problem-solving cycle. Through all the succeeding levels in the onion decried above, this multi-abstractional view of the environment is inherited. In the succeeding section about the flow knowledge we will meet still another argument for why this is important.

The control layer (flow keys)

Control, contrôle, F.A counter-roll, or check-book kept by another officer, that each may be examined by the other; and hence restraint, check authority. From the Latin contra and rotula.

(Robert Sullivan, A Dictionary of Derivations, Dublin 1851)

This is a representation of the flow keys (the softer flow constraints at our disposal), i.e. possible control constraints that can be imposed on the flow. In traffic control lingo, this is mostly called signal plans. For completeness, we must state that a signal plan can be the state of a set of traffic-lights in a junction or a state in a variable sign (such as "don't drive here"). For simplicity, our discussion will consider only signal plans working on traffic-lights. Signal plans in variable signs can, however, be used in a similar manner.

A signal plan in a junction gives the respective partitions of time the traffic-lights can be green during a cycle. The cycle is repeated as long as the signal plan is active. An urban traffic control system typically administers a library of such signal plans,[10] each signal plan in the library having been calculated to cater for specific traffic situations that are thought to occur. A signal plan can be represented as a graph where the compatibility between traffic-lights is given. In some traffic control systems, signal plans are modelled with some extra information giving the reason for its existence (the triggering traffic situations) and dependencies to other signal plans in neighbouring junctions (for the sake of upholding green waves when such exists).

Nothing more needs to be said about the flow keys, except that they must be modelled in such a way as to enable counterparts in the flow layer (below). Again, cost-effectivity judged against expressiveness is the main consideration.

Behaviour (the flow layer)

This layer is a representation of the traffic flow at any time. Traffic flow can be expressed in primarily two ways: quantitatively and qualitatively. A quantitative representation of the traffic situation is important because it maintains the relation with the situation in the real world. However, it is often useful to employ a sparse quantity space to express a real world (and real number value) situation as an analysed entity. A sparse quantity space is usually expressed by a set of qualitative labels a variable can range over.

The reason for imposing a space of qualitative labels is fourfold. First, the different numeric values denoting the traffic situations may not convey important information (in average 43.5 cars and in average 43.3 cars is for all practical purposes the same number). It seems more natural to express a range (say between 40 and 45 cars) as denoting the same number of cars, because it may not be that we can distinguish between 41 and 44 cars in our program. This makes it easier to compare traffic situation values for one network object. Secondly, having between 40 and 45 cars in a link usually means something. It can for instance mean the link is saturated (i.e. this is the number of cars on the dlink when the dlink is saturated). By expressing "between 40 and 45 cars" with the symbolic, quantitative label "saturated", we are thus embedding an analysis in the description of the traffic situation. Thirdly, it may be that the number of cars in two links for those two links to be saturated is different (i.e.. link A is saturated when there are between 40 and 50 cars on it while link B is saturated when it contains between 50 and 60 cars). (We are aware of the existence of quantitative measures expressing this, but the argument holds for other variables as well.)

A fourth reason is that when expressing relationships between the behaviour of traffic on two links, qualitative description makes the relationships more intuitive, and in a sense more user-friendly. This will be discussed further in the section about the empirical layer.

Another aspect about the behaviour is time. Behaviour has extensions in the temporal dimension. One reason for this is that it changes. Trend analysis is an important feature when avoiding temporal suboptimization (solving the current traffic problems even though we should know that it will change drastically in the near future). Secondly, we would like to express knowledge about the temporal aspects of the behaviour, e.g.:

IF the traffic situation is not bettered by signal plan A
THEN stop suggesting signal plan A as a remedy.

The empirical layer (flow knowledge)

A man who has the knowledge but lacks the power clearly to express it is no better off than if he never had any ideas at all.

(Thucydides, *ca.* 400 BC)

Superimposed on the layers described above, we have an empirical level where relationships between traffic behaviour, network structure and control measures (flow keys) can be expressed. This is an important part of the descriptive elements, because this is where rules of thumb expressed by expert traffic engineers, can be expressed and utilized in solving the traffic control problem. Usually, in a traffic control system, empirical knowledge (causal information and relationships between objects in the traffic environment) are not explicitly expressed. This kind of information is hidden in adjustable parameters (cryptic), and in some cases in the program code itself (!).

The information in this layer could be represented as rules, e.g.:

IF traffic-situation in link #2 is-worse-than "saturated"
THEN link #2 is a priority link.

It is important from a modelling and a consistency point of view that the empirical knowledge (for instance expressed as rules as described above) is distributed to where they belong spatially in the layered collection of declarative elements. The reason for this can be argued as follows: as described under the section about the network structure, the topology of the traffic environment can be viewed as a hierarchy of objects, where each level may contain objects of the level beneath. The bottom level consists of atomic elements such as secondary objects (e.g. objects representing the path a vehicle describes when turning left at a junction). The middle level consists of primary (identifiable) objects such as junctions and links. Even higher level are composed of these objects, and so on. Knowledge about each level may constrain the selection of signal plans (which is after all our main task to accomplish), thus all layers should be taken into consideration when searching for solutions. For example, a traffic-light may argue that it needs more green time. However, from the perspective of the layer above, i.e. the whole junction, other problems than the one of the traffic-light may be more important to solve, and hence the demand for increased green time may not be met. The same kind of conflicts can arise between higher levels, e.g. single junctions and areas, and small and big areas. Expressing the empirical knowledge about the different entities and the entities flow (behaviour) and flow keys (signal plans) enables us to reason about the knowledge at a meta-level, e.g.:

IF a rule says that a junction needs anything and this conflicts with what an area containing the junction needs, then discard this rule.

A scheme for implementing such a distributed, hierarchic set of rules have been described elsewhere.[9]

19.5 MODEL-BASED COLLABORATION: ARCHITECTURES

As we have seen, traffic control involves several distinct sub-tasks, all which may be solved by a feasible representation of imperative and declarative elements. All of the sub-tasks are reasonably hard problems both from a modelling and a computational point of view. All are necessary in our chosen solution for the successful control of traffic in a large traffic network. To be able to develop, operate and handle several distinct tasks, we have the choice of several (partly) disjunct architectures.

By analysing the problem-solving elements described above from the point of view of several of these architectures, taking into account the requirements below, we will end up with an architectural synthesis which will be task dependent (traffic control) and have foundations in AI and software engineering.

19.5.1 Requirements

The requirements can be subdivided into two sets: criteria relating to the task that is being performed, and criteria related to the technology enabling the tasks to be performed. Attaching technologies to the criteria domains, we may say that the one set originates from software engineering and the other from traffic engineering. Both are necessary in a solution.

Traffic engineering requirements

1. Experimentation.
2. Portability.
3. Plug-in/pull-out models (easy modification of parts of the models).
4. Parallel development and/or maintenance.
5. Dynamic change between traffic control tactics.
6. Independence between how flow constraints are changed.

Software/knowledge engineering requirements

1. Modularity.
2. Explicit representation of all data. All information, all data about traffic flow, traffic flow knowledge and traffic flow control means (flow constrains) are represented in an explicit model.
3. A layered model with plug-in and pull-out modules.
4. Model-based reasoning: all reasoning is based on the state of the model.
5. Collaborative problem solving: tasks and sub-tasks are identified and represented as independent agents. The agents are working purely on the basis of the state of the model, and do not have direct interaction. The user of the system is regarded as an agent.
6. Explicit represented task control: the choice of which agents that may operate at any time is done by a set of management agents, again based on information in the model.

19.5.2 A Model-based Architecture

Model-based reasoning can be defined as the usage of explicitly represented models of physical structure and/or behaviour in reasoning systems. Reasoning by means of empirical knowledge solely was for long a dominant method within the AI community. However, it is often seen that there is a need for "deeper" domain models than are given by empirical knowledge. As stated in the section about the problem identification antity, the rule-based formalization of knowledge we find in expert systems, is well-suited for problem areas that can be attacked using empirical knowledge. Systems based on empirical cause–effect relations, do not take into account underlying reasons, and as a result of this, gives efficient short-cuts in the solution process — and hence effective systems.

Many traditional systems suffer from such shallow models of knowledge. If you have an empirical model, you get an efficient system when nothing unforeseen happens. In other words your empirical model has (implicitly) labelled the different situations that may arise. Such a "labelling" is often far from exhaustive and creates a problem when we meet situations that are not a part of our background of experience, and hence not a part of the experience base.

Model-based reasoning is a field of study that has risen from the need for undertaking qualitative reasoning over domain models. This need has roots in both traditional modelling, and in AI. As empirics offers effective systems, while deeper models offers the possibility for a more general problem solving in the problem domain, a combination of symbolic reasoning and domain models is often appropriate. Such systems, which have been denoted *second generation expert systems*, or *model-based systems*, represent a way of reasoning where we may have (at least) two abstraction levels: the empirical associations and the explicit models. Model-based reasoning is often called *deep reasoning*, as it focus on reasoning using explicitly represented models. Systems based on this approach can, in addition to providing a method for solving a set of otherwise too difficult problems, have advantages, such as easier construction of informative explanations,[11] separation of the structure of the model from system control procedures, and the possibility of treating new problems that were not foreseen when the system was made.

A general pattern on how to exploit a model in the reasoning process contains three stages:

1. *Model analysis*. Analyse the model on the same abstraction level as the problem domain.
2. *Model usage*. Use the model or the result of the analysis as a support in a reasoning process.
3. *Model updating*. Change the model.

Using our descriptive entities as the models, referable on the various abstraction levels previously described, we have a large program that utilizes the different parts of a model in a pragmatic way, trying on a high abstraction level first, and on a lower if this is required. In the traffic control case, as we have seen, the declarative elements represented in the model, may be layers as seen in the figure above.

19.5.3 A Generic Tasks Architecture

There has been some attention about the concept of generic problem solving taxonomies.[12] We will now put forward a task-based architecture based on some of these ideas.

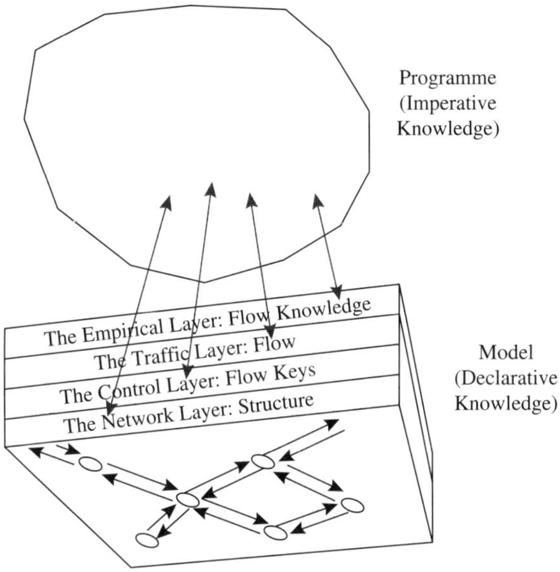

Figure 19.1.

The main tasks in the system are:

1. *Model maintenance*. Maintaining the traffic model
2. *Traffic flow control*. Decide upon flow constraint changes, and execute them.
3. *Management*. Decide upon which tasks that have priority.

Taking Model maintenance down one level we get:

1. *Physical structure maintenance*. Updating the physical network, like closing of links, changing link flow, and changing the flow control hardware.
2. *Flow knowledge maintenance*. Maintenance/updating of flow statistics and empirical knowledge.
3. *Flow constraint maintenance*. Maintaining/updating of the flow constraints (the means for flow control).
4. *Flow updating*. Updating traffic information in the model.

And so on.

Similarly, we may describe the tasks of traffic flow control drawing on the previous descriptions of how the imperative entities in our problem-solving model function.

The main point is that we have a taxonomy of tasks, where each task level is dependent on the level higher up in the hierarchy. Going from a higher level to a lower towards a solution, we need information (knowledge) about the situation to find out which lower level task we can choose (provided that there are more than one). The higher up in the hierarchy, the more generic is the task. The lower in the hierarchy, the more domain specific is the task.

In some sense, making a task-oriented architecture means creating a grammar for all the functions in our system, and placing all these functions at their right level in the taxonomy of tasks along with their activation criteria. Proceeding from a generic task to the accomplishment of a specific task is thus reduced to searching through a general tree structure.

The advantages with this architecture is that devising the task hierarchy is a good exercise in functional design.

The disadvantage is that we do not employ the strong dependency between the declarative and the imperative part of the model. Although abstraction as such can be expressed as different levels in the taxonomy of tasks, this will blur the clean-cut generic versus specific tasks order and make the architecture more hybrid than there is methodological grounding for.

19.5.4 A Blackboard-based Architecture

The blackboard architecture is another scheme for organizing reasoning steps and domain knowledge to construct solutions to formal problems. The blackboard architecture supports a collaborative model for problem solving. Basically, it consists of a set of agents operating on a common data structure, called a blackboard.

The domain knowledge is partitioned into independent modules of knowledge, i.e. the agents. They may consist of algorithmic procedures, declarative statements, functional actions, heuristic rules or whatever. It is determined dynamically which agent to apply next, resulting in the incremental generation of partial solutions on the blackboard. The choice of an agent is based on the state on the blackboard, i.e. the solution state, and on the existence of agents capable of improving the current state of the solution; hence, the blackboard architecture constitutes an opportunistic reasoning model. The agents are applied when, and in a way and at the moment it is most opportune.

A clear advantage, is the support for separating distinct tasks. In the traffic control domain, this means a possibility to treat the different sub-tasks independent of each other. As traffic control is process oriented it would be natural to represent a traffic model on the blackboard. Hence, we may have agents that update the sensor points in the model, agents that disseminate the traffic load over the network as a whole, agents that perform short-term predictions, and agents that take traffic control decisions based on the state of the blackboard, i.e. the model. Another advantage, is the possibility to apply dynamically heuristics and strategies in the process, as how and when the different agents are activated is not predetermined.

A disadvantage is that it is not possible to decouple the predetermined sequence of the agents in a way that conform 100% to the blackboard architecture. In some ways the architecture is too strong, and we cannot utilize it to all its potential.

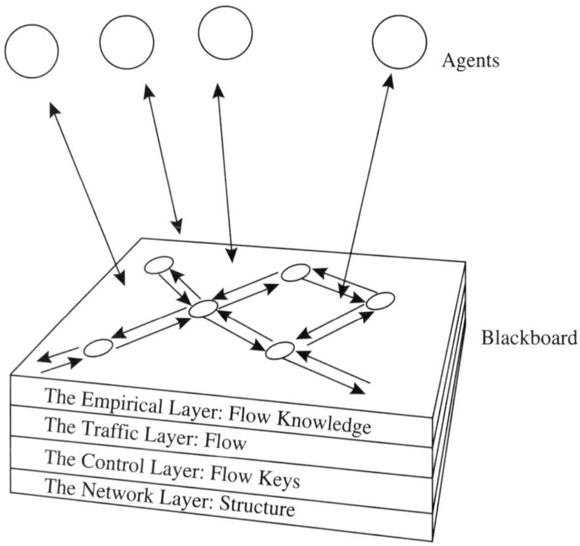

Figure 19.2.

19.5.5 Cooperating Supervised Agents

None of the above-mentioned architectures are optimal enough regarding the domain or the chosen view of how to combine the solution elements. The following architectural suggestion will make a further attempt to tailor a solution to the problem domain, can also draw a synthesis of some of the features in the above architectures.

The applicable features from the previous architectures are:

(1) model-based characteristics — from the model-based architecture;
(2) a language for generic domain problem solving — from the task-based architecture; and
(3) a distributed set of active entities working on the traffic model — from the blackboard architecture.

What we are lacking is an architecture distributed by spatial problem area. The observant reader will of course remember that the blackboard-based architecture is distributed in the functional problem are (i.e. one agent for each of the subprograms solving functional subproblems). What we are introducing now is a distribution in network areas. When solving real life problems in large traffic environments, a "singular go, we'll solve it all" architecture as the one described in the blackboard-based architecture will not be sufficient — even taking into account that we have several cycles to work on. Furthermore, as larger and larger areas as handled, the solution characteristics for similar tasks in different areas may not be overlapping (e.g. traffic prediction may not be performed in the same way for both a highway link and an urban link).

This leads us to introduce two new concepts: actors and supervisors. The functional collection of imperative entities called agents in the blackboard-based architecture will still be the same. The traffic model will also be identical.

1. *Actor*. An actor is an imperative (active, program-like) entity that has as its main task to manage the traffic in one area. An actor has at its command a full set of agents as described under the blackboard-based architecture section.

2. *Supervisor*. A supervisor is an imperative (active, program-like) entity that are in control of and manage the traffic in a larger area than that handled by an actor. A supervisor has at its command a set of actors managing traffic in sub-areas of the area managed by the actor, or a set of supervisors managing spatially smaller areas.

The imperative entities of the system will now be sorted in a hierarchy, determined by what are the area and sub-area divisions that are necessary in the traffic model.

As the imperative entities (now supervisors, actors and agents) no longer work solely by changing the state of the traffic model, but also by communicating and cooperating between them, a language for communication and channel for communication must be perceived. This language draws on some of the ideas from the task-based architecture. As the imperative entities are sorted in a hierarchy where the top nodes refer to general traffic problems in large areas and the bottom nodes

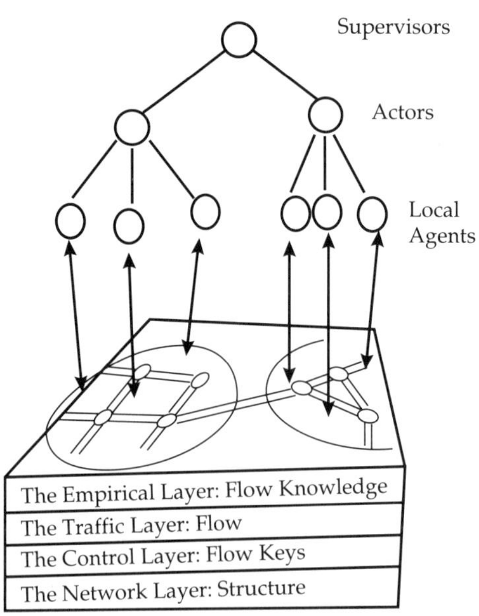

Figure 19.3.

refer to specific traffic problems in small areas, a hierarchy of tasks needs to be set up. The total vocabulary of tasks (the name, or the syntax) of tasks are identical for nodes of the same types i.e. the tasks are generic in our architecture. The semantics of the tasks can, however, be different (the task of traffic prediction may be performed differently by different task handlers (traffic prediction agents)). To complicate the matter, further task requests can be put forward from lower-level entities (e.g. agents) to higher-level entities (e.g. actors) which the higher-level entities in turn address to the right lower-level entities (agents). Although this does not conform with what is traditionally associated with task architectures, we still feel that we have conveyed some of its flavour.

19.6 SOLUTION ELEMENTS INSTANTIATED: IUTCS AND KITS

I must create a system or be enslaved by another man's

(William Blake, ca. 1815)

In 1989 a project was initiated under the DRIVE framework of the European Community. The title of the project was "AI Techniques for Urban Traffic Control". The project consisted of five partners from three European nations (two members of the EEC and one member of EFTA). The collaborating organizations had competence in traffic engineering, AI, architectures for problem solving, operations research and general computer science. In 1991 the project ended successfully as it had brought forth a prototype for traffic control based on methods from the field of AI called the intelligent urban traffic control system (IUTCS). The prototype relied heavily on some of the characteristics of blackboard systems (described above). The system (IUTCS) was tested on a simulation of a traffic environment from Hamburg. The simulation system used was Mission, and the tests were performed in Karlsruhe, Germany under the close supervision of expert traffic engineers. The results from the tests focused in particular on the behaviour of the various agents (subprograms developed for solving subproblems, such as data completion, data analysis, traffic control, traffic predictions, etc.). The conclusions from the tests were that the behaviour of the singular agents were promising, but that some reimplementation would be necessary in the architectural framework if the system were to be handle large real-life problems.

In 1992, a continuation of the described project was established under the DRIVE II framework of the European Community. The title of this project was "Knowledge Based and Intelligent Traffic Control Systems". The main purpose was to refine the existing prototype and undertake a massive experimentation in four European cities: Trondheim, Cologne, Madrid and Genova. The name of the refined prototype was KITS. The four different test cites had different characteristics, and some (especially Madrid and Genova) were so big that the previous architecture implemented for the IUTCS was no longer sufficient. An architecture resembling the one described in Section 19.5.5 above was implemented.

19.7 CONCLUSION

In Carceribus denuo adsumus (We're back to square one)

(Henry Beard.)

A set of solution characteristics for the traffic control problem has been identified. These solution characteristics are associated with the subproblems of traffic control. The solution characteristics have been described and put into several architectural frameworks. These frameworks have been analysed, and arguments for a synthesis of several architectures have been put forward. Practically two of these architectures has been pursued in two different projects (DRIVE v1059: 1989–91 "Blackboard-based Architecture", and DRIVE v3029: 1992–94 "Cooperating, Supervised Agents").

It is our conclusion that the analysis leading up to the final choice of architecture is in fact more important than the resulting system. This analysis is generic, as it contains characteristics of similar analysis required for building many RTI (road traffic informatics) systems. In particular, we would like to point out that all architectures have their methodological background in AI. Thus a second conclusion is that AI techniques are now practically available and feasible in building RTI systems.

As a final point relating to the implemented systems, we would like to quote Nierstrasz:[13]

> *"First we did it, then we did it right, and then we did it fast" The popular wisdom behind this statement is that when building (new) software systems, one should not attempt to do all three things at once. A first prototype can give you the insight to understand how to do the job properly and cleanly, and then you may gain the insight needed to streamline the solution. Try to do them all at once and you are sure to fail.*

We hope that the next implementation will be fast.

REFERENCES

1 Dougherty MS, Montgomery OF. *OECD Workshop on Knowledge-based Expert Systems in Transportation*, Vol. 1 (Heikki J. Ed.). Finland. VTT, Espoo, 1990.

2 C. Krogh: SPGen — an inference engine for signal plan generation. *Proceedings from 2nd OECD Workshop on Knowledge Based Expert Systems in Transportation*, Montreal, Canada, June 1992.

3 *Proceedings from 2nd OECD Workshop on Knowledge Based Expert Systems in Transportation*, Montreal 1992.

4 *Proceedings from Drive Conference*, Bruxelles, 4–6 February 1991.

5 Cuena J. AURA: a second generation expert system for traffic control in urban motorways. *Proceedings of the Congress on Expert Systems and Application*, Avignon, 1989.

6 Ambrosino G, Bielli M, Boero M, Fleishcmann S, Hock R, Irgens M. A blackboard model for traffic control operations. *Proceedings from Drive Conference*, Bruxelles 4–6 February 1991.
7 Ritchie SG. A knowledge-based decision support architecture for advanced traffic management. *Transportation Research* 1990; **24A**: 27–37.
8 Irgens M, Bielli M, Boero M, Fleischman S, Hock R. Towards intelligent urban traffic control. *Proceedings, OECD Workshop on Knowledge Based Expert Systems in Transportation*, 1990.
9 Krogh C, Irgens M, Trætteberg H. A novel architecture for traffic control. *Proceedings from "Vehicle Navigation and Information Systems"*. Oslo: IEEE, 1992.
10 Irgens M, Bielli M, Ambrosino G, Boero M, Fleischman S, Hock R. Artificial intelligence and expert systems: Towards the application to road traffic control; OECD *Workshop on Knowledge-based Expert Systems in Transportation*, Vol. 1 (Heikki J., ed.). Finland: VTT, Espoo, 1990.
11 Brown JS, Burton RR. Multiple representations of knowledge for tutorial reasoning. In *Representation and Understanding: Studies in Cognitive Science* (Bobrow D, Collins A, ed.). New York: Academic Press, 1976.
12 Chandrasekaran B. Towards a taxonomy for problem solving types. *AI Magazine* 1983; **4** (1):
13 Nierstrasz O. In a comp.sotfware.engineering news article. 1992.

20

Application of Artificial Intelligence and Expert Systems to Traffic Control

20.1 INTRODUCTION

During the last years, we observed an impressive acceleration of research efforts and investments in the application of information technology and telecommunications to road transport. The perception of the key role played by transport in the development of world economy, together with the evident and continuous growth of demand of mobility and of road use, have led to the activation of strategic initiatives such as DRIVE (*D*edicated *R*oad *I*nfrastructure for *V*ehicle safety in *E*urope) and PROMETHEUS (*PRO*gra*M*me for *E*uropean *T*raffic with a *H*ighest *E*fficiency and *U*nprecedented *S*afety) in Europe, IVHS (*I*ntelligent *V*ehicle–*H*ighway Systems) in the USA, RACS and VICS in Japan. Central to such research programmes are the development and the implementation of advanced concepts usually referred to as "intelligent road infrastructures" and "intelligent vehicles". This will provide, in the short–medium term, a totally new generation of advanced transport telematic systems and will bring fundamental technological changes in the whole transport environment, with a relevant impact in terms of improved traffic management, increased efficiency and safety.

Within this broader area of research and application, the development and experimentation of artificial intelligence (AI) and knowledge-based architectures to traffic management and control has significantly progressed.[1] Several experimental developments are underway worldwide, including, among others.[2–8] In this framework

Advanced Vehicles and Infrastructure Systems, Edited by C.O. Nwagboso

this chapter describes the application of knowledge-based (KB) techniques to support the operators in monitoring and control activities. In particular, firstly, the chapter introduces some current urban traffic control (UTC) problems which could be faced by the application of knowledge-based systems (KBS) techniques and describes a possible general architecture of AI systems to face these traffic control problems. Then it provides a detailed description of the software prototype developed in a European project carried out during 1989–91 under the programme DRIVE aimed at the application of AI systems to traffic control (DRIVE Project V1055[5]). This prototype KBS, intelligent urban traffic control systems (IUTCS) has been implemented to support traffic control and management operations and has been tested in the laboratory using data from a real urban test case, an area in the city centre of Hamburg, Germany. Finally, the chapter provides a brief description of a real test case adopted for evaluation describes the experimental results obtained.

20.2 KNOWLEDGE-BASED AND EXPERT SYSTEMS APPROACH TO TRAFFIC CONTROL

In this section we give a short description of the main open problems of UTC systems and outline the possible contribution that AI and KB techniques could provide.

20.2.1 Current Problems with Conventional UTC Technology

Present generation urban traffic control systems differ in a number of aspects. Various control strategies can be implemented which may differ with respect to: (i) the spatial organization of control (centralized, distributed systems); (ii) the level of aggregation of control measures (isolated intersection, arterial co-ordination, area co-ordination); (iii) the modality of achieving flow regulation (fixed-time, traffic flow actuated, demand responsive); and (iv) the priorities and combination of criteria and control policies adopted (e.g. maximum average flow, minimum queue length, minimum level of pollution, etc.). However, whatever combination of such options, present-day traffic control methods show clear limitations, especially when traffic conditions reach critical levels. Current systems, based upon algorithms and quantitative models (e.g. optimization procedures) usually cope with free flow or undersaturated situations, but are not able to manage or to prevent oversaturated situations. Furthermore, even in non-critical situations, they fail to adapt to changing conditions and unexpected events which disrupt smooth flow (e.g. accidents, stopped vehicles, etc.).

At present, the gap between the traffic demand and requirements, and such limitations of control systems can only be filled by human experience and ability. As a usual practice, experienced operators in traffic control centres monitor running signal plans, assess to which extent the signal plans meet the evolving traffic situation and eventually force control actions accordingly, by overriding the choice and the timings of the automated system.

The need and the opportunities for improvement of current systems can be individuated at various levels of operations outlined in the following.[1]

Data collection level

Owing to both economical and technical reasons (limited number of sensors due to costs and operational constrains, sensor malfunction, transmission faults, etc.) only an incomplete view of current levels and composition of traffic is available for analysis and decision. What would be desirable, at this level, is an overall enhancement of the traffic data/base used for control and management decisions, with respect to both its reliability and completness.

Several aspects are important:

(1) inference of missing data, to provide plausible values for road stretches where sensors are not present and substitute values for faulty detectors;

(2) estimation of turning traffic, to infer the distribution of measured data with respect to different directions;

(3) estimation of the origin/destination behaviour of the traffic crossing the area.

These appear to be major requirements to achieve a more comprehensive basis of information to understand traffic developments and make control decisions.

Monitoring level

The interpretation of collected traffic data and the identification of the overall traffic situation from a limited set of "local" indications, is in the first instance where improved methods are required.

A skeletal classification process of traffic patterns is already implied, for example, in the mechanisms for signal plan selection used in vehicle-actuated control strategies; either in a state–space description (vector method) or, more empirically, based on a set of threshold rules. However, the algorithm approach needs to be extended to adapt to the changing environmental conditions and disturbances (accidents, etc.) or to include appropriate consideration of traffic behaviour over time (both history and trends).

Analysing congestion and understanding its causes is another problem where improvements are needed and contributions from AI can be obtained. Levels of congestion can be defined according to empirical criteria, and congested sections in the network can be identified and also related to each other using their chronological sequence. Plausible causes should be identified (e.g. critical intersections and road stretches) and priorities in solving congested areas should be set up, in order to make a complete assessment about congestion in the whole area.

Scemama[2] has discussed the main issues and problems of a KB approach to congestion management.

The capability of predicting traffic situations on a short-term basis (5–10 min) is a third desirable enhancement of the monitoring operations. This would enable improved strategies for the management of control action, including, for instance, to anticipate or delay the selection of a given signal plan under strategic considerations.

Given the difficulties of quantitative models for short-term prediction, especially in large networks, the development of qualitative approaches able to predict only relevant changes of traffic state can be an effective alternative.[3]

Control level

The management of control actions is based on the results provided by the above operational levels.

Hence the quality and the effectiveness of control depends largely on the quality and the appropriateness of data provided by such levels. However, besides an enhanched provision of data, "expert" control requires additional capabilities such as, for instance, the ability of continuously evaluating the performance of the traffic system under the control strategies currently implemented, individuating the necessity and the urgency of an intervention, and assessing the merit and the possible effects of the control actions applicable at a time interval.[5]

Key aspects of an improved control, for which contributions from AI can be envisaged, These include:

(1) the capabilty of dealing simultaneously with multiple problems occurring in the network, e.g. congestion and critical situations at different sites;

(2) the ability to cope with and managing conflicting objectives; favouring traffic in an area versus restraining access to reduce pollution;

(3) the capability of making decisions on the basis of temporal analysis of traffic behaviour and developments in networkwide, i.e not simply reacting to actual demand

(4) the ability of managing non-recurring congestions and responding to incidents and unexpected events.

20.2.2 Roles for Knowledge-based Techniques in Urban Traffic Control Systems

Based on this background, the general objective for the application of AI techniques to traffic control is to develop a general software tool which could integrate various KB-based functions within the normal flow of operations of today's traffic control systems to address the above identified problems of UTC systems (Figure 20.1).

At these various problem levels the main contributions of the application of KB techniques could be the following:

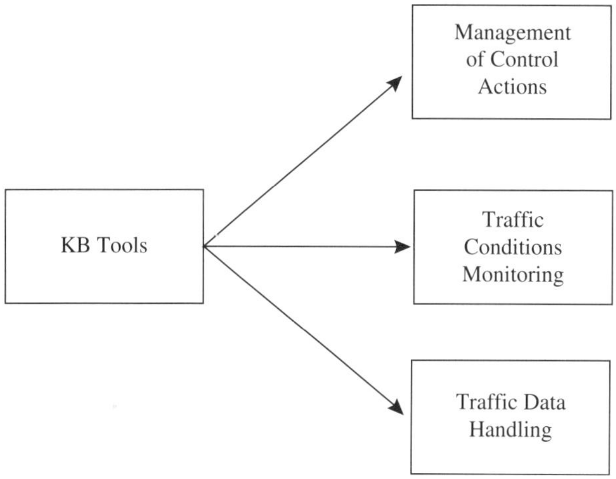

Figure 20.1. Different problem levels addressed by knowledge-based (KB) tools

(1) enhance the set of data collected in real-time and used for traffic monitoring and control decisions, by inferring additional traffic data related to links where traffic information is not available;

(2) analyse collected data and make inferences about traffic conditions to recognize critical traffic patterns and situations;

(3) predict, in a qualitative way, near-future traffic situations in order to recognize in advance possible inconvenient changes of state in the network;

(4) improve the management of available control strategies by relying on an extended base of information, including the knowledge of traffic behaviour over time, of predicted traffic situations and of the different control objectives and constraints related to the particular situation.

20.3 A KNOWLEDGE-BASED ARCHITECTURE FOR TRAFFIC CONTROL

Taking into account the above problems and the possible rules of the KB tools, an AI architecture has been defined to enhance the performances of current UTC systems.

This architecture is based on a blackboard problem-solving paradigm,[15–17] where four knowledge sources are associated with the above different problem levels and traffic control operations. In the following, before describing the details of this architecture (IUTCS prototype) developed in the DRIVE project, we briefly recall the basic aspects of the blackboard problem-solving model for better understanding of the technical approach followed.

20.3.1 The Blackboard Approach

A model for problem solving gives a conceptual framework for organizing domain knowledge and strategy for applying that knowledge (i.e. reasoning steps) to implement a solution to a given problem. A quite "classical" problem-solving model in AI systems is based on the structure "knowledge base/inference engine/working memory". This represents the traditional, core structure of KBS, where the domain-specific knowledge is often represented as rules, what we usually call rules of thumb or heuristics. As in human problem solving, the major function of these heuristics is to reduce the complexity of a problem by limiting the number of alternatives that must be considered in reaching a decision.

The blackboard model for problem solving is an alternative to the classical approach, one which is well-suited for independent problem-solving agents cooperating in a common domain. In this model, different knowledge sources (KS) act upon a common global data/base, called the blackboard. The KS respond to changes occurring in the blackboard. When something happens on the blackboard, one or more KS may be triggered, announcing their interest in performing their actions.

These actions may change the configuration on the blackboard (problem-solving state) and in turn trigger new KS.

The current execution of the KS is regulated by a control component, the action scheduler, which maintains a record of the requests of activation from the KS and manages priorities according to some specified policy.

This way, the global solution is built opportunistically, i.e. step-by-step and cooperatively by the various KS, whenever an opportunity for adding a partial solution arises.

Blackboard architectures are becoming a major trend in AI applications to engineering problems, especially in process monitoring and control. Briefly, the charactristics which make such conceptual models attractive in many applications[15] are:

(1) organization of the domain knowledge into different modular KS;

(2) incorporation of diverse types of knowledge;

(3) integration of different knowledge representation methods and reasoning techniques;

(4) intelligent use of available knowledge in the absence of a known problem-solving algorithm;

(5) implementation in a distributed computing environment for greater efficiency;

(6) support for cooperative system development among multiple system builders;

(7) support of system modification and incremental evolution.

20.3.2 The IUTCS Architecture

The prototype IUTCS, developed in the DRIVE Programme, implements a blackboard-oriented knowledge architecture.

Following this architecture, IUTCS includes several task-oriented modules, associated with different key tasks of traffic control operations. A central blackboard provides a common representation of the traffic environment and a data exchange facility for the various subsystems.

The overall system architecture is outlined in Figure 20.2
It consists of the following components:

(1) a data completion KS, which acts on a real-time database of flow information (traffic data collected from sensors) to infer additional traffic data;

(2) a traffic evaluation KS, which acts on acquired and inferred data, and on a set of qualitative indicators of traffic behaviour in the network, to recognize traffic situations, detect critical events and make a diagnosis;

(3) a traffic prediction KBS, to make qualitative prediction of traffic flow based on collected sensor data;

(4) a traffic control KS, which continuously evaluates the level of performance of the current control strategy and decides about appropriate control actions;

(5) a shared working memory, the central blackboard, accessible to the above independent knowledge modules, including a representation of the traffic

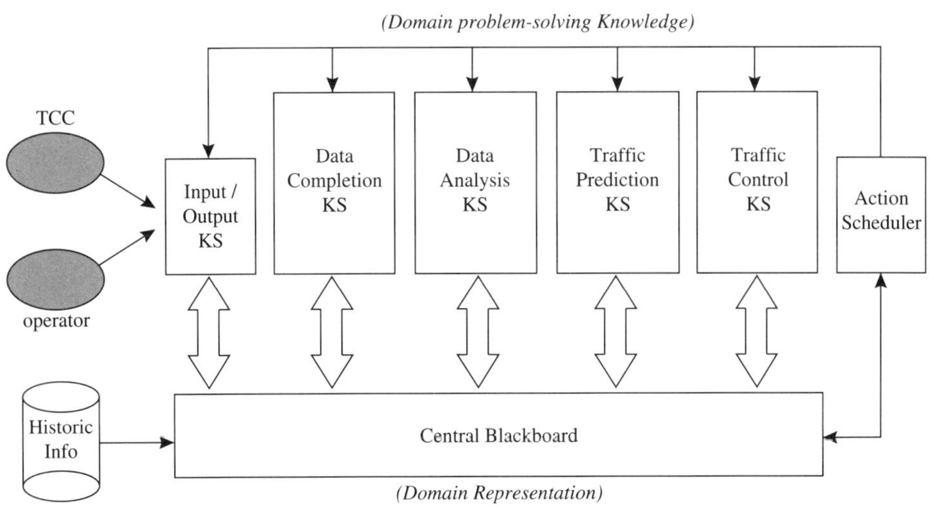

Figure 20.2. Components of the IUTCS prototype. KS, knowledge source(s); TCC, traffic control computer

network, information about the implemented control actions and a qualitative model of current and foreseen traffic behaviour;

(6) an action scheduler, which controls the selection and the execution of the KS as a result of the changes introduced in the date stored in the blackboard.

The major operational assumptions of the IUTCS architecture are the use of a symbolic and qualitative model of the traffic environment and the functional breakdown of traffic management operations into a collection of basic tasks. IUTCS works by reasoning on a model of the real world. A model of the traffic environment is maintained in the global working memory and is updated at each cycle of data acquisition on the basis of traffic data coming from roadside sensors (through the interconnection with the TCC — traffic control computer) and as a result of the reasoning of the various KS.

20.3.3 The knowledge sources

In the following, we introduce the basic aspects and the role of each KB agent of the IUTCS prototype.

The data completion KS

Data completion operates on traffic data collected by the inductive loops available and tries to infer flow data, i.e. traffic volumes and occupancy rates, for the links where detector data are not available. It provides the following services:

(1) plausibility check of data collected from active sensors; and

(2) inference of flow values and occupancy values (faulty detectors or unequipped links).

The data completion KS incorporates a model of traffic flow propagation through the network. This involves both *physical knowledge* about the traffic network (e.g. knowledge of physical characteristics of individual links) and *historical knowledge* describing traffic behaviour at intersections and on links (e.g. turning proportions at intersections, traffic patterns at detectors, etc.).

Essentially, the inference of unknown volumes is carried out by setting up a network of interdependencies relating all the measuring points in the network (active detectors) and the points for which values can only be inferred (virtual detectors). Intersections provide the basic elements to model flow interactions through a set of constraints. A *constraint propagation* approach is used to propagate measured values across this set of contraints in order to infer unknown values. Furthermore, heuristics are applied to infer plausible occupancy values from flows, based on the actual flow values, the capacity of the links and the last measured (or inferred) value of the occupancy rate.

The data analysis KS

The data analysis KS works on collected and inferred data to detect relevant traffic conditions and make global judgments about the overall traffic situation.This KS includes two basic reasoning steps:

(1) qualitative evaluation of a set of indicators of traffic behaviour;

(2) diagnosis of congested states; i.e. identification of probable origins of expanding congestions.

The evaluation of traffic situation on the network is implemented as a quite straightforward rule-based shallow reasoning system. Essentially, two basic indicators are used to evaluate traffic conditions at a local level, i.e. at individual links and intersections.

Traffic flows on each link are classified according to the *degree of saturation v/c*, where *v* is the actual flow in the link and *c* is the capacity. This parameter indicates how near the link is at its maximum capacity and gives a measure of the spare green time available.

Analogously, traffic conditions at any intersection *i* are evaluated by a global indicator given by the *capacity factor* (CF).

For both parameters, qualitative indicators are defined by setting up a classification of load levels through an appropriate definition of a *quantity space* for the corresponding parameter. Table 20.1 shows the instantiation of this load level

Table 20.1. Instance of range spaces for traffic flow analysis in the Hamburg test case

Link load levels		
Flow	Occupancy	Load level
v < 20% C	or < 0.2	very low
20% C < v < 50% C	or < 0.2	low
50% C < v < 80% C	or < 0.2	undersaturation
80% C < v	or < 0.2	saturation
	0.2 < or < 0.3	saturation
	0.3 < or < 0.4	precongestion
	or < 0.4	congestion

Capacity factor	Load level
CF > 5	very low
5 > CF > 2	low
2 > CF > 1.25	undersaturation
1.25 > CF > 1	precongestion
1 > CF	congestion

classification for the test case of Hamburg. Furthermore, as the direction of the changes of the parameters are important as well to describe traffic conditions, trends of flow are also defined based on a three states of distinction: increasing, decreasing and steady. For instance, the complete qualitative description of flow on a link, in a given time instant, may have the following form: ⟨low, +⟩, meaning "low traffic with increasing volume".

The traffic prediction KS

The traffic prediction KS performs a qualitative prediction of traffic on a short-term horizon (e.g. 5–10 min). The approach followed for KB prediction is based on: (i) the knowledge of historic volume information (detector patterns) at any relevant traffic situation of the day, and (ii) on the time series of flow data currently measured by detectors. Current volumes are matched against a database of historic volume profiles conveying a description of average behaviour on the link in order to identify the best match and to extract predicted volumes. Traffic prediction has therefore two steps, one being an interpretation task, the other a symbolic reasoning task on identified flow profile:

(1) identification of the best fitting historic trend
(2) extrapolation of qualitative predicted traffic volume (Figure 20.3)

The prediction reasoning as tackled in the IUTCS prototype is essentially a shallow reasoning approach. The criteria for evaluating and reasoning over historic patterns are basically heuristic and are supported by the knowledge of the specific area under examination. On the whole, the approach relies on the assumption that traffic evolves typically, which is often not the case. However, this intrinsic weakness of the method seems to be compensated by the requirement of operating at a qualitative

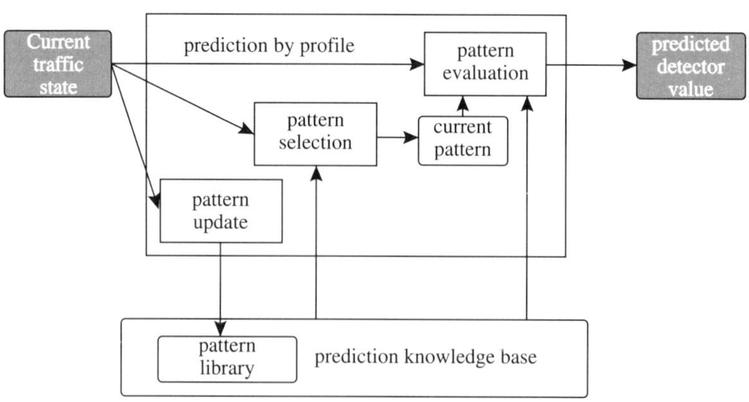

Figure 20.3. Basic structure of the prediction task

level; i.e. at a first level, predicted flows can be individuated according to the classification of qualitative load levels handled by the data analysis KS and previously introduced.

The traffic control KS

The traffic control (TC) KB models the decision process underlying the management of control actions in face of dynamically changing traffic conditions. Control actions are available in the form of a predefined library of signal plans (SP) for intersections. The TC KS provides the following services:

(1) evaluate the performance of the traffic network under the current control settings;

(2) identify required local changes of control for critical intersections;

(3) evaluate the effects of proposed local changes on neighbouring intersections and clusters;

(4) select and propose compatible control changes for the whole area to cope with identified local problems.

The background of this decision process includes several elements, some of which made available by the other KS of IUTCS. These include:

(1) a performance measure, which is continuously evaluated to monitor the performance of current control with respect to the current traffic situation;

(2) a set of quantitative and qualitative traffic parameters, conveying a picture of the traffic behaviour in the medium-term period (as provided by the data completion and data analysis KS);

(3) the focalization of control options on critical links and intersections, i.e. spare capacity at intersections, flow levels, etc. (as provided by the data analysis KS);

(4) the prediction of the traffic situation in the medium-term time span (as provided by the traffic prediction KS).

The overall structure of the TC KB is outlined in Figure 20.4.

The TC KS is based on the evaluation of a *performance measure*, providing a measure of the success the given settings of SP have with respect to the current traffic situation. An object function is used based on the capacity factor parameter provided by the analysis component, introduced in the previous sections.

On the basis of this performance evaluation, the analysis and decision process implemented by the TC source is able to assess current and available SP at

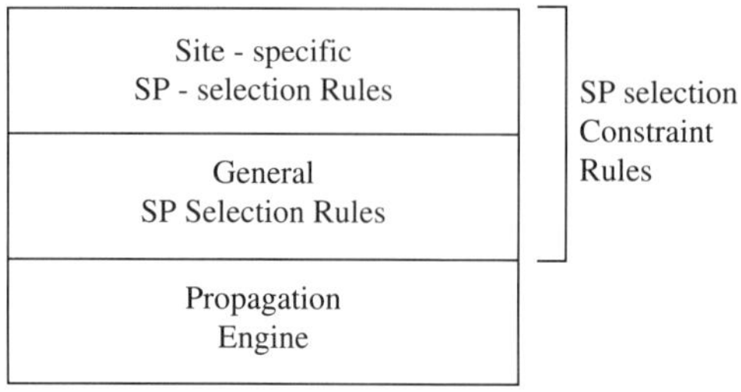

Figure 20.4. Overall structure of the traffic control knowledge source. SP, signal plan(s)

intersections, in order to identify a candidate list of SP. However, to handle the combinatorial explosion arising in the search process some constraining heuristic mechanisms are needed. These provide compatibility constraints to reduce the analysis and selection process further, embodying, respectively, general, domain-independent SP selection criteria and specific, site-dependent rules. Examples of site-independent criteria are provided below, expressing global constraints for a given cluster of junctions and links:

Rule 1: *if* CF > 2 in all key junctions
then restrict candidates SP to those with shorter cycle times

Rule 2: *if* CFi < 2 in all key junctions
then restict candidates SP to those with longer cycle times

Rule 3: *if* CFi < 2 on junction *i* as *t* > *n***T* (*T* = sampling time; *n* = 3)
and $ SPK | (cycle-time(SPk)) = cycle time (current SP)
and direction of coordination (SP) = current direction of coordination
and CFi > 2
then SPk is a candidate
else constrain selection to SP with longer cycle times
Site-independent SP selection constraints

20.3.4 Knowledge Representation

In the IUTCS environment several knowledge representation schemes and inference mechanisms have been implemented on top of LISP. These include, essentially, object-oriented, rule-oriented and constraint-based representation and reasoning. Object-oriented programming has been also extensively used to implement such higher-level representation functionalities and is supported by CLOS, the Common LISP Object System.[14]

Object-oriented representation and reasoning are used mostly to implement the global blackboard, providing a model of the traffic environment accessed by the

```
(defrule MEDIUM-OUTBOUND
 (:agent DA)
 (:ruleset TRAFFIC-SITUATION)
 (if
      (and (load-level j32-j42 (one-of 'precongested' congested)
      (eq (congestion-cause j32-j42) j42-j8042)
      (load-trend (all j22-j32 j13-j14) 'increasing)
      (load-level (all-others (dlinks *envir*)
                         '(j13-j14 j14-j15 j22-j32))
                                   (lower= 'undersaturated)))
(then
      (change-traffic-situation(traf-state *envir*) 'MEDIUM-OUTBOUND))))
```

Figure 20.5. Example rule: traffic situation identification

various KS. Basically, the blackboard implements a three level structure, supporting a representation of:

(1) the *network structure*, including all relevant topological and physical parameters;

(2) the *traffic state* in the network at different time points, including both quantitative traffic data and qualitative indicators;

(3) the *control state*, i.e. the parameters of both active traffic signals and of available SP (SP library at any intersection).

Accessing individual elements of the network, one can obtain information about the local traffic state and about the state of the signals. Link and intersection representations include also a local memory of traffic events. So, accessing any one element, one can obtain a history (of variable length) of local traffic conditions.

Shallow knowledge is embedded in the IUTCS environment through a rule-based representation. A (rather simple) rule language, based on the usual antecedent/consequent syntax, has been developed to express the empirical knowledge used by the various KS. The rule in Figure 20.5 identifies a specific traffic situation for the whole test area.

Rules of this kind are also defined for other KS. For instance, in the traffic control KS, shallow rules are defined to provide indications for SP selection.

20.4 TESTING THE IUTCS PROTOTYPE

From the operational level IUTCS has been developed to run on a "traffic engineer's workstation", connected to the traffic control computer (TCC) operating in the UTC via a serial or a LAN connection. Traffic flow data, specifically, flows and occupancy rates, are provided by the TCC as input to IUTCS, as well as the information about running SP at intersections. Based on these data IUTCS works at a "conceptual level" providing as output indications about the current and predicted traffic situations, as well as suggestions about appropriate control actions to be implemented in the network.

The version tested in this laboratory runs on standard hardware, either SUN4 or Macintosh workstations. LISP was adopted as a basic language to develop the whole representation and reasoning environment. To guarantee a good portability across different hardware systems, Common LISP[14] has been chosen.

IUTCS has been tested in laboratory using both detectors data recorded in the test site and a microsimulation environment implementing a model of the test area, fine tuned with collected data. The test site selected was the Hammerbrook area, a part of the inner city of Hamburg, Germany and the road network used in the simulation test environment is an abstraction of the original test site and consists of 11 major intersections and 42 directed links, including some feeder links.

The tests performed focused mainly on the traffic data provision capabilities of IUTCS and on the control aspects. The first level of assessment entailed essentially testing the data completion functionality of IUTCS, while the second level involved the whole IUTCS prototype.

In order to give an idea of the level of testing reached, the following two sections report on two lab tests, concerning, respectively, the traffic data provision and the traffic controlling functionalities of IUTCS.

20.4.1 Testing Traffic Data Inference Agent

In this test, the data completion module of IUTCS was run in open-loop, reading flow data from a file dumped from the traffic control computer in the UTC of Hamburg. This input file consisted of traffic counts and occupancies collected by detectors at an acquisition rate of 90 s, which is typically operated in the Hammerbrook area.

The test presented in the following is based on the traffic data recorded in the Hamburg/Hammerbrook area in June 1991, during two collection periods of 2 h each. In addition to the traffic data recorded by detectors, on-street surveys and manual traffic counts were carried out by operators in order to observe traffic behaviour and to estimate other parameters, such as, e.g. the turning proportions at intersections, during the data collection interval.

The analysis of the test focused on the completion of traffic data from two uninterrupted road sections, the Amsinckstraße arterial road (inbound) and the Nordkanalstraße arterial road (outbound). The remaining detectors were located on minor roads; some of them only covered a single lane.

The strategy to assess the performance of data completion was quite simple to implement. For test purposes, one or two detectors were "disactivated" in the model of the area used by IUTCS; i.e. in the representation of the Hammerbrook area maintained in the blackboard. This allowed running the data completion module on a subset of active detectors to complete the values corresponding to the disactivated detectors, comparing afterwards the inferred flow data for "blinded" detectors with collected known values.

The tests performed this way showed the traffic data completion inference mechanisms of IUTCS work, in principle, correctly. Inferred data for unequipped road stretches were always within reasonable approximations, even if with some difference in accuracy between the results obtained for traffic volumes and those produced for occupancy rates. The inference of traffic volumes obtained through the constraint satisfaction approach was more than acceptable. The deviations between measured and inferred occupancy rates were slightly more pronounced. However, these were related mostly to the accuracy of the data collected from the test area (especially, the turning proportions) and to the tuning of some parameters in the heuristics used.

The results obtained are exemplified by the case presented in Figure 20.6 where 10 detected and 10 completed detection intervals are compared.

20.4.2 Testing the Traffic Control Agent

The performance of the traffic controlling functionalities of IUTCS were tested in the laboratory using a model of the test site, calibrated with real world data collected in the Hammerbrook area. The MISSION miscrosimulation model[20] developed by the University of Karlsruhe, Germany, was used as an evaluation tool, providing both simulated traffic behaviour and a variety of measures of effectiveness and performance parameters. This second kind of testing entailed assessing the whole IUTCS, i.e. the whole cycle of opeation starting from the data completion, to data analysis and TC functionalities.

Both IUTCS and MISSION run on a SUN4 workstation under the Unix operating system as independent asyncronous processes. A two-way communication link was set up between the two processes, using the socket interprocess communication mechanism of Unix. Via a "detector interface", at the scanning time of detectors, IUTCS receives simulated detector readings from MISSION. Via a "control action

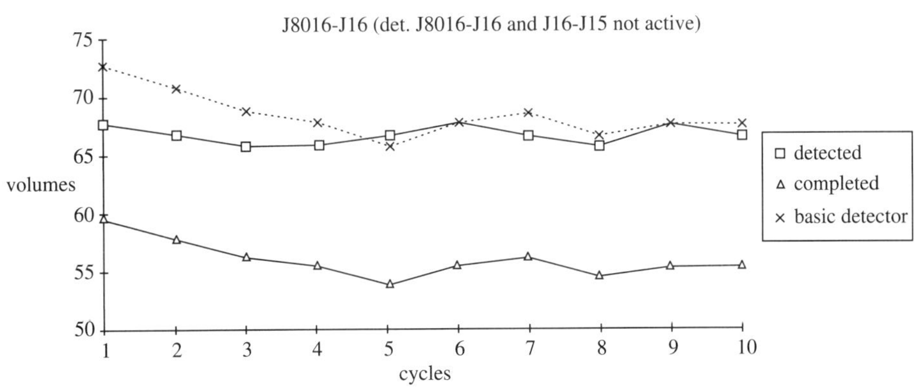

Figure 20.6. Completion of traffic volumes on link J8016-J16; one disactivated detector, one supporting detector

interface", the KBS prototype sends back signal plan change suggestions to MISSION, identifying the interested signal groups and the relevant signal plans. Like scanning time for detector values, changes of signal plans are transmitted at a predefined time. This enables synchronizing of the two processes during the whole testing time interval.

In order to evaluate the performance of the IUTCS prototype, two performance evaluation measures were defined based on: (i) number of stops and waiting time in the network, and on (ii) cruise speed over some defined routes in the network.

As a first performance measure, a performance index P of stops and waiting time was adopted, corresponding to the approach used in the TRANSYT model.[21] This is defined according to equation (20.1).

$$P = \sum_{j=1}^{m} \sum_{i=1}^{n} (W_{ij} + kH_{ij}) \tag{20.1}$$

where

m = number of links

n = number of events of delay

W_{ij} = delay time on link j at event i

H_{ij} = number of stops on link j at event i

k = factor converting stop in delay time ($k = 10$ s suggested).

The second performance measure is based on the evaluation of cruise speed along key routes. In the test case network, six main routes have been defined as sample itineraries to evaluate control performances.

Taking into account these routes, the average cruise speed v can be straightforwardly defined according to equation (20.2).

$$\nabla = \frac{\sum_{r=1}^{n} \nabla_r F_r}{\sum_{r=1}^{n} F_r} \tag{20.2}$$

where

v_r = average cruise speed of route r (km/h)

F_r = traffic flow on route r (vh/h)

n = number of routes (here 6).

Based on these evaluation criteria various assessment runs were carried out. During each testing session, IUTCS read simulated detector values from MISSION and then suggested (when required), SP changes to be implemented in the network. For each operation cycle, the operations of IUTCS were recorded on a log-book to trace the computation done and to explain the decisions taken.

Table 20.2. Evaluation of test cases by performance index *P*

Waiting time (10^6 s)	Fixed-time strategy	IUTCS strategy
Morning rush hour	1.09	1.05
Noon	1.00	1.17
Afternoon	1.12	0.95

Average cruise speed (km/h)	Fixed-time strategy	IUTCS strategy
Morning rush hour	20	22

Two test cases are presented below, addressing three different traffic situations in the Hammerbrook area, specifically, the morning, noon and afternoon peak hours, respectively. In all these cases, IUTCS was assessed against a fixed-time control strategy.

The performance index *P* for the investigated test cases is presented in Table 20.2. As can be seen, in the morning rush hour test case, the IUTCS strategy produced 1.05×10^6 s average delay time per vehicle, while the fixed-time strategy resulted in 1.09×10^6 s average delay. In the test case of noon traffic, the fixed-time strategy takes only 1.00×10^6 s in contrast to 1.17×10^6 s of the IUTCS strategy. The situation turns over again when considering a test case after noon, about 17.30. Then, IUTCS produces less delay time (0.95×10^6 s) than fixed-time control (1.12×10^6 s).

The evaluation of the average cruise speed also showed a slight improvement of the fixed-time control by IUTCS. The average cruise speed has been computed for the test case of morning rush hour traffic. For the six routes defined in the test site, the IUTCS strategy performs better than the fixed-time control strategy, yelding to an average 6.02 m/s cruise speed while fixed-time control resulted in 5.56 m/s average speed.

20.5 CONCLUSIONS

In this chapter we described a possible application of KB techniques to support the operators in monitoring and surveillance activities of UTC system.

To face some current UTC problems an advanced KB architecture was designed based on the blackboard paradigm. Its essential features are the use of a symbolic and qualitative model of the traffic environment updated over time as traffic data come from roadside sensors and functional breakdown of traffic management operations into a collection of basic tasks addressing central problems of traffic control. The approach was developed in a European project carried out during period 1989–91 under the DRIVE Programme. The work is now being continued in a follow-up project, the DRIVE 2039 project (KITS) undertaken within the new phase of DRIVE (1992–94), with the goal of achieving extensive validation in this field.

ACKNOWLEDGEMENTS

The work described in this chapter has been partially funded by the Commission of European Communities, Directorate General XIII, in the framework of the DRIVE Programme, under the Contract DRIVE V1055. Partners of the V1055 Consortium are: AUTOMA, Genova (Italy); FZI, Karlsruhe (Germany), HB Gmbh, Aachen (Germany); CNR-IASI, Rome (Italy); SI, Oslo (Norway). The Authors wish to thank Daniele Tezza for his helpful encouragement throughout the writing of this chapter.

REFERENCES

1 Ambrosino G, Bielli M, Boero M, Mastretta M. Expert system approach to road traffic control. In *Concise Encylopedia of Transportation Systems* (M. Papageorgiou, ed.) Pergamon Press, 1991.

2 Scemama G, Forasté B. Surveillance and congested traffic control in Paris by expert system. In *Proceedings of the 2nd International Conference on Road Traffic Control*. London: Institution of Electrical Engineers, 1986.

3 Cuena J. AURA: a second generation expert system for traffic control in urban motorways. In *Proceedings of the 9th International Workshop on Expert Systems and their Applications, EC2*, Avignon, May 1989.

4 Ritchie SG, Prosser NA. Development of a prototype real-time expert system for managing non-recurrent congestion. In *Proceedings of the 1st OECD Workshop on Knowledge-Based Expert Systems in Transportation*, Espoo, Finland, June 1990.

5 Ambrosino G, Bielli M, Boero M, Fleischmann S, Hock R, Irgens M. A blackboard model for traffic control operations. In *Advanced Telematics in Road Transport, Proceedings of the DRIVE Conference*, Bruxelles, February 4–6 1991. Elsevier.

6 Scemama G, Bell M, Ibbetson L. CLAIRE: an expert system for congestion management. In *Advanced Telematics in Road Transport Proceedings of the DRIVE Conference*, Bruxelles, February 4–6, 1991, Elsevier.

7 Sellam S, Boulmakoul A, Pierrelée JC. Intelligent intersection: a monitoring and control integrated system. In *Advanced Telematics in Road Transport, Proceedings of the DRIVE Conference*, Bruxelles, February 4–6, 1991, Elsevier.

8 Cuena J, Molina M, Martin G. An architecture for knowledge-based traffic management for the Expo-92 Sevilla urban ring. In *Proceedings of the 2nd OECD Workshop on Knowledge-Based Expert Systems in Transportation*, Montreal, Canada, June 15–17, 1992.

9 Commission of the European Communities. *Advanced Telematics in Road Tansport, Proceedings of the DRIVE Conference*, Bruxelles, February 4–6, 1991, Elsevier.

10 Bell MC. The fundamental issues of an expert system for urban traffic control. In *Proceedings of the IEE Colloquium on Applications of Expert Systems in Road Transportation*, Digest No. 9, 7/1-6. London: Institution of Electrical Engineers, January 1988.

11 Forschungsgesellshaft fur das Strassenvesen, Richtlinen fur Lichtsignalanlagen *(RILSA)* Cologne, 1981.

12 *Traffic Control in Saturated Conditions*. OECD Report, Paris, 1981.

13 Philipps P. Modelling on-line signal network control. In *Proceedings of Engineering Foundation Conference on Traffic Control Methods*, Santa Barbara, CA, February 1989.

14 Steele G.*Common LISP*, 2nd ed. Digital Press, 1990.

15 Engelmore R, Morgan T. *Blackboard Systems*. Wokingham, UK: Addison Wesley, 1988.

16 Nii HP. Blackboard systems: the blackboard model of problem solving and the evolution of blackboard architectures. *AI Magazine* 1986; 7: 38–53.

17 Nii HP. Blackboard systems: blackboard application systems, blackboard systems from knowledge engineering perspective. *AI Magazine* 1986; 7: 82–106.

18 Jansen B, Kirschfink H, Richter M. Traffic data support systems via AI-based methods. In *Proceedings of the 1st International Conference on Artificial Intelligence Applications in Transportation Engineering*, San Buenaventura, CA, June 20–24, 1992.

20 Hubschneider H. *Mikroskopisches Simulationsystem für Individualverkehr und offentlichen Personennahverkehr*, Schriftenreihe Institut für Verkehrswesen, Heft 26, 1983.

21 Robertson DI. *TRANSYT: a Traffic Network Study Tool*. TRRL Report No. LR253. Crowthorne, UK: Transport and Road Research Laboratory, 1969.

22 Moreno S, Toledo F, Rosich F, Martin G. Qualitative simulation for temporal reasoning in urban traffic control. In *Proceedings of the 10th International Conference on Expert Systems and their Applications, EC2*, Avignon, France.

21

Application of a Neural Network with Computer Vision for a Vehicle Identification and Detection System

21.1 INTRODUCTION

There is a growing demand for road traffic data (i.e. traffic volume, speeds, lane occupancy, vehicle classification, inter-vehicle gaps, road congestion, travelling time). These traffic data are required by local and central governments for traffic surveillance and control, for studies of traffic management and road safety, and for the development of transport policy.

Recently, image processing technology has been used in applications in such diverse industries as product manufacturing and inspection. The transport industry is no exception, and a general review and practical experience of image processing applied to traffic have been reported,[1,2] and a variety of systems using image processing (for example, measuring traffic flow,[3] queue detection and congestion monitoring,[4] incident detection in urban roads,[5] and measuring travel time[6]) have been developed for traffic data collection or monitoring during the past decade.

For these traffic application systems using image processing, the most important is extracting an object to be recognized from the picture with accuracy in which basic image processing methods have been used. However, these systems are used outdoors, so changes in illumination or reflectance of objects within the scene occurs repeatedly. Therefore, a powerful recognition method is needed for practical use. Under these conditions, the neural network (NN) based on a human vision system has the feature of robustness for noise or illumination and a lot of good results have

Advanced Vehicles and Infrastructure Systems, Edited by C.O. Nwagboso

been reported. Therefore, we have tried to use NN in image processing procedures for traffic applications.

This chapter shows two traffic application systems using NN and shows the result of simulations and implications for practical use.

21.2 IMAGE PROCESSING FOR TRAFFIC APPLICATIONS

Generally, the image processing algorithm for traffic applications is separated into the image preprocessing stage for extracting an object to be recognized in a picture, the image analysis stage for recognizing an object, and the data processing stage for measuring road traffic data.

Among these three stages, at the image preprocessing stage, image processing algorithms are:

1. *Interframe differencing*. This method extracts only moving objects, by subtracting between two different incoming frames.

2. *Background frame differencing*. This method extracts the object, by subtracting a reference image in which no object exists from the image in which an object exists.

3. *Space differencing*. This method extracts the outline of an object, by edge operations such as Gradient and Laplacian.

At the image analysis stage, the following operations have been given to the above-mentioned preprocessed image data.

1. *Segmentation*. This is a way of distinguishing an object and other variables from the binarized image data.

2. *Projection*. This is a way of distinguishing an object and other variables using characters of projection to the *X*-axis or *Y*-axis.

3. *Template matching*. This is a way of clustering an object by investigating the degree of mismatching between the standard pattern and the input pattern.

However, the image processing system for a traffic application is generally used outdoors, so there are some cases that an object cannot be accurately analysed because of the influence of shadows from a building or a vehicle on a road, or changes in illumination.

Therefore, we have been investigating more intelligent recognition systems using NN, which can recognize the influence of the changes in the surroundings.

In this chapter, we describe the vehicle identification system which recognizes vehicle licence numbers in Section 21.3 and the vehicle detection system which recognizes the vehicle in the input picture in Section 21.4, using NN at the image analysis stage.

21.3 VEHICLE IDENTIFICATION SYSTEM

The vehicle identification system, which is introduced in this section, is in three steps: (i) the extraction of plate character string and cut-out of character stage; (ii) the normalization and image transfer stage; and (iii) the character recognition stage by NN.

21.3.1 Extraction of Plate Character String and Cut-out of Character

The vehicle licence plate has many vertical edge constituents, while the general front view of a vehicle has many horizontal edge constituents. So the plate image can be extracted by horizontal space differencing. After horizontal space differencing operation to the whole image data, vertical edge points are investigated from the bottom right of the picture, taking into consideration that the vehicle licence plate is positioned on the lower part of a vehicle. The first detected vertical edge is regarded as the right position of the plate and the scope corresponding to the width and height of the plate is checked. If the distance does not agree with the inferred width and height of a plate, another image picture is inputted and the above-mentioned extraction operation starts again.

Next, the image data of the extracted plate is binarized and the background of the plate and characters are distinguished. Each character is cut out from the plate image using the projection method and the plate size, which indicates vehicle type, and is determined to be large or standard from the height of the cut-out characters. Moreover, the cut-out characters are examined by the number, heights and intervals, and if these data are proved to be characters, this system proceeds to the next normalization and image transfer step.

21.3.2 Normalization and Image Transfer

As the image of the vehicle licence plate is taken by the camera installed above the road, the input pattern is vertically shrinked or elongated in comparison with the standard pattern. Therefore, the height of the input pattern is normalized as follows. The input pattern is horizontally split into "$n + 1$" parts, where "n" is the number of pixcels calculated from comparison of the input pattern to the standard pattern. If the input pattern is shrinked, a linc of compensating parts made from the neighbouring part is inserted in each split part. If the input pattern is vertically elongated, the split pattern is partly deleted to reduce and modify the input pattern. Moreover image data of normalized input patterns are transferred by the following linear transfer operation. This translation is done to normalize the image data for the next recognition stage by NN.

$$g(x, y) = s/a - b \times \{f(x, y) - b\} + C$$

$f(x, y)$ is image data of input pattern

$g(x, y)$ is transfered image data of $f(x, y)$

$a>b$ (a is maximum value of image data, b is minimum value of image data)

s is constant

c is offset

21.3.3 Character Recognition by the Neural Network

Figure 21.1 shows an example of input patterns into NN. In order to measure travelling time, four numerals in the licence number plate are needed for recognition.

This time, we have used the back-propagation network model, which was proposed in 1986 Rumelhart *et al*[7] as a NN.

This model is a multilayer model which improves the accuracy of its association by computing an error signal for each new input which is proportional to the total mean square error between the desired and actual system outputs. This error signal is propagated back through the network to adjust the interconnection weights of the processing elements in the internal layer.[8]

Structure of the multilayer network model

The network model which is used as the NN in the vehicle licence number recognition in this paper is shown in Figure 21.2. There are three layers (input, internal and output), and the information going to the input layer is recoded as an internal representation and the output layer is generated by the internal representation rather

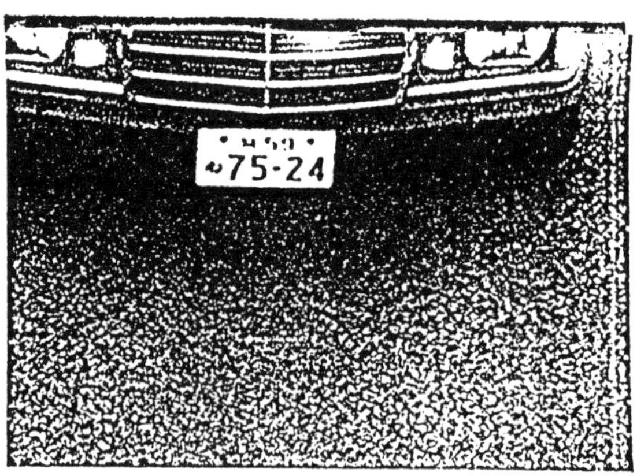

Figure 21.1. Input image into the neural network

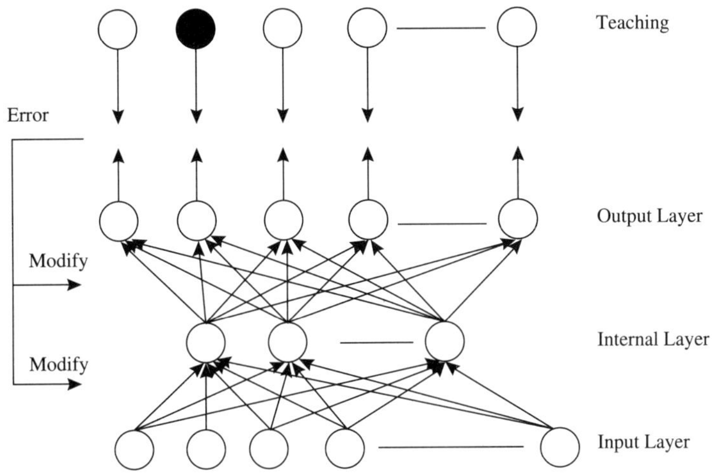

Figure 21.2. Structure of the multilayer network model

than by the original pattern. Input elements are 165 units, internal elements are 40 units and output elements are 10 units.

Results

We did a proof test for recognizing characters in vehicle licence plates according to the above-mentioned algorithm by inputting 100 vehicle licence plates.

Table 21.1 shows the result of the test. Numbers "0", "1", "2", "6" and "8" are clearly recognized but and the number of the others are recognized with a few mistakes. Moreover, we compared the accuracy of recognition by the template matching method which we have already used in the vehicle licence number recognition system with accurate of recognition by the NN.

Table 21.1. Result of recognition by NN

Character to be recognized	Recognition rate by NN (%)
0	100.0
1	100.0
2	100.0
3	97.1
4	97.1
5	96.1
6	100.0
7	97.3
8	100.0
9	97.2
Average	98.4

Table 21.2. Comparison of total recognition rate

	Template matching method	NN method
Recognition rate	99.3%	98.4%
Execution time	28.0 ms	90.0 ms

Table 21.3. Comparison with the mismatching degree and the engaged degree

Input pattern	Template maching method Judged character and (the mismatching degree)		NN method Judged character and (the engaged degree)	
	No. 1	No. 2	No. 1	No. 2
5	5 (10)	9 (10)	5 (0.994)	1 (0.034)
5	5 (10)	9 (11)	5 (0.992)	6 (0.010)
8	8 (7)	4 (8)	8 (0.762)	0 (0.042)
2	2 (7)	7 (8)	2 (0.996)	3 (0.010)
5	5 (10)	9 (12)	5 (0.831)	9 (0.040)
7	7 (14)	3 (16)	7 (0.926)	2 (0.019)
5	5 (6)	9 (9)	5 (0.993)	3 (0.007)
9	9 (1)	3 (4)	9 (0.628)	3 (0.256)

Table 21.2 and Table 21.3 show the comparison. Table 21.2 shows that the template matching method is superior to the NN method with total recognition rate. However, Table 21.3 shows that a lot of patterns which are judged with the same degree or almost the same degree of accuracy using the template matching method, can be distinguished with high engaged degrees by NN.

Consequently, it is said that the NN method is capable of increasing the accuracy of recognition of input pattern in which the mismatching degree of the template matching method is the same or very similar.

21.4 VEHICLE DETECTION SYSTEM

The vehicle detection system which has been developed in this chapter, consists of two steps. The first step is a preprocessing stage where an original image is transformed into the image in which the outline of a car is extracted as input image data for the NN. The second step is the image analysis stage and a car is recognized by the NN.

21.4.1 Pre-processing for a Vehicle Detection System

The original image is a picture taken by a surveillance camera installed on a road (see Figure 21.3). The picture is structured by 512 × 512 pixels, and the number

Figure 21.3. The original image taken by a surveillance camera

of its gradients is 256. From this image, a sub-image, which consists of 128 × 128 pixels is selected. Moreover this sub-image is transformed into a binary image with 64 × 64 pixels.

These preprocessing operations used for getting the sub-image, are as follows.

1. *Sobel transform.* A method to extract the edges of an object skilfully from a picture with noise.

2. *Adaptive binarization.* A method to produce the segmented image in which a car and others can be suitably distinguished, by using the grey-level histogram information of a picture.

3. *Thinning.* A method to reduce the noise including the image after adaptive binarization and extracting the main outline of a object.

After preprocessing, the outline of a vehicle in a picture is extracted, and the 64 × 64-pixel binary image is produced as input data to the NN.

21.4.2 Image Analysis by the Neural Network for a Car Detection System

The NNs judge whether a car exists or not in the preprocessed image. Figure 21.4 shows an example of input patterns to the NN. These images consist of 64 × 64 pixels, and each pixel corresponds to the input cell of the NN. A black pixel

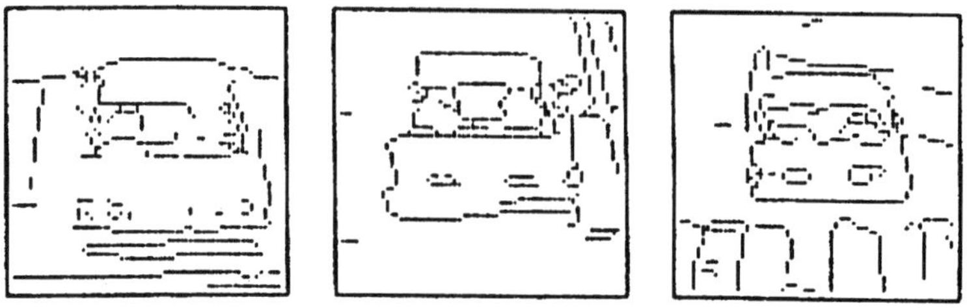

Figure 21.4. Preprocessed images consisting of 64 × 64 pixels

represents an active cell whose output value is given as "1", and the white pixel represents a passive cell whose output is given as "0".

This time, we have used a modified Neocognitron, which is designed based on a mechanism of visual pattern recognition by Fukushima,[9] and used as the NN for a car detection system.[10,11]

This method is dedicated to image clustering. Its most interesting property is its ability to recognize patterns which have shifted in position, deformed or changed in size compared with to the learned representative of the cluster.

Outline of the Neocognitron

The Neocognitron is a multilayer feedforward NN. Each layer is made of two kinds of sub-layers called a *Us* layer and a *Uc* layer. The *Uc* layer is put behind the *Us* layer, as shown in Figure 21.5.

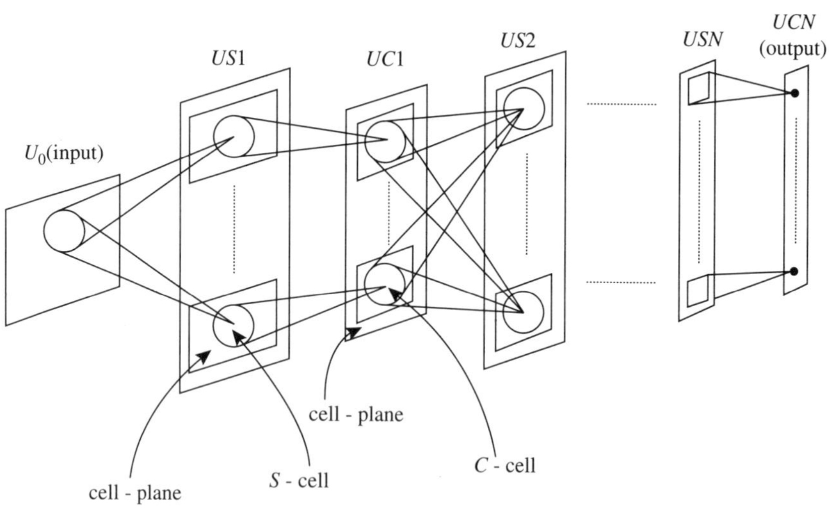

Figure 21.5. Schematic diagram illustrating synaptic connections between the layers in the Neocognitron

The properties are mainly due to its particular, hierarchical structure. The first level extracts local features such as lines of different orientations. The connections are partial and local. There is not only one cell per feature, but a plane of identical cells and each of them are looking at a pixel and its neighbourhood. Hence each feature is not only recognized in a particular position, but anywhere in the image. Planes in higher levels extract bigger and more complex features using the same method.

Design of the Neocognitron

As mentioned above, the Neocognitron can recognize input patterns robustly, which is barely affected by deformations, changes in size, and/or shifts in position. Car size and position in a picture are not definite, so these properties of the Neocognitron perform effectively. The Neocognitron can be trained either by supervised or unsupervised learning. In this chapter, the supervised learning is used. In the case of training the Neocognitron by supervised learning, a training pattern to respond to each *S*-plane must be determined by network designers according to input patterns. Therefore, the capability of recognition which the learned Neocognitron has, depends on a useful selection of training patterns by designers of the Neocognitron.

We have introduced the following assumption for designing the Neocognitron in car detection system: "The car back (or front) view in a picture is formed from combinations of vertical and horizontal lines". From this assumption, we used the training patterns of the *S*-planes in the *Us*1 layer, which are shown in Figure 21.6. These figures are vertical and horizontal lines in 3×3 matrices. To confirm the probability of the above-mentioned assumption, we also made the cell planes which are trained by the patterns shown in Figure 21.7, and compared the responses of the *C*-cells which respond to vertical and horizontal lines with the responses of the *C*-cells which respond to oblique lines. In Figure 21.8, the frame (a) corresponds to an input plane in the *UO* layer, and other frames correspond to *C*-planes in the *Uc*1 layer. The frame (a) consists of 64×64 pixels, and the others consist of 62×62 pixels. Planes (b) and (c) detect the horizontal and vertical lines, respectively. Planes (d) and (e) detect the oblique lines. One pixel in each frame corresponds to one *C*-cell. The *C*-cells corresponding to black and white pixels indicate that their

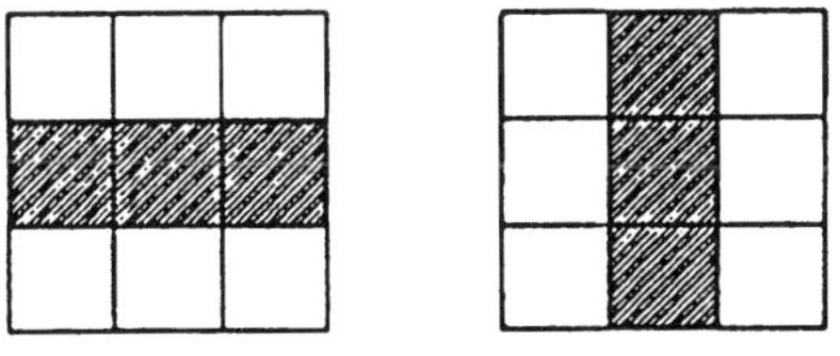

Figure 21.6. Training patterns of the *Us*1 layers (vertical and horizontal lines in 3×3 matrixes)

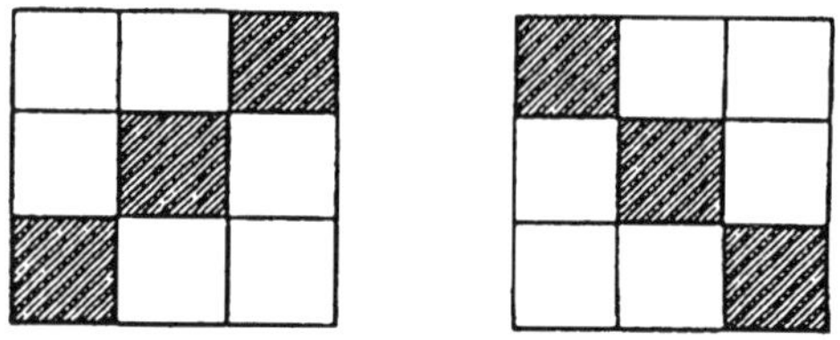

Figure 21.7. Training patterns of the *Us*1 layers (oblique lines in 3 × 3 matrixes)

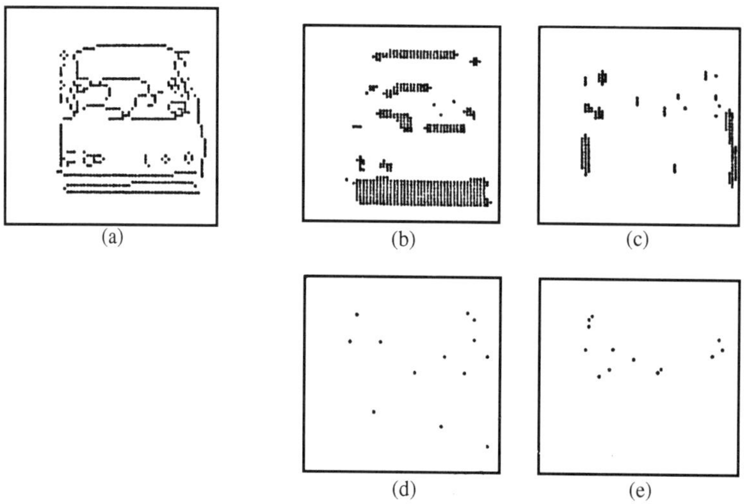

Figure 21.8. (a) Input pattern into the *U*0 layer. (b) and (c) Responses of the *S*-cells to dectect the horizontal and vertical lines. (d) and (e) Responses to detect oblique lines

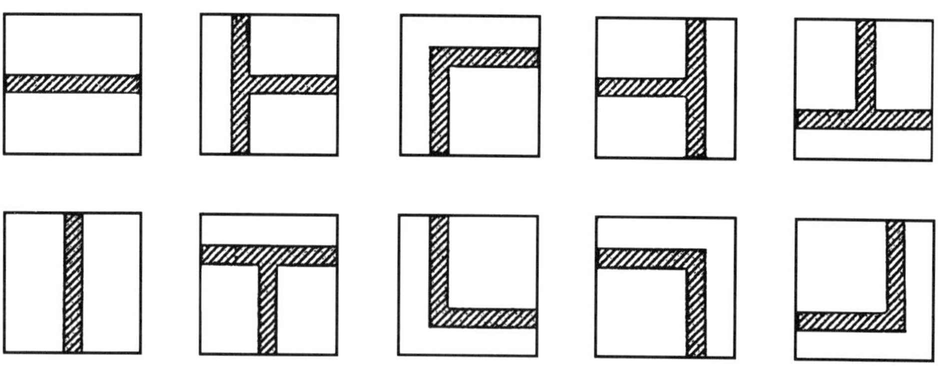

Figure 21.9. Training patterns of the *Us*2 layer consisting of 9 × 9 matrixes

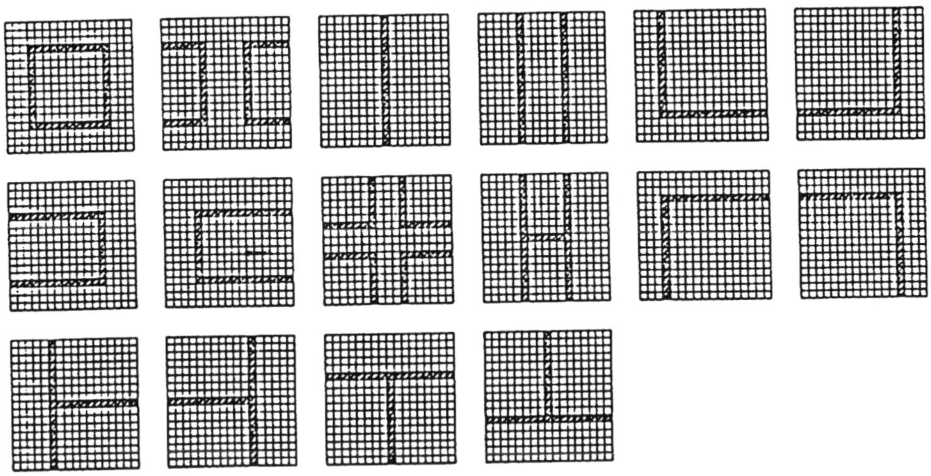

Figure 21.10. Training patterns of the *Us*3 layer consisting 17 × 17 matrixes

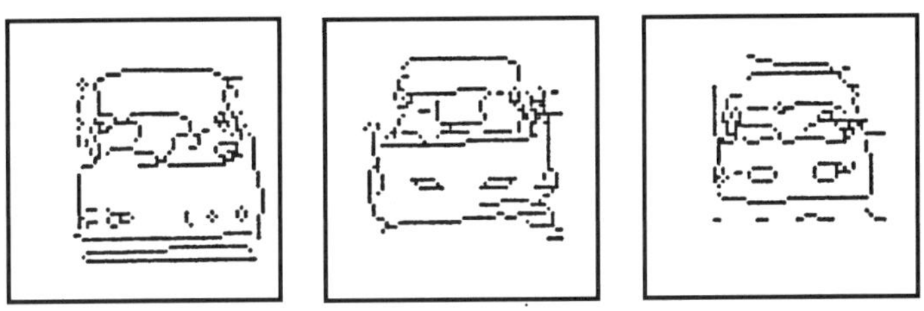

Figure 21.11. Training patterns of the *Us*4 layer consisting of 64 × 64 matrixes

output values are more than and less than 0.3, respectively. From this figure, the oblique-line detection cells respond little to car images.

Training patterns of the *S*-cells in the *Us*2 and *Us*3 layers are the combinations of vertical and horizontal lines shown in Figures 21.9 and 21.10. The *Uc*4 is the output layer, so the training patterns of the *S*-cells in the *Us*4 are preprocessor car images, as shown in Figure 21.11.

Results

To confirm whether our car detection system using the NN performs properly or not, we have simulated some pictures in Figures 21.12–21.14 using a Sun Workstation. In these figures, the number in parentheses under a group of frames corresponds to an index of the *Uc* layer. For example, the frame of (0) is a cell plane in the *U*0 layer; in other words, this is an input pattern. The frame in group (1) is a cell plane in the *Uc*1 layer. The pixel in the frames corresponds to a *C*-cell. The black pixel indicates an active cell and the white pixel is not an active cell.

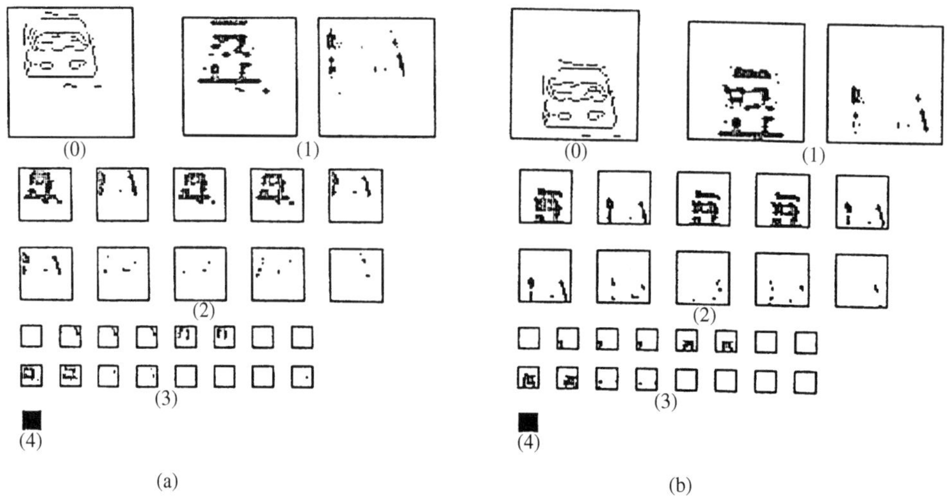

Figure 21.12. Responses of each *C*-cell. (a) and (b) are different training patterns in the position

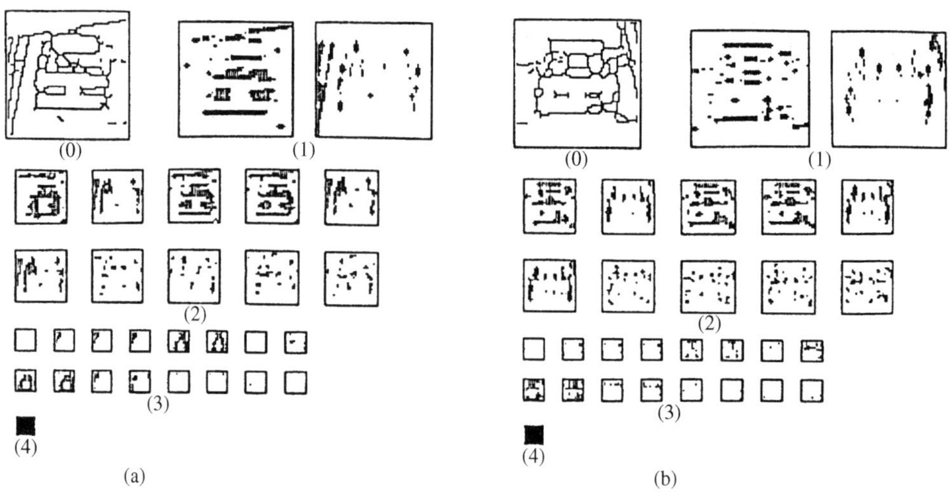

Figure 21.13. Responses of each *C*-cell. (a) and (b) are untrained patterns

The input image in the Figure 21.12 (a) and (b) are the same car and they are used as one of the training patterns. But the position of the car in the image is different. Nevertheless, the active cell patterns appearing in the *Uc*3 layer in the figures (a) and (b) indicate a similar response each other.

Figure 21.13 (a) and (b) are results of inputting the images which are not training patterns. Responses of *C*-cells indicate the same active cell patterns.

Figure 21.14 (a) is a result of inputting the image in which there is shadow from the guidepost on a road, and Figure 21.14 (b) is a result of inputting the image in

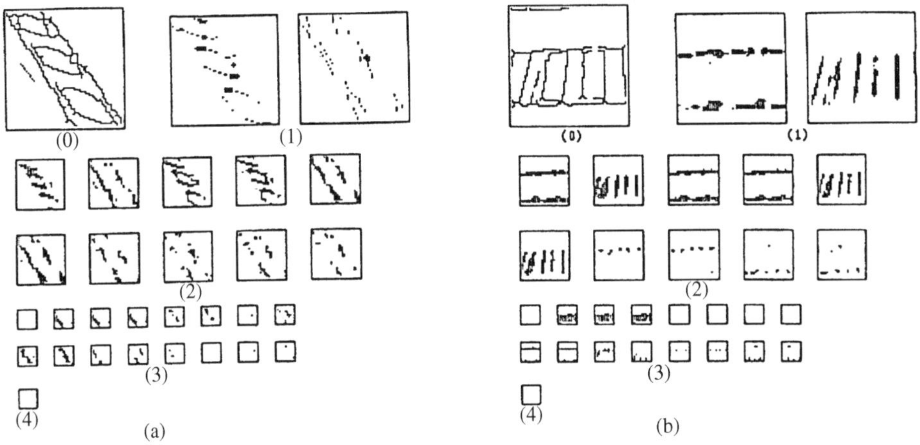

Figure 21.14. Responses of each *C*-cell. (a) Shadow on the road. (b) A pedestrian crossing

which there is a pedestrian crossing. The responses of *Uc*3 indicate the different patterns from those of Figures 21.12 and 21.13.

As a result of the above-mentioned simulation, we can confirm that the method of image analysis by the NN can distinguish a car and other variables (e.g. shadow, a pedestrian crossing, etc.).

21.5 CONCLUSIONS

This chapter has introduced two application systems of the NN in traffic engineering. First, in the vehicle identification system where a three-layer network model with backpropagation was used at the image analysis stage, we confirmed that our proposed system is capable of increasing the accuracy of recognition of the input patterns which the mismatching degree of template matching method is same or very similar.

Next, in the vehicle detection system where the Neocognitron was used, we confirmed that our proposed system consisting of preprocessing which extracts the outline of objects and the image analysis by the NN is capable of distinguishing between one car and another; also we have confirmed that the system can recognize vehicle images successfully without being influenced by the different types of car and shifts in position. As a result of these simulations, we can say that NN is useful for traffic application systems.

Recently, various NN models have been reported and the merits are being introduced. However, the image processing problem means that we have to customize the algorithm to each application. Therefore, the final aim of using the NN in image processing is that the image processing algorithm should be adaptable (we do not need to rewrite algorithms according to variable conditions), robust and not dependent on illumination and noise. This is particularly so in the case of image processing technology for traffic applications; with this in mind there is a great expectation for it to succeed.

In future, in order to use NN in real-time image processing, we feel that parallel processing technology (VLSI or semiconductor technology) will be very important as well as using algorithms for recognizing variable images.

REFERENCES

1 Dickinson KW, Waterfall RC. Image processing applied to traffic: 1. a general review. *Traffic Engineering Control* 1984; **25**: 6–13.
2 Dickinson KW, Waterfall RC. Image processing applied to traffic: 2. Practical experience. *Traffic Engineering Control* 1984; **25**(2): 60–67.
3 Takaba S *et al.* Measurement of traffic flow using real-time processing of moving pictures. *Proceedings of the 32nd Conference on Vehicle Technology*, San Diego, CA U.S. May 1982; 488–494.
4 Rourke A, Bell MGH. Queue detection and congestion monitoring using image processing. *Traffic Engineering Control* 1991; **32**(9): 412–421.
5 Hoose N, Vicencio M, Zhang X. Incident detection in urban roads using computer image processing. *Traffic Engineering Control* 1992; **33**(4): 236–244.
6 Kitamura T, Kanayama K, Tanaka R, Miyamoto K. Travel time measurement system using image-processing technology. *Paper of the Technical Meeting on Transportation and Electric Railway*, IEE, Japan TER. 1987; 87–35.
7 Rumelhart D, Hinton G, Williams R. Learning internal representations by error propagation. In *Parallel Distributed Processing*. 318–362 MIT Press, Cambridge, Mass U.S. (1986).
8 Miller RK. *Neural Networks*. The Fairmont Press, 2–12 Lilburn Press, GA.
9 Fukushima K, MIyake S, Ito T. Neocognitron: a neural network model for a mechanism of visual pattern recognition. *IEEE Transactions on System, Man, and Cybernetics* 1983; **SMC-13**: 826–834.
10 Wang D. An extended model of the Neocognitron for pattern partitioning and pattern composition. *Proceedings, IJCNN-89*. 1989; **II**: 267–274.
11 Minnix JI, Mcvey ES, Inigo RM. Modified Neocognitron with position normalizing preprocessor for translation invariant shape recognition. *Proceedings, IJCNN-90* 1990; **I**: 395–400.

Subject Index